Electrical Machines

Mathematical Engineering

For further volumes:
http://www.springer.com/series/8445

Dieter Gerling

Electrical Machines

Mathematical Fundamentals of Machine Topologies

Springer

Dieter Gerling
Fakultät Elektrotechnik und Informationstechnik
Universität der Bundeswehr München
Neubiberg, Germany

ISSN 2192-4732 ISSN 2192-4740 (electronic)
ISBN 978-3-642-17583-1 ISBN 978-3-642-17584-8 (eBook)
DOI 10.1007/978-3-642-17584-8
Springer Heidelberg New York Dordrecht London

Library of Congress Control Number: 2014950816

Printed on acid-free paper

Springer is part of Springer Science+Business Media (www.springer.com)

Preface

Calculation and design of electrical machines and drives remain challenging tasks. However, this becomes even more and more important as there are increasing numbers of applications being equipped with electrical machines. Some recent examples well-known to the public are wind energy generators and electrical traction drives in the automotive industry. To realize optimal solutions for electrical drive systems it is necessary not only to know some basic equations for machine calculation, but also to deeply understand the principles and limitations of electrical machines and drives.

To foster the know-how in this technical field, this book *Electrical Machines* starts with some basic considerations to introduce the reader to electromagnetic circuit calculation. This is followed by the description of the steady-state operation of the most important machine topologies and afterwards by the dynamic operation and control methods. Continuously giving detailed mathematical deductions to all topics guarantees an optimal understanding of the underlying principles. Therefore, this book contributes to a comprehensive expert knowledge in electrical machines and drives. Consequently, it will be very useful for academia as well as for industry by supporting senior students and engineers in conceiving and designing electrical machines and drives.

After introducing Maxwell's equations and some principles of electromagnetic circuit calculation, the first part of the book is dedicated to the steady-state operation of electrical machines. The detailed description of the brushed DC-machine is followed by the rotating field theory, which in particular explains in detail the winding factors and harmonics of the magneto-motive force of distributed windings. On this basis, induction machines and synchronous machines are described. This first part of the book is completed by regarding permanent magnet machines, switched reluctance machines, and small machines for single-phase use.

Dynamic operation and control of electrical machines are the topics of the second part of this book, starting with some fundamental considerations. Next, the dynamic operation of brushed DC-machines and their control is described (in particular cascaded control using PI-controllers and their adjustment rules). A very important concept for calculating the dynamic operation of rotating field machines is the space vector theory; this is deduced and explained in detail in the following chapter. Then, the dynamic behavior of induction machines and synchronous machines follows, including the description of important control methods like field-oriented control (FOC) and direct torque control (DTC). The permanent magnet machine with surface mounted magnets (SPM) or interior magnets (IPM) is explained concerning the differences of both in torque control and concerning the maximum torque per ampere (MTPA) control method. The last chapter gives an overview of latest research results concerning concentrated windings.

In spite of being a new contribution to a comprehensive understanding of electrical machines and the respective actual developments the reader may find parts of the contents even in different literature, as this book explains the fundamentals of electrical machines (steady-state and dynamic operation as well as control). Concerning these fundamentals it is nearly impossible to list all relevant literature during the text layout. Therefore, the most important references are given at the end of each chapter. In addition, parts of the lectures of Prof. H. Bausch (Universitaet der Bundeswehr Munich, Germany) and Prof. G. Henneberger (RWTH Aachen, Germany) were used as a basis.

The author deeply wishes to express his grateful acknowledgment to all team members of his Chair of Electrical Drives and Actuators at the Universitaet der Bundeswehr Munich and of the spin-off company FEAAM GmbH for their most valuable discussions and support. In particular this holds for (in alphabetical order) Dr.-Ing. Gurakuq Dajaku, Mrs. Lara Kauke, and most notably Dr.-Ing. Hans-Joachim Koebler. Without their beneficial contributions this book would not have been possible in such a high quality.

Last, but not least the author exceedingly thanks his wife and his daughters for their respectfulness and understanding not only concerning the effort being accompanied by writing this book, but even concerning the expenditure of time the author dedicates to professional activities.

Munich, April 2014 *Dieter Gerling*

Contents

Preface .... **v**

Contents .... **vii**

1 Fundamentals **1**

1.1 Maxwell's Equations 1

1.1.1 The Maxwell's Equations in Differential Form 1

1.1.2 The Maxwell's Equations in Integral Form 2

1.1.2.1 Ampere's Law (First Maxwell's Equation in Integral Form) 2

1.1.2.2 Faraday's Law, Law of Induction (Second Maxwell's Equation in Integral Form) 3

1.1.2.3 Law of Direction 4

1.1.2.4 The Third Maxwell's Equation in Integral Form 4

1.1.2.5 The Fourth Maxwell's Equation in Integral Form 4

1.1.2.6 Examples for the Ampere's Law (First Maxwell's Equation in Integral Form) 5

1.1.2.7 Examples for the Faraday's Law (Second Maxwell's Equation in Integral Form) 7

1.2 Definition of Positive Directions 14

1.3 Energy, Force, Power 16

1.4 Complex Phasors 28

1.5 Star and Delta Connection 30

1.6 Symmetric Components 31

1.7 Mutual Inductivity 32

1.8 Iron Losses 34

1.9 References for Chapter 1 35

2 DC-Machines **37**

2.1 Principle Construction 37

2.2 Voltage and Torque Generation, Commutation 38

2.3 Number of Pole Pairs, Winding Design 41

2.4 Main Equations of the DC-Machine 45

2.4.1 First Main Equation: Induced Voltage 45

2.4.2 Second Main Equation: Torque 47

2.4.3 Third Main Equation: Terminal Voltage 48

2.4.4 Power Balance 48
2.4.5 Utilization Factor 49
2.5 Induced Voltage and Torque, Precise Consideration 51
2.5.1 Induced Voltage 51
2.5.2 Torque 53
2.6 Separately Excited DC-Machines 56
2.7 Permanent Magnet Excited DC-Machines 62
2.8 Shunt-Wound DC-Machines 69
2.9 Series-Wound DC-Machines 73
2.10 Compound DC-Machines 77
2.11 Generation of a Variable Terminal Voltage 78
2.12 Armature Reaction 80
2.13 Commutation Pole 84
2.14 References for Chapter 2 88

3 Rotating Field Theory 89
3.1 Stator of a Rotating Field Machine 89
3.2 Current Loading 90
3.3 Alternating and Rotating Magneto-Motive Force 93
3.4 Winding Factor 104
3.5 Current Loading and Flux Density 114
3.5.1. Fundamentals 114
3.5.2. Uniformly Distributed Current Loading in a Zone 114
3.5.3. Current Loading Concentrated in the Middle of Each Slot 115
3.5.4. Current Loading Distributed Across Each Slot Opening 116
3.5.5. Rotating Air-Gap Field 116
3.6 Induced Voltage and Slip 120
3.7 Torque and Power 126
3.8 References for Chapter 3 134

4 Induction Machines 135
4.1 Construction and Equivalent Circuit Diagram 135
4.2 Resistances and Inductivities 141
4.2.1 Phase Resistance 141
4.2.2 Main Inductivity 142
4.2.3 Leakage Inductivity 142
4.2.3.1 Harmonic Leakage 142
4.2.3.2 Slot Leakage 143
4.2.3.3 End Winding Leakage 145
4.3 Operating Characteristics 145
4.3.1 Heyland-Diagram (Stator Phase Current Locus Diagram) 145
4.3.2 Torque and Power 151
4.3.3 Torque as a Function of Slip 154
4.3.4 Series Resistance in the Rotor Circuit 157

4.3.5 Operation with Optimum Power Factor 159
4.3.6 Further Equations for Calculating the Torque 164
4.4 Squirrel Cage Rotor 166
4.4.1 Fundamentals 166
4.4.2 Skewed Rotor Slots 170
4.4.3 Skin Effect 175
4.5 Possibilities for Open-Loop Speed Control 179
4.5.1 Changing (Increasing) the Slip 179
4.5.2 Changing the Supply Frequency 179
4.5.3 Changing the Number of Pole Pairs 181
4.6 Star-Delta-Switching 182
4.7 Doubly-Fed Induction Machine 183
4.8 References for Chapter 4 188

5 Synchronous Machines 189
5.1 Equivalent Circuit and Phasor Diagram 189
5.2 Types of Construction 195
5.2.1 Overview 195
5.2.2 High-Speed Generator with Cylindrical Rotor 196
5.2.3 Salient-Pole Generator 196
5.3 Operation at Fixed Mains Supply 196
5.3.1 Switching to the Mains 196
5.3.2 Torque Generation 198
5.3.3 Operating Areas 200
5.3.4 Operating Limits 203
5.4 Isolated Operation 205
5.4.1 Load Characteristics 205
5.4.2 Control Characteristics 207
5.5 Salient-Pole Synchronous Machines 209
5.6 References for Chapter 5 217

6 Permanent Magnet Excited Rotating Field Machines 219
6.1 Rotor Construction 219
6.2 Linestart-Motor 220
6.3 Electronically Commutated Rotating Field Machine with Surface Mounted Magnets 220
6.3.1 Fundamentals 220
6.3.2 Brushless DC-Motor 222
6.3.3 Electronically Commutated Permanent Magnet Excited Synchronous Machine 228
6.4 Calculation of the Operational Characteristics; Permanent Magnet Excited Machines with Buried Magnets 230
6.5 References for Chapter 6 230

7 Reluctance Machines 231
7.1 Synchronous Reluctance Machines 231
7.2 Switched Reluctance Machines 232
7.2.1 Construction and Operation 232
7.2.2 Torque 234
7.2.3 Modes of Operation 239
7.2.4 Alternative Power Electronic Circuits 242
7.2.5 Main Characteristics 244
7.3 References for Chapter 7 244

8 Small Machines for Single-Phase Operation 247
8.1 Fundamentals 247
8.2 Universal Motor 247
8.3 Single-Phase Induction Machine 250
8.3.1 Single-Phase Operation of Three-Phase Induction Machine 250
8.3.2 Single-Phase Induction Motor with Auxiliary Phase 252
8.3.3 Shaded-Pole (Split-Pole) Motor 253
8.4 References for Chapter 8 254

9 Fundamentals of Dynamic Operation 255
9.1 Fundamental Dynamic Law, Equation of Motion 255
9.1.1 Translatory Motion 255
9.1.2 Translatory / Rotatory Motion 255
9.1.3 Rotatory Motion 256
9.1.4 Stability 257
9.2 Mass Moment of Inertia 258
9.2.1 Inertia of an Arbitrary Body 258
9.2.2 Inertia of a Hollow Cylinder 259
9.3 Simple Gear-Sets 260
9.3.1 Assumptions 260
9.3.2 Rotation / Rotation (e.g. Gear Transmission) 260
9.3.3 Rotation / Translation (e.g. Lift Application) 261
9.4 Power and Energy 262
9.5 Slow Speed Change 264
9.5.1 Fundamentals 264
9.5.2 First Example 264
9.5.3 Second Example 265
9.6 Losses during Starting and Braking 267
9.6.1 Operation without Load Torque 267
9.6.2 Operation with Load Torque 270
9.7 References for Chapter 9 271

10 Dynamic Operation and Control of DC-Machines 273
10.1 Set of Equations for Dynamic Operation 273
10.2 Separately Excited DC-Machines 277
10.2.1 General Structure 277
10.2.2 Response to Setpoint Changes 278
10.2.3 Response to Disturbance Changes 283
10.3 Shunt-Wound DC-Machines 286
10.4 Cascaded Control of DC-Machines 288
10.5 Adjusting Rules for PI-Controllers 291
10.5.1 Overview 291
10.5.2 Adjusting to Optimal Response to Setpoint Changes (Rule "Optimum of Magnitude") 292
10.5.3 Adjusting to Optimal Response to Disturbances (Rule "Symmetrical Optimum") 293
10.5.4 Application of the Adjusting Rules to the Cascaded Control of DC-Machines 294
10.6 References for Chapter 10 295

11 Space Vector Theory 297
11.1 Methods for Field Calculation 297
11.2 Requirements for the Application of the Space Vector Theory 298
11.3 Definition of the Complex Space Vector 299
11.4 Voltage Equation in Space Vector Notation 303
11.5 Interpretation of the Space Vector Description 305
11.6 Coupled Systems 306
11.7 Power in Space Vector Notation 309
11.8 Elements of the Equivalent Circuit 313
11.8.1 Resistances 313
11.8.2 Inductivities 314
11.8.3 Summary of Results 316
11.9 Torque in Space Vector Notation 317
11.9.1 General Torque Calculation 317
11.9.2 Torque Calculation by Means of Cross Product from Stator Flux Linkage and Stator Current 318
11.9.3 Torque Calculation by Means of Cross Product from Stator and Rotor Current 319
11.9.4 Torque Calculation by Means of Cross Product from Rotor Flux Linkage and Rotor Current 319
11.9.5 Torque Calculation by Means of Cross Product from Stator and Rotor Flux Linkage 320
11.10 Special Coordinate Systems 321
11.11 Relation between Space Vector Theory and Two-Axis-Theory 322
11.12 Relation between Space Vectors and Phasors 323
11.13 References for Chapter 11 324

12 Dynamic Operation and Control of Induction Machines 325
12.1 Steady-State Operation of Induction Machines in Space Vector Notation at No-Load 325
12.1.1 Set of Equations 325
12.1.2 Steady-State Operation at No-Load 326
12.2 Fast Acceleration and Sudden Load Change 328
12.3 Field-Oriented Coordinate System for Induction Machines 334
12.4 Field-Oriented Control of Induction Machines with Impressed Stator Currents 344
12.5 Field-Oriented Control of Induction Machines with Impressed Stator Voltages 356
12.6 Field-Oriented Control of Induction Machines without Mechanical Sensor (Speed or Position Sensor) 358
12.7 Direct Torque Control 360
12.8 References for Chapter 12 367

13 Dynamic Operation of Synchronous Machines 369
13.1 Oscillations of Synchronous Machines, Damper Winding 369
13.2 Steady-State Operation of Non Salient-Pole Synchronous Machines in Space Vector Notation 374
13.3 Sudden Short-Circuit of Non Salient-Pole Synchronous Machines 381
13.3.1 Fundamentals 381
13.3.2 Initial Conditions for t = 0 381
13.3.3 Set of Equations for t > 0 383
13.3.4 Maximum Voltage Switching 390
13.3.5 Zero Voltage Switching 394
13.3.6 Sudden Short-Circuit with Changing Speed and Rough Synchronization 395
13.3.7 Physical Explanation of the Sudden Short-Circuit 400
13.4 Steady-State Operation of Salient-Pole Synchronous Machines in Space Vector Notation 402
13.5 Sudden Short-Circuit of Salient-Pole Synchronous Machines 410
13.5.1 Initial Conditions for t = 0 410
13.5.2 Set of Equations for t > 0 412
13.6 Transient Operation of Salient-Pole Synchronous Machines 415
13.7 References for Chapter 13 423

14 Dynamic Operation and Control of Permanent Magnet Excited Rotating Field Machines 425
14.1 Principle Operation 425
14.2 Set of Equations for the Dynamic Operation 426
14.3 Steady-State Operation 432
14.3.1 Fundamentals 432
14.3.2 Base Speed Operation 432

14.3.3 Operation with Leading Load Angle and without Magnetic Asymmetry 434
14.3.4 Operation with Leading Load Angle and Magnetic Asymmetry 436
14.3.5 Torque Calculation from Current Loading and Flux Density 438
14.4 Limiting Characteristics and Torque Control 440
14.4.1 Limiting Characteristics 440
14.4.2 Torque Control 442
14.5 Control without Mechanical Sensor 446
14.6 References for Chapter 14 447

15 Concentrated Windings 449
15.1 Conventional Concentrated Windings 449
15.2 Improved Concentrated Windings 453
15.2.1 Increased Number of Stator Slots from 12 to 24 453
15.2.2 Increased Number of Stator Slots from 12 to 18 460
15.2.3 Main Characteristics of the Improved Concentrated Windings 461
15.3 References for Chapter 15 462

16 Lists of Symbols, Indices and Acronyms 463
16.1 List of Symbols 463
16.2 List of Indices 467
16.3 List of Acronyms 470

Index 471

1 Fundamentals

1.1 Maxwell's Equations

1.1.1 The Maxwell's Equations in Differential Form

The basis for all following considerations are the Maxwell's equations. In differential form these are (the time-dependent variation of the displacement current $\vec{D}$ can always be neglected against the current density $\vec{J}$ for all technical systems regarded here):

1. Maxwell's equation

$$\mathrm{rot}\vec{H} = \vec{J} + \frac{d\vec{D}}{dt} \approx \vec{J} \tag{1.1}$$

2. Maxwell's equation

$$\mathrm{rot}\vec{E} = -\frac{d\vec{B}}{dt} \tag{1.2}$$

3. Maxwell's equation

$$\mathrm{div}\vec{B} = 0 \tag{1.3}$$

4. Maxwell's equation

$$\mathrm{div}\vec{D} = \rho \tag{1.4}$$

The material equations are:

$$\vec{B} = \mu\vec{H} \tag{1.5}$$

$$\vec{D} = \varepsilon\vec{E} \tag{1.6}$$

$$\vec{J} = \gamma\vec{E} \tag{1.7}$$

The used variables have the following meaning:

$\vec{H}$ the vector field of the magnetic field strength;
$\vec{J}$ the vector field of the electrical current density;
$\vec{D}$ the vector field of the displacement current;
$\vec{E}$ the vector field of the electric field strength;
$\vec{B}$ the vector field of the magnetic flux density;
ρ the scalar field of the charge density;
μ the scalar field of the permeability (in vacuum or air there is: $\mu = \mu_0$);
ε the scalar field of the dielectric constant (in vacuum or air there is: $\varepsilon = \varepsilon_0$);
γ the scalar field of the electric conductivity.

The expression "vector field" means that the vector quantity depends on all (usually three) geometric coordinates; the expression "scalar field" means that scalar quantity depends on all geometric coordinates.

In the case of homogeneous, isotropic materials the scalar fields μ, ε and γ are reduced to space-independent material constants.

1.1.2 The Maxwell's Equations in Integral Form

1.1.2.1 Ampere's Law (First Maxwell's Equation in Integral Form)

The first Maxwell's equation in integral form is

$$\oint \vec{H}d\vec{\ell} = \int_A \vec{J}d\vec{A} \tag{1.8}$$

The line integral of the magnetic field strength $\vec{H}$ on a closed geometric integration loop $\vec{\ell}$ ("magnetic circulation voltage") is equal to the total electric current flowing through the area A limited by this loop ("magneto-motive force", "ampere-turns"), if the displacement current is neglected.

For graphical explanation see Fig. 1.1.

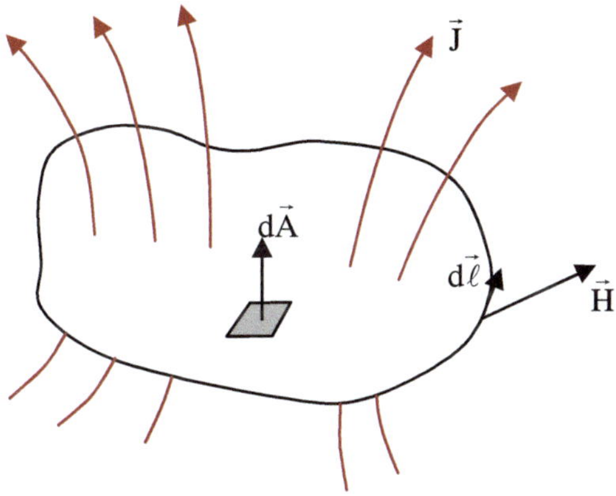

Fig. 1.1. Explanation of Ampere's Law.

1.1.2.2 Faraday's Law, Law of Induction (Second Maxwell's Equation in Integral Form)

The second Maxwell's equation in integral form is

$$\oint \vec{E} d\vec{\ell} = -\frac{d}{dt} \int_A \vec{B} d\vec{A} \tag{1.9}$$

with the magnetic flux being

$$\int_A \vec{B} d\vec{A} = \Phi \tag{1.10}$$

The line integral of the electric field strength $\vec{E}$ on a closed geometric integration loop $\vec{\ell}$ ("electric circulation voltage") is equal to the negative time-dependent variation of the total magnetic flux, that penetrates the area A limited by this loop.

For graphical explanation see Fig. 1.2.

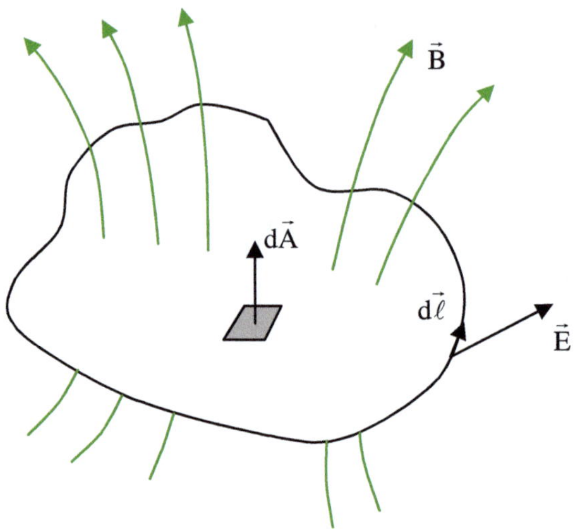

Fig. 1.2. Explanation of Faraday's Law.

1.1.2.3 Law of Direction

The positive direction of the vectors $d\vec{\ell}$ and $\vec{A}$ are defined according to a right-handed screw.

1.1.2.4 The Third Maxwell's Equation in Integral Form

The third Maxwell's equation in integral form is

$$\oint_A \vec{B} d\vec{A} = 0 \tag{1.11}$$

The total magnetic flux penetrating a closed surface of any volume is zero, i.e. there are no single magnetic poles.

1.1.2.5 The Fourth Maxwell's Equation in Integral Form

The fourth Maxwell's equation in integral form is

$$\oint_A \vec{D} d\vec{A} = \int_V \rho dV \tag{1.12}$$

The reason for the total electric field penetrating a closed surface of any volume are the electric charges inside this volume.

1.1.2.6 Examples for the Ampere's Law (First Maxwell's Equation in Integral Form)

The Ampere's Law is an integral law. It shows the dependency of the magneto-motive force (ampere-turns) and the magnetic circulation voltage, but in general it cannot be used to calculate the magnetic field strength. For the calculation of the magnetic field strength $\vec{H}$ at given magneto-motive force additional knowledge of the field is necessary (e.g. symmetry characteristics or simplifying assumptions).

1. Example:

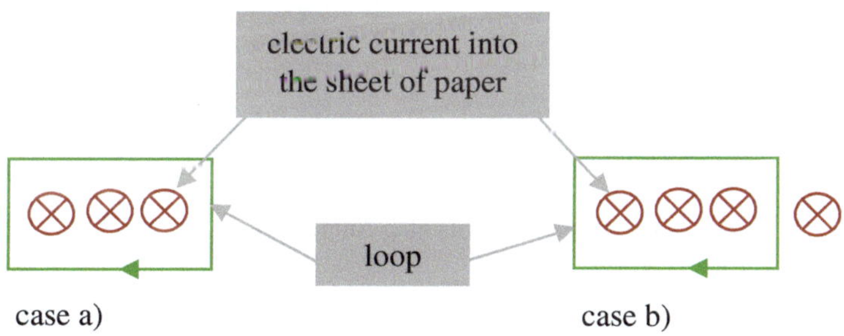

Fig. 1.3. Example for explaining Ampere's Law.

- The magneto-motive force, the integration loop and the magnetic circulation voltage are the same in both cases (Fig. 1.3).
- But the distribution of the magnetic field strength on the integration loop is different (because of the additional current in case b)).
- The calculation of the magnetic field strength is not possible in both cases without additional information.

2. Example:

Calculation of the magnetic field of a straight, current carrying conductor (with the radius R) in air.

Because of the symmetry the magnitude of the field strength is constant at constant distance r from the center of the conductor.

a) Solution outside the conductor:

$$\oint \vec{H}d\vec{\ell} = H_{out}\,2\pi r_{out} = J\pi R^2 \quad \Rightarrow \quad H_{out} = \frac{J}{2}\frac{R^2}{r_{out}} \tag{1.13}$$

b) Solution inside the conductor (J is assumed being equally distributed across the conductor cross section):

$$\oint \vec{H} d\vec{\ell} = H_{in} 2\pi r_{in} = J\pi r_{in}^2 \quad \Rightarrow \quad H_{in} = \frac{J}{2} r_{in} \tag{1.14}$$

3. Example:

The following magnetic circuit is given (Fig. 1.4):

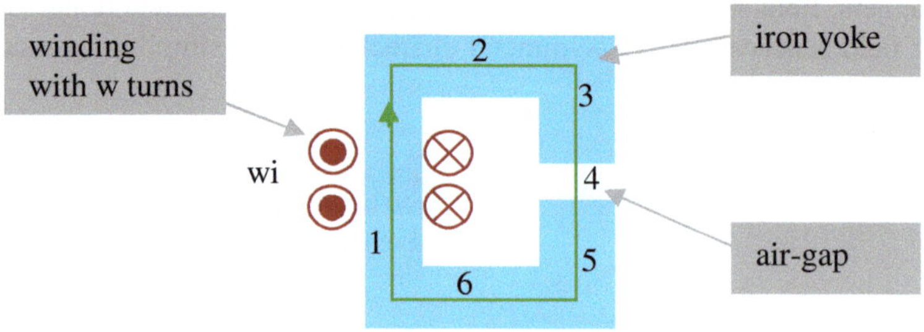

Fig. 1.4. Example for explaining Ampere's Law.

The following is assumed:

- The magnetic circuit may be separated in a finite number of parts ($\nu = 1 \ldots 6$).
- H_ν is constant in each part.
- A closed loop may be described by using a mean field line length.
- The leakage flux is negligible: $\Phi_\nu = \Phi = \text{const.}$

Now, the Ampere's Law is:

$$\oint \vec{H} d\vec{\ell} = \sum_{\nu=1}^{6} H_\nu \ell_\nu = wi \quad \text{with} \quad H_\nu = \frac{B_\nu}{\mu_\nu}; \quad B_\nu = \frac{\Phi_\nu}{A_\nu} \tag{1.15}$$

For $\mu_{Fe,\nu} \to \infty$ it is further:

$$\oint \vec{H} d\vec{\ell} = H_4 \ell_4 = wi \tag{1.16}$$

It follows:

$$B_4 = \mu_0 \frac{wi}{l_4} \tag{1.17}$$

and

$$\Phi_4 = \Phi = B_4 A_4 \tag{1.18}$$

Therefore, the flux density in the different parts becomes:

$$B_v = B_4 \frac{A_4}{A_v} \tag{1.19}$$

1.1.2.7 Examples for the Faraday's Law (Second Maxwell's Equation in Integral Form)

In a closed conductor loop (that is used as integration loop) there is an electrical circulation voltage ("magnetic loss"), if the magnetic flux linked with this conductor loop changes with time:

$$\oint \vec{E} d\vec{\ell} = -\frac{d}{dt} \int_A \vec{B} d\vec{A} = -\frac{d}{dt} \Phi \tag{1.20}$$

Regarding a winding with w turns, the Faraday's Law becomes:

$$\oint \vec{E} d\vec{\ell} = -\frac{d}{dt} w \int_A \vec{B} d\vec{A} = -\frac{d}{dt} w\Phi = -\frac{d}{dt} \Psi \tag{1.21}$$

The time-dependent variation of the flux may originate from:

- time-dependent variation of the induction with stationary conductor loop;
- movement of the conductor loop (totally or partly) relative to the stationary magnetic field.

Obviously, the difference comes from the choice of the coordinate system.

1. Example: stationary winding, time-dependent induction

There is:

$$\oint \vec{E}d\vec{\ell} = -\frac{d}{dt}\Psi = -\frac{\partial\Psi}{\partial i}\frac{di}{dt} = -L\frac{di}{dt} \tag{1.22}$$

This voltage is called “transformer voltage”.

2. Example: moved winding, induction constant in time

The following movement of a conductor loop is regarded (Fig. 1.5):

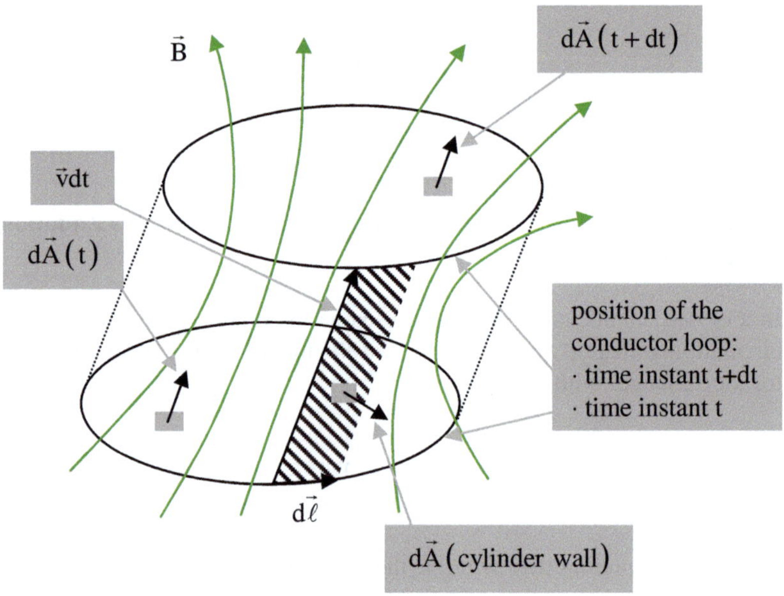

Fig. 1.5. Example for explaining Faraday’s Law.

From $\text{div}\vec{B} = 0$ it follows:

$$\oint_A \vec{B}d\vec{A} = \int_{A(t+dt)} \vec{B}d\vec{A} - \int_{A(t)} \vec{B}d\vec{A} + \int_{\text{cylinder wall}} \vec{B}d\vec{A} = 0 \tag{1.23}$$

The variation of the flux linked with the regarded winding is:

$$d\Psi = \int_{A(t+dt)} \vec{B}d\vec{A} - \int_{A(t)} \vec{B}d\vec{A} \tag{1.24}$$

For the cylinder wall it is:

$$d\vec{A} = -\left(\vec{v}dt \times d\vec{\ell}\right) = -dt\left(\vec{v} \times d\vec{\ell}\right) \tag{1.25}$$

Therefore:

$$\begin{aligned}\int\limits_{\text{cylinder wall}} \vec{B}d\vec{A} &= -dt\oint \vec{B}\left(\vec{v} \times d\vec{\ell}\right) = -dt\oint\left(\vec{B} \times \vec{v}\right)d\vec{\ell} \\ &= dt\oint\left(\vec{v} \times \vec{B}\right)d\vec{\ell}\end{aligned} \tag{1.26}$$

and further:

$$d\Psi + dt\oint\left(\vec{v} \times \vec{B}\right)d\vec{\ell} = 0 \quad \Rightarrow \quad -\frac{d\Psi}{dt} = \oint\left(\vec{v} \times \vec{B}\right)d\vec{\ell} \tag{1.27}$$

In total this results in:

$$\oint \vec{E}d\vec{\ell} = -\frac{d}{dt}\Psi = \oint\left(\vec{v} \times \vec{B}\right)d\vec{\ell} \tag{1.28}$$

This voltage is called "voltage of movement".

3. Example: short circuit of a conductor loop

A conductor loop (cross section A_{wire}, conductivity γ and resistance R) is penetrated by a time-dependent magnetic field, see Fig. 1.6.

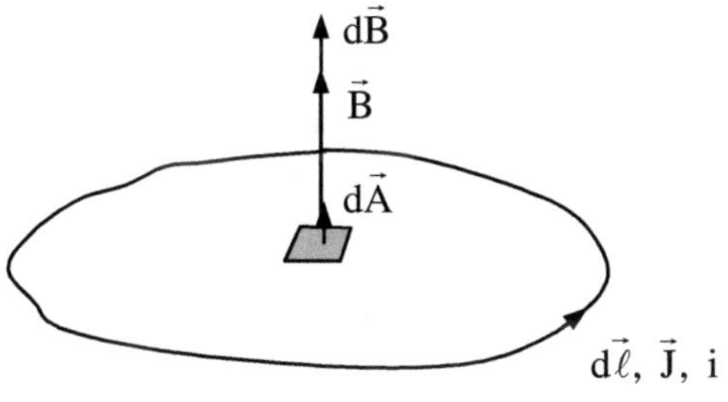

Fig. 1.6. Example for explaining Faraday's Law.

From $\oint \vec{E}d\vec{\ell} = -\frac{d}{dt}\Psi$ and $\vec{J} = \gamma\vec{E}$ it follows:

$$\oint \frac{\vec{J}}{\gamma}\, d\vec{\ell} = -\frac{d}{dt}\Psi \tag{1.29}$$

Because on each point of the conductor the directions of $d\vec{\ell}$ und $\vec{J}$ are identical, with $J = \frac{i}{A_{wire}}$ the following is true:

$$i\oint \frac{d\ell}{\gamma A_{wire}} = iR = -\frac{d}{dt}\Psi \quad \Rightarrow \quad 0 = iR + \frac{d}{dt}\Psi \tag{1.30}$$

If $\frac{\partial \vec{B}}{\partial t} > 0$ is true, $\vec{J} = \gamma\vec{E}$ will flow against the direction of $d\vec{\ell}$.

Lenz's Law: The current caused by induction variation (induced current) always flows in that direction that its magnetic field opposes the generating induction variation.

4. Example: conductor loop in open-circuit

Opening the above conductor loop the situation shown in Fig. 1.7 is obtained.

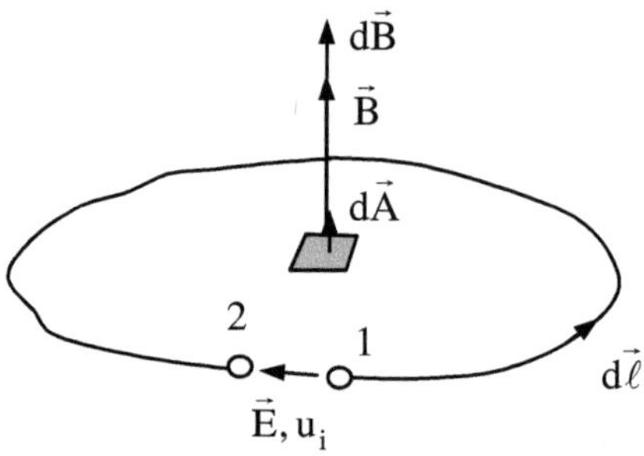

Fig. 1.7. Example for explaining Faraday's Law.

There is

$$\oint \vec{E}d\vec{\ell} = \int_2^1 \vec{E}d\vec{\ell} + \int_1^2 \vec{E}d\vec{\ell} \tag{1.31}$$

where the direction of $d\vec{\ell}$ determines the execution of the integral. As on the path from "1" to "2" the conductivity γ is limited, but the current (and therefore even the current density) is zero because of the open terminals, from $\vec{J} = \gamma\vec{E}$ even $\vec{E} = 0$ is obtained. It remains:[1]

$$\oint \vec{E} d\vec{\ell} = \int_2^1 \vec{E} d\vec{\ell} = -\frac{d}{dt}\Psi = -u_i \tag{1.32}$$

Note: The negative sign is valid only for the positive directions shown in Fig. 1.7!

5. Example: Electrical Circuit

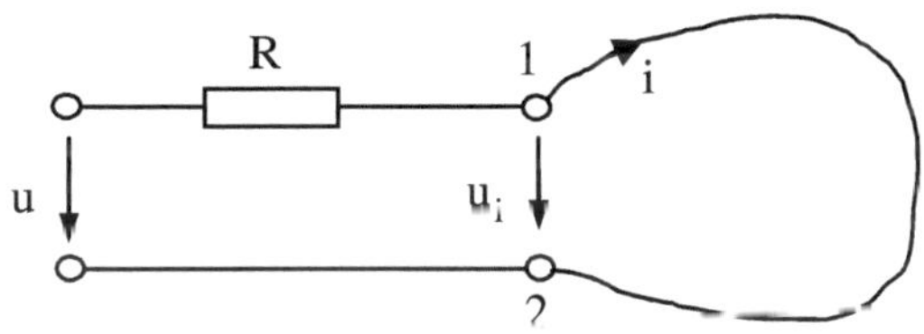

Fig. 1.8. Example for explaining Faraday's Law.

In Fig. 1.8 the voltage at the outside terminals should be fixed by a voltage source to a certain value u; the current i is flowing. It follows:

$$u = Ri + u_i = Ri + \frac{d}{dt}\Psi \tag{1.33}$$

Even here the signs are used according to the defined positive directions.

6. Example: moved coil in a stationary magnetic field

There is a flat rectangular coil with a single turn (the extension in x-direction is τ, the extension in z-direction is ℓ_z; please refer to Fig. 1.9). This coil is moved in x-direction with the speed v, penetrating a stationary stepwise magnetic field being constant in time with

---

[1] In the field theory often $\oint \vec{E} d\vec{\ell} = u_i$ is defined; the definition of u_i used here turned out to be appropriate for electrical machines and therefore will be used further. Sometimes the induced voltage (also called "back electromotive force", "back emf" or "counter emf") is nominated with "e". As it has the nature of a voltage, here the name "u_i" is preferred.

$$B(x) = -B(x - \tau) \tag{1.34}$$

The following holds true:

$$\begin{aligned} u_i &= -\oint \vec{E}_i d\vec{\ell} = -\oint (\vec{v} \times \vec{B}) d\vec{\ell} \\ &= 2v|B(x)|\ell_z \end{aligned} \tag{1.35}$$

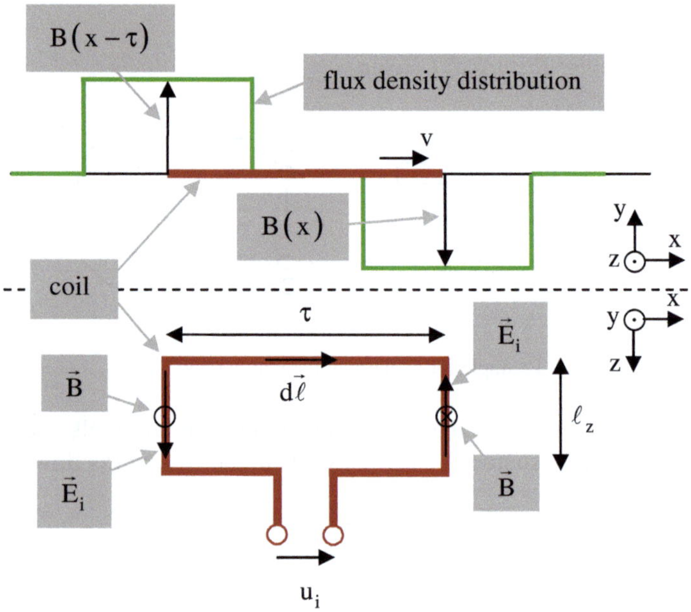

Fig. 1.9. Example for explaining Faraday's Law.

Even here the signs are valid regarding the positive directions of circulation and voltage.

For constant speed v the time-dependent characteristic of the induced voltage corresponds to the space-dependent characteristic of the induction.

7. Example: stationary coil in a moved magnetic field:

There is a flat, stationary, rectangular coil with w turns and a magnetic travelling field

$$B(x,t) = \hat{B}\cos\left(\omega t + \varphi - \frac{\pi}{\tau}x\right), \qquad \omega = 2\pi f \tag{1.36}$$

with the angular frequency ω, the phase angle φ and the pole pitch τ (half wave length). The extension of the coil in the direction of movement of the travelling wave (x-direction) is s (effective width), the extension perpendicular to the direction of movement (z-direction) is ℓ_z (effective length), see Fig. 1.10.

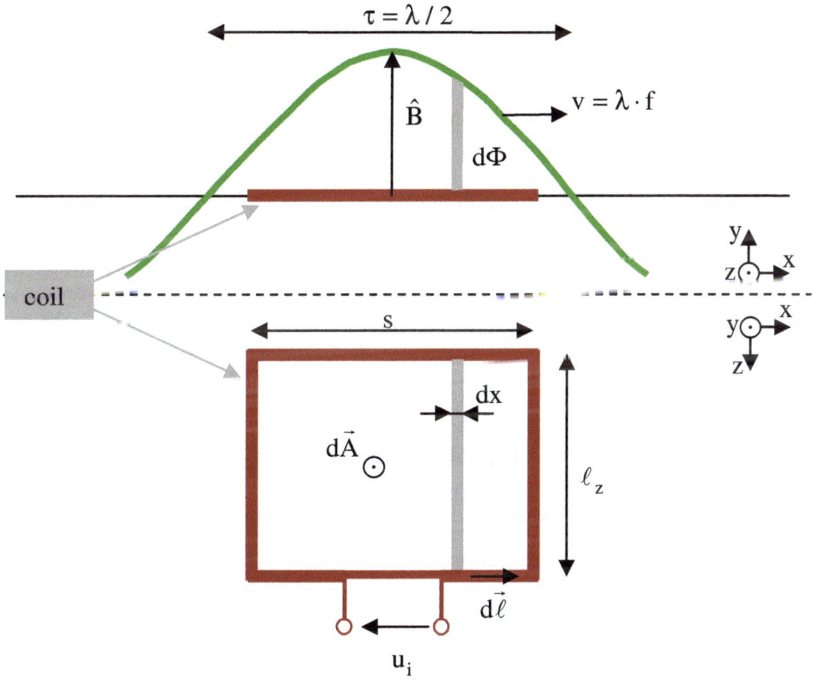

Fig. 1.10. Example for explaining Faraday's Law.

The flux linked with the coil is:

$$\Psi = w\int_A B(x,t)dA = w\hat{B}\ell_z \int_{-s/2}^{s/2} \cos\left(\omega t + \varphi - \frac{\pi}{\tau}x\right) dx$$

$$= w\hat{B}\ell_z \int_{-s/2}^{s/2} \left[\cos(\omega t + \varphi)\cos\left(-\frac{\pi}{\tau}x\right) - \sin(\omega t + \varphi)\sin\left(-\frac{\pi}{\tau}x\right)\right] dx \tag{1.37}$$

This integral can be solved like follows:

$$\begin{aligned}
\Psi &= w\hat{B}\ell_z\left[\cos(\omega t+\varphi)\left(-\frac{\tau}{\pi}\right)\sin\left(-\frac{\pi}{\tau}x\right)\right. \\
&\qquad\left. -\sin(\omega t+\varphi)\left(-\frac{\tau}{\pi}\right)\left(-\cos\left(-\frac{\pi}{\tau}x\right)\right)\right]_{-s/2}^{s/2} \\
&= w\hat{B}\ell_z\cos(\omega t+\varphi)\left(-\frac{\tau}{\pi}\right)\left[\sin\left(-\frac{\pi}{\tau}\frac{s}{2}\right)-\sin\left(\frac{\pi}{\tau}\frac{s}{2}\right)\right] \\
&= w\hat{B}\ell_z\cos(\omega t+\varphi)\frac{\tau}{\pi}2\sin\left(\frac{s}{\tau}\frac{\pi}{2}\right) \\
&= w\xi\frac{2}{\pi}\hat{B}\tau\ell_z\cos(\omega t+\varphi), \qquad \xi=\sin\left(\frac{s}{\tau}\frac{\pi}{2}\right)\le 1
\end{aligned} \tag{1.38}$$

This equation can be interpreted like follows:

- $\hat{B}\frac{2}{\pi}$ is the mean value of a half harmonic flux density wave.
- $\hat{B}\frac{2}{\pi}\tau\ell_z$ is then the mean flux penetrating through the area $\tau\ell_z$.
- The factor ξ is called "short-pitch factor" and it reduces this flux to that amount penetrating through the area $s\ell_z$.
- The number of turns w transforms the flux to the flux linkage.
- The $\cos$ -term shows the time dependency and the phase shift.

The induced voltage is (the positive directions of $d\vec{A}$ and $\vec{B}$ or $d\vec{B}$ are identical):

$$u_i = \frac{d}{dt}\Psi = -\omega\hat{\Psi}\sin(\omega t+\varphi), \qquad \hat{\Psi} = w\xi\frac{2}{\pi}\hat{B}\tau\ell_z \tag{1.39}$$

1.2 Definition of Positive Directions

For the unambiguous description of electrical circuits, directions have to be assigned to voltages, currents, and power. The definition of the direction may be chosen arbitrarily. In principle there are two different possibilities as it is shown in Fig. 1.11:

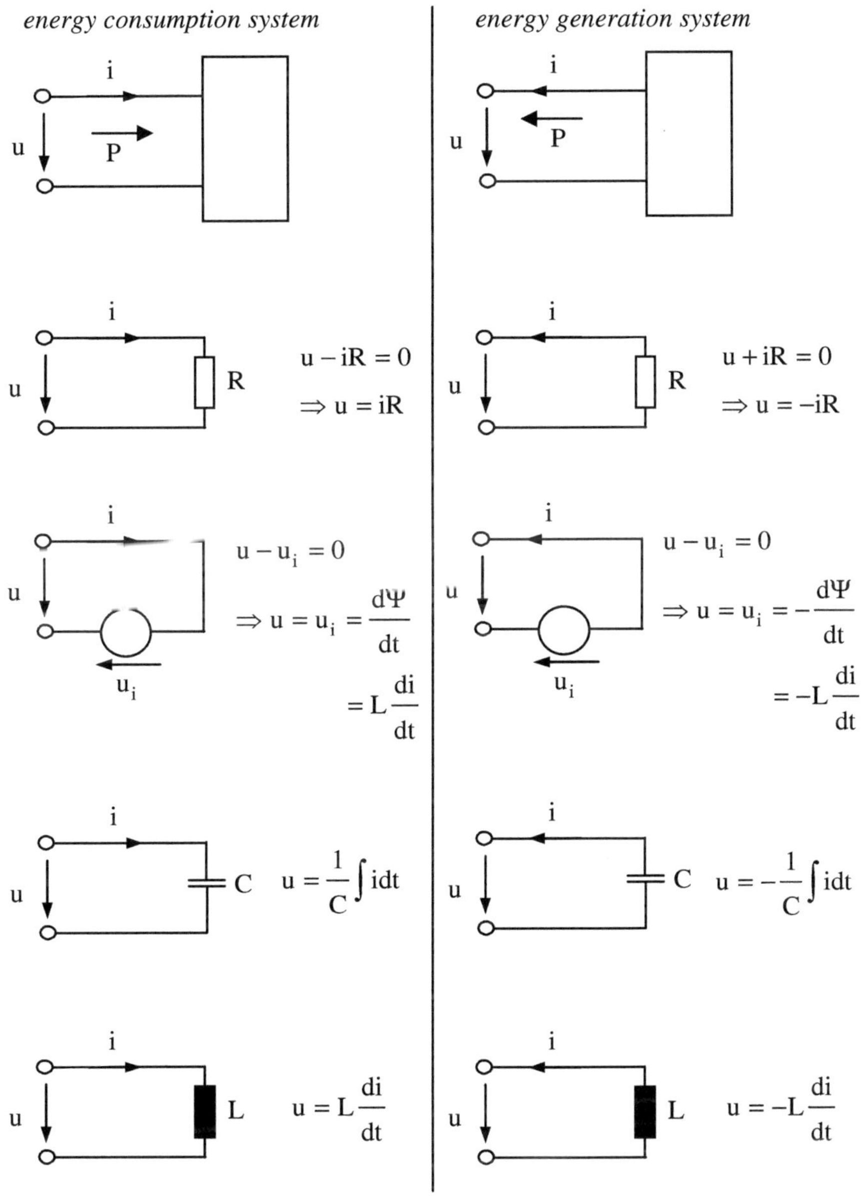

Fig. 1.11. Examples for the energy consumption system (*left*) and the energy generation system (*right*).

Poynting's vector describes the power density in the electromagnetic field, please refer to Fig. 1.12:

$$\vec{S} = \vec{E} \times \vec{H} \tag{1.40}$$

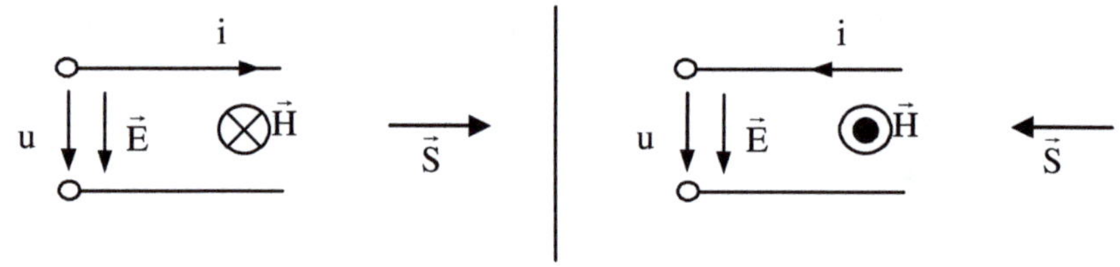

Fig. 1.12. Poynting's vector for the energy consumption system (*left*) and the energy generation system (*right*).

1.3 Energy, Force, Power

The principle of electromechanical energy conversion will be explained using the example of a simple lifting magnet, see Fig. 1.13. The occurring energies can be calculated as follows.

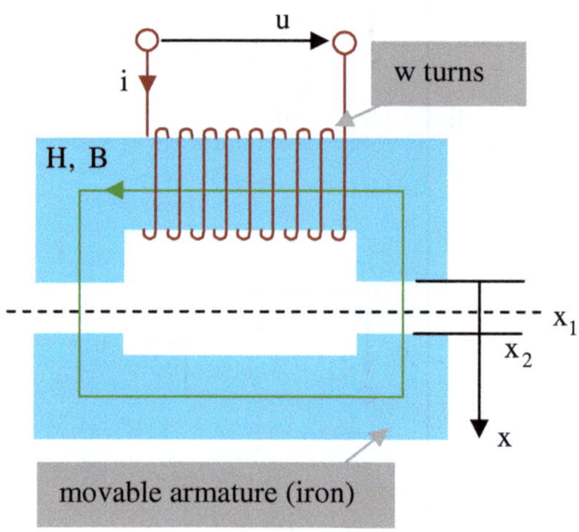

Fig. 1.13. Electromagnetic system with movable armature.

The different possible kinds of energy are electrical energy, electrical losses, magnetic energy, and mechanical energy:

$$W_{el} = \int uidt \tag{1.41}$$

$$W_{loss} = \int i^2 Rdt \tag{1.42}$$

$$W_{mag} = \iiint_V \int HdB\, dV = \int id\Psi \tag{1.43}$$

$$W_{mech,lin} = \int Fdx \qquad W_{mech,rot} = \int Td\alpha \tag{1.44}$$

The B-H-curve of the iron and the Ψ -i-curve of the magnetic circuit in general (i.e. considering magnetic saturation) have the characteristics shown in Fig. 1.14.

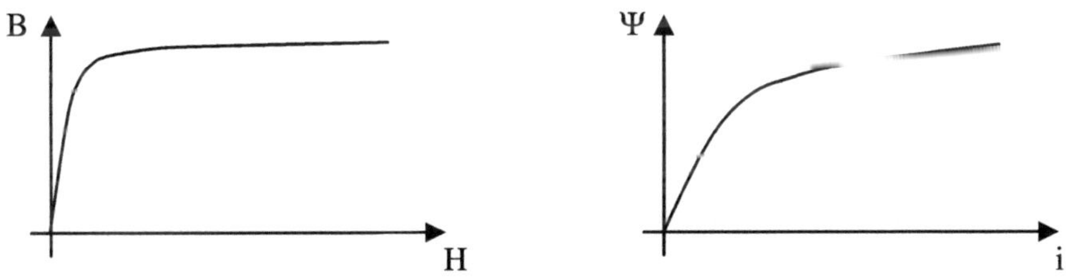

Fig. 1.14. Principle B-H- and Ψ -i-characteristics.

The energy density of the magnetic field in air is:

$$w_{mag} = \frac{1}{2}HB = \frac{B^2}{2\mu_0} \tag{1.45}$$

With $B = 0.5T = 0.5\frac{Vs}{m^2}$ (typical value) and $\mu_0 = 4\pi \cdot 10^{-7}\frac{Vs}{Am}$ it follows:

$$w_{mag} = \frac{0.25\frac{V^2s^2}{m^4}}{8\pi \cdot 10^{-7}\frac{Vs}{Am}} = 0.995 \cdot 10^5 \frac{VAs}{m^3} \approx 1 \cdot 10^5 \frac{N}{m^2} \tag{1.46}$$

The energy density of the electrical field in air is:

$$w_{el} = \frac{1}{2}ED = \frac{1}{2}\varepsilon_0 E^2 \tag{1.47}$$

With $E = 3\frac{kV}{mm}$ (breakdown field strength in air) and $\varepsilon_0 = 8.854 \cdot 10^{-12}\frac{As}{Vm}$ it follows:

$$w_{el} = \frac{1}{2}8.854 \cdot 10^{-12}\frac{As}{Vm}\left(3 \cdot 10^6\frac{V}{m}\right)^2 = 39.8\frac{VAs}{m^3} \approx 40\frac{N}{m^2} \tag{1.48}$$

Because of the considerable lower energy density there are virtually no electrostatic machines, except for extremely small geometries (please refer to equations (1.46) and (1.48)). In the following, the energy stored in the electrical field will be neglected against the energy stored in the magnetic field.

The *energy balance* of the lifting magnet is:

$$dW_{el} = dW_{loss} + dW_{mag} + dW_{mech} \tag{1.49}$$

The electrical energy supplied via the terminals is equal to the sum of losses, change of magnetic energy and change of mechanical energy.

Case 1: fixed armature ($x = \text{const.}$, see Fig. 1.15):

$$\begin{aligned}
& dW_{mech} = 0 = d\left(W_{el} - W_{loss}\right) - dW_{mag} \\
\Rightarrow \quad & d\left(W_{el} - W_{loss}\right) = dW_{mag} \\
\Rightarrow \quad & \left(ui - i^2R\right)dt = dW_{mag} \\
\Rightarrow \quad & u_i i dt = dW_{mag} \\
\Rightarrow \quad & \frac{d\Psi}{dt} i dt = dW_{mag} \\
\Rightarrow \quad & dW_{mag} = i d\Psi
\end{aligned} \tag{1.50}$$

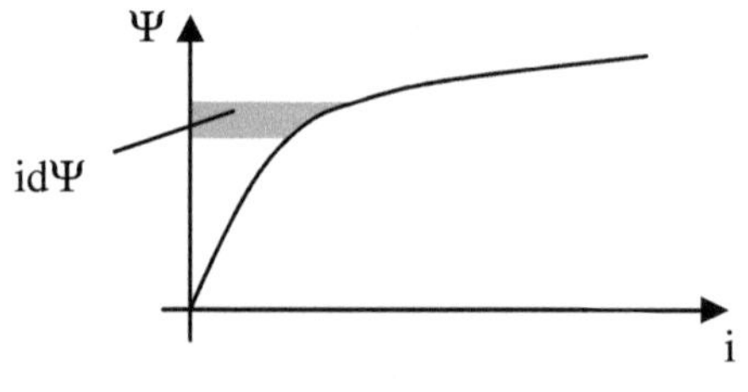

Fig. 1.15. Ψ -i-characteristic for case 1.

The total magnetic energy is:[2]

$$W_{mag} = \int_0^{\Psi} i d\tilde{\Psi} \tag{1.51}$$

Case 2: movable armature, constant current:

The armature is moved from $x = x_1$ to $x = x_2$ with $i_0 = \text{const.}$. Doing this the flux linkage changes from $\Psi = \Psi_1$ to $\Psi = \Psi_2$.

$$\begin{aligned} d\left(W_{el} - W_{loss}\right) &= \left(ui_0 - i_0^2 R\right) dt = u_i i_0 dt \\ &= \frac{d\Psi}{dt} i_0 dt = i_0 d\Psi \end{aligned} \tag{1.52}$$

This equals area ① plus area ② in Fig. 1.16.

$$dW_{mag} = W_{mag,1} - W_{mag,2} = \int_0^{\Psi_1} i d\Psi - \int_0^{\Psi_2} i d\Psi \tag{1.53}$$

This equals area ②+③ minus area ③+④, being equivalent to area ② minus area ④ (please refer to Fig. 1.16). Therefore:

$$dW_{mech} = d\left(W_{el} - W_{loss}\right) - dW_{mag} \tag{1.54}$$

which equals area ① plus area ④ (see Fig. 1.16). Consequently:

$$\begin{aligned} dW_{mech} &= \int_0^{i_0} \Psi\left(x_1\right) di - \int_0^{i_0} \Psi\left(x_2\right) di \\ &= dW'_{mag} \end{aligned} \tag{1.55}$$

W'_{mag} is called the magnetic co-energy. The force can be calculated like follows:

[2] The tilde serves for the differentiation between integration limit and integration variable.

$$F = \frac{dW_{mech}}{dx} = \left.\frac{dW'_{mag}}{dx}\right|_{i=const.} \tag{1.56}$$

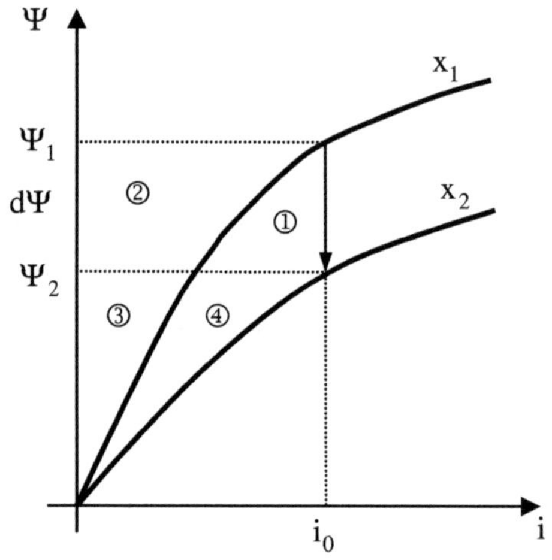

Fig. 1.16. Ψ -i-characteristics for case 2.

Case 3: movable armature, constant flux linkage:

From $\Psi = \Psi_0 = \text{const.}$ it follows $id\Psi = 0$ and therefore $d(W_{el} - W_{loss}) = 0$ (the electrical input power is used only for covering the losses). Consequently:

$$dW_{mech} = -dW_{mag} \tag{1.57}$$

$$dW_{mag} = W_{mag,1} - W_{mag,2} = \int_0^{\Psi_0} id\Psi(x_1) - \int_0^{\Psi_0} id\Psi(x_2) \tag{1.58}$$

This equals area ① minus area ①+②. This is equivalent to being equal to minus area ②. Therefore dW_{mech} equals area ② and consequently (see Fig. 1.17):

$$dW_{mech} = \int_0^{i_1} \Psi(x_1)\,di + \int_{i_1}^{i_2} \Psi_0 \; di - \int_0^{i_2} \Psi(x_2)\,di$$
$$= \int_0^{i_2} \Psi_1(i)\,di - \int_0^{i_2} \Psi_2(i)\,di = dW'_{mag} \tag{1.59}$$

The force is calculated as follows:

$$F = \frac{dW_{mech}}{dx} = \left.\frac{dW'_{mag}}{dx}\right|_{\Psi=const.} \tag{1.60}$$

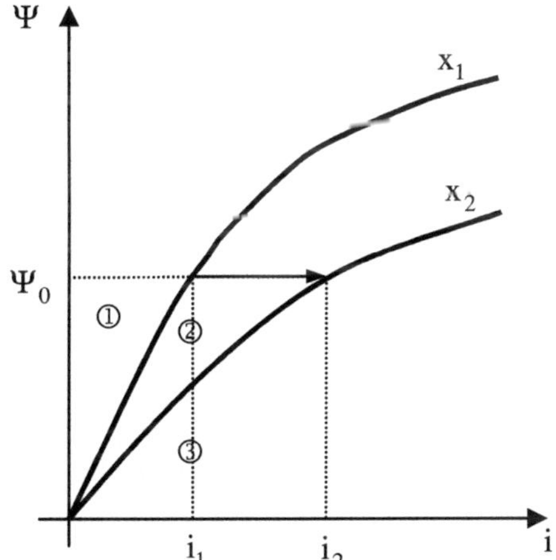

Fig. 1.17. Ψ -i-characteristics for case 3.

Case 4: arbitrary case; movable armature, current and flux linkage are variable (see Fig. 1.18):

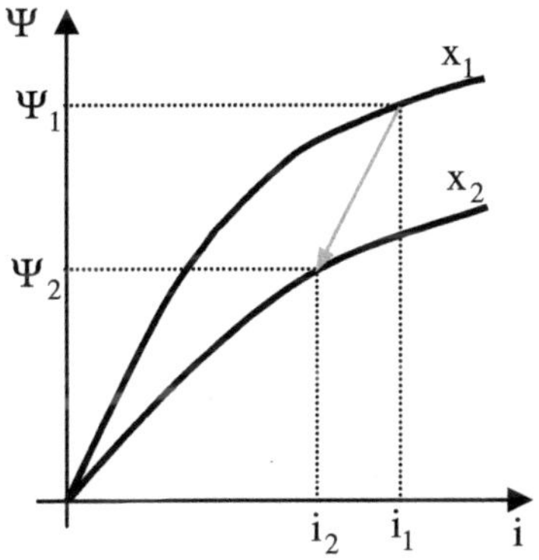

Fig. 1.18. Ψ -i-characteristics for case 4.

The changes of energies calculated in the following for this case are based on the solutions of the above cases 1 and 2.

Case 4.1 (see Fig. 1.19):

a) firstly the armature is fixed, change of current and flux linkage
b) secondly the current is constant, movable armature and change of flux linkage

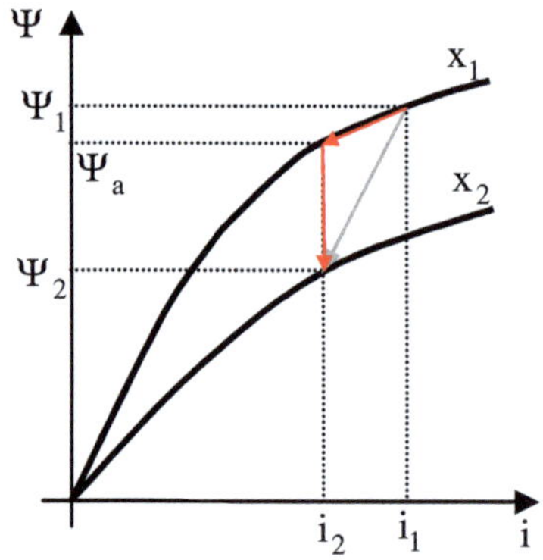

Fig. 1.19. Ψ -i-characteristics for case 4.1.

a) $x_1 = \text{const.}$;

i is changed from i_1 to i_2; Ψ is changed from Ψ_1 to Ψ_a.

$$dW_{mech,1a} = 0$$

$$dW_{mag,1a} = i d\Psi = \int_0^{\Psi_1} i(x_1)\, d\Psi - \int_0^{\Psi_a} i(x_1)\, d\Psi \tag{1.61}$$

$$d\left(W_{el,1a} - W_{loss,1a}\right) = dW_{mag,1a}$$

b) $i_2 = \text{const.}$;

x is changed from x_1 to x_2; Ψ is changed from Ψ_a to Ψ_2.

$$dW_{mech,1b} = dW'_{mag,1b} = \int_0^{i_2} \Psi(x_1)\,di - \int_0^{i_2} \Psi(x_2)\,di$$

$$dW_{mag,1b} = \int_0^{\Psi_a} i(x_1)\,d\Psi - \int_0^{\Psi_2} i(x_2)\,d\Psi \tag{1.62}$$

$$d\left(W_{el,1b} - W_{loss,1b}\right) = i_2 d\Psi = i_2\left(\Psi_a - \Psi_2\right)$$

Case 4.2 (see Fig. 1.20)*:*

a) firstly the current is constant, movable armature and change of flux linkage
b) secondly the armature is fixed, change of current and flux linkage

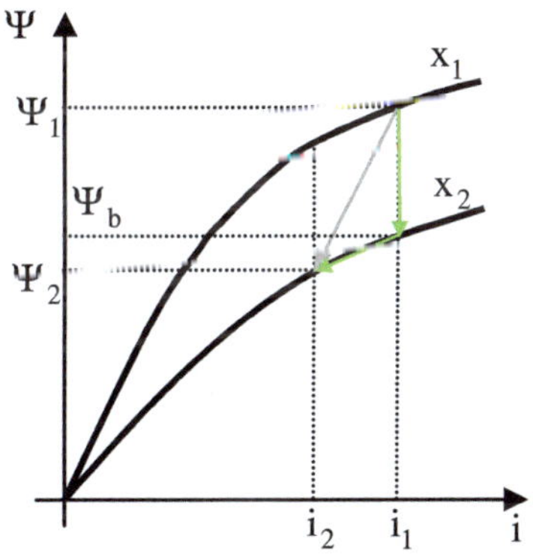

Fig. 1.20. Ψ -i-characteristics for case 4.2.

a) $i_1 = \text{const.}$;

x is changed from x_1 to x_2; Ψ is changed from Ψ_1 to Ψ_b.

$$dW_{mech,2a} = dW'_{mag,2a} = \int_0^{i_1} \Psi(x_1)\,di - \int_0^{i_1} \Psi(x_2)\,di$$

$$dW_{mag,2a} = \int_0^{\Psi_1} i(x_1)\,d\Psi - \int_0^{\Psi_b} i(x_2)\,d\Psi \tag{1.63}$$

$$d\left(W_{el,2a} - W_{loss,2a}\right) = i_1 d\Psi = i_1\left(\Psi_1 - \Psi_b\right)$$

b) $x_2 = \text{const.}$;

i is changed from i_1 to i_2; Ψ is changed from Ψ_b to Ψ_2.

$$dW_{mech,2b} = 0$$
$$dW_{mag,2b} = id\Psi = \int_0^{\Psi_b} i(x_2)\,d\Psi - \int_0^{\Psi_2} i(x_2)\,d\Psi \tag{1.64}$$
$$d\left(W_{el,2b} - W_{loss,2b}\right) = dW_{mag,2b}$$

Comparison of cases 4.1 and 4.2:

a) change of mechanical energy dW_{mech} (Fig. 1.21)

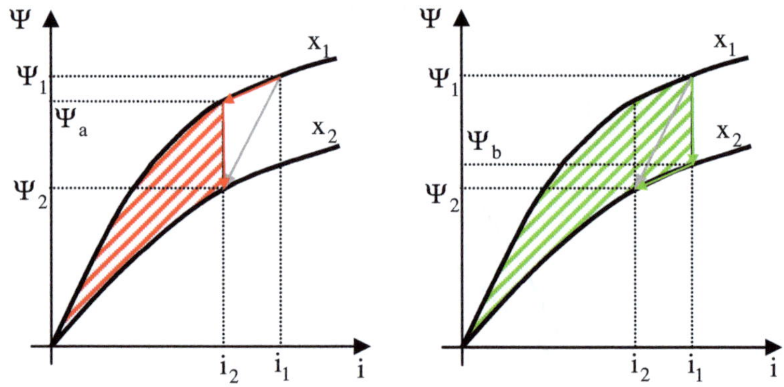

Fig. 1.21. Ψ -i-characteristics: different change of mechanical energy in both cases.

b) change of magnetic energy dW_{mag} (Fig. 1.22)

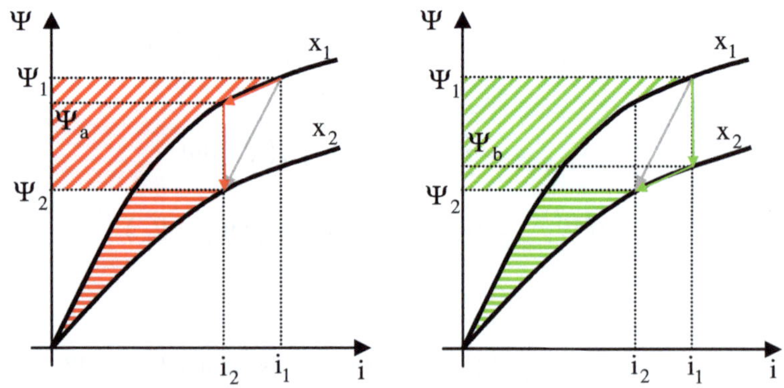

Fig. 1.22. Ψ -i-characteristics: equal change of magnetic energy in both cases.

c) change of difference: electrical energy and losses $d\left(W_{el} - W_{loss}\right)$ (Fig. 1.23)

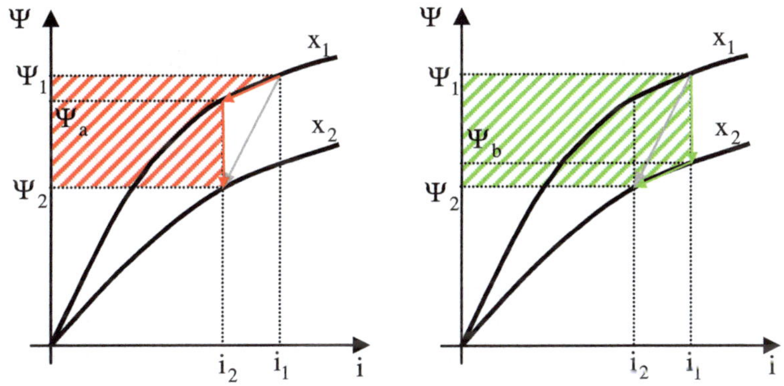

Fig. 1.23. Ψ -i-characteristics: different change of difference between electrical energy and losses in both cases.

For *linear materials* $W_{mag} = W'_{mag}$ holds true. This means that in this case (and only in this case!) the force may be calculated from the magnetic energy.

The *magnetic pulling force* on the surface area of flux carrying iron parts can be calculated as follows (Fig. 1.24):

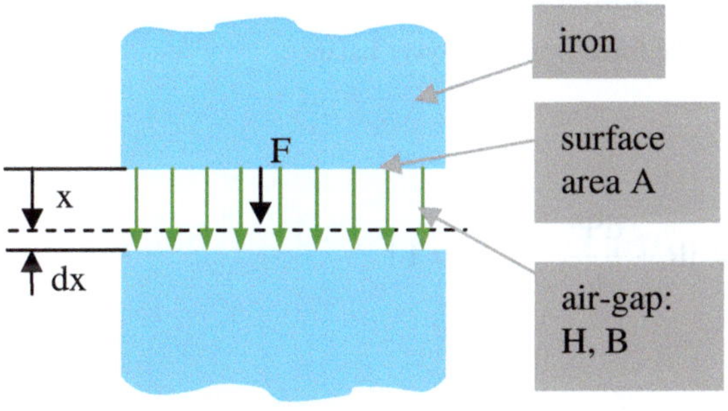

Fig. 1.24. Explanation of the magnetic pulling force.

Because of $\mu_{r,Fe} \to \infty$ and $\mu_{r,air} = 1$ the used materials are linear. Consequently the force may be calculated from the change of the magnetic energy.

Because of $H_{Fe} \to 0$ the iron paths may be neglected. Therefore, the force will be calculated from the change of magnetic energy in the air-gap.

$$F = \frac{dW_{mag}}{dx} = \frac{w_{mag} A dx}{dx} = \left(\frac{1}{2} HB\right) A = \frac{B^2}{2\mu_0} A \quad (1.65)$$

The *specific force* (force per cross section unit, “Maxwell's attractive force“) is:

$$f = \frac{F}{A} = \frac{B^2}{2\mu_0} \quad (1.66)$$

Calculating the force from the power balance
A cylindrical coil shall have the Ohmic resistance R and an armature movable only in x-direction. The inductivity of that coil depends on the position of the armature: $L = L(x)$. Saturation will be neglected: $L \neq L(i)$ (Fig. 1.25).

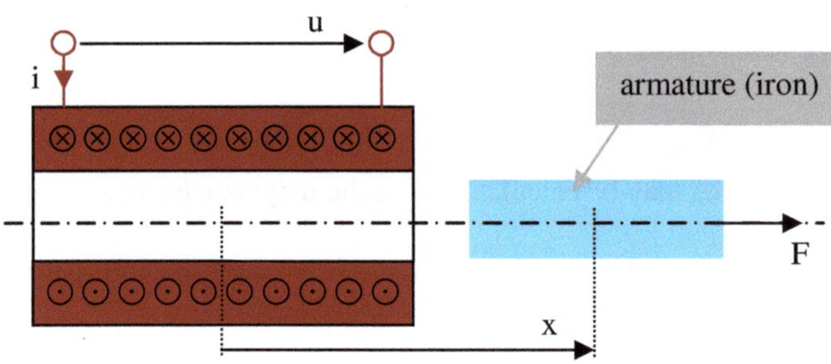

Fig. 1.25. Explanation of calculating the force from the power balance.

The voltage equation is:

$$u = iR + \frac{d\Psi}{dt}, \qquad \Psi = Li \quad (1.67)$$

Case 1: armature is fixed at position x (then L is constant)

From the voltage equation (1.67) the power balance follows by multiplication with the current i:

$$ui = i^2 R + Li\frac{di}{dt} \quad (1.68)$$

From the magnetic energy $W_{mag} = \frac{1}{2}Li^2$ it follows:

$$\frac{d}{dt}W_{mag} = \frac{d}{dt}\left(\frac{1}{2}Li^2\right) = L\frac{d}{dt}\left(\frac{1}{2}i^2\right) = Li\frac{di}{dt} \tag{1.69}$$

From equations (1.68) and (1.69) it follows further:

$$ui = i^2R + \frac{d}{dt}\left(\frac{1}{2}Li^2\right) \tag{1.70}$$

Therefore, the electrical input power is equal to the sum of electrical losses and change of magnetic energy.

Case 2: movable armature ($L = L(x)$)

In this case the voltage equation becomes:

$$u = iR + L\frac{di}{dt} + i\frac{dL}{dt} \tag{1.71}$$

and consequently the power balance:

$$ui = i^2R + Li\frac{di}{dt} + i^2\frac{dL}{dt} \tag{1.72}$$

From the magnetic energy $W_{mag} = \frac{1}{2}Li^2$ it follows:

$$\frac{d}{dt}W_{mag} = \frac{d}{dt}\left(\frac{1}{2}Li^2\right) = Li\frac{di}{dt} + \frac{1}{2}i^2\frac{dL}{dt} \tag{1.73}$$

From equations (1.71) and (1.72) it follows further:

$$ui = i^2R + \frac{d}{dt}\left(\frac{1}{2}Li^2\right) + \frac{1}{2}i^2\frac{dL}{dt} \tag{1.74}$$

The additional term in the power balance compared with case 1 must be the mechanical power. Therefore the mechanical power is

$$F\frac{dx}{dt} = \frac{1}{2}i^2\frac{dL}{dt} = \frac{1}{2}i^2\frac{\partial L}{\partial x}\frac{dx}{dt} \tag{1.75}$$

and the force can be calculated like follows:

$$F = \frac{1}{2}i^2\frac{\partial L}{\partial x} \tag{1.76}$$

1.4 Complex Phasors

Alternating voltages and currents with sinusoidal time dependency are described in the electrical power engineering as complex phasors of the rms values (Fig. 1.26):

$$\begin{aligned} u(t) &= \sqrt{2}U\cos(\omega t) = \mathrm{Re}\left\{\sqrt{2}Ue^{j\omega t}\right\} \\ &= \mathrm{Re}\left\{\sqrt{2}\underline{U}e^{j\omega t}\right\}, \qquad \underline{U} = Ue^{j0} \end{aligned} \tag{1.77}$$

$$\begin{aligned} i(t) &= \sqrt{2}I\cos(\omega t - \varphi) = \mathrm{Re}\left\{\sqrt{2}Ie^{j\omega t}e^{-j\varphi}\right\} \\ &= \mathrm{Re}\left\{\sqrt{2}\underline{I}e^{j\omega t}\right\}, \qquad \underline{I} = Ie^{-j\varphi} \end{aligned} \tag{1.78}$$

The non time-dependent components $\underline{U}$ and $\underline{I}$ are called (complex) phasors. Phasors describe the amplitude of the respective variable with their length; the direction of the phasor shows the position of the maximum of this variable. The instantaneous value of the physical magnitude (voltage and current) results from the projection of the rotating phasors onto the real axis of the complex plane. The phasors rotate mathematically positive (anti-clockwise).

The choice of the phase angle φ is arbitrary as well, but usually the phase angle of the voltage is chosen being zero. Defining the phase angle of the current like shown above, for resistive-inductive impedances (which are mostly relevant for electrical drives) positive values for the phase angle φ are obtained.

The orientation of the complex plane is arbitrary, but in the electrical power engineering usually the positive real axis is oriented vertically upright, the negative imaginary axis to the right.

The complex impedance is:

$$\underline{Z} = \frac{\underline{U}}{\underline{I}} = \frac{U}{I} e^{j\varphi} = Ze^{j\varphi} = Z\cos(\varphi) + jZ\sin(\varphi) = R + jX$$
$$Z = \sqrt{R^2 + X^2} \qquad \tan(\varphi) = \frac{X}{R} \tag{1.79}$$

The complex apparent power is the product of the complex rms-value of the voltage and the conjugate complex rms-value of the current:

$$\underline{S} = \underline{U}\,\underline{I}^* = UIe^{j\varphi} = P + jQ \tag{1.80}$$

The different kinds of power are the

- active power (real power)

$$P = \text{Re}\{\underline{S}\} = UI\cos(\varphi) \tag{1.81}$$

- reactive power (wattless power)

$$Q = \text{Im}\{\underline{S}\} = UI\sin(\varphi) \tag{1.82}$$

- and apparent power

$$S = |\underline{S}| = UI = \sqrt{P^2 + Q^2} \tag{1.83}$$

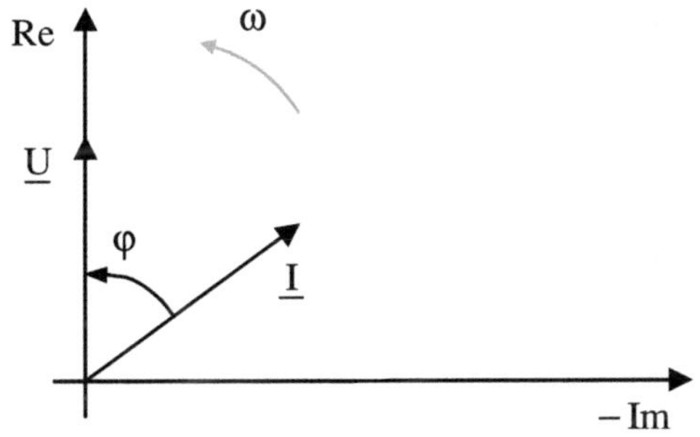

Fig. 1.26. Phasor diagram.

1.5 Star and Delta Connection

Regarding symmetric three-phase systems without neutral line there are the possibilities illustrated in Fig. 1.27:

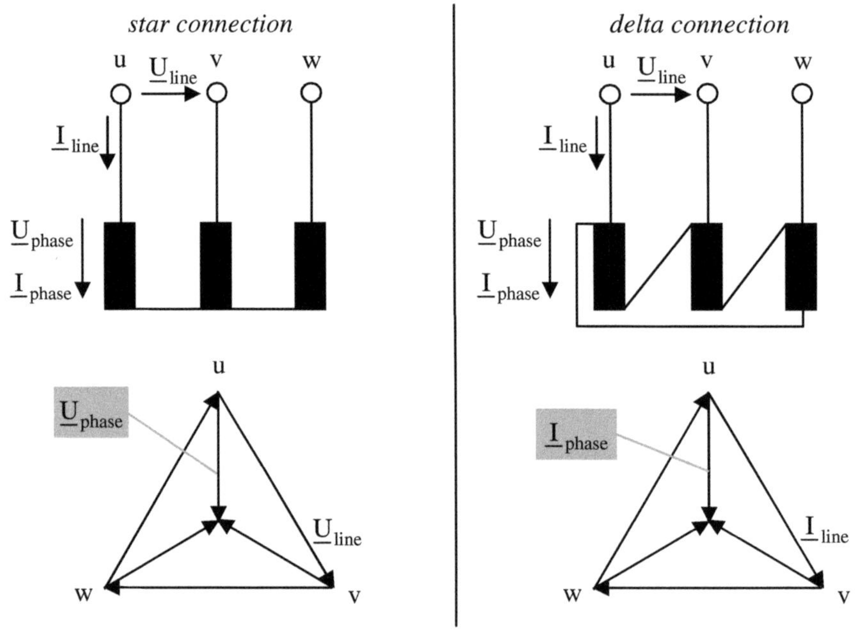

Fig. 1.27. Star and Delta connection.

For the phase voltages U_{phase} and currents I_{phase} it holds:

- for star connection: $\sum I_{phase} = 0$
- for delta connection: $\sum U_{phase} = 0$

The terminal voltages U_{line} and terminal currents I_{line} are:

- for star connection: $U_{line} = \sqrt{3}\ U_{phase}$; $\quad I_{line} = I_{phase}$
- for delta connection: $U_{line} = U_{phase}$; $\quad I_{line} = \sqrt{3}\ I_{phase}$

The electrical power is:

- for star connection: $S = 3U_{phase}I_{phase} = 3\frac{U_{line}}{\sqrt{3}}I_{line}$
- for delta connection: $S = 3U_{phase}I_{phase} = 3U_{line}\frac{I_{line}}{\sqrt{3}}$

Therefore, it is always:

$$S = \sqrt{3}\ U_{line}\ I_{line} = \sqrt{3}\ U\ I \tag{1.84}$$

Usually the index "line" is omitted. The values on the name plate of electrical machines are always the terminal values!

1.6 Symmetric Components

A symmetric three-phase system may be operated asymmetrically, e.g. by:

- supplying with asymmetric voltages or
- single-phase load between two phases or between one phase and the neutral line.

Describing these asymmetric (unknown) operating conditions by symmetric ones, a simplified calculation method is gained. The method of symmetric components is qualified for this: An asymmetric three-phase-system is separated into three symmetric systems (positive, negative, and zero system), the circuit calculated and the results superposed.

The preconditions are:

- The three currents or voltages have the same frequency and they are sinusoidally in time (i.e. there is no harmonic content); phase shift and amplitude are arbitrary.
- Because of the superposition of the results the system must be linear.

In the following the complex phasor $\underline{a} = e^{j\frac{2\pi}{3}}$ will be used. There is:

$$\underline{a}^2 = e^{j\frac{4\pi}{3}} = e^{-j\frac{2\pi}{3}}; \qquad 1 + \underline{a} + \underline{a}^2 = 0 \tag{1.85}$$

The following asymmetric current system $\underline{I}_u, \underline{I}_v, \underline{I}_w$ (Fig. 1.28) will be represented by the components $\underline{I}_p, \underline{I}_n, \underline{I}_0$ (Fig. 1.29).

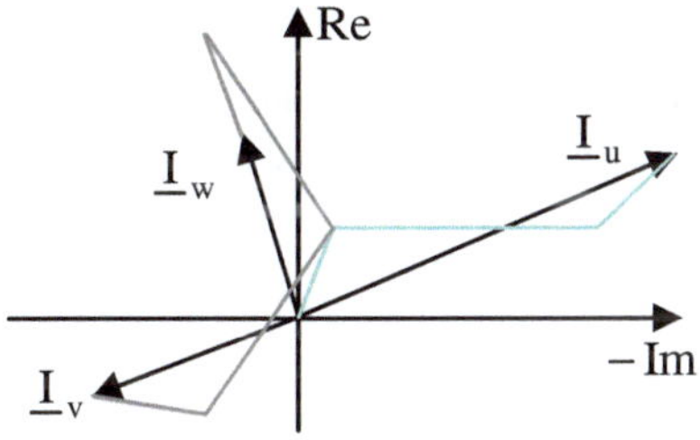

Fig. 1.28. Asymmetric current system.

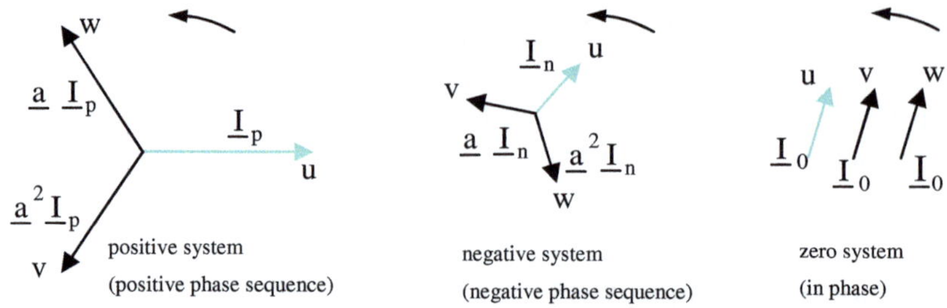

Fig. 1.29. Three symmetric current systems.

The following holds true:

$$\begin{aligned} \underline{I}_u &= \underline{I}_p + \underline{I}_n + \underline{I}_0 \\ \underline{I}_v &= \underline{a}^2 \underline{I}_p + \underline{a}\ \underline{I}_n + \underline{I}_0 \\ \underline{I}_w &= \underline{a}\ \underline{I}_p + \underline{a}^2 \underline{I}_n + \underline{I}_0 \end{aligned} \quad \Leftrightarrow \quad \begin{pmatrix} \underline{I}_u \\ \underline{I}_v \\ \underline{I}_w \end{pmatrix} = \begin{pmatrix} 1 & 1 & 1 \\ \underline{a}^2 & \underline{a} & 1 \\ \underline{a} & \underline{a}^2 & 1 \end{pmatrix} \begin{pmatrix} \underline{I}_p \\ \underline{I}_n \\ \underline{I}_0 \end{pmatrix} \tag{1.86}$$

Solving this results in:

$$\begin{pmatrix} \underline{I}_p \\ \underline{I}_n \\ \underline{I}_0 \end{pmatrix} = \frac{1}{3} \begin{pmatrix} 1 & \underline{a} & \underline{a}^2 \\ 1 & \underline{a}^2 & \underline{a} \\ 1 & 1 & 1 \end{pmatrix} \begin{pmatrix} \underline{I}_u \\ \underline{I}_v \\ \underline{I}_w \end{pmatrix} \tag{1.87}$$

Now, the asymmetric system $\underline{I}_u, \underline{I}_v, \underline{I}_w$ can be separated into three symmetric systems $\underline{I}_p, \underline{I}_n, \underline{I}_0$ according to the above equation; these three systems can be calculated easily and the solution is gained by inverse transformation (superposition of the three single results).

1.7 Mutual Inductivity

There are two coils, each generating a magnetic field. Both magnetic fields shall penetrate both coils, see Fig. 1.30. As an example, one coil produces a homogeneous field, the other coil an inhomogeneous field.

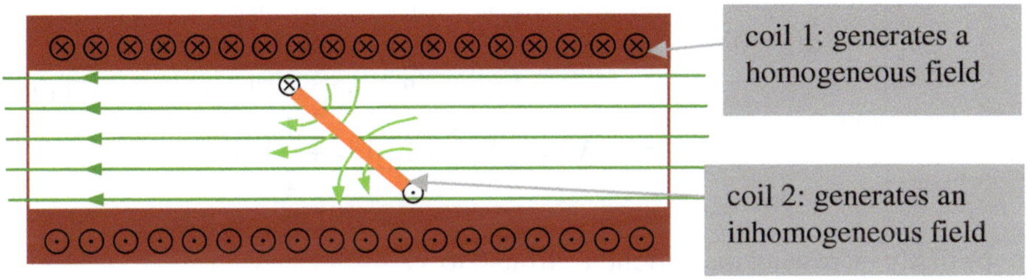

Fig. 1.30. Magnetic fields of a system made of two coils.

Calculation of the magnetic energy (the tilde is introduced to distinguish between integration limit and integration variable) if supplying

a) only coil 1:

$$dW_1 = i_1 d\Psi_1 \quad \Rightarrow \quad W_1 = \int_0^{\Psi_1} i_1 d\tilde{\Psi}_1 = \int_0^{i_1} \tilde{i}_1 L_1 d\tilde{i}_1 = \frac{1}{2} L_1 i_1^2 \tag{1.88}$$

b) only coil 2:

$$dW_2 = i_2 d\Psi_2 \quad \Rightarrow \quad W_2 = \int_0^{\Psi_2} i_2 d\tilde{\Psi}_2 = \int_0^{i_2} \tilde{i}_2 L_2 d\tilde{i}_2 = \frac{1}{2} L_2 i_2^2 \tag{1.89}$$

c) coils 1 and 2:

$$\begin{aligned} dW &= dW_1 + dW_2 = i_1 d\Psi_1 + i_2 d\Psi_2, \qquad \text{with} \\ &\Psi_1 = L_1 i_1 + L_{12} i_2, \qquad d\Psi_1 = L_1 di_1 + L_{12} di_2 \\ &\Psi_2 = L_2 i_2 + L_{21} i_1, \qquad d\Psi_2 = L_2 di_2 + L_{21} di_1 \\ dW &= i_1 \left(L_1 di_1 + L_{12} di_2 \right) + i_2 \left(L_2 di_2 + L_{21} di_1 \right) \end{aligned} \tag{1.90}$$

Assuming $\mu = \text{const.}$ it follows:

a) firstly increasing the current $\tilde{i}_1$ from 0 to i_1

$$i_2 = 0,\ di_2 = 0 \quad \Rightarrow \quad W = \int_0^{i_1} \tilde{i}_1 L_1 d\tilde{i}_1 = \frac{1}{2} L_1 i_1^2 \tag{1.91}$$

then increasing the current $\tilde{i}_2$ from 0 to i_2

$$i_1 = i_1,\ di_1 = 0 \qquad \Rightarrow \quad W = \frac{1}{2}L_1 i_1^2 + \int_0^{i_2} i_1 L_{12} d\tilde{i}_2 + \int_0^{i_2} \tilde{i}_2 L_2 d\tilde{i}_2$$
$$\Rightarrow \quad W = \frac{1}{2}L_1 i_1^2 + L_{12} i_1 i_2 + \frac{1}{2}L_2 i_2^2 \tag{1.92}$$

b) firstly increasing the current $\tilde{i}_2$ from 0 to i_2

$$i_1 = 0,\ di_1 = 0 \qquad \Rightarrow \quad W = \int_0^{i_2} \tilde{i}_2 L_2 d\tilde{i}_2 = \frac{1}{2}L_2 i_2^2 \tag{1.93}$$

then increasing the current $\tilde{i}_1$ from 0 to i_1

$$i_2 = i_2,\ di_2 = 0 \qquad \Rightarrow \quad W = \frac{1}{2}L_2 i_2^2 + \int_0^{i_1} \tilde{i}_1 L_1 d\tilde{i}_1 + \int_0^{i_1} i_2 L_{21} d\tilde{i}_1$$
$$\Rightarrow \quad W = \frac{1}{2}L_2 i_2^2 + \frac{1}{2}L_1 i_1^2 + L_{21} i_2 i_1 \tag{1.94}$$

Independent from the sequence of increasing the currents (switching on the coils) the magnetic energy must always have the same value. Therefore, the following is true:

$$L_{12} = L_{21} \tag{1.95}$$

1.8 Iron Losses

In addition to the copper losses (caused by current flow in wires having a resistance) iron losses are known in electrical machines. These iron losses mainly are composed of two parts:

According to Lenz's Law the flux change in the electrical conducting iron material causes eddy currents that oppose their generating induction variation. The eddy current losses are proportional to the squared frequency, the squared flux density and the iron volume:

$$P_{Fe,edd} \sim f^2\ \hat{B}^2\ V_{Fe} \tag{1.96}$$

These eddy current losses can be reduced by using isolated lamination sheets and by using iron laminations with low electrical conductivity.

Because of ever changing magnetizing direction inside the iron hysteresis losses are generated that are proportional to the area of the hysteresis loop enclosed during each cycle; these losses are proportional to the frequency, the squared flux density and the iron volume:

$$P_{Fe,hys} \sim f\ \hat{B}^2\ V_{Fe} \tag{1.97}$$

The hysteresis losses can be reduced by using iron material with a narrow hysteresis loop.

Mostly, the iron losses are calculated according to the following Steinmetz equation:

$$P_{Fe} = \left(a_{edd} \left(\frac{f}{50\text{Hz}} \right)^2 + a_{hys} \frac{f}{50\text{Hz}} \right) \left(\frac{\hat{B}}{1\text{T}} \right)^2 \rho_{Fe}\ V_{Fe} \tag{1.98}$$

where ρ_{Fe} is the specific iron weight. The material specific loss factors (eddy current loss factor a_{edd} and hysteresis loss factor a_{hys}, both in $\mathrm{W/kg}$) are given by the iron material suppliers.

1.9 References for Chapter 1

Küpfmüller K, Kohn G (1993) Theoretische Elektrotechnik und Elektronik. Springer-Verlag, Berlin

Lehner G (1994) Elektromagnetische Feldtheorie. Springer-Verlag, Berlin

Müller G, Ponick B (2005) Grundlagen elektrischer Maschinen. Wiley-VCH Verlag, Weinheim

Müller G, Vogt K, Ponick B (2008) Berechnung elektrischer Maschinen. Wiley-VCH Verlag, Weinheim

Richter R (1967) Elektrische Maschinen I. Birkhäuser Verlag, Basel

Schwab AJ (1993) Begriffswelt der Feldtheorie. Springer-Verlag, Berlin

Simonyi K (1980) Theoretische Elektrotechnik. VEB Deutscher Verlag der Wissenschaften, Berlin

Veltman A, Pulle DWJ, DeDoncker RW (2007) Fundamentals of electrical drives. Springer-Verlag, Berlin

2 DC-Machines

2.1 Principle Construction

Figure 2.1 shows a photograph of an open cut DC-machine, where all relevant parts are described.

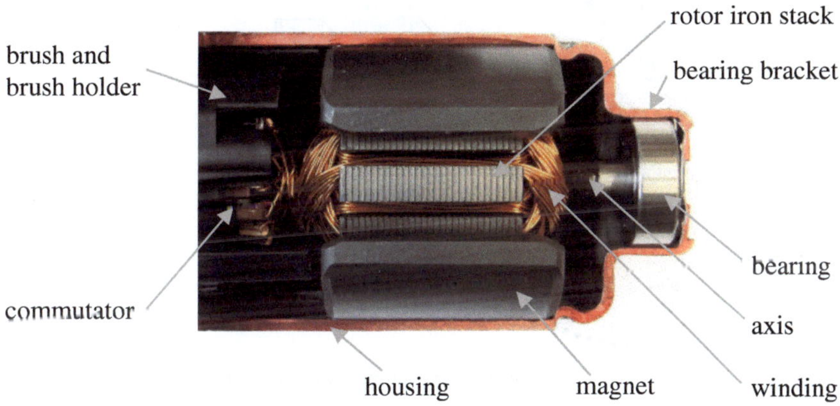

Fig. 2.1. Photograph of a DC-machine.

The principle construction of a DC-machine is like follows:

The stationary part (called "stator") mostly is composed of massive iron (to lead the magnetic flux). A stationary magnetic field with changing polarity is generated, either by permanent magnets (see Fig. 2.1) or by salient poles having coils with DC-currents.

The movable part (called "rotor") - separated from the stator by an air-gap - is composed of an iron stack made from laminations, in whose slots coils made from copper are placed. These coils are connected with the clamps of the commutator segments. On the commutator the carbon brushes are sliding, so that the current is supplied from the stationary terminals to the rotating coils.

By this commutator the supplied DC-current permanently changes direction in the rotor in such a way, that the current in the rotor coils below a permanent magnet pole of the stator always flows in the same direction (under the magnet pole with opposite polarity the current flows in opposite direction). By this changing of current flow direction in the rotor coils an alternating current arises.

2.2 Voltage and Torque Generation, Commutation

In principle each electrical machine can be operated as motor or as generator. In generator operation usually voltage production constant in time is required, in motor operation usually torque production constant in time is asked for.

In a rotating coil a voltage is induced according to the induction law, see Fig. 2.2.

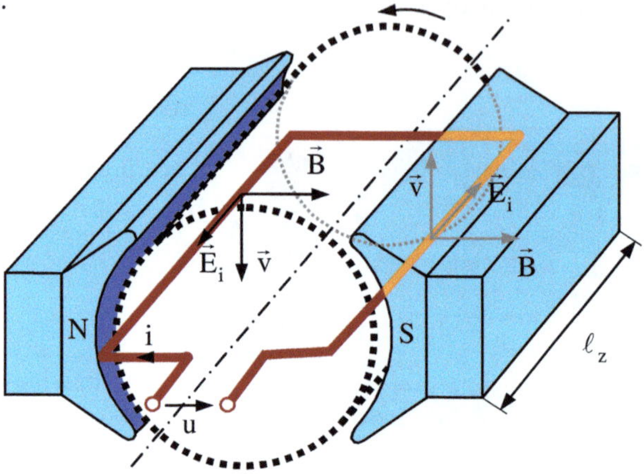

Fig. 2.2. Principle sketch of voltage induction in rotating representation.

The induced voltage is:

$$\begin{aligned} u_i &= -\oint \vec{E}_i d\vec{\ell} = -\oint (\vec{v} \times \vec{B})\, d\vec{\ell} = 2|B|\, v\ell_z \\ u &= Ri + u_i \\ &\Rightarrow \quad u = Ri + 2|B|\,\ell_z v \end{aligned} \tag{2.1}$$

For the signs the following is true:

- $u_i = -\oint \vec{E}_i d\vec{\ell}$ is defined like this in Sect. 1.1, see Eq. 1.32;
- $\vec{E}_i$ and $d\vec{\ell}$ (in the direction of the current i) are opposite to each other.

Figure 2.3 shows the same situation in a "wound-off" representation. The induced voltage for this situation can be calculated like follows:

$$\begin{aligned} u_i &= \frac{d\Psi}{dt} = w\frac{d\Phi}{dt}, \qquad w = 1 \\ \Rightarrow \quad u_i &= \frac{|B|\, dA}{dt} = \frac{|B|\, 2\,\ell_z\, v dt}{dt} = 2|B|\,\ell_z v, \qquad v = \omega_{mech} r = 2\pi n r \end{aligned} \tag{2.2}$$

For the signs the following is true:

- $u_i = \frac{d\Psi}{dt}$ is deduced in Sect. 1.1;
- $d\vec{\ell}$ (in the direction of the current i) and $\vec{B}$ in the left part of Fig. 2.3 (increase of $\vec{B}$) are linked together like a right-handed screw.

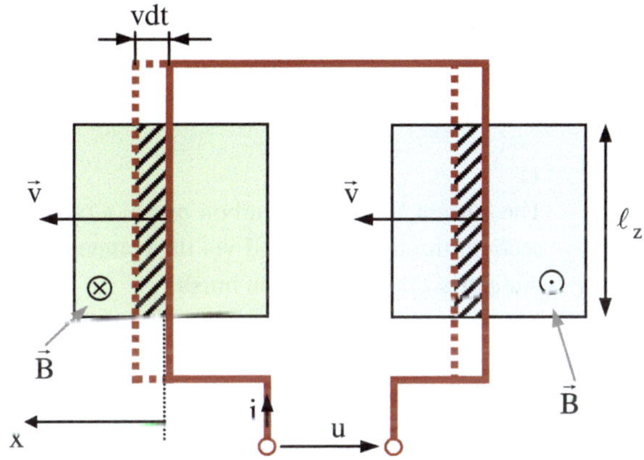

Fig. 2.3. Principle sketch of voltage induction in "wound-off" representation.

The spatial characteristic of the flux density and the time-dependent characteristic of the voltage are like follows (Fig. 2.4):

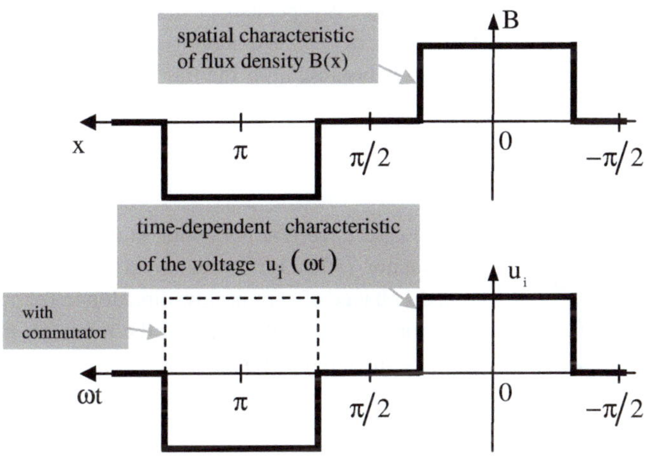

Fig. 2.4. Flux density and induced voltage characteristics.

Between the electrical angular frequency ω and the mechanical angular frequency ω_{mech} the following relation is true (p being the number of pole pairs):

$$\omega = p\omega_{mech}$$
$$\Rightarrow \quad 2\pi f = p\ 2\pi n \tag{2.3}$$
$$\Rightarrow \quad f = pn$$

The commutator converts the AC-voltage in the coil into a DC-voltage (with harmonics) at the terminals. By series connection of several coils evenly distributed along the rotor circumference a higher DC-voltage with lower harmonic content is obtained.

The procedure of commutation is explained in Fig. 2.5:

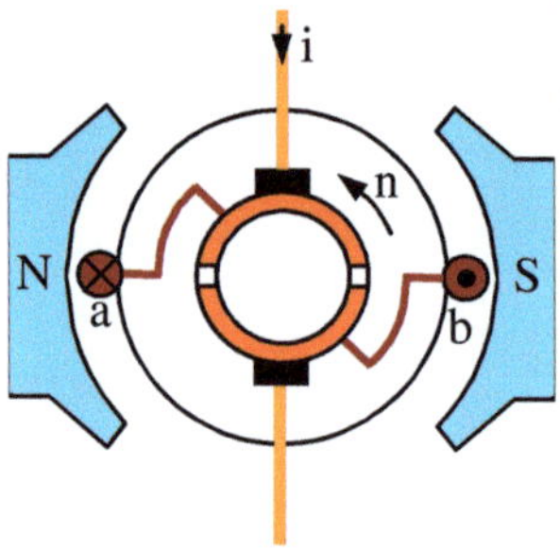

1)
The current flows via a carbon brush, a commutator section, through a coil and via the counterpart commutator section and carbon brush.
For motor operation a torque in the direction of motion occurs.

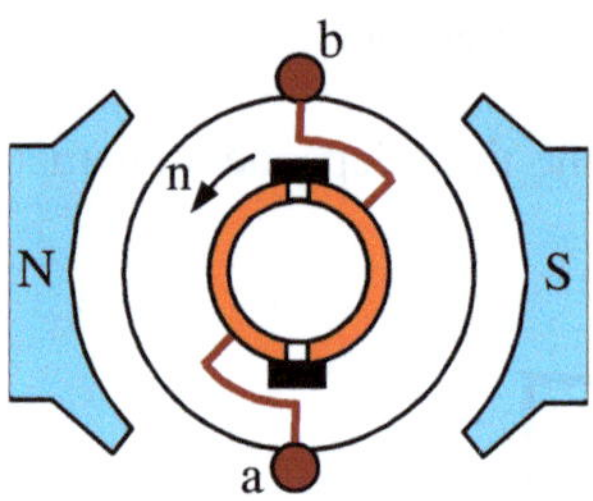

2)
Under each carbon brush both commutator sections are located; there is no current in the coil (begin and end of the coil are short-circuited via the brushes) and no torque is generated.
The rotor of the DC-motor stays in rotational movement because of its inertia.

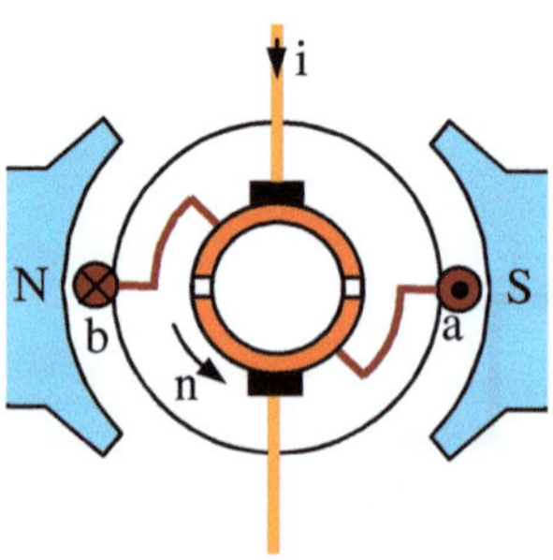

3)
Like in case 1) current is flowing in the coil, but the commutator (after 180° rotation of the rotor) has forced a change of current flow direction in the coil.
Therefore, torque and current at the terminals have the same direction like in case 1).

Fig. 2.5. Procedure of commutation.

As a first approximation it may be assumed that the current in the coil changes linearly from its maximum to its minimum value (maximum absolute value, negative sign). In the time period between two commutation events the current in the coil is (approximately) constant.

The force onto a current conducting wire is: $\vec{F} = i(\vec{\ell} \times \vec{B})$. From this speed direction and torque of the DC-machine in motor operation follow.

In generator operation the voltage $u = -Ri + 2B\ell v$ is produced (here the energy generation system is assumed, see Fig. 2.6).

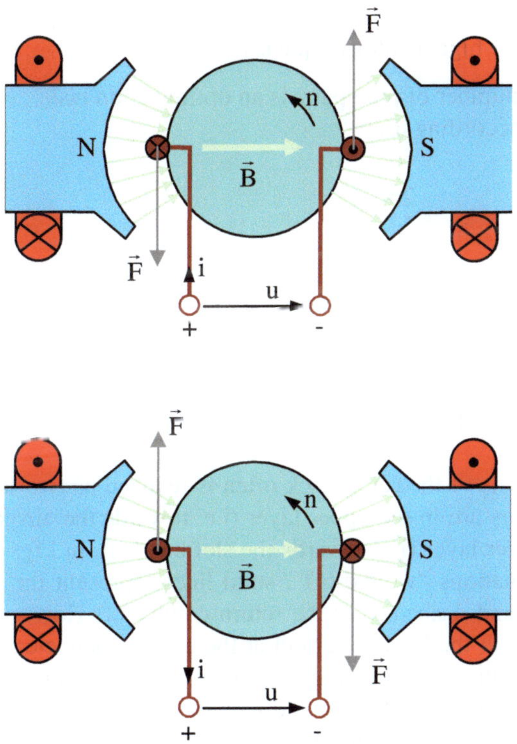

Fig. 2.6. Motor operation (*above*) and generator operation (*below*).

2.3 Number of Pole Pairs, Winding Design

Up to now two-pole machines were presented. Nevertheless, DC-machines with even more poles are possible. For these constructions the arrangement is repeated p times along the circumference (e.g. for $p = 2$ there are 4 carbon brushes and 4 magnets or excitation poles).

The advantages of a high number of poles are:

- The total flux is divided into 2p part fluxes. By this the cross section of the yokes in stator and rotor can be chosen smaller (material savings).
- A smaller pole pitch results in shorter end windings (with smaller resistance and lower losses).

The disadvantages are:

- By having smaller distances between the poles the leakage between the poles is increased.
- The losses are increased by the higher rotor frequency.

Therefore, the choice of the number of pole pairs is an optimization task.
The pole pitch is calculated according to:

$$\tau_p = \frac{2\pi r}{2p} \tag{2.4}$$

Between the mechanical angle α and the electrical angle β the following relation is true:

$$\beta = p\alpha \tag{2.5}$$

The winding placed in the slots of the rotor stack often is realized as two-layer winding: The forward conductors are in the upper layer (i.e. towards the air-gap), the return conductors in the lower layer (i.e. towards the slot bottom). In Figs. 2.7 to 2.9, showing the general situations "wound-off", solid lines represent the forward conductors (upper layer) and dashed lines the return conductors (lower layer). For DC-machines each coil at the beginning and at the end is connected to a commutator section, i.e. the number of coils and the number of commutator sections are identical; in the following this will be named with the variable K.

For DC-machines the following nominations are introduced:

K	number of commutator sections (equal to number of coils)
u	number of coils sides side-by-side in a single slot
N	number of rotor slots
w_S	number of turns per coil (number of conductors per coil side)
z	total number of conductors in all slots

The distance between two carbon brushes (i.e. between the positive brush and the negative brush) is:

$$y_B = \frac{K}{2p} \tag{2.6}$$

Moreover, the following relations are true:

$$K = Nu$$
$$z = 2w_S K \tag{2.7}$$

Mostly $u > 1$ is true, then the number of rotor slots is smaller than the number of commutator sections ($N < K$).

Examples (lap winding):

- In Fig. 2.7 the three upper sketches show the conductors in a rotor slot for different winding layout.
- The lower sketches illustrate the according winding layout (in each sketch on the left side only the upper layer and on the right side only the lower layer is shown).

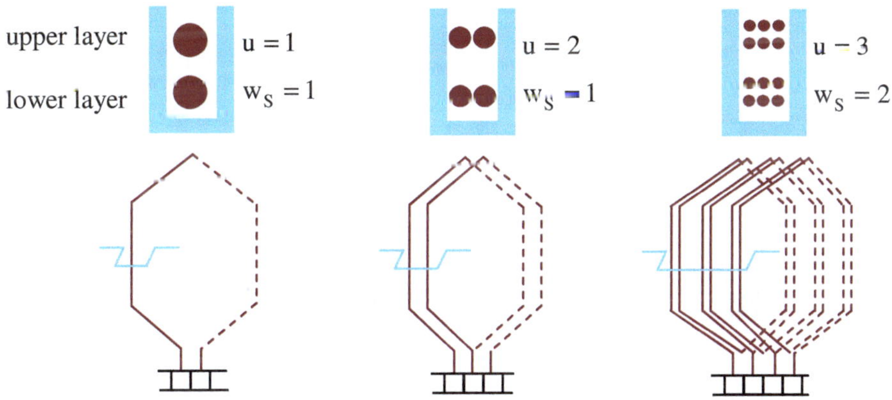

Fig. 2.7. Sketches of conductors in a slot (*above*) and the according winding layout (*below*).

The coils can be connected to the commutator in two different ways:

Having a *lap winding* (Fig. 2.8) the end of a coil is connected directly to the beginning of the next coil of the same pole pair at the commutator. Between two commutator sections only one coil is placed. All p pole pairs are connected in series by the carbon brushes; the number of parallel paths in the rotor is: $2a = 2p$. The total rotor current is therefore divided into $2p$ parallel conductor currents.

Naming the coil width with y_1 (in numbers of rotor slots) and the connection step with y_2 (in numbers of rotor slots), then the commutator step y for the lap winding is (see Fig. 2.8): $y = y_1 - y_2 = 1$.

Having a *wave (or series) winding* (Fig. 2.9) the end of a coil is connected with the beginning of a corresponding coil of the next pole pair, so that - until reaching the neighboring commutator section - a path along the circumference of the rotor with p coils is completed. Between positive and negative carbon brush all p pole pairs are connected in series; the number of parallel paths in the rotor is: $2a = 2$. The total rotor current is therefore divided into 2 parallel conductor currents.

Naming the coil width with y_1 (in numbers of rotor slots) and the connection step with y_2 (in numbers of rotor slots), then the commutator step y for the wave winding is (see Fig. 2.9): $y = y_1 + y_2 = \frac{K-1}{p}$.

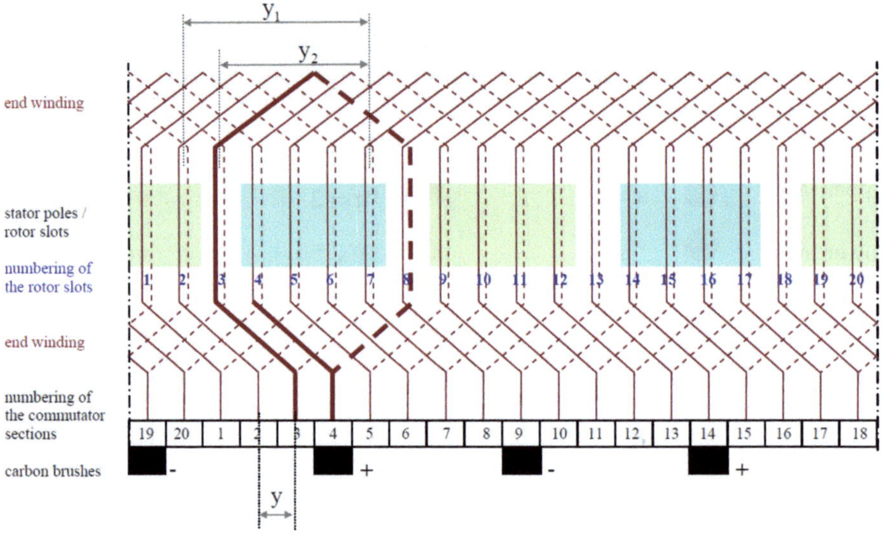

Fig. 2.8. Winding layout of a DC-machine with lap winding.

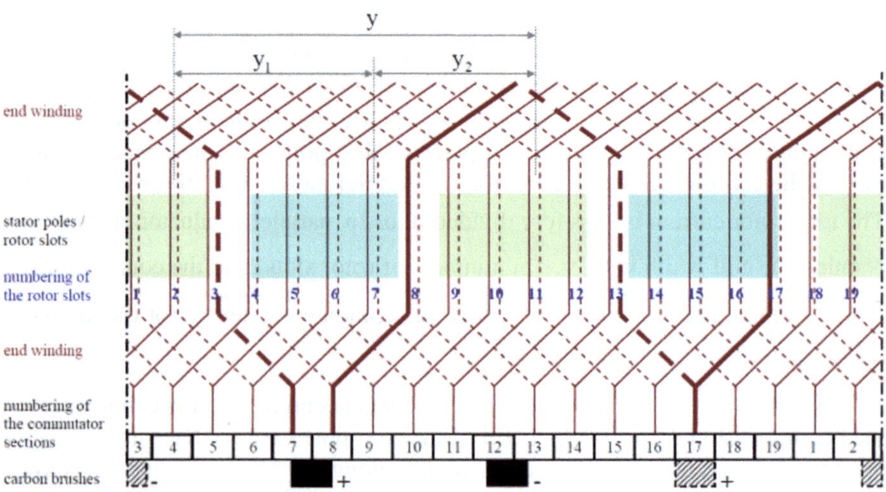

Fig. 2.9. Winding layout of a DC-machine with wave winding.

For the wave winding an arbitrary number of pole pairs can be realized with just two carbon brushes, because all positive brushes and all negative brushes are connected in series, respectively. This is illustrated in Fig. 2.9 by hatching one positive brush and one negative brush.

2.4 Main Equations of the DC-Machine

A two-pole DC-machine is regarded in the following (please refer to Fig. 2.10). Here α_i is the pole arc in parts of the pole pitch (α_i is a dimensionless number that gives the ratio between pole arc and pole pitch).

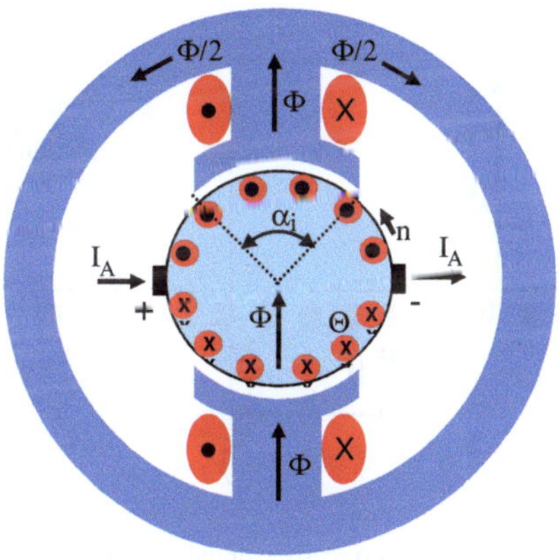

Fig. 2.10. Cross-sectional sketch of a two-pole DC-machine.

The number of turns of the rotor winding (armature winding) is:

$$w_A = \frac{z/2}{2a} = \frac{z}{4a} \tag{2.8}$$

Under the $2p$ poles $2\alpha_i w_A$ turns are effective.

2.4.1 First Main Equation: Induced Voltage

The induced voltage in a *single* rotor conductor is: $u_i = B\ell v$. For every rotor path there are $\frac{z}{2a}$ conductors in series; in each element $d\alpha$ (see Fig. 2.11) there are

$\frac{z}{2a}\frac{d\alpha}{2\pi p}$ conductors in series, if all z conductors are distributed evenly along the circumference.

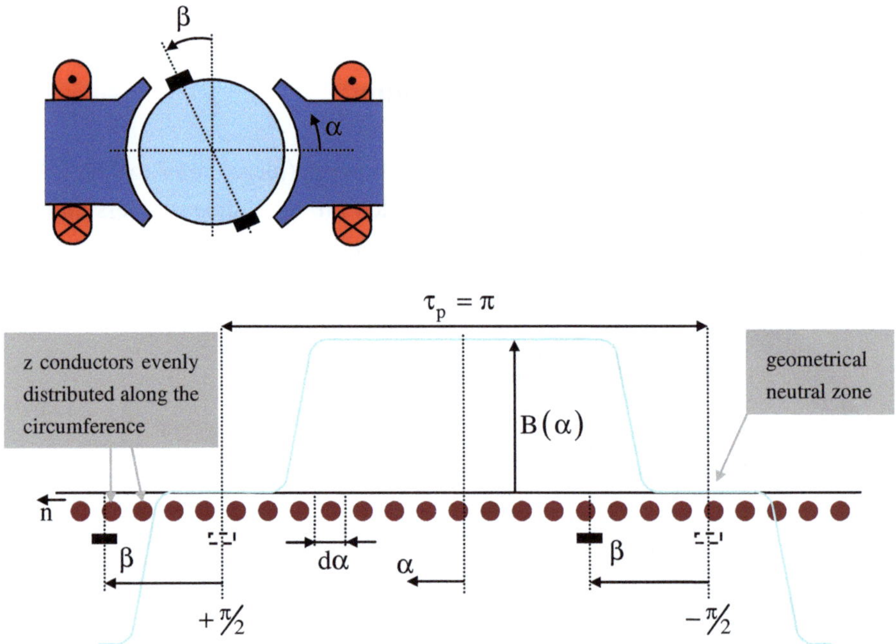

Fig. 2.11. Sketch of a two-pole DC-machine (*above*) and respective flux density distribution in "wound-off" representation (*below*).

The induced voltage in the conductors of a circumference element dα becomes:

$$u_i(d\alpha) = B(\alpha)\,\ell v \frac{z}{2a}\frac{d\alpha}{2\pi p} \tag{2.9}$$

Because of the parallel connection of the paths the total voltage of a path is equal to the induced voltage. This voltage is gained by integration between the limits given by the carbon brush position (shifted brushes). For $2p$ poles there is:

$$u_i = 2p \int_{-\frac{\pi}{2}+\beta}^{\frac{\pi}{2}+\beta} B(\alpha)\,\ell v \frac{z}{2a}\frac{d\alpha}{2\pi p} \tag{2.10}$$

With $v = 2\pi r n$ it follows:

$$u_i = z\frac{p}{a}n \int_{-\frac{\pi}{2}+\beta}^{\frac{\pi}{2}+\beta} B(\alpha)\ell\frac{r}{p}\,d\alpha \tag{2.11}$$

Here the integral describes the flux Φ, which is enclosed by the brushes. With

$$k = z\frac{p}{a} = 4pw_A \tag{2.12}$$

(the constant k is called motor constant or rotor constant) it follows:[3]

$$u_i = k\Phi n \tag{2.13}$$

For $\beta = 0$ (i.e. no shift of the brushes, brushes in neutral position) and the air-gap flux density below the excitation poles B_δ it follows:

$$\Phi = \int_{-\frac{\pi}{2}}^{\frac{\pi}{2}} B(\alpha)\ell\frac{r}{p}\,d\alpha = \int_{-\alpha_i\frac{\pi}{2}}^{\alpha_i\frac{\pi}{2}} B_\delta\ell\frac{2\pi r}{2\pi p}\,d\alpha = \int_{-\alpha_i\frac{\pi}{2}}^{\alpha_i\frac{\pi}{2}} B_\delta\ell\frac{\tau_p}{\pi}\,d\alpha$$
$$= \alpha_i\tau_p B_\delta\ell \tag{2.14}$$

2.4.2 Second Main Equation: Torque

The torque can be calculated from the force on current conducting wires (here for $\beta = 0$):

$$T = \alpha_i w_A i\, B_\delta\ 2\,\ell\, r$$
$$= \frac{4pw_A}{2\pi}\alpha_i\tau_p\,\ell B_\delta i \tag{2.15}$$

Therefore, it follows:

---

[3] Sometimes the induced voltage (also called back electromotive force, back emf, counter emf) is nominated with "e". As it has the nature of a voltage, here the name " u_i " is preferred.

$$T = \frac{k}{2\pi}\Phi i \tag{2.16}$$

2.4.3 Third Main Equation: Terminal Voltage

For the terminal voltage there is (in the energy consumption system), see Fig. 2.12:

$$u = u_i + Ri + L\frac{di}{dt} \tag{2.17}$$

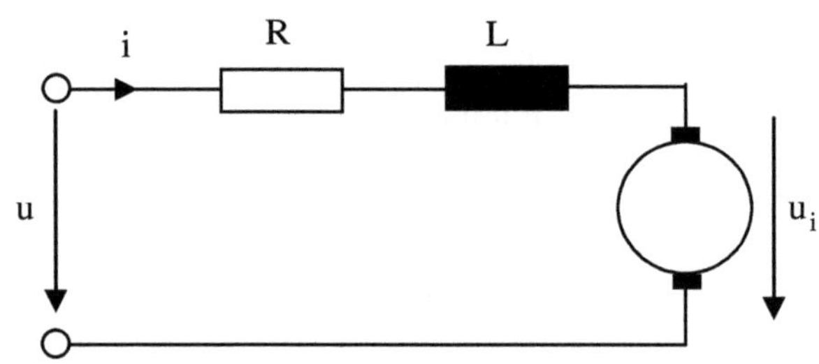

Fig. 2.12. Equivalent circuit diagram of the DC-machine.

For steady-state operation it follows:

$$U = U_i + RI \tag{2.18}$$

2.4.4 Power Balance

By means of the voltage equation a power balance can be made (multiplication of the voltage equation with the current i):

$$ui = u_i i + i^2 R + Li\frac{di}{dt} \tag{2.19}$$

From this can be deduced: The electrical input power equals the internal power of the DC-machine plus the electrical losses plus the change of magnetic energy.

Neglecting the iron and friction losses the internal power of the DC-machine equals the mechanical power. Therefore:

$$u_i i = P_i = P_{mech} = \omega_{mech} T = 2\pi n T \tag{2.20}$$

Consequently:

$$T = \frac{u_i i}{2\pi n} = \frac{k\Phi n i}{2\pi n} = \frac{k}{2\pi}\Phi i \tag{2.21}$$

2.4.5 Utilization Factor

Decisive for the design of DC-machines is the internal power $P_i = u_i i$. This internal power is limited by the material characteristics of copper (losses) and iron (magnetic saturation). These limits can be described by the values B (flux density) and A (current loading). The "current loading" is a theoretical concept, that simplifies the winding placed in the slots: It is assumed that the conductors are distributed infinitely thin on the rotor surface (please refer to Sect. 3.2). The following relationships are true:

$$\begin{aligned} u_i &= k\Phi n = z\frac{p}{a}\alpha_i \tau_p \ell B n \\ i &= 2\pi r \frac{2a}{z} A \end{aligned} \tag{2.22}$$

Consequently:

$$\begin{aligned} P_i = u_i i &= \frac{zp}{a}\frac{2a}{z}\alpha_i \tau_p \ell B n 2\pi r A \\ &= 2p\,\alpha_i \frac{2\pi r}{2p} 2\pi r \ell\, n\, AB \\ &= \alpha_i 4\pi^2 r^2 \ell\, n\, AB \\ &= C 4 r^2 \ell\, n \qquad \text{with} \quad C = \alpha_i \pi^2 AB \end{aligned} \tag{2.23}$$

The value C is called utilization factor (Esson's number); the internal power is now described by the geometry, the speed, and the utilization factor.

Example:

Some typical values are: $\alpha_i = 0.65$, $A = 500\,\text{A/cm}$ and $B = 0.8\text{T}$. From this the utilization factor $C = 4.28\,\text{kW min/m}^3$ can be obtained. Now a DC-machine with $P_i = 100\text{kW}$, $n = 2000\,\text{min}^{-1}$ and $p = 2$ shall be designed.

Choosing $\tau_p = \ell$, it follows:

$$\begin{aligned} &\frac{P_i}{Cn} = 4r^2\ell = 4r^2\tau_p = 4r^2\frac{2\pi r}{2p} \quad \Rightarrow \quad \frac{P_i}{Cn} = 2\pi r^3 \\ &\Rightarrow \quad r = \sqrt[3]{\frac{P_i}{2\pi Cn}} \approx 0.123\text{m} \end{aligned} \tag{2.24}$$

and

$$\ell = \frac{1}{4r^2}\frac{P_i}{Cn} = \frac{1}{4r^2}2\pi r^3 = \frac{1}{2}\pi r = 0.193\text{m} \tag{2.25}$$

A transformation gives:

$$C' = \frac{2}{\pi^2}C = \frac{2}{\pi^2}\frac{P_i}{4r^2\ell\,n} = \frac{2}{\pi^2}\frac{2\pi nT}{4r^2\ell\,n} = \frac{T}{\pi r^2\ell} \tag{2.26}$$

Consequently, the utilization factor C is proportional to the torque divided by the bore volume.

Further it follows:

$$C' = \frac{T}{\pi r^2\ell} = \frac{Fr}{\pi r^2\ell} = 2\frac{F}{2\pi r\ell} = 2f \tag{2.27}$$

The utilization factor C also is proportional to the (tangential) force divided by the bore surface area.

With $C = \alpha_i\pi^2 AB$ it is true:

$$\begin{aligned} C' &= 2\alpha_i AB \\ f &= \alpha_i AB \end{aligned} \tag{2.28}$$

2.5 Induced Voltage and Torque, Precise Consideration

2.5.1 Induced Voltage

Up to now the calculation of the induced voltage and the torque was performed assuming that the conductors of the rotor are lying in the air-gap field. But the conductors of the rotor are placed in the rotor slots; the magnetic field is guided around the rotor winding by means of the surrounding iron (Fig. 2.13).

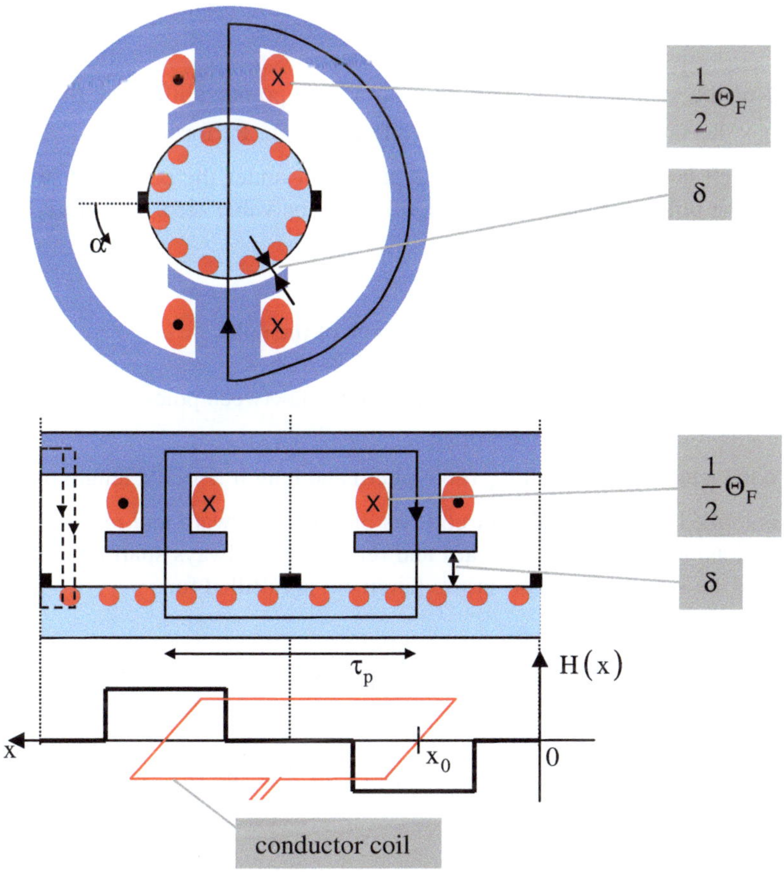

Fig. 2.13. Sketch of the DC-machine in rotatory presentation (*above*) and "wound-off" representation (*below*).

With

$$x = r\,\alpha \tag{2.29}$$

both coordinate systems (Cartesian and cylindrical) can be transformed to each other.

For calculation of the induced voltage Ampere's Law is used with a circulation path over one pole pitch (for the permeability of iron $\mu_{Fe} \to \infty$ is assumed; the rotor current is zero; r is the rotor radius):

$$\begin{aligned}\oint \vec{H}d\vec{\ell} &= \Theta_F = H(x)\delta + \left[-H(x+\tau_p)\right]\delta \\ &= 2H(x)\delta = \frac{B(x)}{\mu_0}2\delta\end{aligned} \tag{2.30}$$

Because of the symmetry the field strength at two points, shifted by the pole pitch τ_p, has the same absolute value, but different sign. Performing the circulation path (which has the width of one pole pitch) not under the poles, but in the gap between the poles, the circulation integral gives the value zero. Therefore, the air-gap flux density becomes:

$$B(x) = \begin{cases} \pm B_\delta = \pm\dfrac{\mu_0}{2\delta}\Theta_F & \text{in the area of the poles} \\ 0 & \text{in the gap area between poles} \end{cases} \tag{2.31}$$

Now a conductor coil of width τ_p (pole pitch) and length ℓ (axial length of the machine) is looked at. This conductor coil has placed its forward and return wires in the rotor slots. Moreover, the forward and return wires always shall be located in the areas of the stator poles. This last requirement is fulfilled for

$$\frac{\tau_p}{2}(1-\alpha_i) \le x_0 \le \frac{\tau_p}{2}(1+\alpha_i) \tag{2.32}$$

if x_0 describes the beginning of the conductor coil. For the flux surrounded by this conductor coil it follows:

$$\phi(x_0) = -B_\delta \ell \tau_p \left(\frac{\tau_p}{2} - x_0\right)\frac{2}{\tau_p} = -B_\delta 2\ell\left(\frac{\tau_p}{2} - x_0\right) \tag{2.33}$$

Shifting now the conductor coil by the value Δx (i.e. rotating the rotor by $\Delta\alpha = \Delta x/r$), but this coil remains under the stator poles, the surrounded flux is:

$$\phi(x_0 + \Delta x) = -B_\delta 2\ell\left(\frac{\tau_p}{2} - (x_0 + \Delta x)\right) \tag{2.34}$$

The induced voltage equals the change of flux with respect to time (see Sect. 2.2); therefore it follows:

$$\begin{aligned} \Delta\phi &= \phi(x_0 + \Delta x) - \phi(x_0) \\ &= -B_\delta 2\ell(-\Delta x) \\ \Rightarrow \quad U_i &= \frac{\Delta\phi}{\Delta t} = -B_\delta 2\ell \frac{-\Delta x}{\Delta t} = B_\delta 2\ell v \end{aligned} \tag{2.35}$$

The sign of the induced voltage depends on the direction of movement of the conductor coil (i.e. depending on the direction of movement of the rotor, because the conductor coil is placed inside the rotor slots).

In total the following can be stated: *The induced voltage of wires placed in slots can be calculated as if these wires would lie in the air-gap field.*

2.5.2 Torque

In the preceding section the calculation of the induced voltage was performed using the "wound-off" representation, in this section the computation will be done using the original rotatory geometry (of course, both calculations can be performed using the other alternative).

A conductor coil placed inside the rotor slots is assumed having a rotor (armature) current $I_A > 0$ (see Fig. 2.14). At time instant $t = t_1$ the rotor has the position shown in the upper part of Fig. 2.14, at time instant $t = t_2$ the rotor has the position shown in the lower part of Fig. 2.14. For both cases the shown circulation path along one pole pitch (which is identical for both cases and which is illustrated by the black solid line) is evaluated. At time instant $t = t_1$ the circulation path includes the excitation magneto-motive force (of the stator) and the return wire of the current conducting rotor coil, at time instant $t = t_2$ the circulation path includes the excitation magneto-motive force and the forward wire of this coil.

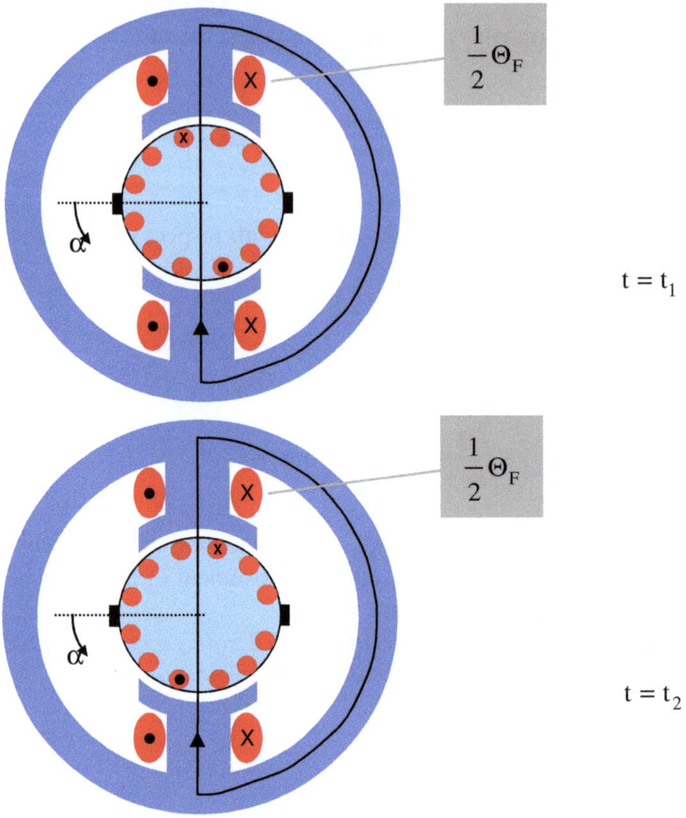

Fig. 2.14. Sketch of the DC-machine in rotatory presentation for two different rotor positions at two different points in time: t_1 (*above*) and t_2 (*below*); these different rotor positions are noticeable from the different locations of the current conducting rotor coil.

By means of symmetry conditions it follows for time instant $t = t_1$, if α_0 describes the mid-point of a stator pole:

$$\Theta_F - I_A = H(\alpha_0, t_1)\, 2\delta = \frac{B(\alpha_0, t_1)}{\mu_0} 2\delta$$
$$\Rightarrow \quad B(\alpha_0, t_1) = \frac{\mu_0}{2\delta}\Theta_F - \frac{\mu_0}{2\delta} I_A = B_\delta - \frac{\mu_0}{2\delta} I_A \tag{2.36}$$

The same circulation path at time instant $t = t_2$ gives, because the forward wire of the conductor coil is included:

$$\Theta_F + I_A = H(\alpha_0, t_2)2\delta = \frac{B(\alpha_0, t_2)}{\mu_0}2\delta$$
$$\Rightarrow \quad B(\alpha_0, t_2) = \frac{\mu_0}{2\delta}\Theta_F + \frac{\mu_0}{2\delta}I_A = B_\delta + \frac{\mu_0}{2\delta}I_A \tag{2.37}$$

Rotating the rotor by $\Delta\alpha$ changes the magnetic energy in the volume element $\Delta V = 2\ell\delta r\Delta\alpha$ like follows (the magnetic energy outside this volume element does not have to be regarded, because the magnetic field outside the space described by the moved conductor coil does not change):

$$\begin{aligned}\Delta W_{mag} &= W_{mag}(t = t_2) - W_{mag}(t = t_1) \\ &= \frac{B^2(t_2, \alpha_0) - B^2(t_1, \alpha_0)}{2\mu_0}2\ell\delta r\Delta\alpha \\ &= \frac{2\frac{2B_\delta}{2\delta}\mu_0 I_A}{2\mu_0}2\ell\delta r\Delta\alpha = 2B_\delta I_A \ell r\Delta\alpha\end{aligned} \tag{2.38}$$

The force onto a *single* conductor is calculated from the change of magnetic energy with respect to movement (please note that up to now *two* conductors, forward and return conductor of the coil, were regarded):

$$F = \frac{1}{2}\frac{\Delta W_{mag}}{r\Delta\alpha} = I_A B_\delta \ell \tag{2.39}$$

The sign of the force is – at constant stator field – depending on the direction of the current in the coil (i.e. depending on the direction of the voltage switched to the conductor coil).

In total the following can be stated: *The force onto wires placed in slots can be calculated as if these wires would lie in the air-gap field.*

The force does not act directly onto the wires, but it acts onto the iron teeth because of different flux densities. From the force calculation the torque generated by the machine can be deduced. Therefore, even the torque direction depends on the current direction in the rotor (i.e. depending on the direction of the DC-voltage switched to the rotor coils).

2.6 Separately Excited DC-Machines

The excitation winding of a separately excited DC-machine is supplied by an additional voltage source, therefore this machine topology in the steady-state operation can be described by the following equivalent circuit diagram (Fig. 2.15):

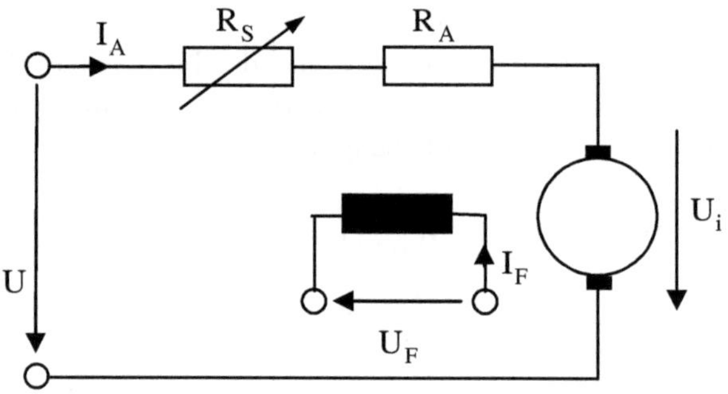

Fig. 2.15. Equivalent circuit diagram of the separately excited DC-machine.

The terminal voltage U and the excitation voltage U_F are independently adjustable. By the variable series resistance R_S the total resistance in the rotor circuit $R = R_S + R_A$ can be increased.

From the three main equations (here for steady-state operation)

$$U_i = k\Phi n \tag{2.40}$$

$$T = \frac{k}{2\pi}\Phi I_A \tag{2.41}$$

$$U = U_i + RI_A \tag{2.42}$$

the following speed characteristic is deduced:

$$n = \frac{U_i}{k\Phi} = \frac{U}{k\Phi} - \frac{RI_A}{k\Phi} \tag{2.43}$$

In no-load operation ($I_A = 0$) there is:

$$n = n_0 = \frac{U}{k\Phi} \tag{2.44}$$

At stand-still ($n = 0$) the so-called stall current (also called short-circuit current) is:

$$I_A = I_{stall} = \frac{U}{R} \tag{2.45}$$

This stall current has to be limited by the series resistance R_S. The stall torque amounts to:

$$T_{stall} = \frac{k}{2\pi} \Phi I_{stall} \tag{2.46}$$

At operation with (positive) nominal voltage $U = U_N$, nominal flux $\Phi = \Phi_N$ (at $U_F = U_{F,N}$) and $R_S = 0$ the fundamental characteristic of the separately excited DC-machine becomes (Fig. 2.16):

$$n = n_0 - \frac{R_A I_A}{k\Phi_N} \tag{2.47}$$

$$T = \frac{k}{2\pi} \Phi_N I_A \tag{2.48}$$

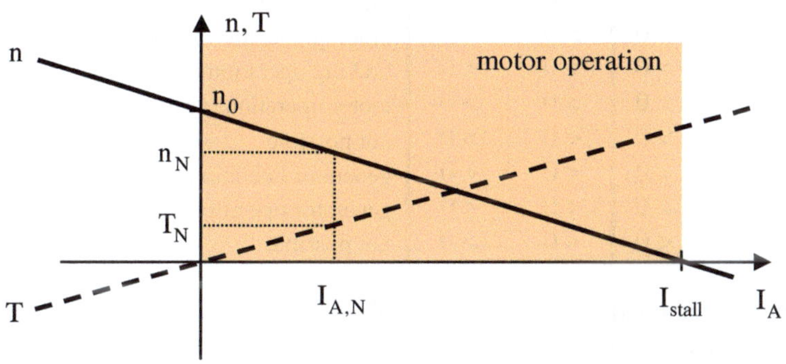

Fig. 2.16. Torque and speed versus current of the separately excited DC-machine.

In the regarded energy consumption system motor operation is for

$$n > 0,\ T > 0,\ I_A > 0 \text{ (i.e. in the first quadrant for } I_A < I_{stall}\text{)};$$

generator operation is for

$$n > 0, T < 0,\ I_A < 0 .$$

For $I_A > I_{stall}$ it is true: $n < 0,\ T > 0,\ I_A > 0$. This is the braking operation of the machine.

The power flow in such a DC-machine is depicted in Fig. 2.17, assuming the energy consumption system for the definition of positive directions.

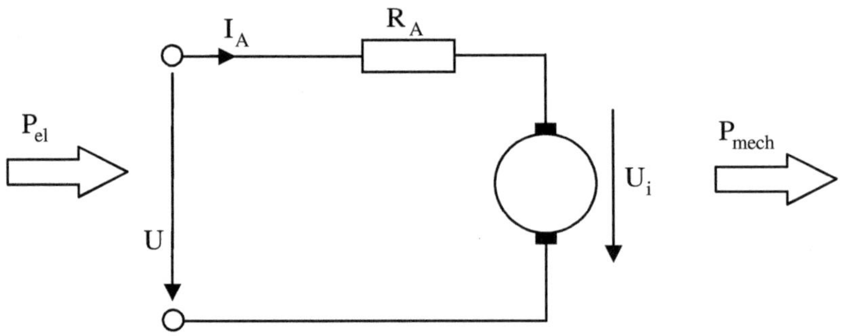

Fig. 2.17. Equivalent circuit diagram of the general DC-machine with power flow.

Generally, the following operational conditions are possible (assumed is the energy consumption system and a positive terminal voltage $U > 0$) (Table 2.1):

Table 2.1. Possible operational conditions (assuming positive terminal voltage and energy consumption system)

I_A	n	T	P_{el}	P_{mech}	
>0	>0	>0	>0	>0	motor operation (@ positive speed)
>0	>0	<0	>0	<0	braking operation (@ positive speed)
>0	<0	>0	>0	<0	braking operation (@ negative speed)
>0	<0	<0	>0	>0	motor operation (@ negative speed)
<0	>0	>0	<0	>0	not possible
<0	>0	<0	<0	<0	generator operation (@ positive speed)
<0	<0	>0	<0	<0	generator operation (@ negative speed)
<0	<0	<0	<0	>0	not possible

Two operational conditions are not possible, because these would mean that the DC-machine delivers electrical power as well as mechanical power (having a perpetuum mobile). For negative terminal voltage $U < 0$ an additional, similar table can be deduced.

There are three general possibilities for speed control:

1. Reduction of the terminal voltage:
With setting the voltage $U/U_N < 1$ the following equations are true:

$$n = n_0 \frac{U}{U_N} - \frac{R_A I_A}{k\Phi_N} \tag{2.49}$$

$$T = \frac{k}{2\pi} \Phi_N I_A \tag{2.50}$$

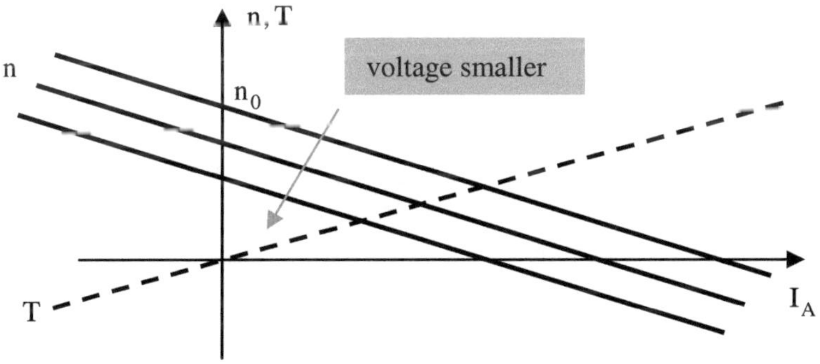

Fig. 2.18. Torque and speed versus current of the separately excited DC-machine for voltage variation.

Therefore, the no-load speed is decreased, but the slope of the speed characteristic does not change. The relation between torque and rotor current remains unchanged. The stall (short-circuit) current is decreased proportional to the terminal voltage; the direction of rotation can be changed by reversing the polarity of the terminal voltage (Fig 2.18).

The speed control by changing the terminal voltage is lossless. Because of the small rotor time constant this alternative for speed control is highly dynamic.

2. Reduction of the flux:
Reducing the excitation flux (by reducing the excitation current) and neglecting the saturation, it follows:

$$\frac{I_F}{I_{F,N}} = \frac{\Phi}{\Phi_N} = \frac{1}{f} < 1 \tag{2.51}$$

$$n = fn_0 - f\frac{R_A I_A}{k\Phi_N} \tag{2.52}$$

$$T = \frac{k}{2\pi}\frac{\Phi_N}{f} I_A \tag{2.53}$$

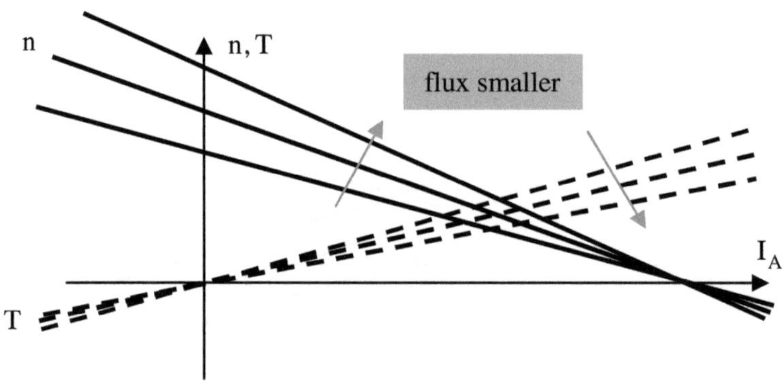

Fig. 2.19. Torque and speed versus current of the separately excited DC-machine for flux variation.

With increasing field weakening (field weakening factor $f > 1$) the no-load speed is increased and the slope of the speed characteristic rises (i.e. loading the machine means an exceeded speed reduction). At constant rotor current the torque decreases with increasing field weakening (Fig. 2.19).

The speed control by field weakening is lossless, but (because of the usually large time constant of the excitation winding) it is less dynamic compared with the speed control by changing the terminal voltage.

3. Increasing the rotor resistance (series resistance):

By inserting a series resistance R_S into the rotor circuit the total resistance is increased. The following equations are obtained:

$$n = n_0 - \frac{(R_A + R_S) I_A}{k\Phi_N} \tag{2.54}$$

$$T = \frac{k}{2\pi} \Phi_N I_A \tag{2.55}$$

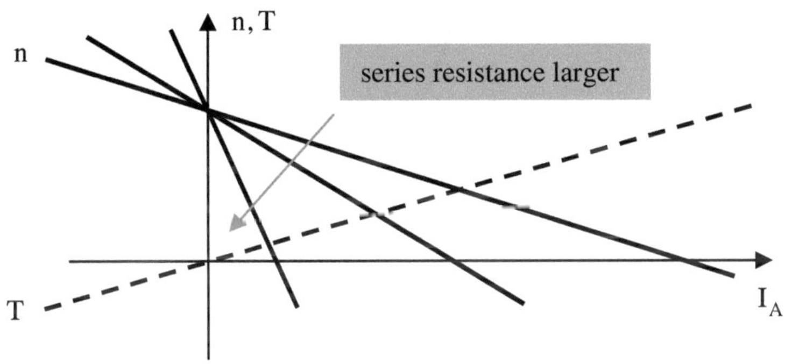

Fig. 2.20. Torque and speed versus current of the separately excited DC-machine for series resistance variation.

The no-load speed n_0 remains unchanged, the same holds true for the relation between torque and rotor current. The slope of the speed characteristic increases, i.e. loading the machine means an exceeded speed reduction. The stall (short-circuit) current is reduced by increasing the series resistance, therefore this alternative is used for starting the motor (Fig. 2.20).

Because of the additional losses in the series resistance ($R_S I_A^2$) this method is not lossless.

The following general operational limits have to be obeyed:

- In continuous operation the acceptable heating may not be exceeded: $I_A \le I_{A,N}$.
- Even in field weakening operation the machine may not be overloaded: $2\pi nT \le P_N$.
- Because of the centrifugal forces onto the rotating parts the maximum acceptable speed may not be exceeded: $n \le n_{max}$.

In literature the characteristics of DC-machines are often shown versus the rotor current I_A (see Figs. 2.18 to 2.20). Describing rotating field machines, mainly the speed is used as horizontal axis. To make the comparison of the characteristics easier, Fig. 2.21 shows the torque of a separately excited DC-machine as a function of the speed (this is the same relation like in the preceding figures, it is just illustrated in a different way).

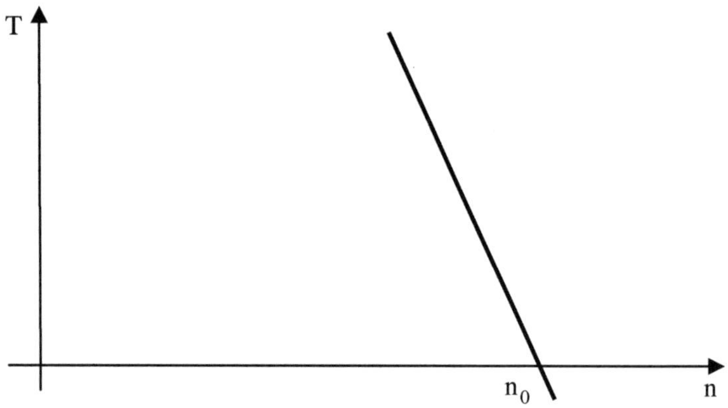

Fig. 2.21. Torque versus speed of the separately excited DC-machine.

2.7 Permanent Magnet Excited DC-Machines

The permanent magnet excited DC-machine can be regarded as being a special case of the separately excited DC-machine. Even for this alternative the speed control can be performed by changing the terminal voltage and by introducing a series resistance. But field weakening is not possible (because the permanent magnets impress the magnetic flux). The advantages of the permanent magnet excitation against the electrical excitation are (please refer to Fig. 2.22):

- smaller outer diameter
- smaller volume and weight
- more simple construction (no stator winding, less terminal contacts)
- higher efficiency (no excitation losses)
- more cost-effective production
- better dynamics (no field increase by excitation current increase)

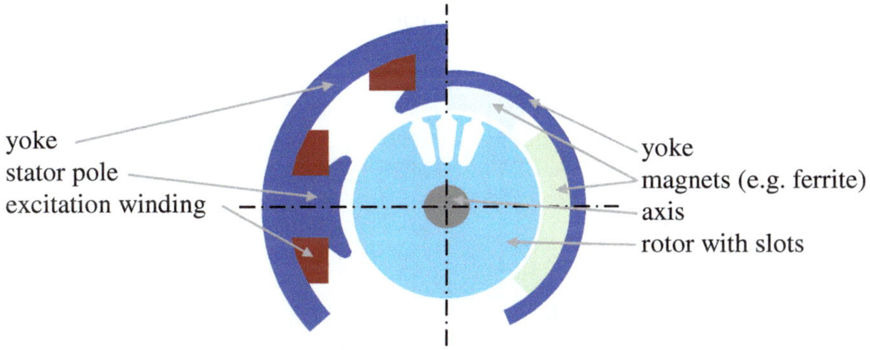

Fig. 2.22. Separately excited DC-machines: electrical excitation (*left*) and magnetic excitation (*right*).

The technical characteristics of a permanent magnet material is described by the hysteresis loop in the second quadrant (Fig 2.23):

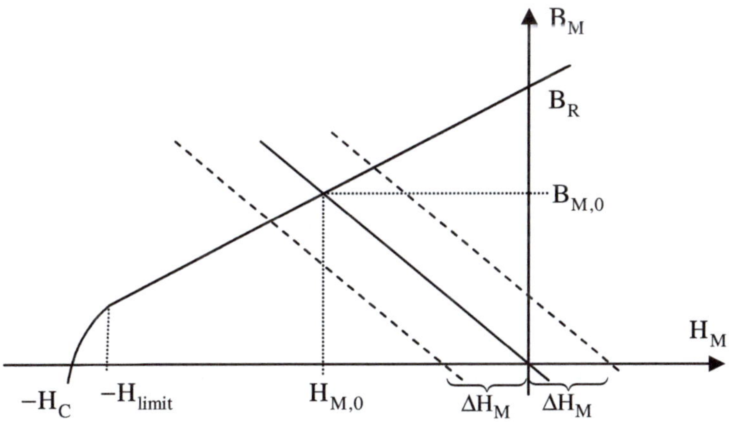

Fig. 2.23. Permanent magnet characteristic.

Flux density and field strength are linked by remanent flux density and permeability:

$$B_M = B_R + \mu_0 \mu_r H_M$$

$$B(H = 0) = B_R \qquad \text{remanent flux density}$$

$$H(B = 0) = -H_C \qquad \text{coercive force} \tag{2.56}$$

$$\mu_r \approx 1,0 \cdots 1,1 \qquad \text{reversible permeability}$$

$$-H_{limit} \qquad \text{limiting field strength}$$

The development of different magnet materials according to time is illustrated in principle in Fig. 2.24:

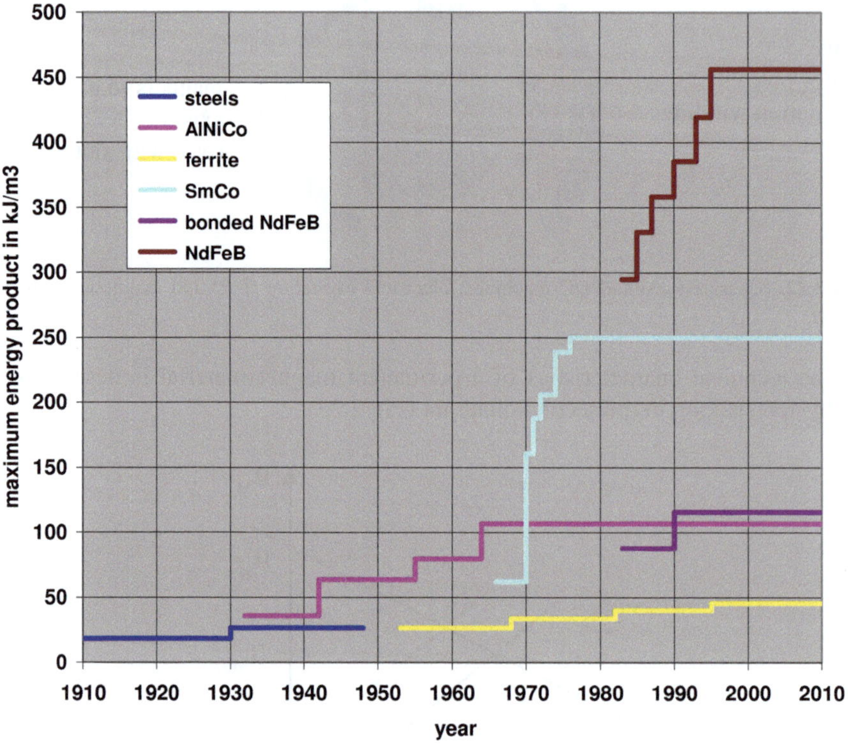

Fig. 2.24. Different magnet materials: maximum energy product versus time.

A permanent magnet can only be operated in the linear section of this characteristic, otherwise it will be (at least partly) demagnetized. If higher opposite field strength than H_{limit} is applied, an irreversible flux loss occurs.

Naming the radial magnet height h_M it follows from Ampere's Law with "circulation path 1" (please refer to Fig. 2.25):

$$\frac{B_{\delta,1}}{\mu_0} 2\delta + H_{M,1} 2h_M = 0 \tag{2.57}$$

The flux densities can be calculated from the according cross sections:

$$B_\delta = B_M \frac{A_M}{A_\delta} \tag{2.58}$$

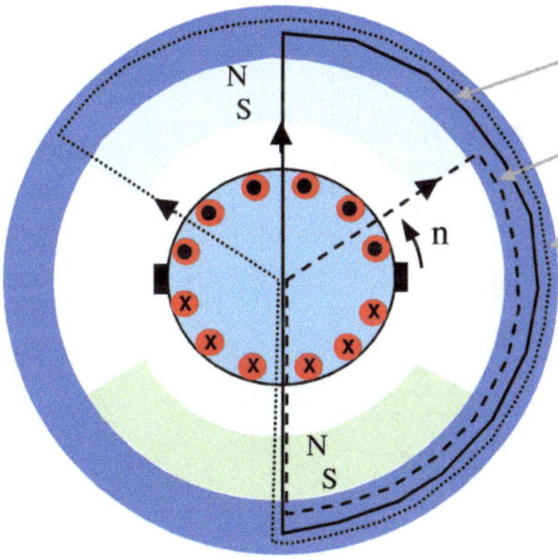

Fig. 2.25. Permanent magnet DC-Machine with different circulation paths.

The field strength in the magnet is:

$$H_M = \frac{B_M - B_R}{\mu_0 \mu_r} \tag{2.59}$$

Consequently the flux density in the magnet is:

$$\begin{aligned}
&\frac{B_{M,1}}{\mu_0}\frac{A_M}{A_\delta}2\delta + \frac{B_{M,1} - B_R}{\mu_0\mu_r}2h_M = 0 \\
\Rightarrow \quad & B_{M,1}\left(\frac{A_M}{A_\delta}\delta + \frac{h_M}{\mu_r}\right) = B_R\frac{h_M}{\mu_r} \\
\Rightarrow \quad & B_{M,1} = B_R \frac{1}{\left(\frac{A_M}{A_\delta}\delta + \frac{h_M}{\mu_r}\right)\frac{\mu_r}{h_M}} \\
& = B_R \frac{1}{1 + \mu_r \frac{A_M}{A_\delta}\frac{\delta}{h_M}}
\end{aligned} \tag{2.60}$$

and the field strength in the magnet:

$$
\begin{aligned}
H_{M,1} &= \frac{B_R}{\mu_0\mu_r}\left(\frac{1}{1+\mu_r\dfrac{A_M}{A_\delta}\dfrac{\delta}{h_M}}-1\right) \\
&= -\frac{B_R}{\mu_0\mu_r}\frac{\mu_r\dfrac{A_M}{A_\delta}\dfrac{\delta}{h_M}}{1+\mu_r\dfrac{A_M}{A_\delta}\dfrac{\delta}{h_M}} \\
&= -\frac{B_R}{\mu_0\mu_r}\frac{1}{1+\dfrac{1}{\mu_r}\dfrac{h_M}{\delta}\dfrac{A_\delta}{A_M}}
\end{aligned} \tag{2.61}
$$

These calculated values for $B_{M,1}$ and $H_{M,1}$ are valid in the symmetry plane, and during no-load operation ($I_A = 0$) for the entire magnet.

Now the air-gap flux density at the run-on edge will be calculated ("circulation path 2"). It has to be considered that along the entire circumference of the rotor $2w_A$ conductors are placed, the current loading is $A = \dfrac{2w_A I_A}{\pi D}$. The current under one pole is: $A\alpha_i\tau_p$.

$$
\begin{aligned}
\oint \vec{H}d\vec{\ell} &= 0-\left(-\frac{1}{2}A\alpha_i\tau_p\right) \\
&= \frac{1}{2}\frac{2w_A I_A}{\pi D}\alpha_i\frac{\pi D}{2p} \\
&= \frac{1}{2p}\alpha_i w_A I_A = \Delta I
\end{aligned} \tag{2.62}
$$

Consequently:

$$\frac{B_{\delta,1}}{\mu_0}\delta + H_{M,1}h_M + \frac{B_{\delta,2}}{\mu_0}\delta + H_{M,2}h_M = \Delta I$$

$$\Rightarrow \quad \frac{B_{M,1}}{\mu_0}\frac{A_M}{A_\delta}\delta + H_{M,1}h_M + \frac{B_{M,2}}{\mu_0}\frac{A_M}{A_\delta}\delta + \frac{B_{M,2}-B_R}{\mu_0\mu_r}h_M = \Delta I \tag{2.63}$$

$$\Rightarrow \quad \frac{B_{M,2}}{\mu_0}\left(\frac{A_M}{A_\delta}\delta + \frac{1}{\mu_r}h_M\right) = -\frac{B_{M,1}}{\mu_0}\frac{A_M}{A_\delta}\delta - H_{M,1}h_M + \frac{B_R}{\mu_0\mu_r}h_M + \Delta I$$

and further

$$B_{M,2} = \mu_0 \frac{1}{\frac{A_M}{A_\delta}\delta + \frac{1}{\mu_r}h_M}\left[-\frac{B_R}{\mu_0}\frac{\frac{A_M}{A_\delta}\delta}{1+\mu_r\frac{A_M}{A_\delta}\frac{\delta}{h_M}}\right.$$

$$\left. + \frac{B_R}{\mu_0\mu_r}\frac{1}{1+\frac{1}{\mu_r}\frac{h_M}{\delta}\frac{A_\delta}{A_M}}h_M + \frac{B_R}{\mu_0\mu_r}h_M + \Delta I\right]$$

$$\Rightarrow \quad B_{M,2} = B_R\frac{\frac{\mu_r}{h_M}}{1+\mu_r\frac{A_M}{A_\delta}\frac{\delta}{h_M}}\left[\frac{h_M}{\mu_r} + \frac{1}{\mu_r}\frac{h_M}{1+\frac{1}{\mu_r}\frac{h_M}{\delta}\frac{A_\delta}{A_M}}\right.$$

$$\left. - \frac{\frac{A_M}{A_\delta}\delta}{1+\mu_r\frac{A_M}{A_\delta}\frac{\delta}{h_M}}\right] + \mu_0\mu_r\Delta I\frac{1}{\mu_r\frac{A_M}{A_\delta}\delta + h_M}$$

$$\Rightarrow \quad B_{M,2} = B_R\frac{\frac{\mu_r}{h_M}}{1+\mu_r\frac{A_M}{A_\delta}\frac{\delta}{h_M}}\left(\frac{h_M}{\mu_r} + \delta\frac{A_M}{A_\delta}\right)\frac{1}{1+\mu_r\frac{A_M}{A_\delta}\frac{\delta}{h_M}} + \tag{2.64}$$

$$\mu_0\mu_r\Delta H_M, \qquad \Delta H_M = \Delta I\frac{1}{\mu_r\frac{A_M}{A_\delta}\delta + h_M}$$

Further it follows:

$$B_{M,2} = B_R \frac{1}{1+\mu_r \frac{A_M}{A_\delta} \frac{\delta}{h_M}} + \mu_0 \mu_r \Delta H_M$$
$$= B_{M,1} + \mu_0 \mu_r \Delta H_M \quad (2.65)$$

Under load the flux density in the magnet B_M at the run-on edge is higher than under no-load condition.

For the run-off edge the flux density under load ("circulation path 3", using an analogous deduction like above) becomes:

$$B_{M,3} = B_{M,1} - \mu_0 \mu_r \Delta H_M \quad (2.66)$$

Therefore, under load the flux density in the magnet B_M at the run-off edge is smaller than under no-load condition. The danger of demagnetization occurs (please refer to "armature reaction", Sect. 2.12)!

Especially for applications with a wide temperature range the demagnetization must be checked in the entire temperature range as the magnet characteristic changes with temperature.

Typical values of remanent flux density (B_R) and coercive field strength (H_C) and the respective temperature coefficients for different magnet materials are shown in Table 2.2.

Table 2.2. Main characteristic values of important permanent magnet materials.

	remanent flux density		coercive field strength	
	value @ 20°C	temperature coefficient	value @ 20°C	temperature coefficient
ferrite	0.4 T	-0.190 %/K	170 kA/m	+0.30 %/K
NdFeB	1.2 T	-0.090 %/K	1900 kA/m	-0.60 %/K
SmCo	1.1 T	-0.032 %/K	1800 kA/m	-0.19 %/K

Figure 2.26 principally illustrates the influence of the temperature onto the limiting field strength H_{limit} and the value ΔH_M.

As counter-measure against the danger of demagnetization the magnet height h_M can be chosen larger.

To illustrate the effect of different demagnetization danger at the run-on edge and the run-off edge, Fig. 2.27 shows the field lines of a 2-pole 12 slots DC-motor with ferrite magnets with and without rotor currents.

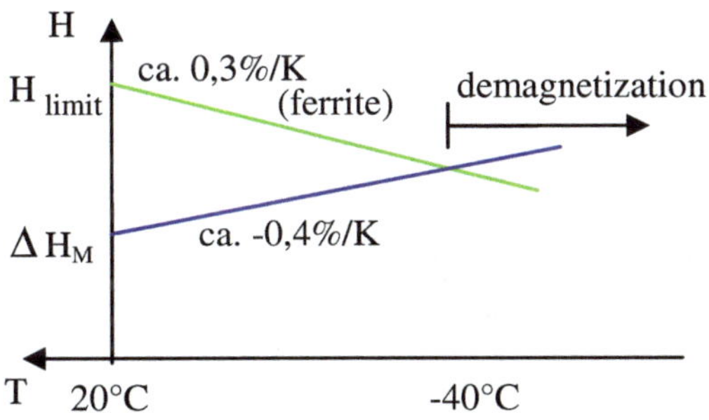

Fig. 2.26. Field strength and limiting field strength versus temperature.

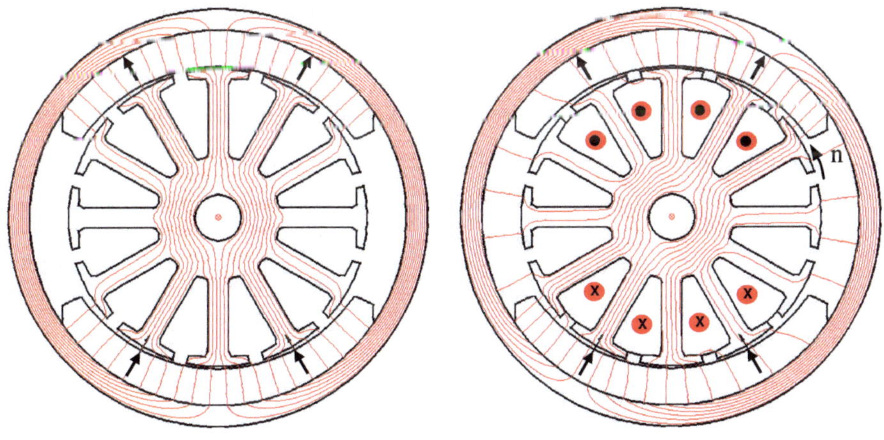

Fig. 2.27. Field lines of a 2-pole 12 slots DC-motor with ferrite magnets without (*left*) and with (*right*) rotor currents.

2.8 Shunt-Wound DC-Machines

For the shunt-wound DC-machine the excitation winding is connected in parallel to the rotor circuit. This results in the equivalent circuit diagram shown in Fig. 2.28. In this figure the resistance R_F is drawn with the symbol of an inductivity. This makes obvious that

- in the excitation circuit an inductivity is present, but
- in the steady-state operation regarded here only the resistance is effective.

Similar depictions are even chosen in Figs. 2.30, 2.32, and 2.37.

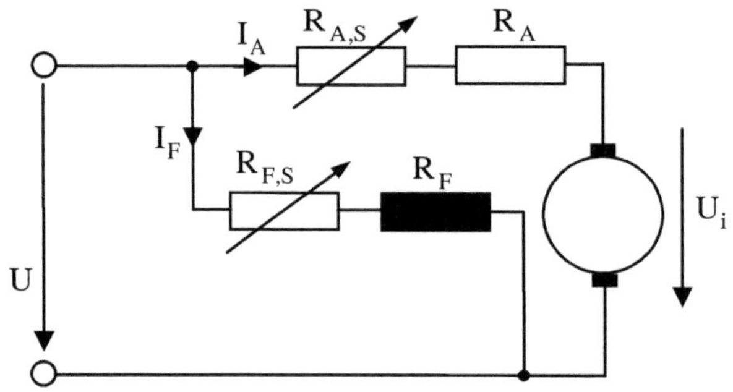

Fig. 2.28. Equivalent circuit diagram of the shunt-wound DC-machine.

Rotor and excitation circuit are switched to the line with constant voltage. In each circuit a series resistance may be introduced.

Is this shunt-wound DC-machine switched to constant voltage, the speed characteristic is equal to that of a separately excited DC-machine. For $R_{A,S} = 0$, $R_{F,S} = 0$ it follows:

$$n = n_0 - \frac{R_A I_A}{k\Phi_N}, \qquad n_0 = \frac{U_N}{k\Phi_N} \tag{2.67}$$

$$T = \frac{k}{2\pi}\Phi_N I_A, \qquad I_{stall} = \frac{U_N}{R_A} \tag{2.68}$$

For speed control or change of speed direction the change of the terminal voltage is without effect, because the excitation current and the rotor current are changed or reversed simultaneously. The speed can only be changed by using the series resistances:

- in the rotor circuit: lossy speed reduction
- in the excitation circuit: nearly lossless speed increase

The characteristics are analogously to those of the separately excited DC-machine.

Self-excitation (generator operation):
Switching the excitation winding via a series resistance parallel to the rotor and driving the machine with constant speed, a remanent voltage U_R is induced by

the always present residual magnetism in every iron circuit. This voltage generates an excitation current I_R, which strengthens the residual field and the induced voltage is increased. This happens as long as the induced voltage is equal to the voltage drop across the resistances in the excitation circuit. Then a stable operating point is reached. This is called "dynamo-electrical principle" (Fig. 2.29) (Werner von Siemens, 1866).

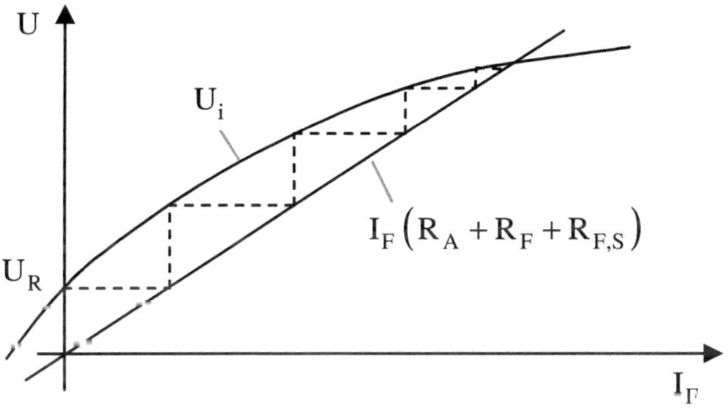

Fig. 2.29. Voltage versus excitation current during self-excitation.

Switching on the excitation winding in opposite direction, the excitation current demagnetizes the circuit and the self-excitation does not take place.

The series resistance in the excitation circuit $R_{F,S}$ has two effects:

- the generator voltage can be set and
- the time constant of the excitation winding can be reduced to speed-up the self-excitation.

This series resistance may not exceed a critical value, because otherwise the self-excitation does not take place.

Loading the generator:

For the separately excited DC-machine the load characteristic $U = f(I)$ is a straight line: $U = U_i - R_A I_A$ (energy generation system).

For the shunt-wound DC-machine it follows in the energy generation system (Fig. 2.30):

$$\begin{aligned} U_i &= R_A I_A + R_F I_F \\ &= R_A (I + I_F) + R_F I_F \\ &= R_A I + (R_A + R_F) I_F \\ \Rightarrow \quad I &= \frac{1}{R_A}(U_i - (R_A + R_F) I_F) \end{aligned} \tag{2.69}$$

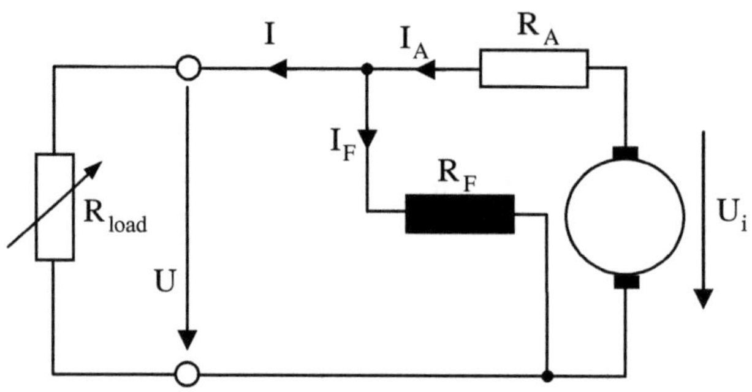

Fig. 2.30. Equivalent circuit diagram of the loaded shunt-wound DC-generator.

By means of the no-load characteristic the load characteristic can be deduced (Fig. 2.31).

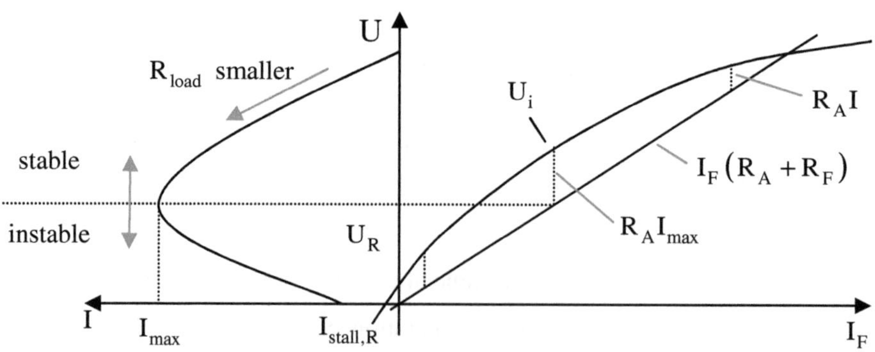

Fig. 2.31. Load characteristic of the shunt-wound DC-generator.

The shunt-wound DC-generator only can be loaded until the current I_{max}, at further loading the voltage collapses. Then only the stall current $I_{stall,R}$ flows, which is generated by the residual voltage.

2.9 Series-Wound DC-Machines

For the series-wound DC-machine the excitation winding is switched in series to the rotor circuit, see Fig. 2.32. The series resistance in the rotor circuit and the resistance parallel to the excitation winding can be used for speed control. As the excitation current of this alternative depends on the load, the speed characteristic of the series-wound DC-machine is principally different to the DC-machines discussed so far.

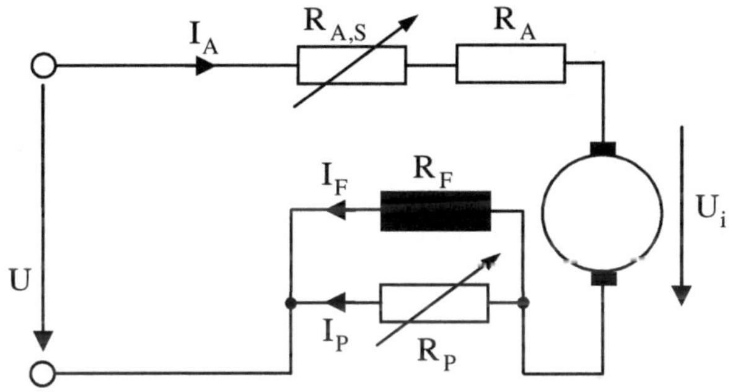

Fig. 2.32. Equivalent circuit diagram of the series-wound DC-machine.

Neglecting the saturation and for $R_P \to \infty$ it follows for steady-state operation:

$$\begin{aligned} w_F \Phi &= L_m I_F = L_m I_A \\ k\Phi &= 4pw_A \frac{L_m I_A}{w_F} = L_m' I_A \end{aligned} \tag{2.70}$$

For the speed and torque equation it follows:

$$\begin{aligned} n &= \frac{U_N}{k\Phi} - \frac{(R_A + R_F) I_A}{k\Phi} = \frac{U_N}{L_m' I_A} - \frac{R_A + R_F}{L_m'} \\ T &= \frac{k}{2\pi} \Phi I_A = \frac{L_m'}{2\pi} I_A^2 \end{aligned} \tag{2.71}$$

Loading this machine the speed decreases drastically (“series characteristic”); the characteristics during motor operation are shown in Fig. 2.33.

In no-load operation ($I_A = 0$) it follows $n \to \infty$, the machine “runs away“. The speed is only limited by the friction torque (carbon brushes, bearings). The stall current ($n = 0$) is:

$$I_{stall} = \frac{U_N}{R_A + R_F} \tag{2.72}$$

This stall (short-circuit) current has to be limited by the series resistance $R_{A,S}$ during starting operation of the motor.

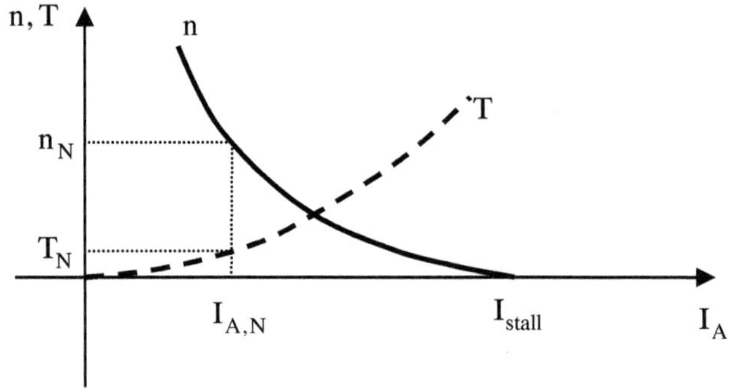

Fig. 2.33. Torque and speed versus current of the series-wound DC-machine.

By reversing the polarity of the terminal voltage no speed reversal can be obtained, because rotor current and excitation current are changed simultaneously.

For speed control the following possibilities can be used (in Figs. 2.34 to 2.36 always the motor operation is illustrated):

1. Reduction of the terminal voltage:
The speed and the short circuit current are reduced lossless, the relation between torque and rotor current remains unchanged:

$$n = \frac{U}{L'_m I_A} - \frac{R_A + R_F}{L'_m} \tag{2.73}$$

$$I_{stall} = \frac{U}{R_A + R_F} \tag{2.74}$$

$$T = \frac{L'_m}{2\pi} I_A^2 \tag{2.75}$$

The respective characteristics are illustrated in Fig. 2.34.

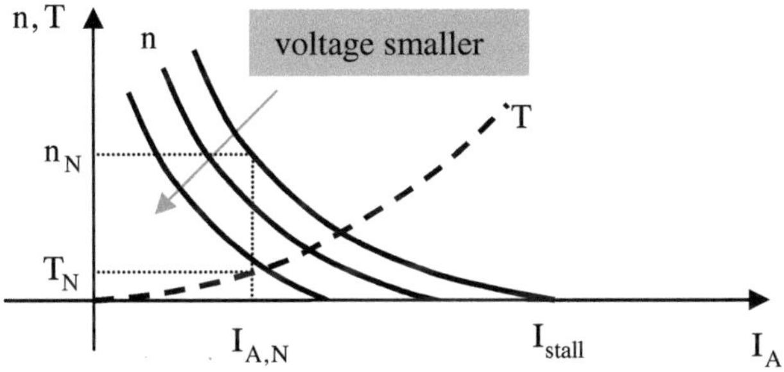

Fig. 2.34. Torque and speed versus current of the series-wound DC-machine for voltage variation.

2. Reduction of the flux:
By switching a resistance parallel to the excitation winding the excitation can be reduced (Fig. 2.35). With

$$\begin{aligned} R_P I_P &= R_F I_F \\ I_A &= I_F + I_P \end{aligned} \tag{2.76}$$

it follows:

$$I_F = \frac{1}{1 + \frac{R_F}{R_P}} I_A = \frac{1}{f} I_A \tag{2.77}$$

where f again is the field weakening factor. The characteristics are:

$$n = f \frac{U_N}{L'_m I_A} - f \frac{R_A + \frac{R_F}{f}}{L'_m} \tag{2.78}$$

$$I_{stall} = \frac{U_N}{R_A + \frac{R_F}{f}} \tag{2.79}$$

$$T = \frac{L'_m}{2\pi} \frac{I_A^2}{f} \tag{2.80}$$

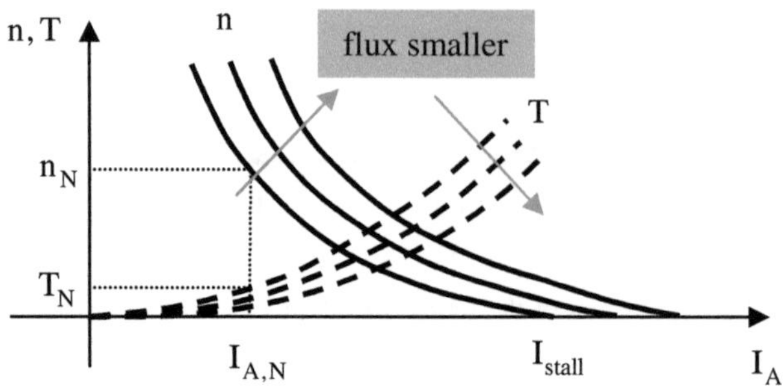

Fig. 2.35. Torque and speed versus current of the series-wound DC-machine for flux variation.

3. Increasing the resistance in the rotor circuit (series resistance):
The speed reduction when loading the machine is increased, the stall current becomes smaller, see Fig. 2.36. The relation between torque and rotor current remains unchanged:

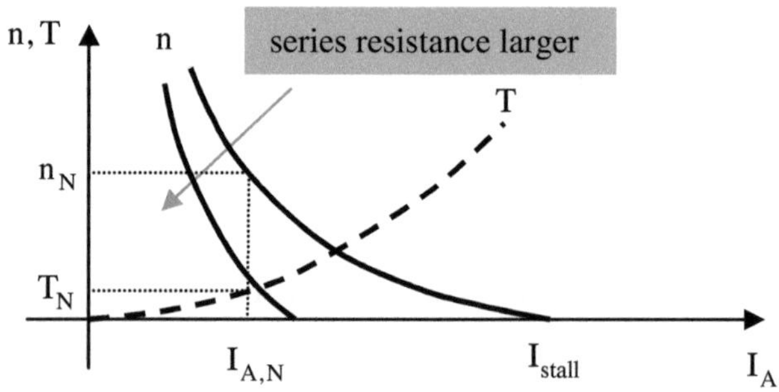

Fig. 2.36. Torque and speed versus current of the series-wound DC-machine for series resistance variation.

These characteristics can be deduced from the following mathematical descriptions:

$$n = \frac{U_N}{L'_m I_A} - \frac{R_A + R_{A,S} + R_F}{L'_m} \tag{2.81}$$

$$I_{stall} = \frac{U_N}{R_A + R_{A,S} + R_F} \tag{2.82}$$

$$T = \frac{L'_m}{2\pi} I_A^2 \tag{2.83}$$

2.10 Compound DC-Machines

The compound (double shunt-wound) DC-machine contains two excitation windings (Fig. 2.37).

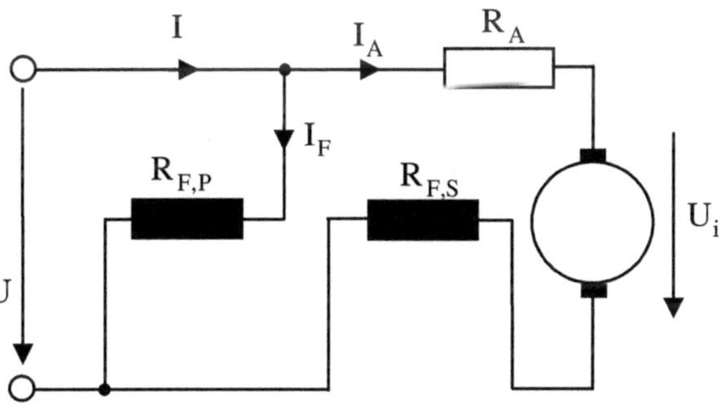

Fig. 2.37. Equivalent circuit diagram of the compound DC-machine.

By dividing the excitation winding into a shunt-wound part and a series-wound part the compound DC-machine has shunt-wound characteristic near the no-load operation and series-wound characteristic when loading the machine.

The torque-speed-characteristics of the three DC-machine alternatives are qualitatively shown in Fig. 2.38.

The most relevant characteristics are:

- series-wound DC-machine:
  - weak characteristic (high starting torque and relatively low nominal torque)
  - danger of "run away " at low load
- shunt-wound DC-machine:
  - stiff characteristic (the speed depends only marginally on the load)
  - fixed no-load speed

- compound DC-machine:
  - weak characteristic (high starting torque and relatively low nominal torque)
  - fixed no-load speed (no danger of "run away" at low load)

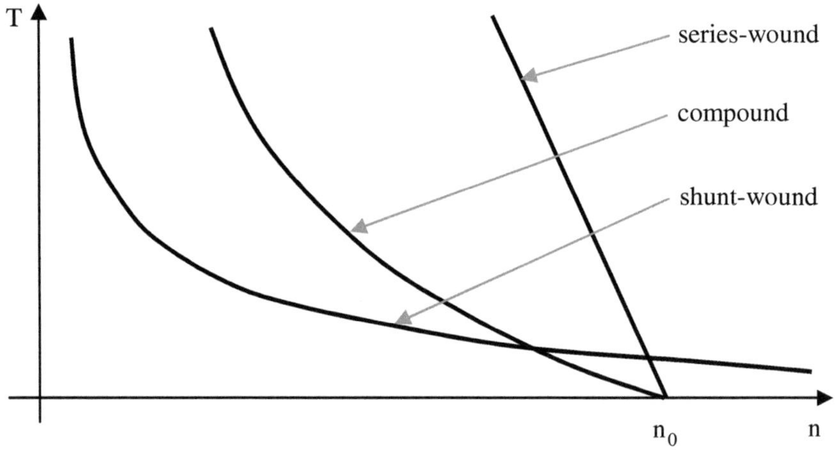

Fig. 2.38. Torque versus speed of the series-wound, shunt-wound and compound DC-machine.

2.11 Generation of a Variable Terminal Voltage

A formerly frequently used method for generating a variable terminal voltage was the usage of the so-called "Leonard machine set" (Fig. 2.39):

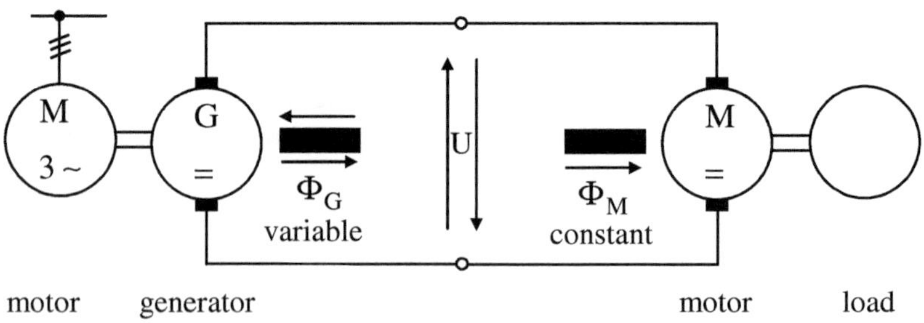

Fig. 2.39. Block diagram of the Leonard machine set.

A rotating field machine (induction or synchronous machine) drives a DC-generator with variable flux; this generator supplies the DC-motor with a variable terminal voltage. With the Leonard-converter operation in all four quadrants is possible.

Today most often the inverter supply is used to generate a variable terminal voltage.

With a controlled three-phase bridge (see left-hand side in Fig. 2.40) an operation in two quadrants is possible (no energy recovery); switching two controlled three-phase bridges in parallel and opposite to each other (see right-hand side in Fig. 2.40) an operation in all four quadrants is possible.

Principle circuit diagram, switching diagram, and the voltages of a controlled three-phase bridge are shown in Fig. 2.41.

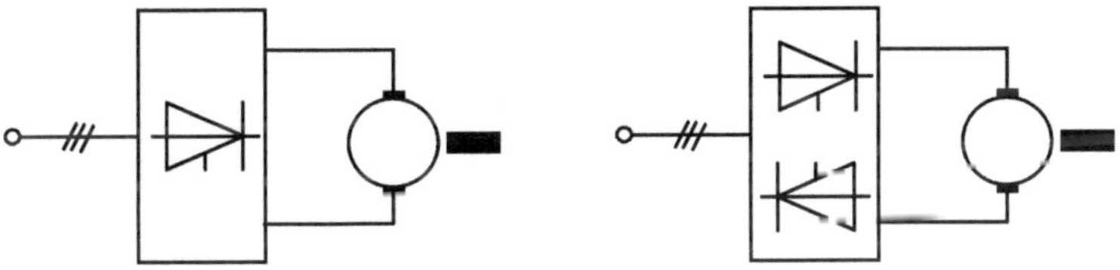

Fig. 2.40. Block diagram of inverter supply: controlled three-phase bridge (*left*) and two opposite three-phase bridges (*right*).

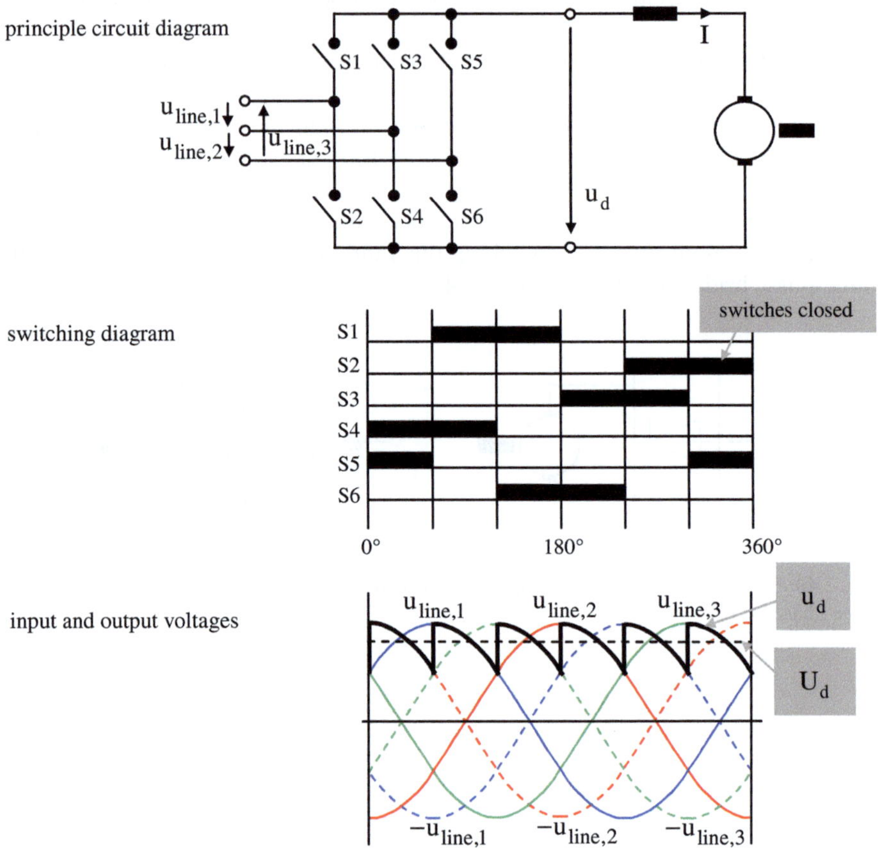

Fig. 2.41. Principle circuit diagram, switching diagram, and voltages of a controlled three-phase bridge.

2.12 Armature Reaction

Up to now it was assumed that the magnetic field in the air-gap of the DC-machine is generated solely by the excitation winding. But this is only the case for no-load operation ($I_A = 0$). Loading the machine (i.e. if $I_A \neq 0$ is true), the rotor current generates a magneto-motive force as well (Fig. 2.42); this is oriented perpendicular to the magneto-motive force of the excitation winding. By superposition of these magneto-motive forces the resulting field is generated ("armature reaction").

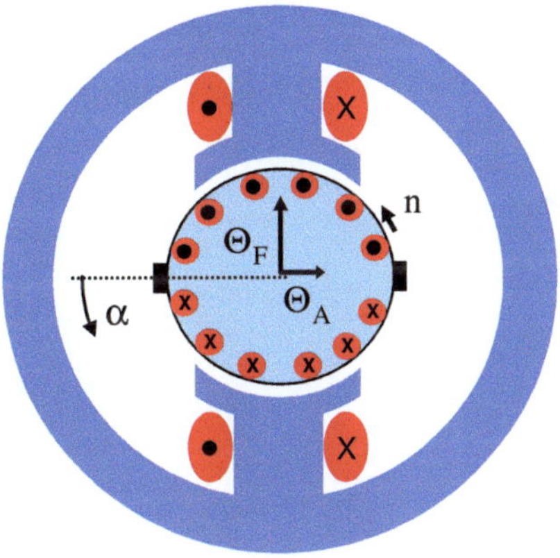

Fig. 2.42. Sketch of the DC-machine illustrating the armature reaction.

Neglecting the saturation and the magnetic voltage drops in the iron, there is by means of Ampere's Law (with α being the angle on the circumference of the rotor).

$$\frac{B(\alpha)}{\mu_0} 2\delta(\alpha) = \Theta_F(\alpha) + \Theta_A(\alpha) \tag{2.84}$$

The resulting air-gap field $B_{\delta,res}$ qualitatively can be deduced from Fig. 2.43.

The magnetic neutral zone is shifted against the geometric neutral zone depending on the load (for motor operation against the direction of rotation, for generator operation in the direction of rotation). If saturation occurs the peaks of the resulting field are rounded. With this even the mean value of the flux density decreases and consequently even the torque.

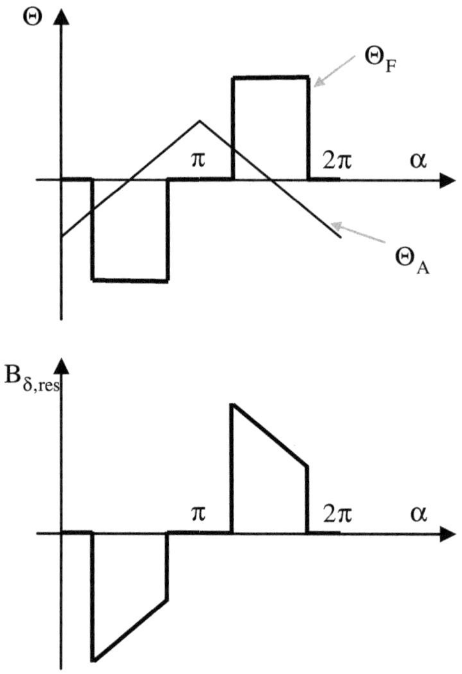

Fig. 2.43. Magneto-motive forces of stator and rotor current (*above*) and resulting air-gap flux density (*below*).

By this field deformation even the voltage between two commutator segments no longer is distributed evenly along the circumference of the rotor. The voltage between two commutator segments can be locally increased by far (but the mean value remains unchanged, if saturation is neglected).

In field weakening operation it may happen that the field under the run-off edge of the rotor (in motor operation) becomes negative, because the main field decreases and the armature reaction remains constant.

To compensate the armature reaction and their negative effects a *compensation winding* may be introduced to the DC-machine. To realize this the main poles are equipped with slots; the conductors placed in these slots have to be supplied with the rotor current in opposite direction. The number of turns of this compensation winding has to be chosen in such a way that the magneto-motive force of the rotor under the poles is compensated. Then the field distribution under the poles is equal to that in no-load operation (as the compensation winding is supplied with the rotor current, this is true for arbitrary operating conditions, i.e. independent from the load) (Fig. 2.44).

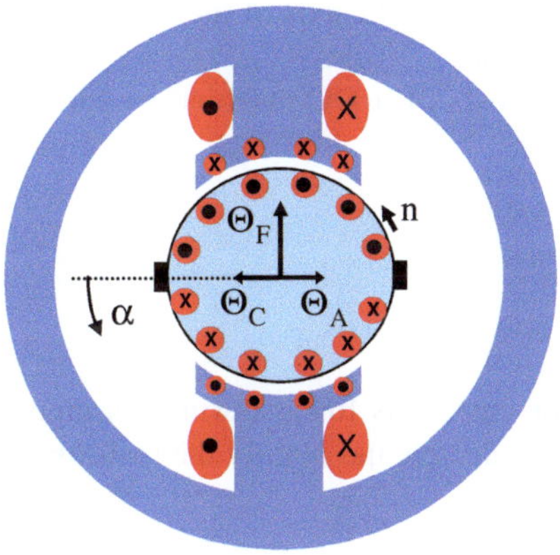

Fig. 2.44. Sketch of the DC-machine with compensation winding in the main poles.

The magneto-motive force distribution and the resulting air-gap field is shown in Fig. 2.45.

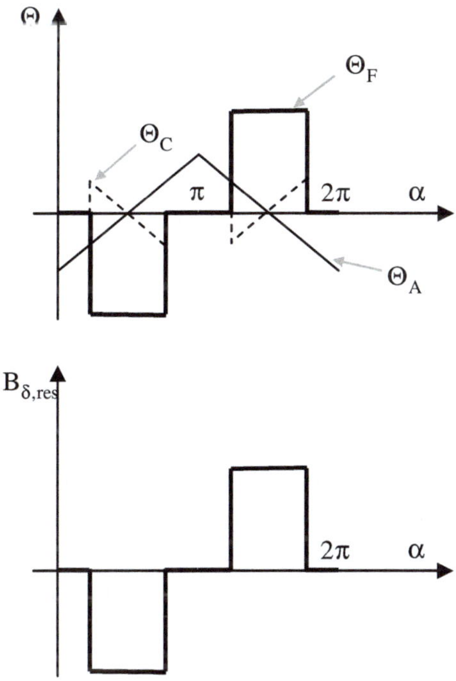

Fig. 2.45. Magneto-motive forces of stator and rotor current and compensation winding (*above*) and resulting air-gap flux density (*below*).

The axis of the resulting field is now identical to the axis of the excitation field; the commutation takes place in the neutral zone.

A compensation winding increases the costs of a DC-machine by far. Therefore, this is realized only for large machines.

2.13 Commutation Pole

In Sect. 2.2 it was assumed that the rotor current is linear during commutation. This is the case if the inductivities can be neglected and mainly the resistances determine the commutation behavior (Fig. 2.46).

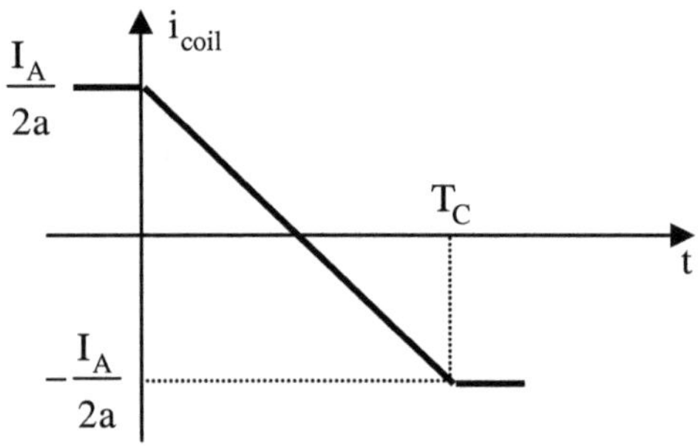

Fig. 2.46. Current characteristic during ideal commutation.

Here T_C is the commutation time,

$$T_C = \frac{b_B}{v_C} \tag{2.85}$$

with b_B being the width of the carbon brush and $v_C = \pi D_C n$ being the circumferential speed of the commutator (D_C is the diameter of the commutator).

In reality there is a non-negligible inductivity because of the slot and end winding leakage. By current change in the coil which is short-circuited by the carbon brushes a voltage of self-induction is generated. With the coil inductivity L_{coil} this voltage becomes:

$$u_s = -L_{coil}\frac{di_{coil}}{dt} \approx L_{coil}\frac{I_A/a}{T_C} = L_{coil}\frac{\pi D_C}{ab_B}I_A n \tag{2.86}$$

Therefore, this voltage during current commutation is proportional to the rotor current and the speed. According to Lenz's Law this voltage has such a direction that it acts against its generating reason (current change). This results in a delayed commutation, producing sparks at the run-off edges of the carbon brushes. This spark generation results in increased wear of the brushes and the commutator (Fig. 2.47).

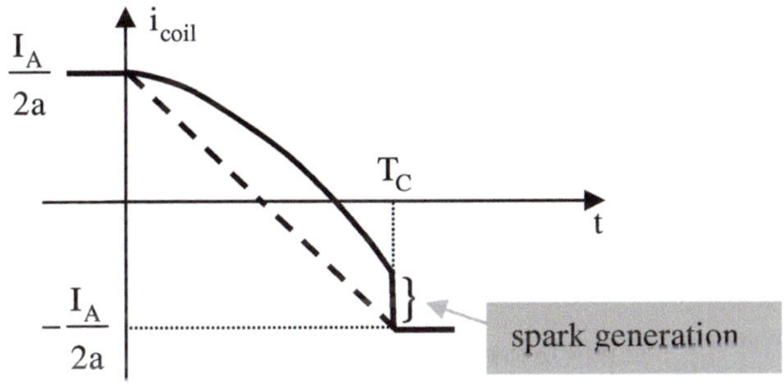

Fig. 2.47. Current characteristic during non-ideal commutation.

Compensating now this voltage coming from current commutation by a rotatory induced voltage, a linear commutation is reached. To realize this so-called commutation poles are inserted in the area between the main poles (commutation zones). The windings of these commutation poles are switched in series with the rotor winding. The principle sketch (Fig. 2.48) shows the cross section of a DC-machine with compensation windings and commutation poles.

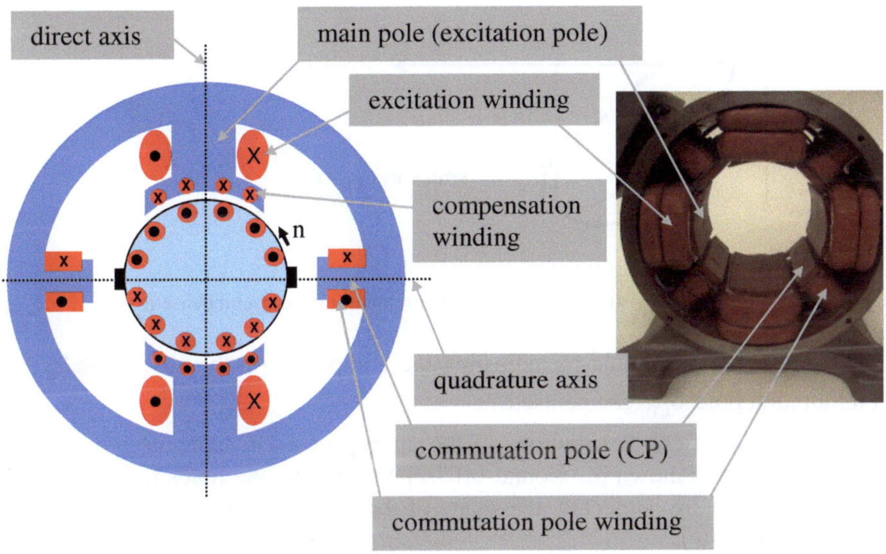

Fig. 2.48. Sketch (p=1) and photograph (p=2) of a DC-machine with compensation winding and commutation poles.

Applying Ampere's Law for one pole pitch (having the integration path through the commutation poles) the following can be deduced:

$$\Theta_{CP} - \Theta_A + \Theta_C = \frac{B_{CP}}{\mu_0} 2\delta_{CP}$$

$$\Rightarrow \quad \Theta_{CP} - \Theta_A (1-\alpha_i) = \frac{B_{CP}}{\mu_0} 2\delta_{CP} \tag{2.87}$$

$$\Rightarrow \quad B_{CP} = \frac{\mu_0}{2\delta_{CP}} \left(\Theta_{CP} - \Theta_A (1-\alpha_i)\right) \quad \sim \quad I_A$$

For the rotary induced voltage it follows:

$$u_{CP} = \oint (\vec{v} \times \vec{B}) d\vec{\ell} \quad \sim \quad nI_A \tag{2.88}$$

Consequently, the voltage coming from current commutation can be compensated. At overcompensation the current is commutating too fast, again resulting in spark generation (Fig. 2.49).

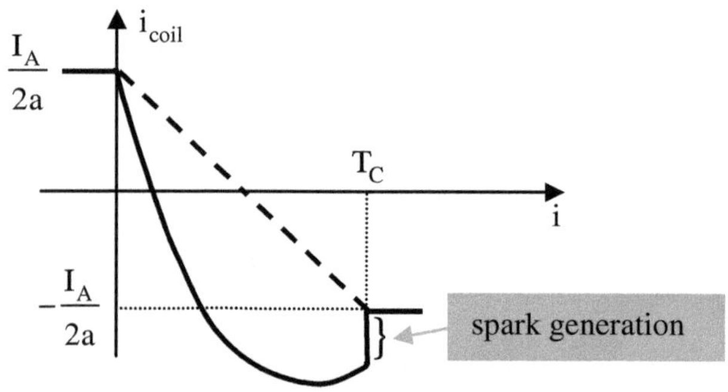

Fig. 2.49. Current characteristic during non-ideal commutation (over-compensation).

As independent variable for adjusting the compensation the air-gap under the commutation pole δ_{CP} can be used.

The principle field distributions in the air-gap of a DC-machine are summarized in Fig. 2.50 (shown in "wound-off" representation, the fields in the gaps between two poles are neglected, the shape of the fields is illustrated to a large extend schematically).

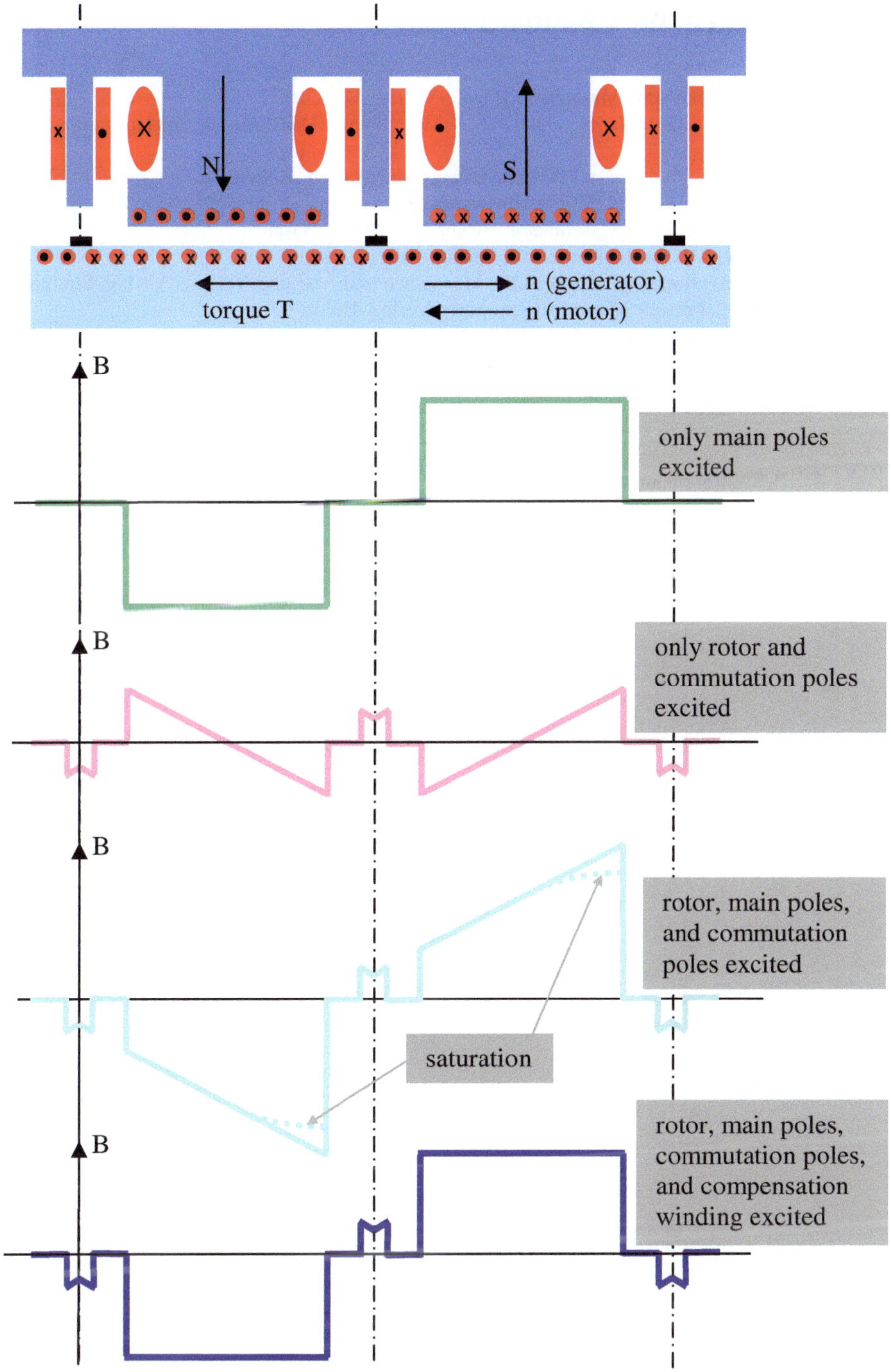

Fig. 2.50. Principle field distribution in the air-gap of a DC-machine.

2.14 References for Chapter 2

Fischer R (2006) Elektrische Maschinen. Hanser-Verlag, München

Kleinrath H (1975) Grundlagen elektrischer Maschinen. Akademische Verlagsgesellschaft, Wiesbaden

Mohr A (1987a) Kleinmotoren mit Permanentmagneterregung 1. Robert Bosch GmbH, Bühlertal

Mohr A (1987b) Kleinmotoren mit Permanentmagneterregung 2. Robert Bosch GmbH, Bühlertal

Richter R (1967) Elektrische Maschinen I. Birkhäuser Verlag, Basel

Schröder D (1994) Elektrische Antriebe 1. Springer-Verlag, Berlin

Seinsch HO (1993) Grundlagen elektrischer Maschinen und Antriebe. Teubner Verlag, Stuttgart

Spring E (1998) Elektrische Maschinen. Springer-Verlag, Berlin

Taegen F (1970) Einführung in die Theorie der elektrischen Maschinen I. Vieweg Verlag, Braunschweig

3 Rotating Field Theory

3.1 Stator of a Rotating Field Machine

Induction machines and synchronous machines are rotating field machines. The stator construction of these machine types are principally the same, they differ in the rotor design. The similarity of these machine types are discussed first, later a detailed consideration of induction machine and synchronous machine will follow.

In the simplest alternative the stator is composed of a stack with electrically isolated laminations to reduce eddy currents. Windings are placed in the slots of the lamination stack. Usually there are $m = 3$ phases, shifted against each other by a spatial angle of

$$\alpha_0 = \frac{2\pi}{pm} = \frac{1}{p}\frac{2\pi}{3} \tag{3.1}$$

with p being the number of pole pairs of the machine. These three phases are supplied by three sinusoidal currents with the same amplitude and frequency, shifted in time by an angle of $\beta_0 = 2\pi/m = 2\pi/3$. Figure 3.1 schematically shows this assembly with $p = 1$; the phases are named u (with return wires in slot x), v (with return wires in slot y) and w (with return wires in slot z):

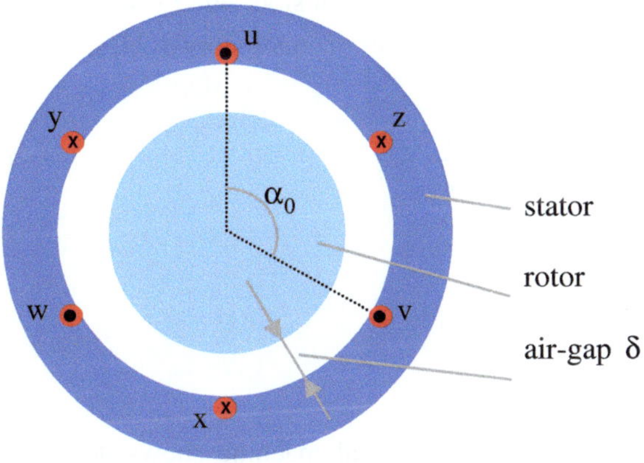

Fig. 3.1. Schematic cross-section of a rotating field machine.

Having a star connection of the phases it schematically looks like it is shown in Fig. 3.2 ("wound-off" representation):

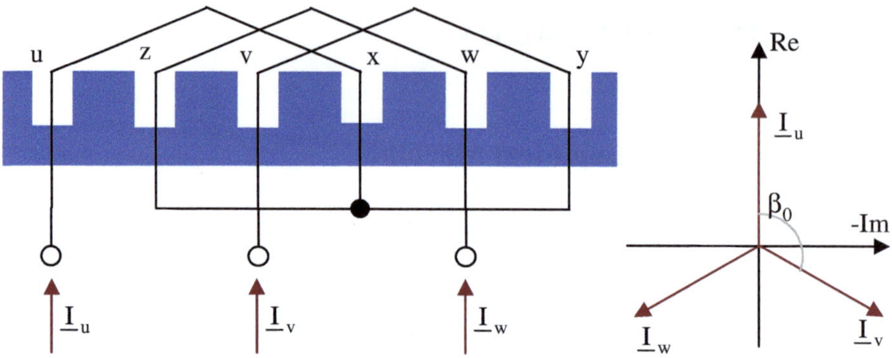

Fig. 3.2. Current system of a rotating field machine.

The mechanical angle α and the electrical angle $\beta = p\alpha$ are linked by the number of pole pairs. The pole pitch is:

$$\tau_p = \frac{\pi D}{2p} \tag{3.2}$$

Consequently, two pole pitches are ($2\tau_p$):

- 2π electrical or
- $2\pi/p$ mechanical.

The coils of a phase may be even placed in several slots side by side. The number of slots per pole per phase is (the total number of stator slots is N_1):

$$q = \frac{N_1}{2pm} \tag{3.3}$$

3.2 Current Loading

The magnetic field in the air-gap of an electrical machine shows decisive importance on the characteristics of the machine, e.g. concerning torque generation. Therefore, the air-gap field has to be calculated precisely.

The reason causing this air-gap field are the current containing conductors in the slots of the stator. Because of the complicated geometry (even the description in the last section is very simplified) this is not possible easily. Therefore, in the following the air-gap field will be separated from the field in the slots by means of the idealized assumption of a homogeneous field in the slot openings, please refer to Fig. 3.3 (in other words: after separating both parts of the field – field in the air-gap and field in the slots – not the entire field considering the complicated lamination contour has to be calculated, but both parts may be calculated separately).

For this a single slot is regarded, whereupon the relative permeability of the surrounding iron is assumed being very large against that of air ($\mu_{Fe} \to \infty$). Then, in the slot there will be a so-called slot leakage field, that can be easily deduced from Ampere's Law ($\oint \vec{H} d\vec{\ell} = \Theta$).

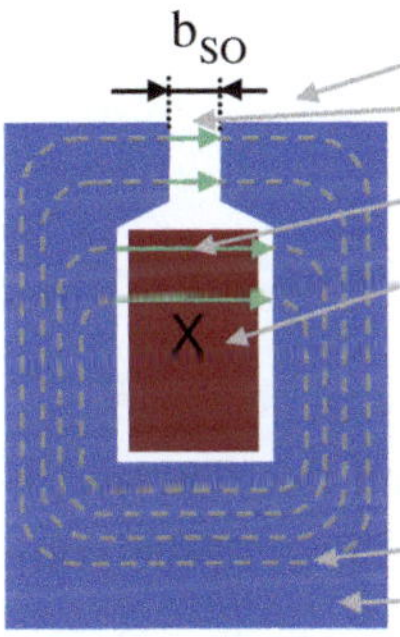

Fig. 3.3. Detailed view of a stator slot.

This slot leakage field joins to the air-gap field at the border between the slot opening and the air-gap. As the air-gap field is to be calculated the slot leakage field is a boundary condition for this calculation. The field in the slot opening (H_{SO}), which can be simplified being tangential concerning the machine geometry, can be computed easily from Ampere's Law ($\mu_{Fe} \to \infty$ is still assumed):

$$H_{SO} = \frac{\Theta_S}{b_{SO}} \tag{3.4}$$

This boundary condition for calculating the air-gap field can be generated even differently. It is assumed that the magneto-motive force of the slot Θ_S is distributed evenly in the area of the slot opening b_{SO} and infinitely thin onto the smooth iron surface (the light blue color in Fig. 3.4 means that the slot notionally is filled with iron). This conception can be described by using the current loading

$$A = \frac{\Theta_S}{b_{SO}} \tag{3.5}$$

This idea is illustrated in Fig. 3.4.

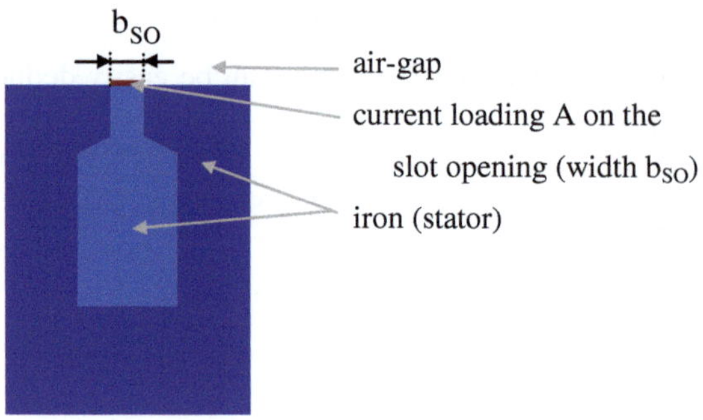

Fig. 3.4. Alternative concept of a stator slot.

Calculating Ampere's Law using an integration loop around this current loading, the tangential field strength H_t on the iron surface in the area of the current loading is obtained (because of $\mu_{Fe} \rightarrow \infty$ the integration path in the iron does not contribute to the integral).

$$\oint \vec{H} d\vec{\ell} = \Theta_S \quad \Rightarrow \quad H_t b_{SO} = A\, b_{SO}$$
$$\Rightarrow \quad H_t = A = \frac{\Theta_S}{b_{SO}} \tag{3.6}$$

This proves that the boundary condition for calculating the air-gap field is unchanged, if the conductors in the stator slots are replaced by an according current loading on the smooth iron surface (in other words: the effect of the magneto-motive force of the slot onto the air-gap field is represented with sufficient accuracy by a current loading distributed in the area of the slot opening). With this it is successfully accomplished to separate the air-gap field from the field in the slots.

For the calculation of the air-gap field the field in the slots is not to be considered. In addition, a quite simple geometry can be regarded: The boundary contours of stator and rotor can be assumed being smooth. This results in a machine with constant air-gap and therefore even a constant magnetic air-gap reluctance.

For calculating the magneto-motive force distribution an additional simplification will be introduced in the next section: The current loading A will be assumed being concentrated in the center of the slot in tangential direction (and not distributed across the area of the slot opening). With this the entire magneto-motive force of a slot Θ_S is concentrated in a single line on the smooth stator iron surface in axial direction[4].

3.3 Alternating and Rotating Magneto-Motive Force

Supplying the winding (with w turns) of a single slot pair with the current i results in magneto-motive forces of these slots of $+wi$ and $-wi$.

Now the space-dependent magneto-motive force $\Theta(\alpha)$ shall be calculated. As the magneto-motive force is defined as an integral value ($\Theta = \oint \vec{H} d\vec{\ell}$), the calculation of $\Theta(\alpha)$ only makes sense, if the integration loop is defined simultaneously. The circulation integral will be defined as follows: The air gap always is crossed radially and the loop always is going through the middle of the slot „u". The field intensity in the iron is neglected ($\mu_{Fe} \rightarrow \infty$), so that the exact path of the integration loop in the iron is not relevant (see Fig. 3.5).

Is the angle α larger than zero, the integration loop includes half of the slot "u" (direction of the loop and direction of the current are assigned positive); then $\Theta(\alpha)$ is equal to ½ wi. This is the case for $0 < \alpha < \pi$. At $\alpha = \pi$ the magneto-motive force steps by the amount of wi into the negative direction. Therefore, for $\pi < \alpha < 2\pi$ there is $\Theta(\alpha)$= -½ wi (Fig. 3.5).

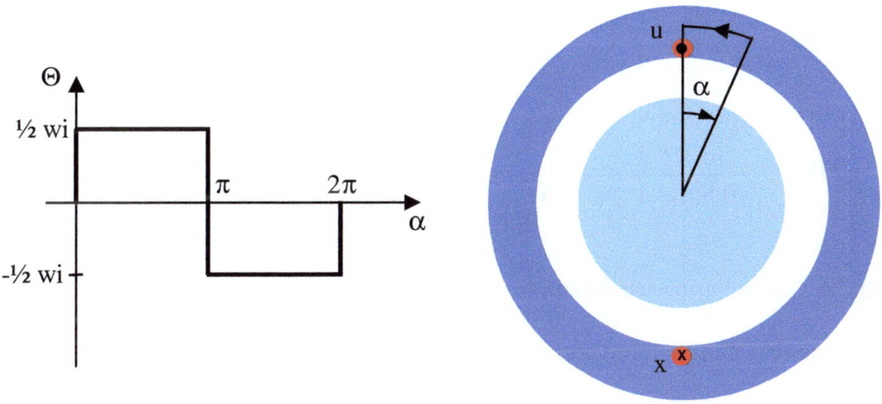

Fig. 3.5. Magneto-motive force distribution of a single slot pair and integral definition.

For the magnetic flux density it holds:

[4] In Sect. 3.5 the effect of different alternatives for the current loading distribution on the magnetic field characteristics will be discussed.

$$B(\alpha) = \mu_0 H(\alpha) = \mu_0 \frac{\Theta(\alpha)}{\delta} \tag{3.7}$$

If the current i is an alternating current with

$$i = \sqrt{2} I \cos(\omega t) \tag{3.8}$$

the function for different points in time looks like it is shown in Fig. 3.6.

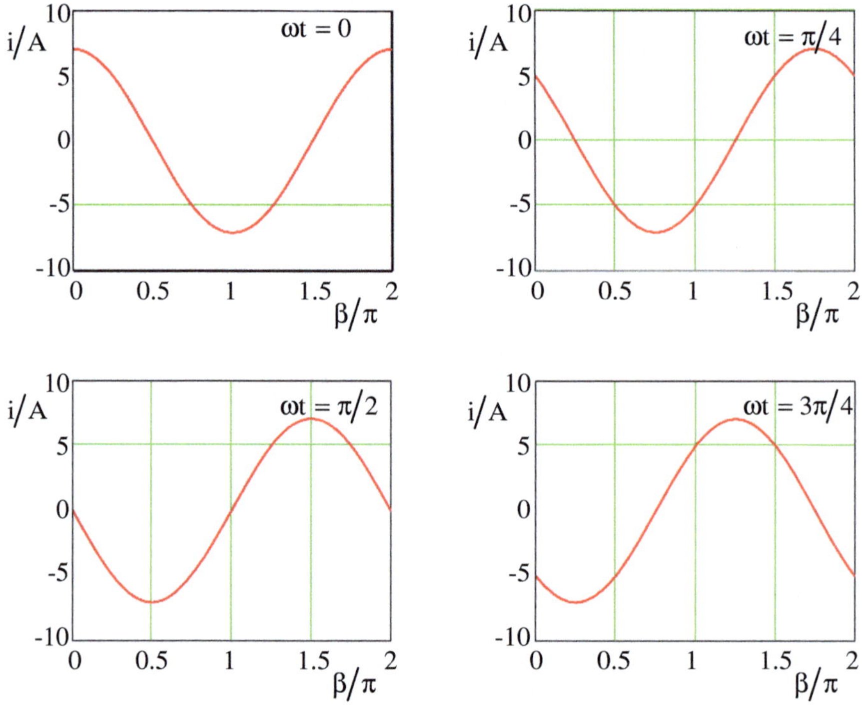

Fig. 3.6. Current versus angle for different points in time.

The magneto-motive force becomes:

$$\Theta(\alpha, t) = \begin{cases} +\dfrac{w}{2}\sqrt{2} I \cos(\omega t) & 0 < \alpha < \pi \\ -\dfrac{w}{2}\sqrt{2} I \cos(\omega t) & \pi < \alpha < 2\pi \end{cases} \tag{3.9}$$

This magneto-motive force distribution is characterized by the facts that the zero crossings are constant in space and time. The amplitude changes in time proportional to the current. This distribution is an alternating magneto-motive force.

If the machine contains more than one pole pair the magneto-motive force $\Theta(\alpha, t)$ is repeated p-times on the circumference. Distributing the number of turns on the p pole pairs the alternating magneto-motive force (now characterized by the index "alt") becomes:

$$\Theta_{alt}(\alpha, t) = \begin{cases} +\dfrac{w}{2p}\sqrt{2}I\cos(\omega t) & 0 < \alpha < \dfrac{\pi}{p} \\ -\dfrac{w}{2p}\sqrt{2}I\cos(\omega t) & \dfrac{\pi}{p} < \alpha < \dfrac{2\pi}{p} \end{cases} \tag{3.10}$$

The function of the alternating magneto-motive force is shown in Fig. 3.7 (for $p = 1$, $I = 5A$, $w = 200$):

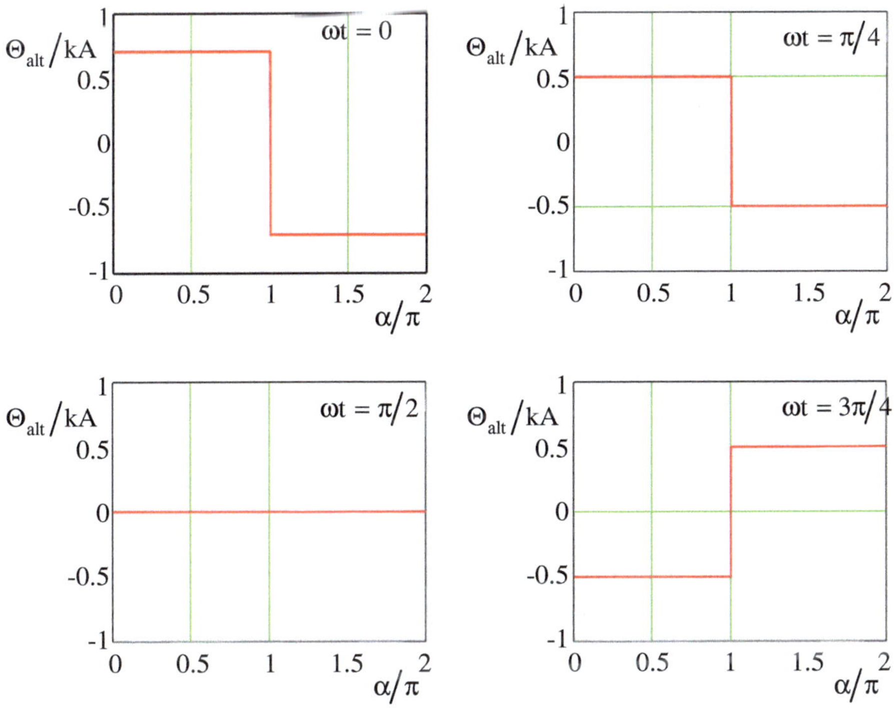

Fig. 3.7. Magneto-motive force versus angle for different points in time.

In the following mathematical description of the spatial distribution of the alternating magneto-motive force a Fourier analysis of the square-wave function will be applied (g' is an integer). With this, the distinction between cases can be replaced by a compact description.

$$\Theta_{alt}(\alpha, t) = \frac{w}{2p}\sqrt{2}I\cos(\omega t)\frac{4}{\pi}\sum_{g'=1}^{\infty}\frac{\sin((2g'-1)\alpha p)}{2g'-1} \tag{3.11}$$

There are an infinite number of waves, each having an odd ordinal number. The fundamental wave is obtained for $g' = 1$:

$$\begin{aligned}\Theta_{alt,1}(\alpha, t) &= \frac{w}{p}\sqrt{2}I\cos(\omega t)\frac{2}{\pi}\sin(\alpha p)\\ &= \Theta_{alt,1}\sin(\alpha p)\cos(\omega t)\end{aligned} \tag{3.12}$$

Convention:
Electrical values (e.g. the current) are described as rms-values, magnetic values (e.g. magneto-motive force, flux density) are described as amplitudes. The differentiation between a function and the amplitude will be done in that way, that for the function the dependencies are explicitly given.

The amplitude of the fundamental wave is:

$$\Theta_{alt,1} = \frac{w}{p}\sqrt{2}I\frac{2}{\pi} \tag{3.13}$$

The amplitudes of the harmonic waves are:

$$\Theta_{alt,2g'-1} = \frac{\Theta_{alt,1}}{2g'-1} \tag{3.14}$$

The amplitudes of the fundamental and all harmonic waves change proportional to the amplitude of the current. Additionally, the amplitudes of the harmonic waves are inversely proportional to their ordinal number. The locations of maxima and zero crossings are constant in time. These are stationary waves. In Fig. 3.8 the fundamental wave $\Theta_{alt,1}(\alpha, t)$ and the first two harmonic waves $\Theta_{alt,3}(\alpha, t)$ and $\Theta_{alt,5}(\alpha, t)$ are shown (for $p = 1$, $I = 5A$, $w = 200$) in red, blue, and magenta, respectively.

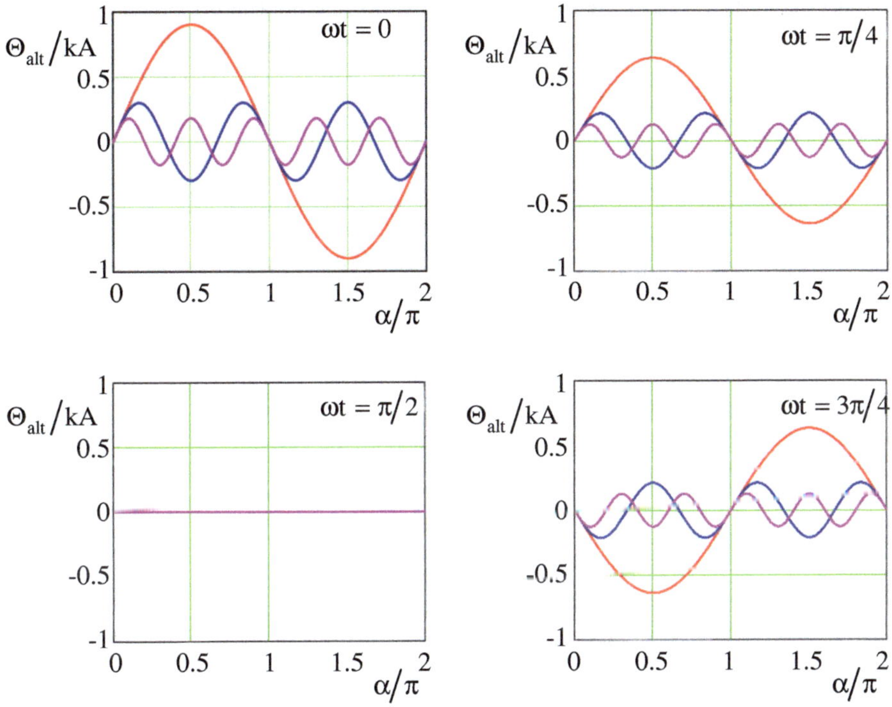

Fig. 3.8. Fourier components of the magneto-motive force versus angle for different points in time (*red*: fundamental; *blue*: first harmonic; *magenta*: second harmonic).

The harmonic waves are generated only from the spatial distribution of the winding, the current is purely sinusoidal (without any harmonic content)!

Convention:

- A *wave* is a space-dependent event.
- An *oscillation* is a time-dependent event.

As it is well-known, a stationary wave can be composed of two waves with half of the amplitude travelling with equal speed in opposite direction. Then the fundamental wave of the alternating magneto-motive force can be described as follows:

$$\begin{aligned}\Theta_{alt,1}(\alpha, t) &= \Theta_{alt,1}\cos(\omega t)\sin(\alpha p) \\ &= \frac{\Theta_{alt,1}}{2}\left[\sin(\alpha p - \omega t) + \sin(\alpha p + \omega t)\right] \\ &= \Theta_{alt,1a}(\alpha, t) + \Theta_{alt,1b}(\alpha, t)\end{aligned} \tag{3.15}$$

This decomposition of the fundamental wave of the alternating magneto-motive force (stationary wave) in two travelling waves is shown in Fig. 3.9. $\Theta_{alt,1}(\alpha,t)$, $\Theta_{alt,1a}(\alpha,t)$, and $\Theta_{alt,1b}(\alpha,t)$ are shown in red, blue, and dashed magenta, respectively.

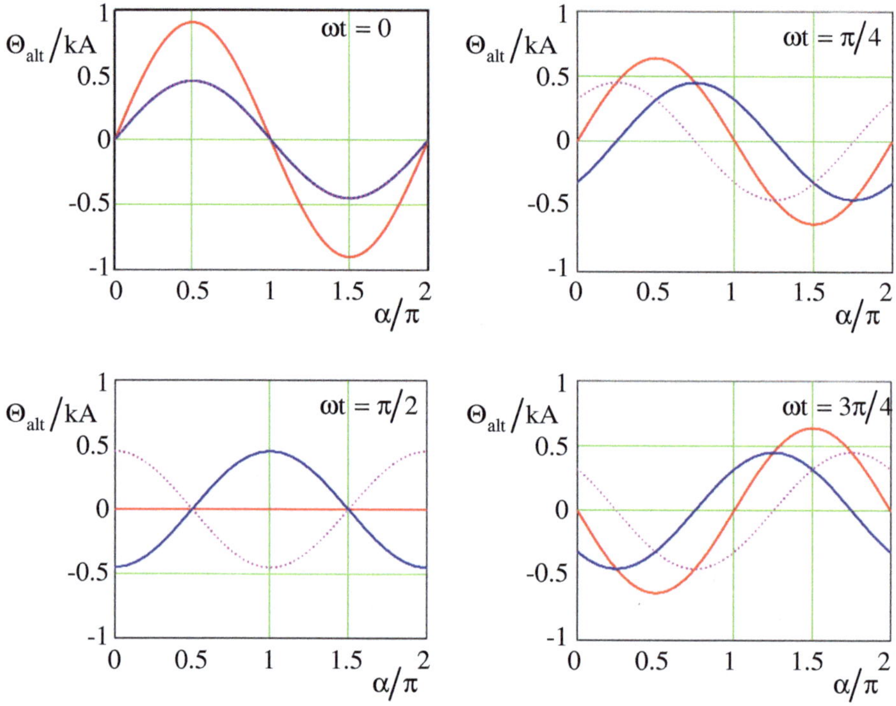

Fig. 3.9. Alternating fundamental wave and decomposition into two travelling waves versus angle for different points in time (*red*: fundamental; *blue*: right travelling wave; *dashed magenta*: left travelling wave).

Consequently, the total alternating magneto-motive force can be described as sum of waves travelling in opposite directions:

$$\begin{aligned}\Theta_{alt}(\alpha,t) &= \Theta_{alt,1}\cos(\omega t)\sum_{g'=1}^{\infty}\frac{\sin((2g'-1)\alpha p)}{2g'-1} \\ &= \frac{\Theta_{alt,1}}{2}\left[\sum_{g'=1}^{\infty}\frac{\sin((2g'-1)\alpha p-\omega t)}{2g'-1}+\right. \\ &\qquad \left.\sum_{g'=1}^{\infty}\frac{\sin((2g'-1)\alpha p+\omega t)}{2g'-1}\right]\end{aligned} \tag{3.16}$$

The travelling magneto-motive force waves are called rotating magneto-motive forces, as these are spreading along the circumference of the machine.

For a fixed point of the rotating magneto-motive force the following is true:

$$(2g'-1)\,\alpha p \mp \omega t = \text{const.} \tag{3.17}$$

Then the mechanical angular frequency of this rotating magneto-motive force wave is:

$$\frac{d\alpha}{dt} = \frac{d}{dt}\left(\frac{\text{const.} \pm \omega t}{(2g'-1)\,p}\right) = \frac{\pm\omega}{(2g'-1)\,p} = \pm\Omega_{2g'-1} \tag{3.18}$$

Consequently, the angular frequency of the fundamental wave is ω/p; the angular frequencies of the harmonic waves are proportional to that of the fundamental wave and inversely proportional to the ordinal number (i.e. with higher ordinal number the angular frequency decreases).

After having analyzed the magneto motive force distribution of the winding of a single slot pair, all three slot pairs of a three-phase machine will be looked at in Fig. 3.10.

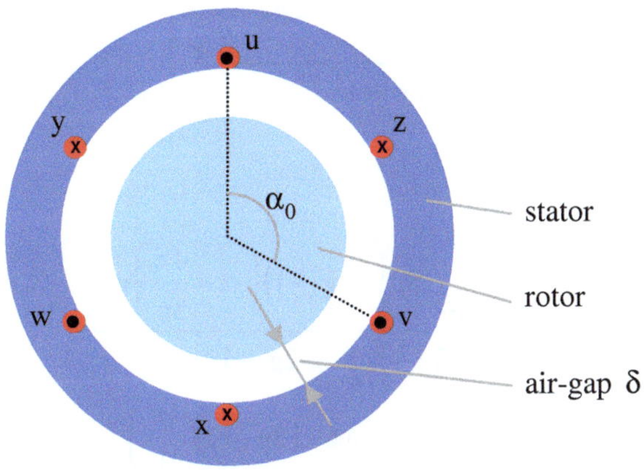

Fig. 3.10. Schematic cross-section of a rotating field machine.

The slot pairs of the different phases are shifted by $\alpha_0 = 2\pi/3p$ in space. The three phases contain w/p turns each and they are supplied by three alternating currents of the same amplitude and frequency shifted by $2\pi/3$ in time.

Each phase generates an alternating magneto-motive force which can be described as follows:

$$\begin{aligned}\Theta_{alt,u}(\alpha,t) &= \Theta_{alt,1}\cos(\omega t)\sum_{g'=1}^{\infty}\frac{\sin((2g'-1)\alpha p)}{2g'-1}\\ &= \frac{\Theta_{alt,1}}{2}\left[\sum_{g'=1}^{\infty}\frac{\sin((2g'-1)\alpha p-\omega t)}{2g'-1}+\right.\\ &\left.\sum_{g'=1}^{\infty}\frac{\sin((2g'-1)\alpha p+\omega t)}{2g'-1}\right]\end{aligned} \tag{3.19}$$

$$\begin{aligned}\Theta_{alt,v}(\alpha,t) &= \Theta_{alt,1}\cos\left(\omega t-\frac{2\pi}{3}\right)\sum_{g'=1}^{\infty}\frac{\sin\left((2g'-1)\left(\alpha-\frac{2\pi}{3p}\right)p\right)}{2g'-1}\\ &= \frac{\Theta_{alt,1}}{2}\left[\sum_{g'=1}^{\infty}\frac{\sin\left((2g'-1)\alpha p-\omega t-(2g'-2)\frac{2\pi}{3}\right)}{2g'-1}+\right.\\ &\left.\sum_{g'=1}^{\infty}\frac{\sin\left((2g'-1)\alpha p+\omega t-2g'\frac{2\pi}{3}\right)}{2g'-1}\right]\end{aligned} \tag{3.20}$$

$$\begin{aligned}\Theta_{alt,w}(\alpha,t) &= \Theta_{alt,1}\cos\left(\omega t-\frac{4\pi}{3}\right)\sum_{g'=1}^{\infty}\frac{\sin\left((2g'-1)\left(\alpha-\frac{4\pi}{3p}\right)p\right)}{2g'-1}\\ &= \frac{\Theta_{alt,1}}{2}\left[\sum_{g'=1}^{\infty}\frac{\sin\left((2g'-1)\alpha p-\omega t-(2g'-2)\frac{4\pi}{3}\right)}{2g'-1}+\right.\\ &\left.\sum_{g'=1}^{\infty}\frac{\sin\left((2g'-1)\alpha p+\omega t-2g'\frac{4\pi}{3}\right)}{2g'-1}\right]\end{aligned} \tag{3.21}$$

The total magneto-motive force of all three phases can be obtained by summing up the three single magneto-motive forces (for this, linearity is necessary, i.e. saturation is neglected):

$$\Theta_\Sigma(\alpha,t) = \Theta_{alt,u}(\alpha,t) + \Theta_{alt,v}(\alpha,t) + \Theta_{alt,w}(\alpha,t) \tag{3.22}$$

Figure 3.11 shows (again for four different points in time) how the total magneto-motive force is composed of three evenly shifted alternating magneto-motive forces of equal amplitude and frequency (the zero crossings are constant in time, but the maximum values are time-dependent). $\Theta_\Sigma(\alpha,t)$, $\Theta_{alt,u}(\alpha,t)$, $\Theta_{alt,v}(\alpha,t)$, and $\Theta_{alt,w}(\alpha,t)$ are shown in red, blue, magenta, and light blue, respectively.

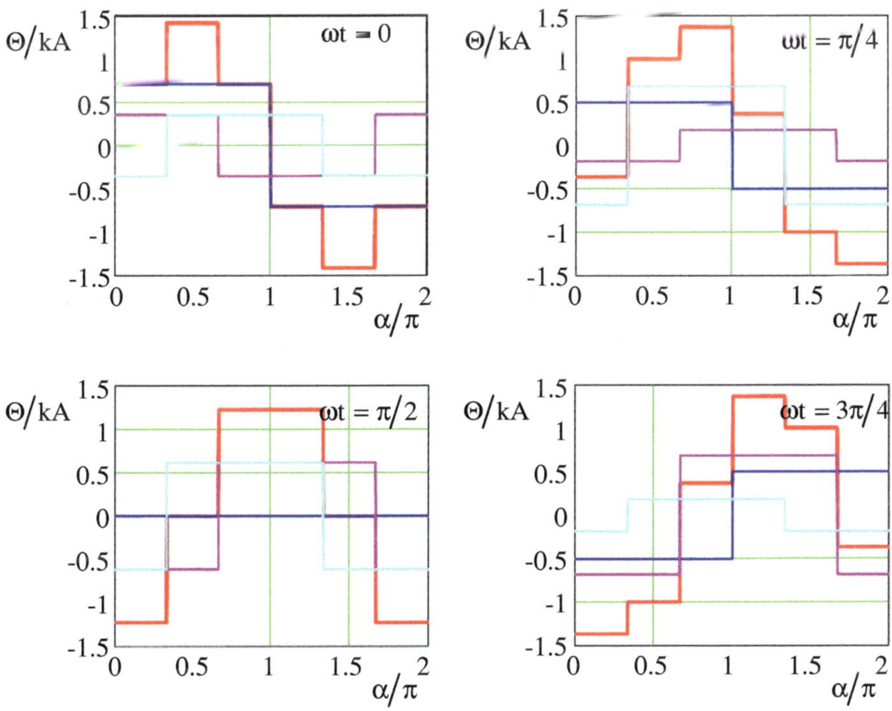

Fig. 3.11. Total magneto-motive force and three evenly shifted alternating magneto-motive forces versus angle for different points in time (*red*: total magneto-motive force; *blue*: alternating magneto-motive force of phase u; *magenta*: alternating magneto-motive force of phase v; *light blue*: alternating magneto-motive force of phase w).

There is:

$$\sin(x)+\sin\left(x-n\frac{2\pi}{3}\right)+\sin\left(x-n\frac{4\pi}{3}\right) = \begin{cases} 0 & \text{if n is not divisible by 3} \\ 3\sin(x) & \text{if n is divisible by 3} \end{cases} \tag{3.23}$$

Consequently, only those travelling waves in opposite direction exist, which fulfill the following condition, where g is a natural number including zero:

$$2g'-2=6g \quad \text{or} \quad 2g'=6g \tag{3.24}$$

$$\Rightarrow \quad 2g'-1=6g+1 \quad \text{or} \quad 2g'-1=6g-1 \tag{3.25}$$

It follows for the total magneto-motive force:

$$\Theta_\Sigma(\alpha,t)=\frac{3}{2}\Theta_{\text{alt},1}\left[\sum_{g=0}^{\infty}\frac{\sin((6g+1)\alpha p-\omega t)}{6g+1}+\sum_{g=1}^{\infty}\frac{\sin((6g-1)\alpha p+\omega t)}{6g-1}\right] \tag{3.26}$$

These two sums can be combined, if negative numbers g are allowed:

$$\Theta_\Sigma(\alpha,t)=\frac{3}{2}\Theta_{\text{alt},1}\sum_{g=-\infty}^{\infty}\frac{\sin((6g+1)\alpha p-\omega t)}{6g+1} \tag{3.27}$$

This total magneto-motive force is a rotating wave (in the following *rotating* waves in the air-gap are labeled with an index "δ"):

$$\Theta_\Sigma(\alpha,t)=\Theta_\delta(\alpha,t) \tag{3.28}$$

with the ordinal numbers

$$6g+1=1,-5,7,-11,13,-17,19,\ldots \tag{3.29}$$

Decomposing the rotating magneto-motive force (red characteristic in Figs. 3.11 and 3.12) into its Fourier components; the fundamental wave (rotating synchronously with the total magneto-motive force, blue) and the harmonic waves with ordinal number -5 (rotating in opposite direction to the fundamental wave, magenta) and 7 (rotating in the same direction as the fundamental wave, light blue) are shown in Fig. 3.12.

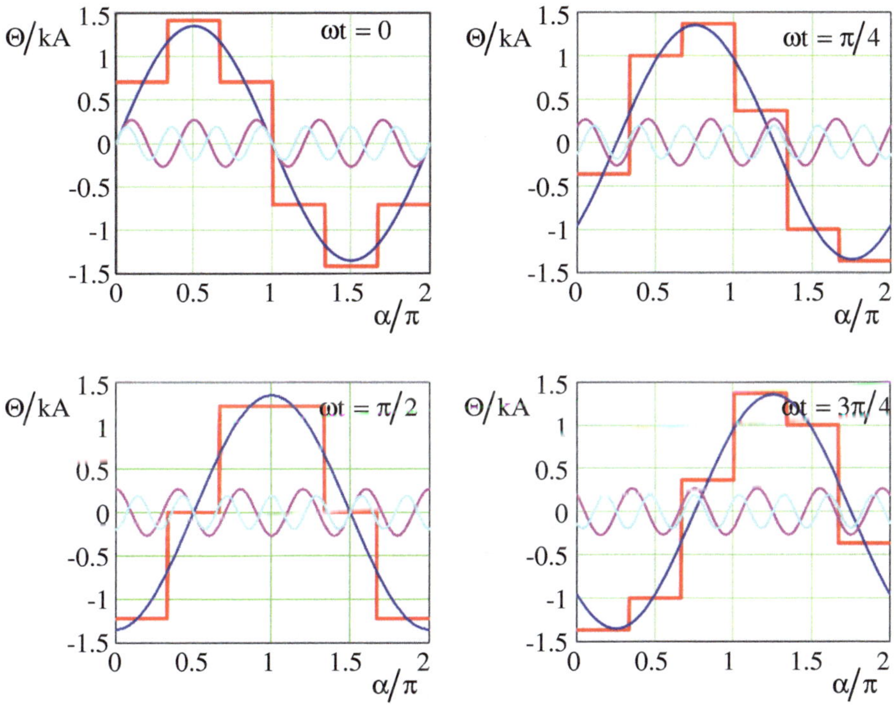

Fig. 3.12. Rotating magneto-motive force and its first Fourier components versus angle for different points in time.

The amplitude of the fundamental wave is:

$$\Theta_{\delta,1} = \frac{3}{2}\Theta_{alt,1} = \frac{3}{2}\frac{2}{\pi}\frac{w}{p}\sqrt{2}I \tag{3.30}$$

The amplitudes of the harmonic waves are:

$$\Theta_{\delta,6g+1} = \frac{\Theta_{\delta,1}}{|6g+1|} \tag{3.31}$$

The rotating speed of the magneto-motive force wave is:

$$\frac{d\alpha}{dt} = \frac{d}{dt}\left(\frac{\text{const.} + \omega t}{(6g+1)p}\right) = \frac{\omega}{(6g+1)p} = \Omega_{6g+1} \tag{3.32}$$

The angular frequency of the fundamental wave is equal to ω/p; the angular frequency of the harmonic waves is proportional to that of the fundamental wave and inversely proportional to the ordinal number (i.e. with higher ordinal number the absolute value of the angular frequency decreases).

3.4 Winding Factor

As demonstrated in the last section the rotating magneto-motive force contains harmonic waves with amplitudes decreasing with the ordinal number. Generally, in rotating field machines only the fundamental wave produces a useful torque, all harmonic waves generate parasitic torques, that disturb the machine operation (e.g. no starting or sticking at low speed).

Therefore, measures are necessary to damp the harmonic waves, but only marginally affect the fundamental wave. Mainly, the following two measures are used:

- Distributing the winding per pole and per phase into several slots (distributed winding).
- Distributing the forward and return conductors of a coil in such a way, that they are no longer shifted by an electrical angle of π (i.e. mechanical angle of π/p), but by an electrical angle less than π (short-pitch winding).

The effects of both measures will be calculated in the following by a common mathematical deduction. Firstly, the magneto-motive force of a single slot pair (known from the proceeding section) is required:

$$\Theta(\alpha, t) = \frac{3}{2}\frac{w}{p}\sqrt{2}I\frac{2}{\pi}\sum_{g=-\infty}^{\infty}\frac{\sin((6g+1)\alpha p - \omega t)}{6g+1} \tag{3.33}$$

The w/p turns per pole and per phase will now be distributed into q equally spaced slots and two layers in radial direction, see Fig. 3.13.

- All $N_1 = 2pqm$ stator slots are equally spaced at the circumference of the machine. The mechanical angle between two slots is then:

$\alpha_{N_1} = \frac{2\pi}{N_1} = \frac{2\pi}{2pqm} = \frac{\pi}{pqm}$, the electrical angle between two slots is: $\beta_{N_1} = p\alpha_{N_1} = \frac{\pi}{qm}$.

- The two layers have a phase shift of $\alpha_S = \left(1 - \frac{s}{\tau_p}\right)\frac{\pi}{p}$ mechanically or $\beta_S = \left(1 - \frac{s}{\tau_p}\right)\pi$ electrically. This phase shift (it is called „short-pitch") can only be realized in integer multiples of a slot pitch (Fig. 3.13).

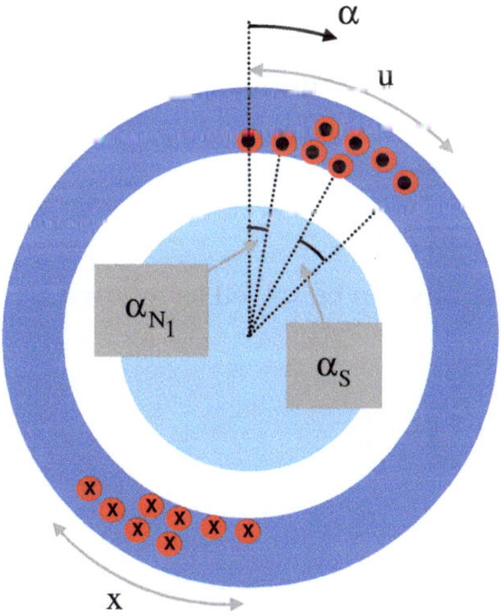

Fig. 3.13. Explanation of distributed and short-pitched winding.

The total magneto-motive force can now be described as follows:

$$\Theta(\alpha, t) = \frac{3}{2}\frac{w}{2pq}\sqrt{2}I\frac{2}{\pi}\cdot \sum_{g=-\infty}^{\infty} \underbrace{\sum_{k=1}^{q}\sum_{\ell=1}^{2}\frac{\sin\left[(6g+1)\left(\alpha-(k-1)\alpha_{N_1}-(\ell-1)\alpha_S\right)p-\omega t\right]}{6g+1}}_{X_1} \tag{3.34}$$

Now, the double sum X_1 will be investigated closely; it can be written as follows:

$$X_1 = \mathrm{Im}\left\{ \frac{1}{6g+1} \sum_{k=1}^{q} \sum_{\ell=1}^{2} e^{j(6g+1)\alpha p} \, e^{-j(6g+1)(k-1)\alpha_{N_1} p} \, e^{-j(6g+1)(\ell-1)\alpha_S p} \, e^{-j\omega t} \right\}$$
$$= \mathrm{Im}\left\{ \frac{e^{j[(6g+1)\alpha p-\omega t]}}{6g+1} \underbrace{\left(\sum_{k=1}^{q} e^{-j(6g+1)(k-1)\alpha_{N_1} p} \right)}_{X_2} \underbrace{\left(\sum_{\ell=1}^{2} e^{-j(6g+1)(\ell-1)\alpha_S p} \right)}_{X_3} \right\} \tag{3.35}$$

The two sums X_2 and X_3 will be calculated separately in the following. For X_2 it can be obtained:

$$X_2 = \sum_{k=1}^{q} e^{-j(6g+1)(k-1)\alpha_{N_1} p}, \qquad \alpha_{N_1} p = \frac{\pi}{qm}$$
$$= e^{j(6g+1)\frac{\pi}{qm}} \sum_{k=1}^{q} e^{-j(6g+1)k\frac{\pi}{qm}} \tag{3.36}$$

This is a finite geometric series which generally can be solved as follows:

$$\sum_{\mu=1}^{n} e^{j\mu\gamma} = \frac{\sin\left(n\frac{\gamma}{2}\right)}{\sin\left(\frac{\gamma}{2}\right)} e^{j(n+1)\frac{\gamma}{2}} \tag{3.37}$$

Consequently:

$$
\begin{aligned}
X_2 &= e^{j(6g+1)\frac{\pi}{qm}} \sum_{k=1}^{q} e^{-j(6g+1)k\frac{\pi}{qm}} \\
&= e^{j(6g+1)\frac{\pi}{qm}} \frac{\sin\left[q(6g+1)\dfrac{\pi}{2qm}\right]}{\sin\left[(6g+1)\dfrac{\pi}{2qm}\right]} e^{-j(q+1)(6g+1)\frac{\pi}{2qm}} \\
&= \frac{\sin\left[q(6g+1)\dfrac{\pi}{2qm}\right]}{\sin\left[(6g+1)\dfrac{\pi}{2qm}\right]} e^{-j(q-1)(6g+1)\frac{\pi}{2qm}} \\
&= q\,\xi_{Z,6g+1}\, e^{-j(q-1)(6g+1)\frac{\pi}{2qm}}, \qquad \xi_{Z,6g+1} = \frac{\sin\left[q(6g+1)\dfrac{\pi}{2qm}\right]}{q\sin\left[(6g+1)\dfrac{\pi}{2qm}\right]}
\end{aligned}
\tag{3.38}
$$

For X_3 it follows (even in this case there is a finite geometric series, in addition rules for trigonometric functions are used):

$$
\begin{aligned}
X_3 &= \sum_{\ell=1}^{2} e^{-j(6g+1)(\ell-1)\alpha_S p} \\
&= e^{j(6g+1)\alpha_S p} \sum_{\ell=1}^{2} e^{-j(6g+1)\ell\alpha_S p} \\
&= e^{j(6g+1)\alpha_S p} \frac{\sin\left[2(6g+1)\dfrac{\alpha_S p}{2}\right]}{\sin\left[(6g+1)\dfrac{\alpha_S p}{2}\right]} e^{-j(2+1)(6g+1)\frac{\alpha_S p}{2}}
\end{aligned}
\tag{3.39}
$$

With $\sin(2x) = 2\sin(x)\cos(x)$ there is further

$$
\begin{aligned}
X_3 &= 2\cos\left[(6g+1)\frac{\alpha_S p}{2}\right] e^{-j(6g+1)\frac{\alpha_S p}{2}}, \qquad \alpha_S p = \left(1-\frac{s}{\tau_p}\right)\pi \\
&= 2\cos\left[(6g+1)\left(1-\frac{s}{\tau_p}\right)\frac{\pi}{2}\right] e^{-j(6g+1)\frac{\alpha_S p}{2}}
\end{aligned}
\tag{3.40}
$$

and with $\cos(x-y)=\cos(x)\cos(y)+\sin(x)\sin(y)$ it follows

$$\begin{aligned} X_3 &= 2\Bigg\{\underbrace{\cos\left[(6g+1)\frac{\pi}{2}\right]}_{=0}\cos\left[(6g+1)\frac{s}{\tau_p}\frac{\pi}{2}\right]+ \\ &\quad \sin\left[(6g+1)\frac{\pi}{2}\right]\sin\left[(6g+1)\frac{s}{\tau_p}\frac{\pi}{2}\right]\Bigg\}\,e^{-j(6g+1)\frac{\alpha_S p}{2}} \\ &= 2\sin\left[(6g+1)\frac{\pi}{2}\right]\xi_{S,6g+1}\,e^{-j(6g+1)\frac{\alpha_S p}{2}}, \qquad \xi_{S,6g+1}=\sin\left[(6g+1)\frac{s}{\tau_p}\frac{\pi}{2}\right] \end{aligned} \tag{3.41}$$

In total it follows:

$$\begin{aligned} \Theta(\alpha,t) &= \frac{3}{2}\frac{w}{2pq}\sqrt{2}I\frac{2}{\pi}\sum_{g=-\infty}^{\infty}\mathrm{Im}\Bigg\{\frac{2\,q\,\xi_{Z,6g+1}\,\xi_{S,6g+1}}{6g+1}\sin\left[(6g+1)\frac{\pi}{2}\right]\cdot \\ &\quad e^{j\left[(6g+1)\left[\alpha p-(q-1)\frac{\pi}{2qm}-\frac{\alpha_S p}{2}\right]-\omega t\right]}\Bigg\} \\ &= \frac{3}{2}\frac{w}{p}\sqrt{2}I\frac{2}{\pi}\sum_{g=-\infty}^{\infty}\frac{\xi_{Z,6g+1}\xi_{S,6g+1}}{6g+1}\mathrm{Im}\Bigg\{\frac{\sin\left[(6g+1)\frac{\pi}{2}\right]}{e^{j(6g+1)\frac{\pi}{2}}}\cdot \\ &\quad e^{j\left[(6g+1)\left[\alpha-(q-1)\frac{\pi}{2pqm}-\frac{\alpha_S}{2}+\frac{\pi}{2p}\right]p-\omega t\right]}\Bigg\} \end{aligned} \tag{3.42}$$

With a coordinate transformation of

$$\alpha' = \alpha-(q-1)\frac{\pi}{2pqm}-\frac{\alpha_S}{2}+\frac{\pi}{2p} \tag{3.43}$$

it follows further:

$$\Theta(\alpha',t)=\frac{3}{2}\frac{w}{p}\sqrt{2}I\frac{2}{\pi}\sum_{g=-\infty}^{\infty}\frac{\xi_{Z,6g+1}\xi_{S,6g+1}}{6g+1}\cdot \operatorname{Im}\left\{\frac{\sin\left[(6g+1)\frac{\pi}{2}\right]}{e^{j(6g+1)\frac{\pi}{2}}}e^{j\left[(6g+1)\alpha' p-\omega t\right]}\right\} \tag{3.44}$$

This coordinate transformation is explained in Fig. 3.14.

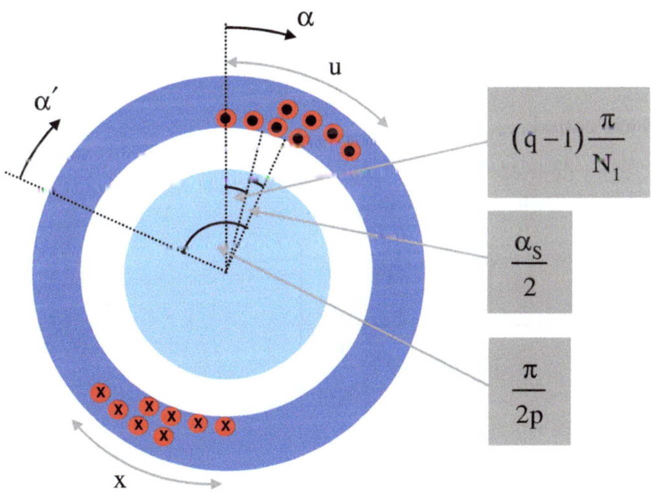

Fig. 3.14. Explanation of the coordinate transformation.

With

$$\frac{\sin\left[(6g+1)\frac{\pi}{2}\right]}{e^{j(6g+1)\frac{\pi}{2}}}=\frac{\sin\left[(6g+1)\frac{\pi}{2}\right]}{\underbrace{\cos\left[(6g+1)\frac{\pi}{2}\right]}_{=0}+j\sin\left[(6g+1)\frac{\pi}{2}\right]}=\frac{1}{j}=-j=e^{-j\frac{\pi}{2}} \tag{3.45}$$

it follows

$$\Theta(\alpha',t)=\frac{3}{2}\frac{w}{p}\sqrt{2}I\frac{2}{\pi}\sum_{g=-\infty}^{\infty}\frac{\xi_{Z,6g+1}\xi_{S,6g+1}}{6g+1}\operatorname{Im}\left\{e^{j\left[(6g+1)\alpha' p-\omega t-\frac{\pi}{2}\right]}\right\} \tag{3.46}$$

A time transformation according to

$$\omega t + \frac{\pi}{2} = \omega t' \quad \Rightarrow \quad t' = t + \frac{\pi}{2\omega} = t + \frac{1}{4f}, \quad \omega = 2\pi f \tag{3.47}$$

delivers:

$$\begin{aligned}\Theta(\alpha', t') &= \frac{3}{2}\frac{w}{p}\sqrt{2}I\frac{2}{\pi}\sum_{g=-\infty}^{\infty}\frac{\xi_{Z,6g+1}\xi_{S,6g+1}}{6g+1}\operatorname{Im}\left\{e^{j[(6g+1)\alpha' p - \omega t']}\right\} \\ &= \frac{3}{2}\frac{w}{p}\sqrt{2}I\frac{2}{\pi}\sum_{g=-\infty}^{\infty}\frac{\xi_{Z,6g+1}\xi_{S,6g+1}}{6g+1}\sin\left[(6g+1)\alpha' p - \omega t'\right]\end{aligned} \tag{3.48}$$

The resulting factors are called

- distribution factor (zoning factor)

$$\xi_{Z,6g+1} = \frac{\sin\left[q(6g+1)\frac{\pi}{2qm}\right]}{q\sin\left[(6g+1)\frac{\pi}{2qm}\right]} \tag{3.49}$$

- short-pitch factor

$$\xi_{S,6g+1} = \sin\left[(6g+1)\frac{s}{\tau_p}\frac{\pi}{2}\right] \tag{3.50}$$

For three-phase windings ($m = 3$) it follows:

$$\xi_{Z,6g+1} = \frac{\sin\left((6g+1)\frac{\pi}{6}\right)}{q\sin\left((6g+1)\frac{\pi}{6q}\right)} \tag{3.51}$$

For a very large number of slots per pole and per phase ($q \to \infty$) there is:

$$\lim_{q\to\infty} \xi_{Z,6g+1} = \lim_{\frac{\pi}{6q}\to 0} \frac{\sin\left((6g+1)\frac{\pi}{6}\right)}{q\sin\left((6g+1)\frac{\pi}{6q}\right)} = \frac{\sin\left((6g+1)\frac{\pi}{6}\right)}{q(6g+1)\frac{\pi}{6q}}$$
$$= \frac{\sin\left((6g+1)\frac{\pi}{6}\right)}{(6g+1)\frac{\pi}{6}} = \mathrm{si}\left((6g+1)\frac{\pi}{6}\right) \tag{3.52}$$

The effect of the distributed winding is that the fundamental waves of the magneto-motive forces of the single slots are added as much as possible undisturbed, but the harmonic waves are nearly compensated.

Table 3.1 reveals this.

Table 3.1. Distribution factor as a function of harmonic number and number of slots per pole and per phase.

$\xi_{Z,6g+1}$		$q=$				
$g=$	$6g+1=$	1	2	3	4	∞
0	1	1,000	0,966	0,960	0,958	0,955
1	7	1,000	-0,259	-0,177	-0,158	-0,136
-1	-5	1,000	0,259	0,218	0,205	0,191
2	13	1,000	-0,966	0,218	0,126	0,073
-2	-11	1,000	-0,966	-0,177	-0,126	-0,087

The effect of the short-pitch winding is that by clever choice of s/τ_p *single harmonic waves of both shifted winding layers are completely (or at least partly) compensated.*

There is e.g.:

$$s/\tau_p = 4/5 \quad \Rightarrow \quad \begin{cases} \xi_{S,1} = 0,951 \\ \xi_{S,7} = 0,588 \\ \xi_{S,-5} = 0,000 \end{cases}$$

$$s/\tau_p = 6/7 \quad \Rightarrow \quad \begin{cases} \xi_{S,1} = 0,975 \\ \xi_{S,7} = 0,000 \\ \xi_{S,-5} = -0,434 \end{cases}$$

For equal damping of both first harmonic waves with a short-pitch winding, the following compromise often is chosen:

$$s/\tau_p = 5/6 \quad \Rightarrow \quad \begin{cases} \xi_{S,1} = 0,966 \\ \xi_{S,7} = 0,259 \\ \xi_{S,-5} = -0,259 \end{cases}$$

The winding factor ξ of a rotating field winding is the product of distribution factor and short-pitch factor:

$$\xi_{6g+1} = \xi_{Z,6g+1}\,\xi_{S,6g+1} \tag{3.53}$$

Without further deduction it shall be mentioned that even the slot opening has an influence onto the winding factor (the influence of the slot opening onto current loading and flux density is explained in the following Sect. 3.5): the total winding factor is then

$$\xi_{6g+1} = \xi_{Z,6g+1}\,\xi_{S,6g+1}\,\xi_{SO,6g+1} \tag{3.54}$$

with the slot opening factor being

$$\xi_{SO,6g+1} = \mathrm{si}\left[(6g+1)\frac{b_{SO}}{\tau_p}\frac{\pi}{2}\right] \tag{3.55}$$

and b_{SO} being the slot opening.

For simplification of writing, the following renaming of the coordinates is used:

$$\alpha' \to \alpha \text{ and}$$

$$t' \to t$$

With this winding factor and renaming of the coordinates, the rotating magneto-motive force of a rotating field winding becomes:

$$\Theta_\delta(\alpha,t) = \frac{3}{2}\frac{w}{p}\sqrt{2}I\frac{2}{\pi}\sum_{g=-\infty}^{\infty}\frac{\xi_{6g+1}}{6g+1}\sin\left((6g+1)\alpha p - \omega t\right) \tag{3.56}$$

For a better understanding of the winding factors, Figs. 3.15 and 3.16 illustrate the periodicity of distribution factor and short-pitch factor.

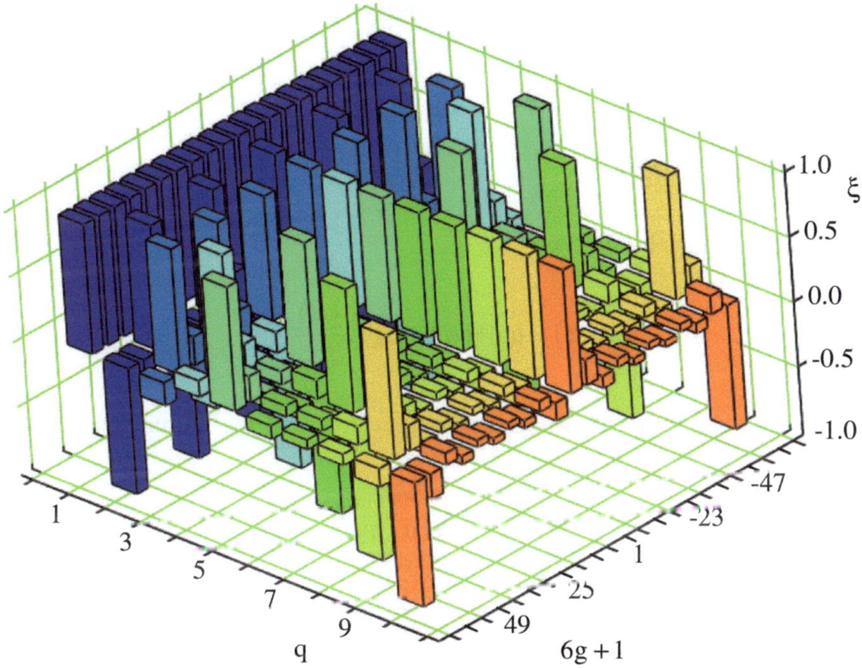

Fig. 3.15. Periodicity of the distribution factor.

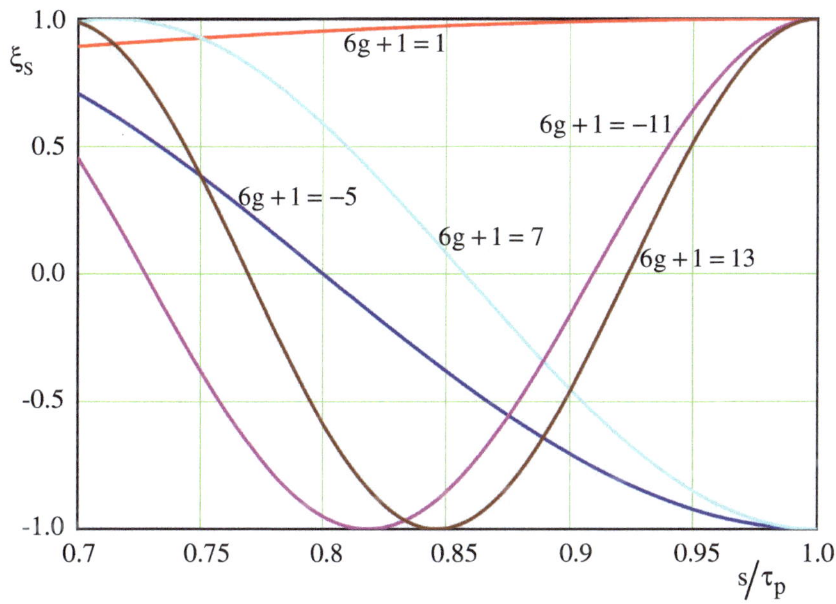

Fig. 3.16. Periodicity of the short-pitch factor.

3.5 Current Loading and Flux Density

3.5.1. Fundamentals

From Ampere's Law it follows:

$$B(\alpha) = \mu_0 H(\alpha) = \mu_0 \frac{\Theta(\alpha)}{\delta} \tag{3.57}$$

The relation between the magneto-motive force and the current loading becomes:

$$\Theta(\alpha) = \int A(\alpha)\, r\, d\alpha \tag{3.58}$$

with r being the bore radius of the machine. In total it can be obtained:

$$B(\alpha) \sim \int A(\alpha)\, d\alpha \tag{3.59}$$

Using this deduction, implicitly it is assumed that there is only a radial component of the air-gap field (see Eq. 3.57). This simplification is allowed as far as the wave length of the regarded field wave is large against the air-gap width (i.e. the equations deduced in the following are not valid for waves with very large ordinal numbers; usually these waves are of minor interest). Furthermore it is assumed that the radial component of the magnetic flux density virtually is unchanged in radial direction inside the air-gap.

By using the concept of electric loading the winding placed in the slots is simplified theoretically. In the following different alternatives of this simplification are examined.

3.5.2. Uniformly Distributed Current Loading in a Zone

The most simple alternative is that all conductors are uniformly distributed infinitely thin on the stator surface (i.e. $q \rightarrow \infty$). Figure 3.17 illustrates such a current loading for a single phase:

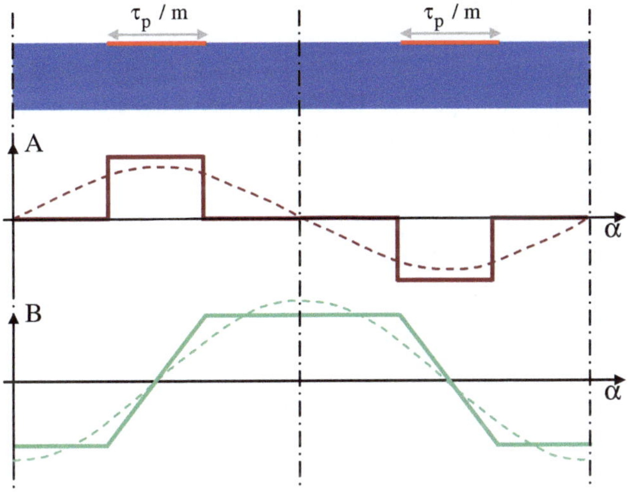

Fig. 3.17. Current loading and flux density of alternative 1.

3.5.3. Current Loading Concentrated in the Middle of Each Slot

Another possibility is to locate the conductors of a slot (theoretically infinitely thin) in the middle of the corresponding slot opening. Analogously to Fig. 3.17 this is shown in Fig. 3.18 for $q = 3$ (for the flux density distribution there is a step curve):

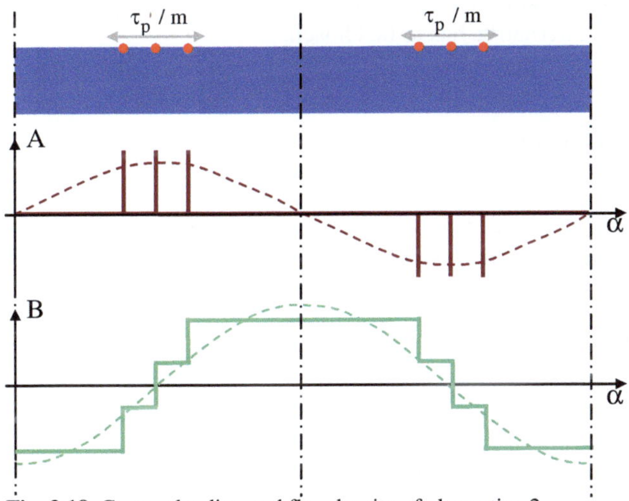

Fig. 3.18. Current loading and flux density of alternative 2.

3.5.4. Current Loading Distributed Across Each Slot Opening

Finally it is possible to distribute the conductors of a slot (theoretically infinitely thin) in the area of the respective slot opening. Then the slope of the steps in the flux density characteristic will be less steep (Fig. 3.19).

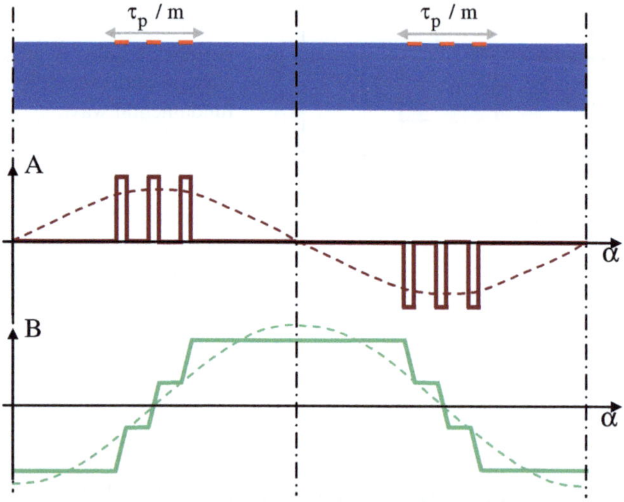

Fig. 3.19. Current loading and flux density of alternative 3.

Which alternative has to be chosen for a specific task inter alia depends on the task itself. For example, if the fundamental torque of the machine has to be calculated usually a relatively rough model is sufficient. Computing characteristics where harmonic waves have a significant influence (typically calculating acoustic noise), a much more precise alternative has to be chosen.

3.5.5. Rotating Air-Gap Field

The rotating air-gap field is deduced from the rotating magneto-motive force. In practice, an air-gap width is used which is enlarged against the geometric air-gap width because of:

- field distortion because of the slotting: $\delta \rightarrow \delta' = \delta\, K_C$ (K_C being Carter's factor) and
- magnetic reluctance (saturation) of the iron: $\delta' \rightarrow \delta''$.

Neglecting this field distortion and the magnetic reluctance of the iron ($\mu_{Fe} \rightarrow \infty$), it can be obtained:

$$B_\delta(\alpha,t) = \frac{\mu_0}{\delta}\Theta_\delta(\alpha,t)$$
$$= \frac{\mu_0}{\delta}\frac{3}{2}\frac{w}{p}\sqrt{2}I\frac{2}{\pi}\sum_{g=-\infty}^{\infty}\frac{\xi_{6g+1}}{6g+1}\sin\left(\left(6g+1\right)\alpha p-\omega t\right) \tag{3.60}$$

The fundamental wave is:

$$B_{\delta,1}(\alpha,t) = B_{\delta,1}\sin(\alpha p-\omega t) \tag{3.61}$$

with the amplitude $B_{\delta,1} = \frac{\mu_0}{\delta}\frac{3}{2}\frac{w\xi_1}{p}\frac{2}{\pi}\sqrt{2}I$ and the angular frequency $\Omega_1 = \frac{\omega}{p}$.

The harmonic waves are:

$$B_{\delta,6g+1}(\alpha,t) = B_{\delta,6g+1}\sin\left(\left(6g+1\right)\alpha p-\omega t\right)$$

with the amplitude $B_{\delta,6g+1} = \frac{\xi_{6g+1}}{\xi_1(6g+1)}B_{\delta,1}$ and the angular frequency $\Omega_{6g+1} = \frac{\Omega_1}{6g+1}$.

The total rotating field can be described as:

$$B_\delta(\alpha,t) = \sum_{g=-\infty}^{\infty} B_{\delta,6g+1}(\alpha,t) \tag{3.62}$$

Figure 3.20 shows this rotating field B_δ (red characteristic) and the rotating current loading A_δ (air-gap width $\delta = 1\text{mm}$; bore diameter $D = 0.1\text{m}$, blue function), together with the respective fundamental waves (dashed lines).

Please consider:

- the phase shift between magnetic field and current loading always is 90° (or $\pi/2$);
- the magnetic field B_δ changes the curve form with respect to time, but the amplitude of the fundamental wave $B_{\delta,1}$ remains unchanged at any point in time.

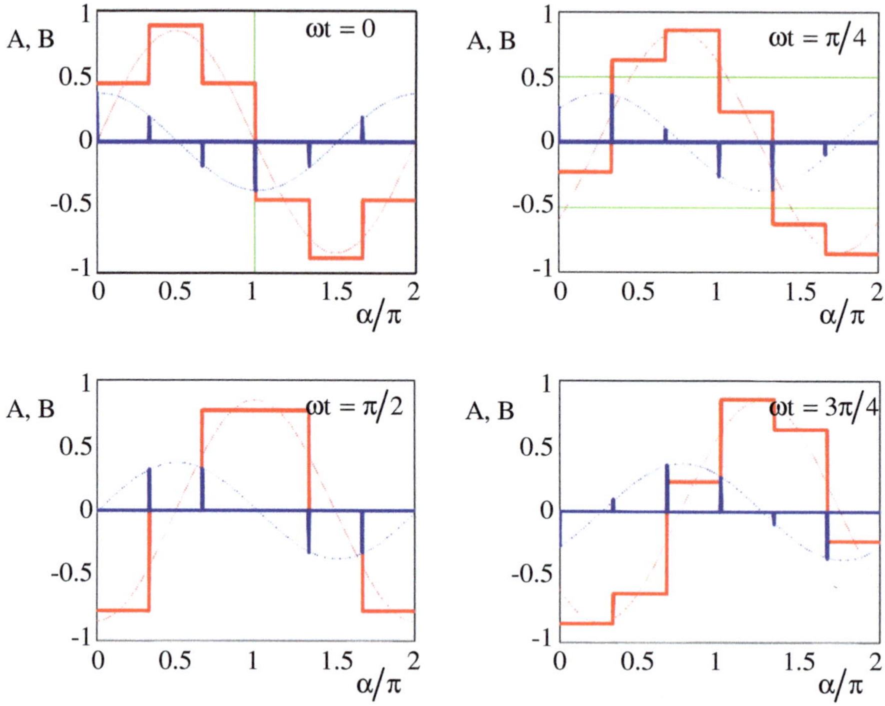

Fig. 3.20. Rotating magnetic field and current loading versus angle for different points in time (*solid red*: rotating field (B_δ in T); *dashed red*: according fundamental wave of rotating field ($B_{\delta,1}$ in T); *solid blue*: rotating current loading (A_δ in $100Am^{-1}$); *dashed blue*: according fundamental wave of rotating current loading ($A_{\delta,1}$ in Am^{-1})).

In Fig. 3.21 the field distribution is illustrated for two different instants in time, obtained by numerical calculation using the Finite Element Method (FEM). The stator contains 4 poles and 36 slots; the rotor is assumed being a cylindrical iron part. In addition, the phasors of the three currents, the current loading and the flux density are shown. The travelling rotating waves can be deduced clearly.

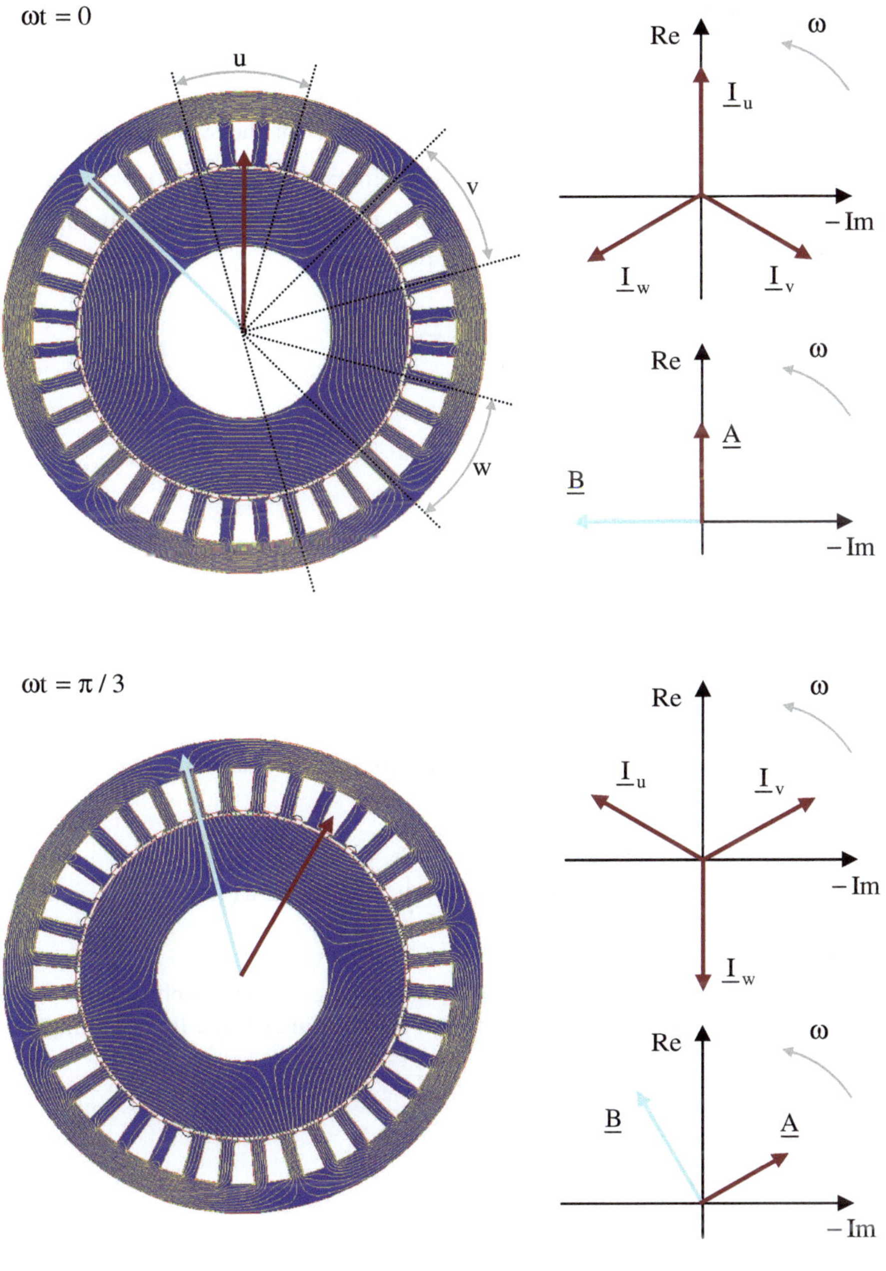

Fig. 3.21. Field distribution (magnetic field lines) and phasors of the three phase currents and the fundamental waves of current loading and air-gap flux density for different points in time.

3.6 Induced Voltage and Slip

In the following the induced voltage will be calculated, which is generated in a rotor coil (winding with w_2 turns) by a rotating field of the stator (winding with w_1 turns). The relative position $\alpha(t)$ is arbitrary (see Fig. 3.22). For this, the flux linkage of the rotor coil has to be calculated firstly. This is done by integration of the air-gap flux density across one pole pitch τ_p:

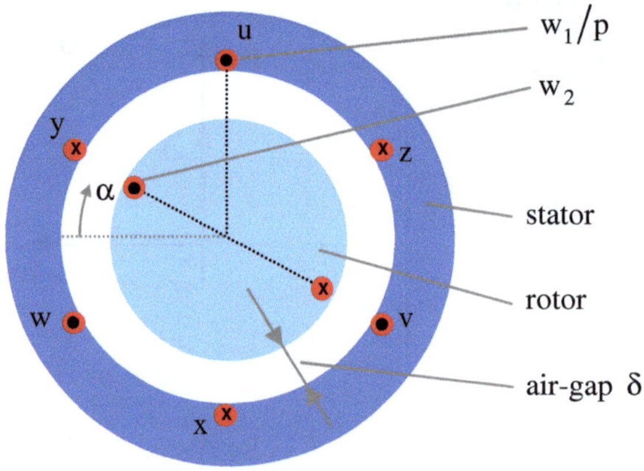

Fig. 3.22. Schematic cross-section of a rotating field machine with a single rotor coil.

$$\psi(\alpha, t) = w_2 \phi(\alpha, t) = w_2 \int_{r\alpha}^{r\alpha+\tau_p} B_\delta(\tilde{\alpha}, t) \ell \, d(r\tilde{\alpha}) \tag{3.63}$$

The variable ℓ describes the effective magnetic length of the machine and r is the bore radius; the tilde is used to distinguish between integration limit and integration variable. It follows:

$$
\begin{aligned}
\psi(\alpha,t) &= w_2 \int_{\alpha}^{\alpha+\frac{\pi}{p}} B_\delta(\tilde{\alpha},t)\,\ell r d\tilde{\alpha} \\
&= w_2 \ell r \int_{\alpha}^{\alpha+\frac{\pi}{p}} \sum_{g=-\infty}^{\infty} B_{\delta,6g+1} \sin\left((6g+1)\tilde{\alpha}p - \omega t\right) d\tilde{\alpha} \\
&= w_2 \ell r \left[\sum_{g=-\infty}^{\infty} B_{\delta,6g+1} \frac{-\cos\left((6g+1)\tilde{\alpha}p - \omega t\right)}{(6g+1)p} \right]_{\alpha}^{\alpha+\frac{\pi}{p}} \\
&= w_2 \frac{2\ell r}{p} \sum_{g=-\infty}^{\infty} \frac{B_{\delta,6g+1}}{6g+1} \cos\left((6g+1)\alpha p - \omega t\right) \\
&= w_2 \sum_{g=-\infty}^{\infty} \phi_{\delta,6g+1}(\alpha,t)
\end{aligned}
\tag{3.64}
$$

The fundamental wave is:

$$
\phi_{\delta,1}(\alpha,t) = \phi_{\delta,1} \cos(\alpha p - \omega t) \tag{3.65}
$$

with the amplitude $\phi_{\delta,1} = \frac{2\ell r}{p} B_{\delta,1} = 2\ell \frac{\tau_p}{\pi} B_{\delta,1} = \frac{2}{\pi} \ell \tau_p B_{\delta,1} = \mu_0 \frac{3\ell r}{\delta} \frac{w_1 \xi_1}{p^2} \frac{2}{\pi} \sqrt{2} I$

and the angular frequency $\Omega_1 = \frac{\omega}{p}$.

The harmonic waves are:

$$
\phi_{\delta,6g+1}(\alpha,t) = \phi_{\delta,6g+1} \cos\left((6g+1)\alpha p - \omega t\right) \tag{3.66}
$$

with the amplitude $\phi_{\delta,6g+1} = \frac{2\ell r}{p} \frac{B_{\delta,6g+1}}{6g+1} = \frac{2}{\pi} \ell \frac{\tau_p}{6g+1} B_{\delta,6g+1}$ and the angular frequency $\Omega_{6g+1} = \frac{\Omega_1}{6g+1}$.

The induced voltage will be calculated from the time-derivative of the magnetic flux linkage. The time dependency of the magnetic flux linkage may be caused by the time-dependent currents as well as by the time-dependent position of the rotor coil.

$$u_i(\alpha,t)=\frac{d\psi(\alpha,t)}{dt}=\frac{\partial\psi(\alpha,t)}{\partial\alpha}\frac{d\alpha}{dt}+\frac{\partial\psi(\alpha,t)}{\partial t}$$
$$=w_2\frac{2\ell r}{p}\sum_{g=-\infty}^{\infty}\frac{B_{\delta,6g+1}}{6g+1}\left(-\sin\left((6g+1)\alpha p-\omega t\right)\right)\left((6g+1)p\frac{d\alpha}{dt}-\omega\right) \tag{3.67}$$
$$=2\ell r\,w_2\sum_{g=-\infty}^{\infty}B_{\delta,6g+1}\sin\left((6g+1)\alpha p-\omega t\right)\left(\frac{\omega}{(6g+1)p}-\frac{d\alpha}{dt}\right)$$

Prior to calculating the induced voltage in detail, the "slip" shall be introduced here:

The *slip* is the difference between the speed of the rotating field wave of the stator and the speed of the rotor (i.e. the mechanical speed), relative to the respective speed of the rotating field wave of the stator.

With the angular frequency of the rotor

$$\frac{d\alpha}{dt}=\Omega=2\pi n \tag{3.68}$$

and the angular frequency of the harmonic field waves

$$\Omega_{6g+1}=\frac{\omega}{(6g+1)p} \tag{3.69}$$

the slip s of such a field wave becomes:

$$s_{6g+1}=\frac{\Omega_{6g+1}-\Omega}{\Omega_{6g+1}} \tag{3.70}$$

The speed of the fundamental wave of the rotating stator field is called synchronous speed. This will be indicated with the index "0":

$$n_0=\frac{\Omega_1}{2\pi}=\frac{\omega/p}{2\pi}=\frac{2\pi f/p}{2\pi}=\frac{f}{p} \tag{3.71}$$

Then the slip of the fundamental wave is:

$$s_1 = \frac{\Omega_1 - \Omega}{\Omega_1} = \frac{n_0 - n}{n_0} \tag{3.72}$$

The following can be deduced easily:

- at synchronous speed ($n = n_0$) there is: $s_1 = 0$;
- at stand-still ($n = 0$) there is: $s_1 = 1$.

The relation between speed and slip is illustrated in Fig. 3.23.

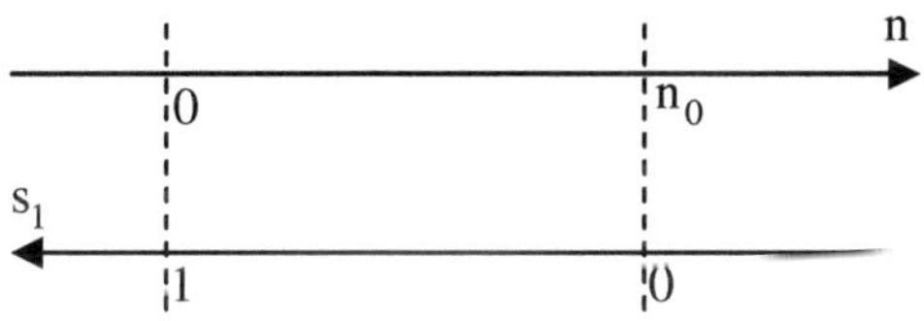

Fig. 3.23. Relation between speed and slip.

The position of the rotor coil changes with the mechanical speed of the rotor. Therefore it follows

$$\alpha(t) = \alpha_R + \Omega t \tag{3.73}$$

Further:

$$\begin{aligned}(6g+1)\alpha p - \omega t &= (6g+1)\alpha_R p + (6g+1)p\Omega t - \omega t \\ &= (6g+1)\alpha_R p - \omega t\left(1 - \frac{(6g+1)p\Omega}{\omega}\right) \\ &= (6g+1)\alpha_R p - \omega t\left(\frac{\dfrac{\omega}{(6g+1)p} - \Omega}{\dfrac{\omega}{(6g+1)p}}\right) \\ &= (6g+1)\alpha_R p - s_{6g+1}\omega t\end{aligned} \tag{3.74}$$

and

$$\frac{\omega}{(6g+1)p}-\frac{d\alpha}{dt}=\left(\frac{\frac{\omega}{(6g+1)p}-\Omega}{\frac{\omega}{(6g+1)p}}\right)\frac{\omega}{(6g+1)p}=s_{6g+1}\Omega_{6g+1} \tag{3.75}$$

For the voltage induced into the rotor coil it follows (by inserting the last two equations into Eq. 3.67):

$$\begin{aligned} u_i(\alpha,t) &= 2\ell r w_2 \sum_{g=-\infty}^{\infty} B_{\delta,6g+1}\sin\left((6g+1)\alpha_R p - s_{6g+1}\omega t\right) s_{6g+1}\Omega_{6g+1} \\ &= 2w_2\ell \sum_{g=-\infty}^{\infty} B_{\delta,6g+1}\sin\left((6g+1)\alpha_R p - s_{6g+1}\omega t\right) s_{6g+1}\Omega_{6g+1} r \\ &= 2w_2\ell \sum_{g=-\infty}^{\infty} B_{\delta,6g+1}(\alpha_R,t)\, v_{6g+1} \\ &= \sum_{g=-\infty}^{\infty} u_{i,6g+1}(\alpha_R,t) \end{aligned} \tag{3.76}$$

Here, $v_{6g+1}=s_{6g+1}\Omega_{6g+1}r$ is the relative speed between the rotor coil and the harmonic wave of the rotating field.

The induced voltage is composed of an infinite number of separate *oscillations* $u_{i,6g+1}(\alpha_R,t)$. For these oscillations the following holds true:

$$\begin{aligned} u_{i,6g+1}(\alpha_R,t) &= 2w_2\ell r s_{6g+1}\Omega_{6g+1}B_{\delta,6g+1}\sin\left((6g+1)\alpha_R p - s_{6g+1}\omega t\right) \\ &= w_2\frac{2\ell r}{(6g+1)p}\omega s_{6g+1}B_{\delta,6g+1}\sin\left((6g+1)\alpha_R p - s_{6g+1}\omega t\right) \\ &= w_2\phi_{\delta,6g+1}\omega s_{6g+1}\sin\left((6g+1)\alpha_R p - s_{6g+1}\omega t\right) \end{aligned} \tag{3.77}$$

The voltages $u_{i,6g+1}(\alpha_R,t)$ induced by the rotating field waves offer the following characteristics:

- The amplitudes of the induced voltages are proportional to the flux of the harmonic wave, to the frequency of the supplying voltage, and to the slip.
- The frequency of all induced voltages is equal to the slip frequency.
- At stand-still of the rotor ($n=0$, i.e. $s_{6g+1}=1$) the frequency of all voltages induced by the harmonic waves is equal to the supply frequency.

- At synchronous speed ($n = n_0$) there is:

$$s_{6g+1} = \frac{\dfrac{\omega}{(6g+1)p} - \dfrac{\omega}{p}}{\dfrac{\omega}{(6g+1)p}} = 1 - (6g+1) = -6g$$

Consequently, no fundamental voltage ($g = 0$) is induced; but the harmonic waves induce harmonic voltage oscillations with different frequencies.

- At other speeds the fundamental wave and each harmonic wave of the rotating stator field induce a voltage of different frequency into the rotor coil.
- The phase shift of the induced voltages in the rotor coil only depends on the spatial position of the coil $\alpha_R (6g+1) p$.

Incorporating a three-phase winding even to the rotor, the three phases will be shifted by an angle of $2\pi/3p$ as well. Having a rotor slot number per pole and per phase $q > 1$, a winding factor for the rotor (index "2" for the rotor) has to be considered. The induced voltage of the kth ($k = 1, 2, 3$) rotor phase becomes ($\alpha_R = (k-1) 2\pi/3p$):

$$u_{i,2,k}(t) = \sum_{g=-\infty}^{\infty} w_2 \xi_{2,6g+1} \phi_{\delta,6g+1} \omega s_{6g+1} \cdot \sin\left((6g+1)\frac{2\pi}{3}(k-1) - s_{6g+1}\omega t \right) \qquad (3.78)$$

The air-gap field will induce voltages even into the stator winding. If in the above equation $w_2 \xi_{2,6g+1}$ is substituted by $w_1 \xi_{1,6g+1}$, and considering that for the stator winding $s_{6g+1} = 1$ is true (because the stator is always in stand-still), the induced voltage into the stator winding becomes:

$$u_{i,1,k}(t) = \sum_{g=-\infty}^{\infty} w_1 \xi_{1,6g+1} \phi_{\delta,6g+1} \omega \sin\left((6g+1)\frac{2\pi}{3}(k-1) - \omega t \right) \qquad (3.79)$$

The ratio of induced stator voltages (maximum values) caused by a harmonic wave and the fundamental wave is:

$$\frac{u_{i,1,6g+1}}{u_{i,1,1}} = \frac{\xi_{1,6g+1}}{\xi_{1,1}} \frac{\phi_{\delta,6g+1}}{\phi_{\delta,1}} = \frac{\xi_{1,6g+1}}{\xi_{1,1}} \frac{B_{\delta,6g+1}}{(6g+1) B_{\delta,1}} = \left(\frac{\xi_{1,6g+1}}{\xi_{1,1} (6g+1)} \right)^2 \quad (3.80)$$

The ratio of induced stator and rotor voltages (maximum values) caused by the same air-gap field wave is:

$$\frac{u_{i,2,6g+1}}{u_{i,1,6g+1}} = \frac{w_2 \xi_{2,6g+1}}{w_1 \xi_{1,6g+1}} s_{6g+1} \quad (3.81)$$

The rms values of the induced stator and rotor voltages caused by the fundamental air-gap field wave are:

$$U_{i,1,1} = w_1 \xi_{1,1} \omega \frac{\phi_{\delta,1}}{\sqrt{2}}$$
$$U_{i,2,1} = w_2 \xi_{2,1} s_1 \omega \frac{\phi_{\delta,1}}{\sqrt{2}} \quad (3.82)$$

3.7 Torque and Power

For the following considerations concerning torque and power only the fundamental waves will be regarded, the influence of the harmonic waves will be neglected.

The stator winding generates a rotating magneto-motive force $\Theta_{\delta,1}$ with the angular frequency ω_1 / p_1:

$$\Theta_{\delta,1}(\alpha, t) = \frac{3}{2} \frac{2}{\pi} \frac{w_1 \xi_1}{p_1} \sqrt{2} I_1 \sin(\alpha p_1 - \omega_1 t) = \Theta_{\delta,1} \sin(\alpha p_1 - \omega_1 t) \quad (3.83)$$

Analogously, the rotor winding generates a rotating magneto-motive force $\Theta_{\delta,2}$ with the angular frequency ω_2' / p_2 and shifted in space by a leading angle of ε. The angular frequency of the rotating magneto-motive force of the rotor is composed of the mechanical angular frequency of the rotor Ω and the angular frequency of the rotor currents ω_2 ($\omega_2' = \Omega + \omega_2$):

$$\Theta_{\delta,2}(\alpha,t) = \frac{3}{2}\frac{2}{\pi}\frac{w_2\xi_2}{p_2}\sqrt{2}I_2 \sin(\alpha p_2 - \omega_2' t - \varepsilon p_2) \\ = \Theta_{\delta,2} \sin(\alpha p_2 - \omega_2' t - \varepsilon p_2) \tag{3.84}$$

At this time the numbers of pole pairs, the angular frequencies and the phase shift between the stator and rotor rotating magneto-motive forces shall be arbitrary. Figure 3.24 schematically illustrates the spatial position of the magneto-motive force distributions by means of phasors: the position of each phasor shows the position of the maximum of the magneto-motive force distribution, the length of the phasor is a quantity for the amplitude (please refer to Sect. 1.4).

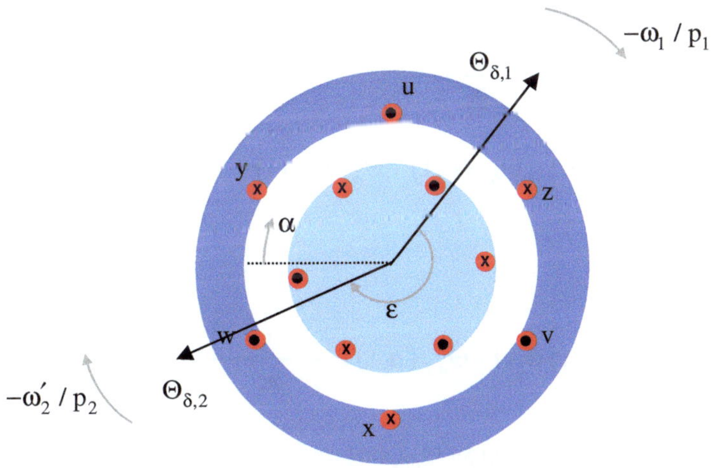

Fig. 3.24. Cross-sectional view of a rotating field machine with winding distribution and illustration of magneto-motive forces by means of phasors.

By means of Ampere's Law the resulting air-gap field is obtained by superposition (this means that saturation is neglected here):

$$H(\alpha,t)\delta = \Theta_{\delta,1}(\alpha,t) + \Theta_{\delta,2}(\alpha,t) \\ \Rightarrow \quad B(\alpha,t) = \frac{\mu_0}{\delta}\left(\Theta_{\delta,1}(\alpha,t) + \Theta_{\delta,2}(\alpha,t)\right) \tag{3.85}$$

The magnetic energy in the air-gap is:

$$W_{mag} = \int_V \frac{B^2(\alpha,t)}{2\mu_0}\, dV, \qquad dV = \ell\, \delta\, r\, d\alpha \tag{3.86}$$

As saturation is neglected (linearity was assumed) the magnetic energy is equal to the magnetic co-energy. Then the torque may be calculated from the derivation of the magnetic energy with respect to the mechanical angle:

$$\begin{aligned} T &= \frac{\partial W_{mag}}{\partial \varepsilon} = \frac{\partial}{\partial \varepsilon} \int_0^{2\pi} \frac{B^2(\alpha,t)}{2\mu_0} \ell\, \delta\, r\, d\alpha \\ &= \frac{\ell \delta r}{2\mu_0} \int_0^{2\pi} \left[\frac{\partial}{\partial \varepsilon} B^2(\alpha,t) \right] d\alpha \end{aligned} \tag{3.87}$$

With

$$\begin{aligned} \frac{\partial}{\partial \varepsilon} B^2(\alpha,t) &= 2B(\alpha,t) \frac{\partial B(\alpha,t)}{\partial \varepsilon} \\ &= 2\left(\frac{\mu_0}{\delta}\right)^2 \left(\Theta_{\delta,1}(\alpha,t) + \Theta_{\delta,2}(\alpha,t)\right) \frac{\partial \Theta_{\delta,2}(\alpha,t)}{\partial \varepsilon} \end{aligned} \tag{3.88}$$

it follows in total:

$$\begin{aligned} T &= \frac{\ell \delta r}{2\mu_0} 2\left(\frac{\mu_0}{\delta}\right)^2 \int_0^{2\pi} \left(\Theta_{\delta,1}(\alpha,t) + \Theta_{\delta,2}(\alpha,t)\right) \frac{\partial \Theta_{\delta,2}(\alpha,t)}{\partial \varepsilon} d\alpha \\ &= \mu_0 \frac{\ell r}{\delta} \int_0^{2\pi} \left(\Theta_{\delta,1} \sin(\alpha p_1 - \omega_1 t) + \Theta_{\delta,2} \sin(\alpha p_2 - \omega_2' t - \varepsilon p_2)\right) \cdot \\ &\qquad \Theta_{\delta,2} \left[-p_2 \cos(\alpha p_2 - \omega_2' t - \varepsilon p_2)\right] d\alpha \\ &= -\mu_0 \frac{\ell r}{\delta} \int_0^{2\pi} \left[p_2 \Theta_{\delta,1} \Theta_{\delta,2} \sin(\alpha p_1 - \omega_1 t) \cos(\alpha p_2 - \omega_2' t - \varepsilon p_2) + \right. \\ &\qquad \left. p_2 \Theta_{\delta,2}^2 \sin(\alpha p_2 - \omega_2' t - \varepsilon p_2) \cos(\alpha p_2 - \omega_2' t - \varepsilon p_2) \right] d\alpha \end{aligned} \tag{3.89}$$

Considering $\int_0^{2\pi} \sin(x)\cos(x)\,dx = 0$ it can be obtained:

$$T = -\mu_0 \frac{\ell r}{\delta} p_2 \Theta_{\delta,1} \Theta_{\delta,2} \int_0^{2\pi} \sin(\alpha p_1 - \omega_1 t) \cos(\alpha p_2 - \omega_2' t - \varepsilon p_2) d\alpha \tag{3.90}$$

With $\sin(x)\cos(y) = \frac{1}{2}[\sin(x+y) + \sin(x-y)]$ it follows:

$$T = -\mu_0 \frac{\ell r}{\delta} p_2 \Theta_{\delta,1} \Theta_{\delta,2} \frac{1}{2} \left[\int_0^{2\pi} \sin((p_1+p_2)\alpha - (\omega_1 + \omega_2')t - \varepsilon p_2) d\alpha + \int_0^{2\pi} \sin((p_1-p_2)\alpha - (\omega_1 - \omega_2')t + \varepsilon p_2) d\alpha \right] \tag{3.91}$$

Generally it is true:

$$\int_0^{2\pi} \sin(nx + \varphi) dx = \begin{cases} 0 & \text{if} \quad n \neq 0 \\ 2\pi \sin(\varphi) & \text{if} \quad n = 0 \end{cases} \tag{3.92}$$

As p_1 and p_2 are the numbers of pole pairs of stator and rotor, respectively, these are positive integers. Therefore, it follows:

$$\int_0^{2\pi} \sin((p_1+p_2)\alpha - (\omega_1 + \omega_2')t - \varepsilon p_2) d\alpha = 0 \tag{3.93}$$

and

$$\int_0^{2\pi} \sin((p_1-p_2)\alpha - (\omega_1 - \omega_2')t + \varepsilon p_2) d\alpha \neq 0 \quad \text{if} \quad p_1 = p_2 = p \tag{3.94}$$

Consequently, torque is only generated if the numbers of pole pairs of stator and rotor are identical. Then it follows:

$$T = -\mu_0 \frac{\ell r}{\delta} p \frac{\Theta_{\delta,1}\Theta_{\delta,2}}{2} 2\pi \sin(-(\omega_1 - \omega_2')t + \varepsilon p) \tag{3.95}$$

From this equation can be deduced that the torque is sinusoidal in time with a mean value (according to time) of zero. It is a pure oscillating torque.

Only if the angular frequencies of the rotating magneto-motive force waves are identical ($\omega_1 = \omega_2' = \omega$) a torque constant in time is generated:

$$T = \mu_0 \frac{\ell r \pi}{\delta} p \Theta_{\delta,1} \Theta_{\delta,2} \sin(-\varepsilon p) \tag{3.96}$$

If both rotating magneto-motive force waves $\Theta_{\delta,1}(\alpha,t)$ and $\Theta_{\delta,2}(\alpha,t)$ are in phase, i.e. if $\varepsilon = 0$ is true, then the torque is equal to zero. The torque is maximum for $\varepsilon p = \pi/2$, i.e. if the rotating magneto-motive force waves have an electrically phase shift of 90° (or $\pi/2$).

The rotating magnetic field has been calculated as follows:

$$B_\delta(\alpha,t) = \frac{\mu_0}{\delta}\left(\Theta_{\delta,1}(\alpha,t) + \Theta_{\delta,2}(\alpha,t)\right) = \frac{\mu_0}{\delta}\Theta_{\delta,tot}(\alpha,t) \tag{3.97}$$

This shall be illustrated in Fig. 3.25 with the help of phasors in the complex plain. With their length phasors describe the amplitude of the respective rotating wave, the direction of the phasor shows the position of the maximum of the rotating wave.

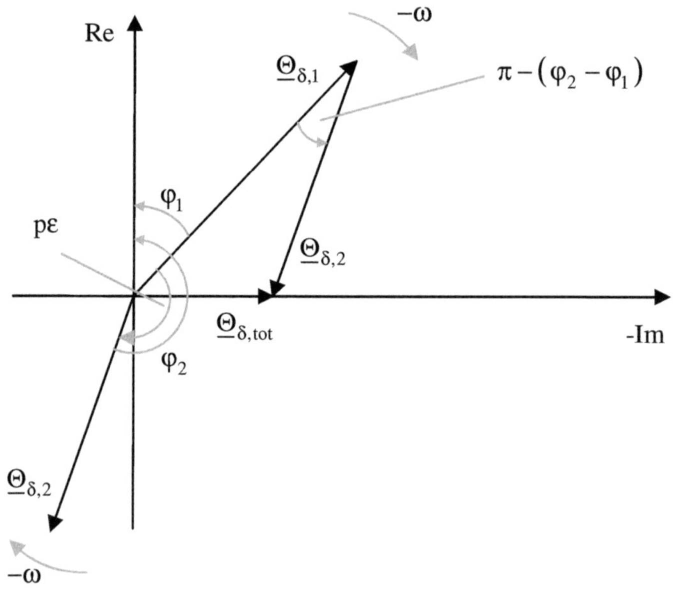

Fig. 3.25. Phasor diagram of the magneto-motive forces.

By means of the law of cosines the absolute value of the total rotating magneto-motive force is obtained:

$$\begin{aligned}\Theta_{\delta,tot} &= \sqrt{\Theta_{\delta,1}^2 + \Theta_{\delta,2}^2 - 2\Theta_{\delta,1}\Theta_{\delta,2}\cos\left[\pi - (\varphi_2 - \varphi_1)\right]} \\ &= \sqrt{\Theta_{\delta,1}^2 + \Theta_{\delta,2}^2 + 2\Theta_{\delta,1}\Theta_{\delta,2}\cos(\varphi_2 - \varphi_1)}\end{aligned} \tag{3.98}$$

The law of sines gives:

$$\begin{aligned}&\frac{\Theta_{\delta,2}}{\sin\left(\dfrac{\pi}{2} - \varphi_1\right)} = \frac{\Theta_{\delta,tot}}{\sin\left(\pi - (\varphi_2 - \varphi_1)\right)} \\ \Rightarrow\quad &\frac{\Theta_{\delta,2}}{\cos(\varphi_1)} = \frac{\Theta_{\delta,tot}}{\sin(\varphi_2 - \varphi_1)} = \frac{\Theta_{\delta,tot}}{\sin(2\pi - p\varepsilon)} = \frac{\Theta_{\delta,tot}}{\sin(-p\varepsilon)} \\ \Rightarrow\quad &\Theta_{\delta,2}\sin(-p\varepsilon) = \Theta_{\delta,tot}\cos(\varphi_1)\end{aligned} \tag{3.99}$$

Therefore, it follows for the torque:

$$\begin{aligned}T &= \mu_0 \frac{\ell r \pi}{\delta} p \Theta_{\delta,1}\Theta_{\delta,tot}\cos(\varphi_1) \\ &= \mu_0 \frac{\ell r \pi}{\delta} p \Theta_{\delta,1} \frac{\delta}{\mu_0} B_\delta \cos(\varphi_1) \\ &= \ell r \pi p \Theta_{\delta,1} B_\delta \cos(\varphi_1) \\ &= \ell r \pi p \frac{3}{2}\frac{2}{\pi}\frac{w_1 \xi_1}{p}\sqrt{2} I_1 B_\delta \cos(\varphi_1) \\ &= 3\ell r w_1 \xi_1 \sqrt{2} I_1 B_\delta \cos(\varphi_1)\end{aligned} \tag{3.100}$$

With the rms-value of the induced voltage

$$U_{i,1} = w_1 \xi_1 \omega \frac{\phi_\delta}{\sqrt{2}} \tag{3.101}$$

and the magnetic flux

$$\phi_\delta = \frac{2\ell r}{p} B_\delta \tag{3.102}$$

the torque can be calculated like this:

$$T = 3U_{i,1}I_1 \frac{p}{\omega} \cos(\varphi_1) = \frac{3U_{i,1}I_1 \cos(\varphi_1)}{\Omega_1} = \frac{3U_{i,1}I_1 \cos(\varphi_1)}{2\pi n_0} \tag{3.103}$$

Here, the expression

$$P_\delta = 3U_{i,1}I_1 \cos(\varphi_1) \tag{3.104}$$

is the rotating air-gap field power. This rotating air-gap field power is equal to the power supplied from the mains minus the copper losses (Ohmic losses) of the stator:

$$P_\delta = P_1 - 3R_1I_1^2 \tag{3.105}$$

The rotating air-gap field power is transferred via the air-gap:

$$P_\delta = 3U_{i,1}I_1 \cos(\varphi_1) = 2\pi n_0 T \tag{3.106}$$

The mechanical power at the shaft of the machine is:

$$P_{mech} = 2\pi n T = 2\pi n_0 T(1-s) = (1-s)P_\delta \tag{3.107}$$

The difference between the rotating air-gap field power and the mechanical power at the axis is transformed to heating losses inside the rotor:

$$P_{loss,rotor} = P_\delta - P_{mech} = P_\delta - (1-s)P_\delta = sP_\delta \tag{3.108}$$

Having synchronous speed of the rotor ($n = n_0$, i.e. $s = 0$), the mechanical power at the shaft equals the rotating air-gap field power; the rotor losses are zero. The power flow inside an electrical machine is illustrated by the co-called Sankey-diagram, see Fig. 3.26 (iron losses and friction is neglected).

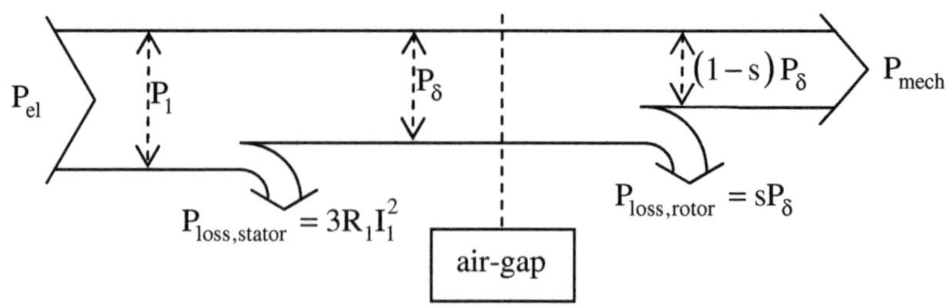

Fig. 3.26. Sankey-diagram.

To get a constant torque in time the condition $\omega_1 = \omega_2' = \omega$ must be fulfilled. The angular frequency of the rotating magneto-motive force wave of the rotor ω_2' can be described like:

$$\begin{aligned}\omega_2' &= \omega = 2\pi f = 2\pi p n_0 = 2\pi p\left(\left(1-s\right)n_0 + s n_0\right) \\ &= 2\pi p\left(n + s\frac{f}{p}\right) = 2\pi p n + 2\pi s f = 2\pi p n + 2\pi f_2 \\ &= \Omega + \omega_2\end{aligned} \tag{3.109}$$

In this equation n is the mechanical speed of the rotor and f_2 is the frequency of the rotor currents. Alternative descriptions are:

$$\begin{aligned}&f_2 = sf = sf_1 = f_1 - pn \\ \Rightarrow \quad &n = \frac{f_1 - f_2}{p}\end{aligned} \tag{3.110}$$

If the rotor of the rotating field machine contains alternating currents $(f_2 > 0)$ there is:

$$n < \frac{f_1}{p} = n_0 \tag{3.111}$$

Consequently, the rotor rotates with a speed less than the synchronous speed n_0. Such a machine is called *asynchronous machine* or *induction machine*. Depending on the frequency of the rotor currents the speed of the machine is achieved.

Supplying the rotor of the rotating field machine with DC current $(f_2 = 0)$ there is:

$$n = \frac{f_1}{p} = n_0 \tag{3.112}$$

Consequently, the rotor rotates with synchronous speed. Such a machine is called *synchronous machine*. The speed of the machine solely depends on the frequency of the *stator* currents.

3.8 References for Chapter 3

Bödefeld T, Sequenz H (1971) Elektrische Maschinen. Springer-Verlag, Wien
Jordan H, Klima V, Kovacs KP (1975) Asynchronmaschinen. Vieweg Verlag, Braunschweig
Kleinrath H (1975) Grundlagen elektrischer Maschinen. Akademische Verlagsgesellschaft, Wiesbaden
Müller G, Ponick B (2009) Theorie elektrischer Maschinen. Wiley-VCH Verlag, Weinheim
Seinsch HO (1992) Oberfelderscheinungen in Drehfeldmaschinen. Teubner Verlag, Stuttgart
Seinsch HO (1993) Grundlagen elektrischer Maschinen und Antriebe. Teubner Verlag, Stuttgart
Taegen F (1971) Einführung in die Theorie der elektrischen Maschinen II. Vieweg Verlag, Braunschweig

4 Induction Machines

4.1 Construction and Equivalent Circuit Diagram

Stator and rotor of an induction machine - separated by an air-gap - contain each a symmetric winding in their slots.

The stator with the number of phases $m_1 = 3$ and the effective number of turns $w_1\xi_1$ is supplied by a symmetric three-phase system with the frequency f_1. Consequently, a rotating field is generated, that rotates with the synchronous speed $n_0 = f_1/p$. To produce a constant torque, the number of pole pairs of stator and rotor are identical ($p_1 = p_2 = p$).

The rotor winding can either be short-circuited directly or it can be passed outside the rotor via slip rings, and there be short-circuited directly or via series resistors. In the m_2 phases of the closed rotor winding with the effective number of turns $w_2\xi_2$ currents are induced by the rotating field. These rotor currents together with the rotating field of the stator generate the torque that rotates the rotor with the mechanical speed $n = (1-s)\,n_0$.

During motor operation the rotor always rotates with a speed lower than the synchronous speed n_0 (i.e. asynchronous to the rotating field of the stator currents), because at synchronous speed no currents are induced into the rotor winding and therefore no torque is generated. This differential speed between rotating field of the stator and the rotor is $n_2 = sn_0$; in the rotor currents with the frequency $f_2 = sf_1$ are induced. These currents generate a rotating field of the rotor that (relative to the rotor) rotates with the speed n_2. Relative to the stator this rotating field has the speed $n_2 + n = n_0$. Consequently, the frequency condition to obtain a constant torque (please refer to Sect. 3.7) is fulfilled.

Figure 4.1 shows the zones and a symbolic representation of the induction machine windings:

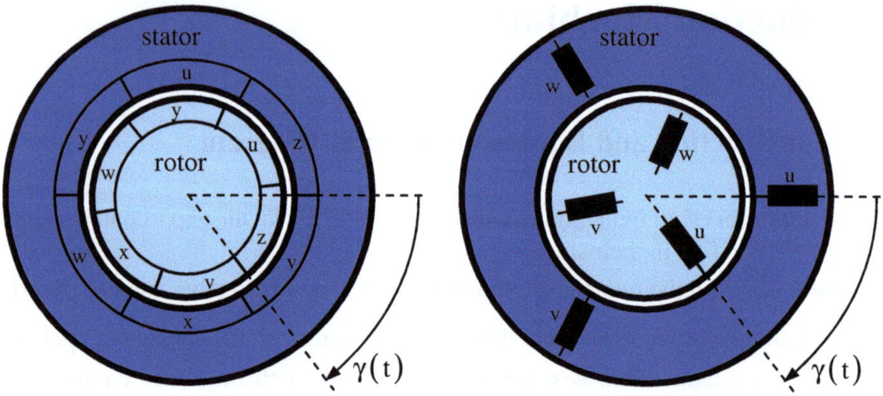

Fig. 4.1. Symbolic representation of the induction machine windings (*left*: distribution of the three stator and rotor windings in zones with 2p=2; *right*: symbolic representation).

To reach an easier mathematical description of this electrical machine firstly the system of coupled and movable coils shall be transformed to a system with stationary elements with the same numbers of phases and turns in stator and rotor.

For this, in the following only the *fundamental waves* will be considered. For calculation of the *steady-state operation* the calculation with complex phasors will be applied. Because of symmetry reasons a single-phase treatment is sufficient.

The windings of stator and rotor can be described with lumped parameters: resistance, leakage and main inductivity. Then the equivalent circuit diagram looks like it is shown in Fig. 4.2 (both partial systems are represented as "energy consumption system"):

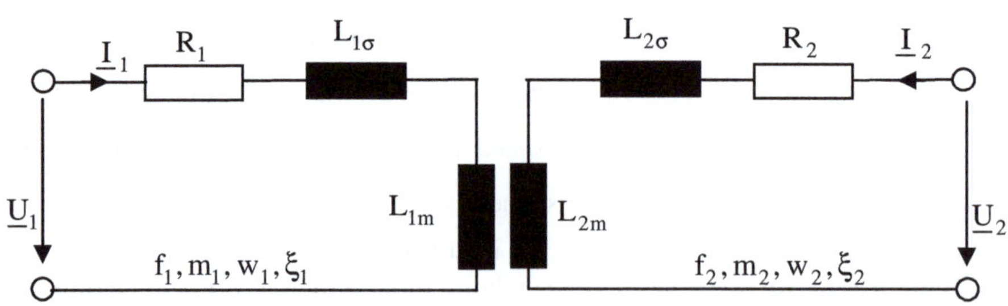

Fig. 4.2. Equivalent circuit diagram of the induction machine with magnetic coupling between stator and rotor.

The voltage equations are (with the mutual inductivities L_{21} and L_{12}):

$$\underline{U}_1 = R_1\underline{I}_1 + j\omega_1 L_1\underline{I}_1 + j\omega_1 L_{21}\underline{I}_2$$
$$\underline{U}_2 = R_2\underline{I}_2 + j\omega_2 L_2\underline{I}_2 + j\omega_2 L_{12}\underline{I}_1 \tag{4.1}$$

with (L_{1m} and L_{2m} being the main or magnetizing inductivities of stator and rotor):

$$L_1 = L_{1\sigma} + L_{1m}$$
$$L_2 = L_{2\sigma} + L_{2m} \tag{4.2}$$

Now voltages and currents in the stationary stator winding are searched, that have the same effect like the actual voltages and currents in the rotating rotor winding. These transformed voltages and currents will be labeled by a prime. The following conditions must be fulfilled:

1. Unchanged magneto-motive force:

$$\frac{m_2}{2}\frac{4}{\pi}\frac{w_2\xi_2}{p}\sqrt{2}I_2 \overset{!}{=} \frac{m_1}{2}\frac{4}{\pi}\frac{w_1\xi_1}{p}\sqrt{2}I_2'$$
$$\Rightarrow \quad I_2' = I_2\frac{m_2 w_2\xi_2}{m_1 w_1\xi_1} \tag{4.3}$$

2. Unchanged main flux:

$$\sqrt{2}U_{i,2} = \omega_2 w_2\xi_2\phi_m$$
$$\sqrt{2}U_{i,2}' \overset{!}{=} \omega_1 w_1\xi_1\phi_m \tag{4.4}$$
$$\Rightarrow \quad \sqrt{2}U_{i,2}' = \omega_1 w_1\xi_1\frac{\sqrt{2}U_{i,2}}{\omega_2 w_2\xi_2}$$

For $\omega_2 = \omega_1$ it is true: $U_{i,2} = U_{2,0}$ (induced voltage is equal to the no-load voltage). Then it follows further:

$$U_{i,2}' = U_{i,2}\frac{\omega_1 w_1\xi_1}{\omega_2 w_2\xi_2} = \frac{1}{s}U_{2,0}\frac{w_1\xi_1}{w_2\xi_2} = \frac{1}{s}U_{2,0}' \tag{4.5}$$

3. Unchanged Ohmic losses:

$$m_2 R_2 I_2^2 \overset{!}{=} m_1 R_2' I_2'^2$$
$$\Rightarrow \quad R_2' = R_2 \frac{m_1 \left(w_1 \xi_1\right)^2}{m_2 \left(w_2 \xi_2\right)^2} \tag{4.6}$$

4. Unchanged magnetic energy (here for the linear, i.e. unsaturated case):

$$\frac{m_1}{2} \Psi_1 I_1 + \frac{m_2}{2} \Psi_2 I_2 \overset{!}{=} \frac{m_1}{2} \left(\Psi_1 I_1 + \Psi_2' I_2'\right)$$
$$\Rightarrow \quad \frac{m_1}{2} \left(L_1 I_1^2 + L_{21} I_1 I_2\right) + \frac{m_2}{2} \left(L_2 I_2^2 + L_{12} I_1 I_2\right) \tag{4.7}$$
$$= \frac{m_1}{2} \left(L_1 I_1^2 + 2 L_{1m} I_1 I_2' + L_2' I_2'^2\right)$$

Further:

$$L_2' = L_2 \frac{m_1 \left(w_1 \xi_1\right)^2}{m_2 \left(w_2 \xi_2\right)^2}$$
$$L_{21} = L_{1m} \frac{m_2 w_2 \xi_2}{m_1 w_1 \xi_1} \tag{4.8}$$
$$L_{12} = L_{1m} \frac{w_2 \xi_2}{w_1 \xi_1}$$

Please note that the following is true:

$$L_{21} = L_{12} \frac{m_2}{m_1} \tag{4.9}$$

Now the voltage equations (Eq. 4.1) can be transformed like follows:

$$\underline{U}_1 = R_1\underline{I}_1 + j\omega_1 L_1 \underline{I}_1 + j\omega_1 \frac{m_1 w_1 \xi_1}{m_2 w_2 \xi_2} L_{21} \frac{m_2 w_2 \xi_2}{m_1 w_1 \xi_1} \underline{I}_2$$

$$\begin{aligned} \underline{U}_2 \frac{\omega_1 w_1 \xi_1}{\omega_2 w_2 \xi_2} &= R_2 \frac{\omega_1 (w_1 \xi_1)^2}{\omega_2 (w_2 \xi_2)^2} \frac{m_1}{m_2} \underline{I}_2 \frac{m_2}{m_1} \frac{w_2 \xi_2}{w_1 \xi_1} \\ &+ j\omega_2 \frac{\omega_1 (w_1 \xi_1)^2}{\omega_2 (w_2 \xi_2)^2} \frac{m_1}{m_2} L_2 \frac{m_2}{m_1} \frac{w_2 \xi_2}{w_1 \xi_1} \underline{I}_2 \\ &+ j\omega_2 \frac{\omega_1 w_1 \xi_1}{\omega_2 w_2 \xi_2} L_{12} \underline{I}_1 \end{aligned} \tag{4.10}$$

Inserting the values transformed to the stator ("primed" values) gives:

$$\begin{aligned} \underline{U}_1 &= R_1 \underline{I}_1 + j\omega_1 L_1 \underline{I}_1 + j\omega_1 L_{1m} \underline{I}_2' \\ \frac{\underline{U}_2'}{s} &= \frac{R_2'}{s} \underline{I}_2' + j\omega_1 L_2' \underline{I}_2' + j\omega_1 L_{1m} \underline{I}_1 \end{aligned} \tag{4.11}$$

With the main reactance, the total stator reactance, the total rotor reactance (transformed to the stator)

$$\begin{aligned} X_{1m} &= \omega_1 L_{1m} \\ X_1 &= X_{1\sigma} + X_{1m} \\ X_2' &= X_{2\sigma}' + X_{1m} \end{aligned} \tag{4.12}$$

and the no-load current

$$\underline{I}_0 = \underline{I}_1 + \underline{I}_2' \tag{4.13}$$

the following set of equations is obtained:

$$\begin{aligned} \underline{U}_1 &= R_1 \underline{I}_1 + jX_{1\sigma} \underline{I}_1 + jX_{1m} \underline{I}_0 \\ \frac{\underline{U}_2'}{s} &= \frac{R_2'}{s} \underline{I}_2' + jX_{2\sigma}' \underline{I}_2' + jX_{1m} \underline{I}_0 \end{aligned} \tag{4.14}$$

From this the following equivalent circuit diagram can be deduced (Fig. 4.3). It becomes obvious that in comparison to Fig. 4.2 a galvanic coupling is introduced that simplifies the calculation of induction machines.

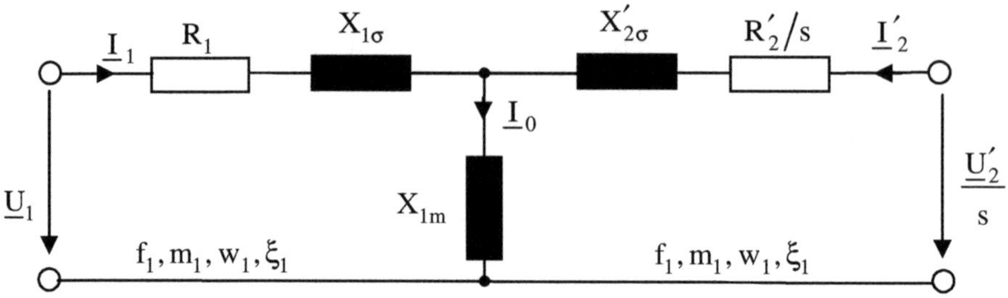

Fig. 4.3. Equivalent circuit diagram of the induction machine with galvanic coupling.

Please note that in this equivalent circuit diagram all values are transformed to stator frequency, phase number of the stator, and effective number of turns of the stator. For $s = 1$ (i.e. stationary rotor) the voltage ratio is equal to the ratio of the effective number of turns:

$$\frac{U_{i,1}}{U_{i,2}} = r_u = \frac{w_1 \xi_1}{w_2 \xi_2} \tag{4.15}$$

Operating the induction machine with constant stator flux linkage most often the following transformation ratio is used:

$$r_u = \frac{w_1 \xi_1}{w_2 \xi_2}(1+\sigma_1) \tag{4.16}$$

Then the leakage reactance of the stator vanishes in the equivalent circuit diagram (see Fig. 4.4).

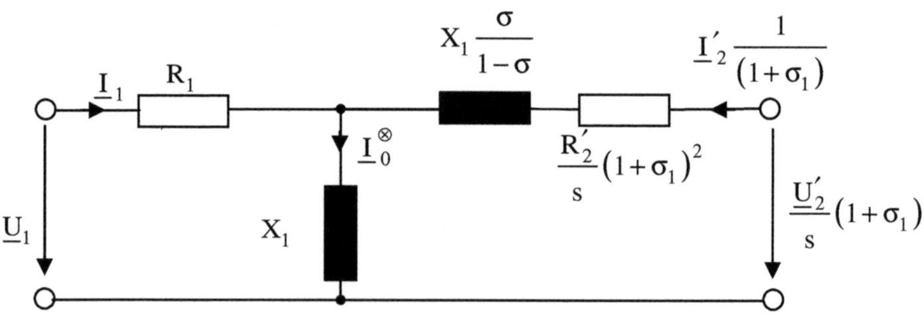

Fig. 4.4. Equivalent circuit diagram of the induction machine without stator leakage reactance.

Operating the induction machine with constant rotor flux linkage most often the following transformation ratio is chosen:

$$r_u = \frac{w_1 \xi_1}{w_2 \xi_2} \frac{1}{1+\sigma_2} \tag{4.17}$$

Then the leakage reactance of the rotor vanishes in the equivalent circuit diagram (see Fig. 4.5).

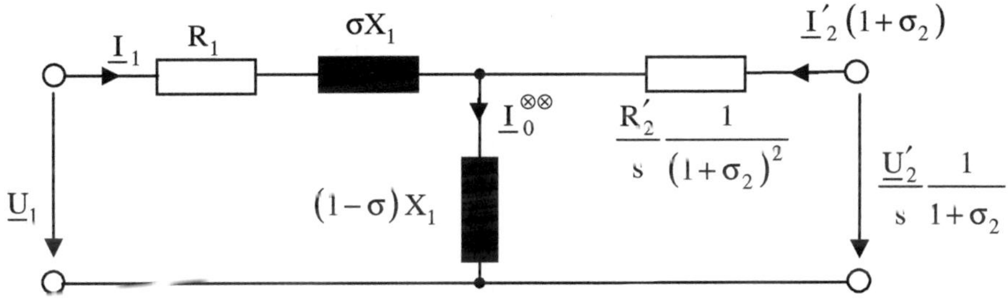

Fig. 4.5. Equivalent circuit diagram of the induction machine without rotor leakage reactance.

4.2 Resistances and Inductivities

4.2.1 Phase Resistance

With the mean length of one turn ℓ_m, the number of turns w of a phase, the cross section of the conductor A_{wire} and the specific resistance ρ_{Cu} the phase resistance becomes:

$$R = \rho_{Cu} \frac{w \ell_m}{A_{wire}} \tag{4.18}$$

The mean length of one turn can be calculated from the stack length of the machine and the length of a turn in the end winding region.

4.2.2 Main Inductivity

Supplying the stator with I_1 and f_1 at no-load operation (i.e. with no currents in the rotor), it follows:

$$L_{1m} = \frac{X_{1m}}{\omega_1} = \frac{U_i}{\omega_1 I_1} \tag{4.19}$$

According to the rotating field theory, the induced voltage is:

$$\sqrt{2}U_i = \frac{3}{2}\mu_0 w_1 \xi_1 \frac{\ell r}{\delta} \frac{w_1 \xi_1}{p^2} \frac{4}{\pi} \omega_1 \sqrt{2} I \tag{4.20}$$

Therefore, it follows in total for the main inductivity:[5]

$$L_{1m} = \frac{3}{2}\mu_0 \left(\frac{w_1 \xi_1}{p} \right)^2 \frac{4}{\pi} \frac{\ell r}{\delta} \tag{4.21}$$

4.2.3 Leakage Inductivity

The leakage is composed of harmonic leakage, leakage of the end windings, and the slot leakage. These three parts are calculated separately and then added up. The separate consideration of single field components is already explained in Sect. 3.2 (for the slot and air-gap regions); even here such a separation of the field regions is performed. The summation of the partial results (i.e. superposition) is allowed because of the assumed linearity.

4.2.3.1 Harmonic Leakage

It has been shown in the rotating field theory that all harmonic waves induce voltages of equal frequencies and phase shift in their generating winding. All these voltages may therefore be added. The sum of these voltages can be interpreted as voltage drop at a leakage reactance, because they do not produce a useful magnetic field.

[5] For the air-gap δ please refer to the remarks in Sect. 3.5.

$$
\begin{aligned}
L_{1,\sigma,\mathrm{harm}} &= \frac{1}{\omega_1}\sum_{g\neq 0}\frac{U_{i,6g+1}}{I_1} \\
&= \frac{3}{2}\mu_0\left(\frac{w_1\xi_1}{p}\right)^2\frac{4}{\pi}\frac{\ell r}{\delta}\sum_{g\neq 0}\left(\frac{\xi_{6g+1}}{\xi_1(6g+1)}\right)^2 = \sigma_{\mathrm{harm}}L_{1m}
\end{aligned}
\tag{4.22}
$$

with:

$$
\sigma_{\mathrm{harm}} = \sum_{g\neq 0}\left(\frac{\xi_{6g+1}}{\xi_1(6g+1)}\right)^2 \tag{4.23}
$$

4.2.3.2 Slot Leakage

A single slot will be regarded in Fig. 4.6.

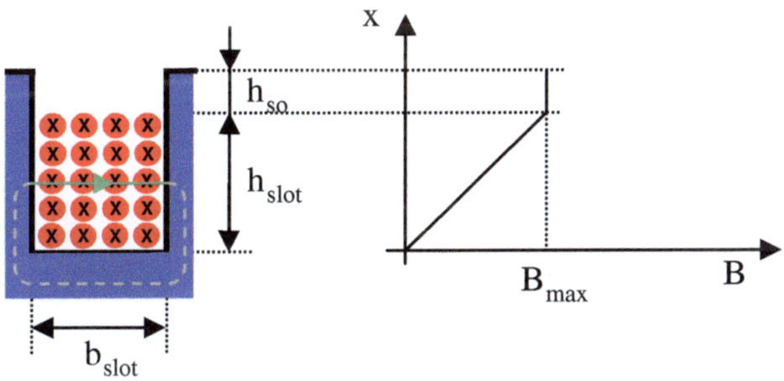

Fig. 4.6. Field distribution in a single slot.

In the slot region filled with conductors Ampere's Law together with the assumption $\mu_{Fe} \to \infty$ gives (z_{slot}: number of conductors in the slot, a: number of parallel paths):

$$
\begin{aligned}
&\Theta(x) = \frac{z_{slot}}{a}\frac{x}{h_{slot}}\sqrt{2}I = \frac{B(x)}{\mu_0}b_{slot} \\
&B(x) = B_{max}\frac{x}{h_{slot}}, \qquad B_{max} = \mu_0\frac{\frac{z_{slot}}{a}\sqrt{2}I}{b_{slot}}
\end{aligned}
\tag{4.24}
$$

In the region of the slot without conductors (slot opening) there is:

$$\Theta(x) = \frac{z_{slot}}{a}\sqrt{2}I = \frac{B(x)}{\mu_0}b_{slot}$$
$$B(x) = B_{max} \tag{4.25}$$

The slot leakage inductivity will be calculated by means of the magnetic energy. For a phase with $2pq$ slots it is true:

$$W_{mag} = \frac{1}{2\mu_0}\int_V B^2 dV = \frac{1}{2}L_{\sigma,slot}\left(\sqrt{2}I\right)^2 \tag{4.26}$$

Therefore:

$$\begin{aligned} W_{mag} &= \frac{2pq}{2\mu_0}\int_0^{h_{slot}}\left(B_{max}\frac{x}{h_{slot}}\right)^2 b_{slot}\ell dx + \frac{2pq}{2\mu_0}\int_{h_{slot}}^{h_{slot}+h_{so}} B_{max}^2 b_{slot} l dx \\ &= 2pq\left(\frac{\mu_0\frac{z_{slot}}{a}\sqrt{2}I}{b_{slot}}\right)^2\frac{b_{slot}\ell}{2\mu_0}\left(\frac{h_{slot}}{3}+h_{so}\right) \\ &= \frac{1}{2}\mu_0 2pq\ell\left(\frac{z_{slot}}{a}\right)^2\left(\frac{h_{slot}}{3b_{slot}}+\frac{h_{so}}{b_{slot}}\right)\left(\sqrt{2}I\right)^2 \end{aligned} \tag{4.27}$$

Consequently:

$$L_{\sigma,slot} = \mu_0 2pq\ell\left(\frac{z_{slot}}{a}\right)^2\left(\frac{h_{slot}}{3b_{slot}}+\frac{h_{so}}{b_{slot}}\right) \tag{4.28}$$

With $\frac{z_{slot}}{a} = \frac{w}{pq}$ it follows further:

$$L_{\sigma,slot} = 2\mu_0\frac{w^2}{p}\ell\frac{\lambda_{slot}}{q}, \qquad \lambda_{slot} = \left(\frac{h_{slot}}{3b_{slot}}+\frac{h_{so}}{b_{slot}}\right) \tag{4.29}$$

Here λ_{slot} is called slot leakage permeance.

4.2.3.3 End Winding Leakage

The end winding leakage is caused by the leakage field of the end winding. Because of the complicated geometry of the end winding a general calculation is not possible. Empirical values lead to the following result:

$$L_{\sigma,endw} = 2\mu_0 \frac{w^2}{p} \ell\lambda_{endw}, \qquad \lambda_{endw} \approx 0.2 \cdots 0.4 \tag{4.30}$$

Here λ_{endw} is called end winding leakage permeance. For simple constructions of the end winding the leakage inductivity of the end winding can be calculated analytically, see the references list.

4.3 Operating Characteristics

4.3.1 Heyland-Diagram (Stator Phase Current Locus Diagram)

For determination of the stator phase current locus the phase resistance will be neglected in a first step: $R_1 = 0$. This approximation, that generally is allowed for machines with high power and supplied by line frequency, will be omitted later.

The equivalent circuit diagram with $r_u = \frac{w_1\xi_1}{w_2\xi_2}(1+\sigma_1)$ is chosen; the rotor winding is short-circuited, either as squirrel cage rotor or as wound rotor via the slip rings and brushes. Then the equivalent circuit diagram is obtained (Fig. 4.7).

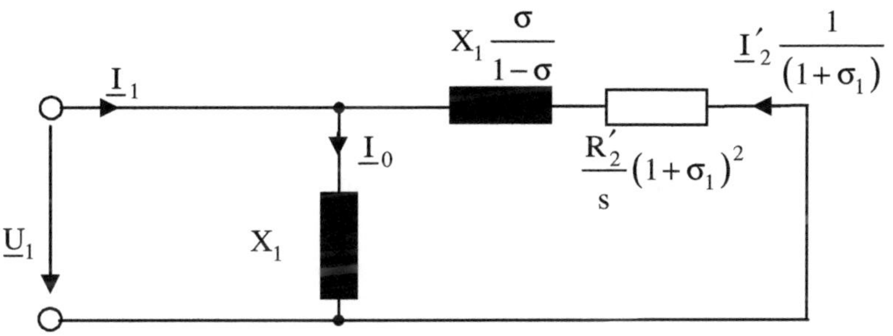

Fig. 4.7. Equivalent circuit diagram of the induction machine.

The following is true:

$$\begin{aligned} &\underline{U}_1 = jX_1\underline{I}_0 \\ &\underline{I}_0 = \underline{I}_1 + \frac{\underline{I}_2'}{1+\sigma_1} \\ &\underline{U}_1 + \frac{\underline{I}_2'}{1+\sigma_1}\left(\frac{R_2'}{s}(1+\sigma_1)^2 + jX_1\frac{\sigma}{1-\sigma}\right) = 0 \end{aligned} \tag{4.31}$$

Defining the voltage $\underline{U}_1$ being real, the phasor diagram is shown in Fig. 4.8:

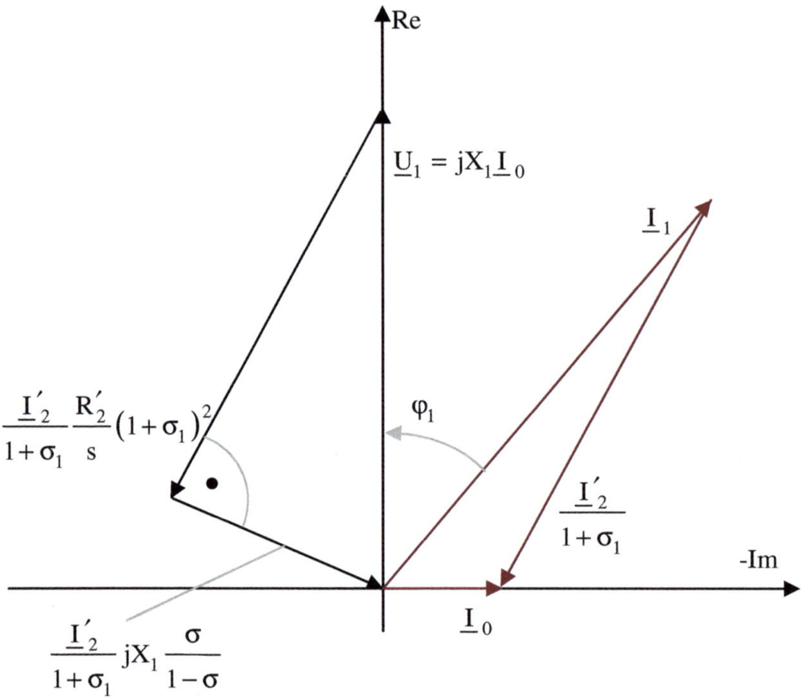

Fig. 4.8. Phasor diagram of the induction motor.

Calculating the stator phase current $\underline{I}_1$ dependent on the slip s, it follows:

$$\underline{I}_1 = \underline{I}_0 - \frac{\underline{I}_2'}{1+\sigma_1} = \frac{\underline{U}_1}{jX_1} + \frac{\underline{U}_1}{\frac{R_2'}{s}(1+\sigma_1)^2 + jX_1\frac{\sigma}{1-\sigma}} \tag{4.32}$$

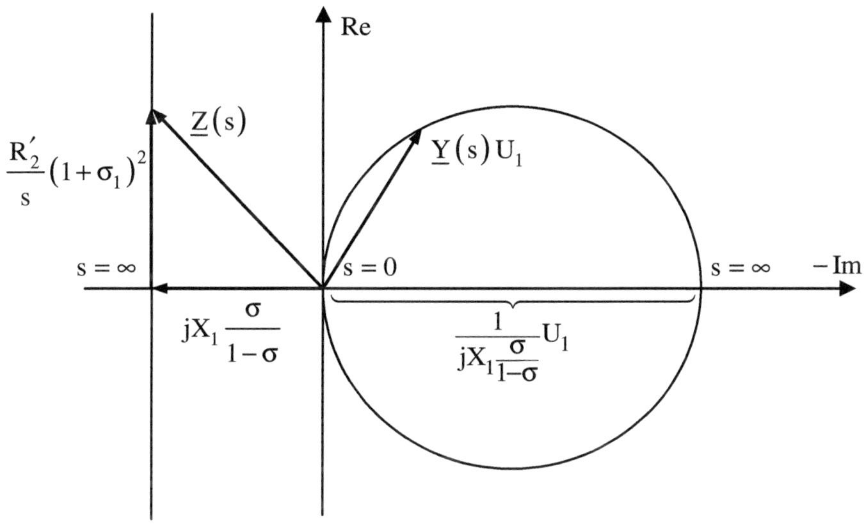

Fig. 4.9. Complex impedance of the induction motor.

The stator phase current locus (see Fig. 4.9) can be deduced like follows:

- The complex resistance $\underline{Z}(s)=\frac{R_2'}{s}(1+\sigma_1)^2+jX_1\frac{\sigma}{1-\sigma}$ is a straight line in the complex plane.
- According to the locus theory, the complex conductance $\underline{Y}(s)=1/\underline{Z}(s)$ is a circle, that
  - reaches the origin of the complex plane for $s=0$,
  - has its midpoint on the negative imaginary axis, and
  - possesses the diameter $\left(X_1\frac{\sigma}{1-\sigma}\right)^{-1}$.
- By multiplication with $\underline{U}_1\equiv U_1$ (the voltage has been defined being real) the locus of $-\frac{\underline{I}_2'}{1+\sigma_1}=U_1\underline{Y}(s)$ is deduced.
- The locus of the stator phase current $\underline{I}_1$ is calculated by adding the current $\underline{I}_0$: $\underline{I}_1=\underline{I}_0-\frac{\underline{I}_2'}{1+\sigma_1}=-j\frac{U_1}{X_1}+U_1\underline{Y}(s)$. Therefore, the locus of $U_1\underline{Y}(s)$ is shifted by the absolute value U_1/X_1 in the direction of the negative imaginary axis.

The no-load current ($s=0$) is

$$\underline{I}_0 = -j\frac{\underline{U}_1}{X_1} \tag{4.33}$$

The rotor current (transformed to the stator system) is

$$-\frac{\underline{I}_2'}{1+\sigma_1} = \frac{\underline{U}_1}{\frac{R_2'}{s}(1+\sigma_1)^2 + jX_1\frac{\sigma}{1-\sigma}} \tag{4.34}$$

The stator current and the "diameter current" are, respectively

$$\underline{I}_1 = \underline{I}_0 - \frac{\underline{I}_2'}{1+\sigma_1} \tag{4.35}$$

$$I_\varnothing = \frac{U_1}{X_1\frac{\sigma}{1-\sigma}} \tag{4.36}$$

The locus of the stator phase current $\underline{I}_1$ is a circle in the complex plane (this is also called "Heyland-diagram", see Fig. 4.10).

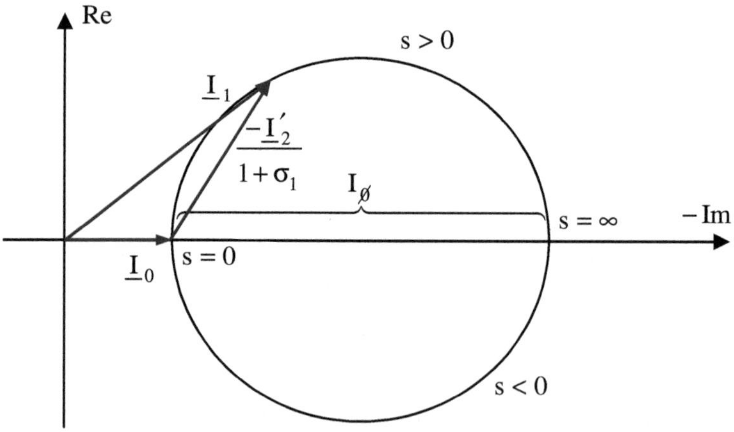

Fig. 4.10. Heyland-diagram of the induction motor: complex currents.

Location and size of this circle are determined only by the main reactance X_1 and the total leakage coefficient σ (if the stator resistance R_1 is neglected).

For parameterization of this circle, the ratio of imaginary part to real part of the rotor current (transformed to the stator system) $-\frac{\underline{I}_2'}{1+\sigma_1}$ is regarded (for explanation please refer to Fig. 4.11):

$$\tan\left(\varphi^*\right) = \frac{\mathrm{Im}\{\underline{I}_2'\}}{\mathrm{Re}\{\underline{I}_2'\}} = \frac{-X_1 \frac{\sigma}{1-\sigma}}{R_2'\left(1+\sigma_1\right)^2} s \tag{4.37}$$

As this function is linear in s, it can be used (most simply) for parameterization. The "slip line" for parameterization can be each arbitrarily chosen parallel line to the $-\mathrm{Im}$-axis:

- The tangent to the circle (stator phase current locus) in the point $\underline{I}_2' = 0$ cuts the slip line at the point $s = 0$.
- Each extension of the current $\frac{\underline{I}_2'}{1+\sigma_1}$ cuts the slip line at the respective slip s, because $\tan\left(\varphi^*\right) \sim s$ is true. To fix the parameterization, the slip of one circle point must be known (the slip of all other circle points then follows from the linear division of the slip line).

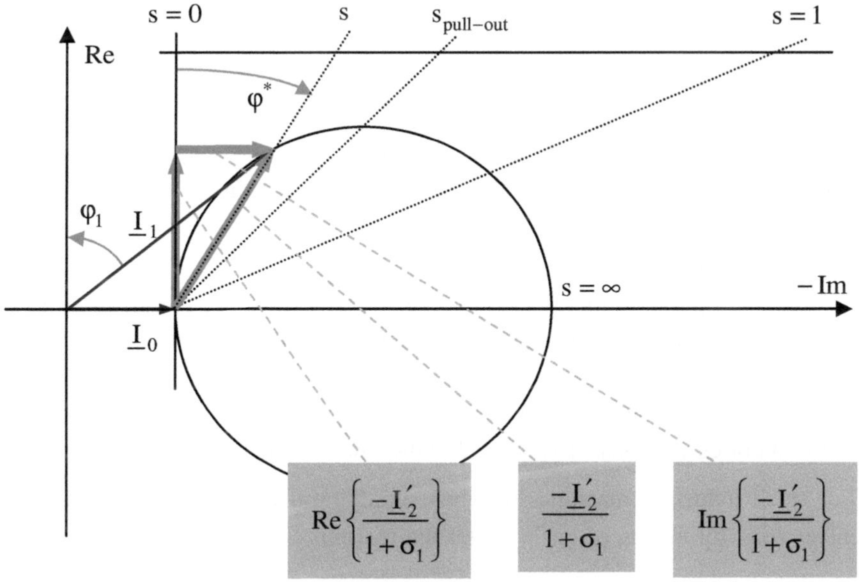

Fig. 4.11. Heyland-diagram of the induction motor: slip line.

The zenith of the Heyland-diagram is called *pull-out point*; the respective slip value is named *pull-out slip*. At this point the induction machine generates its maximum torque (*pull-out torque*). In this point real part and imaginary part of $\dfrac{\underline{I}_2'}{1+\sigma_1}$ are identical, i.e. $\tan\left(\varphi^*\right)=1$ is true. Then it follows:

$$s_{pull-out} = \frac{R_2'\left(1+\sigma_1\right)^2}{X_1 \dfrac{\sigma}{1-\sigma}} \tag{4.38}$$

Further it follows generally:

$$\tan\left(\varphi^*\right) = \frac{s}{s_{pull-out}} \tag{4.39}$$

At stand-still $s = 1$ is true; the stall current (also called short-circuit current) is then $I_{1,stall} \approx 5 \cdots 8 I_{1,N}$. Here, $I_{1,N}$ is the stator phase current in the nominal operation point.

As proven previously, *location and size of the Heyland-diagram* are determined only by X_1 and σ (if R_1 is neglected). From (Eq. 4.38) it can be deduced that the *location of the slip points* depends on R_2'.

The following three operating areas can be distinguished:

- Motor operation:
  For motor operation there is: $0 < s < 1$ (i.e. $n_0 > n > 0$), therefore the arrow-head of the stator current phasor is located in the left part of the upper circle half. The nominal operation point of an induction machine is in the region of: $0 < s_N < s_{pull-out}$.
- Braking operation:
  For braking operation there is: $s > 1$ (i.e. $n < 0$), therefore the arrow-head of the stator current phasor is located in the very right part of the upper circle half.
- Generator operation:
  For generator operation there is: $s < 0$ (i.e. $n > n_0$), therefore the arrow-head of the stator current phasor is located in the lower circle half.

For small induction machines or when supplying the machine with small frequencies (e.g. by means of an inverter) the resistance R_1 generally may not be neglected. The Heyland-diagram then is changed concerning location, diameter, and parameterization. These changes are shown schematically in Fig. 4.12.

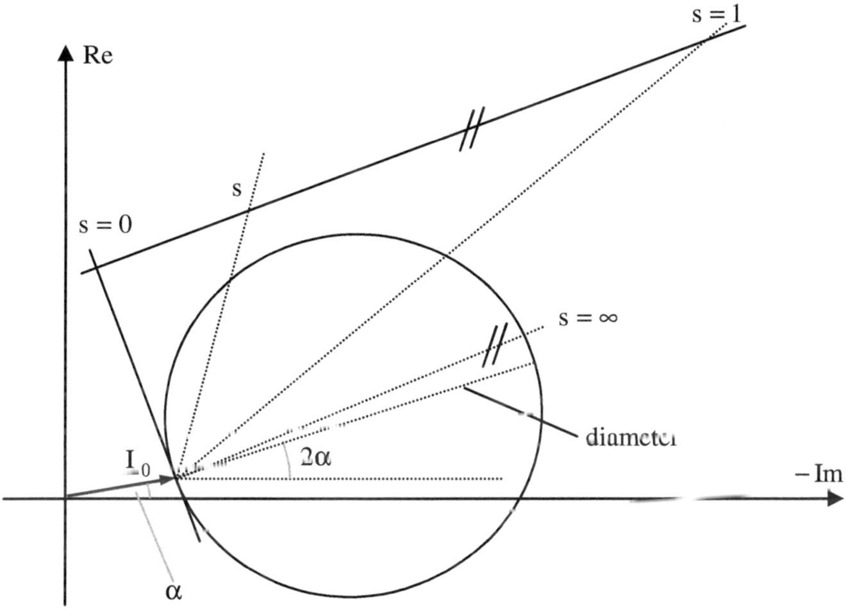

Fig. 4.12. Heyland-diagram of the induction motor: influence of the stator resistance.

For the angle α there is:

$$\tan(\alpha) = \frac{R_1}{X_1} \tag{4.40}$$

4.3.2 Torque and Power

Neglecting the stator resistance R_1 the active power of the induction machine supplied via the terminals is equal to the rotating air-gap field power:

$$P_1 = 3U_1I_1\cos(\varphi_1) = P_\delta = 2\pi n_0 T \tag{4.41}$$

With constant terminal voltage U_1 the rotating air-gap field power P_δ and the torque T are proportional to the real part of the stator current $I_{1,real} = I_1\cos(\varphi_1)$.

Neglecting the stator resistance R_1, the real part of the stator current $I_{1,real}$ of an operating point can be deduced in the Heyland-diagram as projection of $\underline{I}_1$ or $\frac{\underline{I}_2'}{1+\sigma_1}$ onto the real axis.

By means of the notation in Fig. 4.13 there is:

$$I_{1,real} = C\ \overline{AP} \tag{4.42}$$

where C is a constant factor. Then, for the torque it follows:

$$T = \frac{1}{2\pi n_0}\ 3\,U_1\ C\ \overline{AP} \tag{4.43}$$

The joining line between the points $P_{s=0}$ and $P_{s=\infty}$ ($\overline{P_{s=0}P_{s=\infty}}$) is called "torque-line".

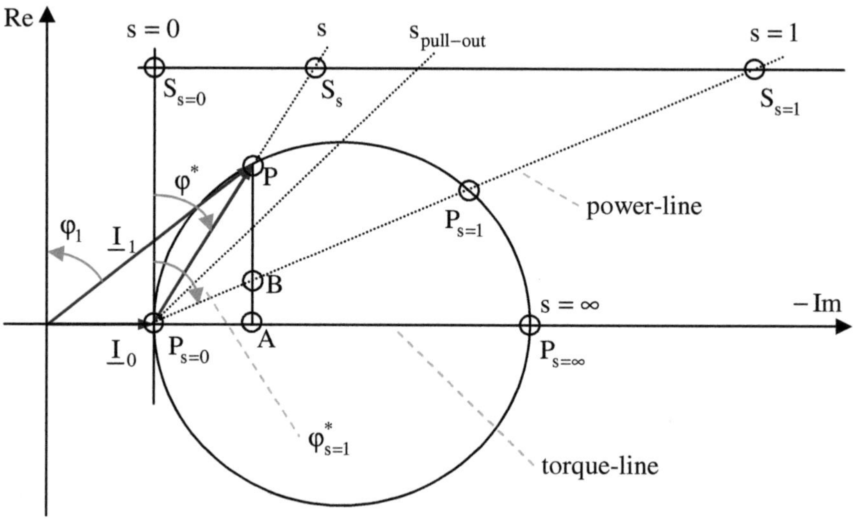

Fig. 4.13. Heyland-diagram of the induction motor: torque line and power line.

For the angles φ^* and $\varphi^*_{s=1}$ the following is true:

$$\frac{\tan(\varphi^*)}{\tan(\varphi^*_{s=1})} = \frac{\overline{S_{s=0}S_s} / \overline{P_{s=0}S_{s=0}}}{\overline{S_{s=0}S_{s=1}} / \overline{P_{s=0}S_{s=0}}} = \frac{\overline{S_{s=0}S_s}}{\overline{S_{s=0}S_{s=1}}} = \frac{s}{1} = s$$
$$\frac{\tan(\varphi^*)}{\tan(\varphi^*_{s=1})} = \frac{\overline{P_{s=0}A} / \overline{AP}}{\overline{P_{s=0}A} / \overline{AB}} = \frac{\overline{AB}}{\overline{AP}} \tag{4.44}$$

Therefore:

$$\frac{\overline{AB}}{\overline{AP}} = s \tag{4.45}$$

and consequently:

$$\frac{\overline{BP}}{\overline{AP}} = 1 - s \tag{4.46}$$

Therefore, the distance $\overline{AP}$ is divided by the so-called "power-line" $\overline{P_{s=0}P_{s=1}}$ by the ratio s to $(1-s)$. With this it can be deduced directly from the Heyland-diagram how the rotating field power (this corresponds to the input power if the stator resistance R_1 is neglected) is divided into the different parts:

- rotating air-gap field power:

$$P_\delta = 3\ U_1\ C\ \overline{AP} \tag{4.47}$$

- electrical power (rotor losses):

$$P_{el} = 3\ U_1\ C\ \overline{AB} \tag{4.48}$$

- mechanical power:

$$P_{mech} = 3\ U_1\ C\ \overline{BP} \tag{4.49}$$

4.3.3 Torque as a Function of Slip

The torque-slip-characteristic shall be deduced in the following by means of the Heyland-diagram (Fig. 4.14). The following relations are already known:

$$\tan\left(\varphi^*\right) = \frac{s}{s_{pull-out}} \qquad \text{see Eq. (4.39)}$$

$$s_{pull-out} = \frac{R_2'\left(1+\sigma_1\right)^2}{X_1 \dfrac{\sigma}{1-\sigma}} \qquad \text{see Eq. (4.38)}$$

$$I_\varnothing = \frac{U_1}{X_1 \dfrac{\sigma}{1-\sigma}} \qquad \text{see Eq. (4.36)}$$

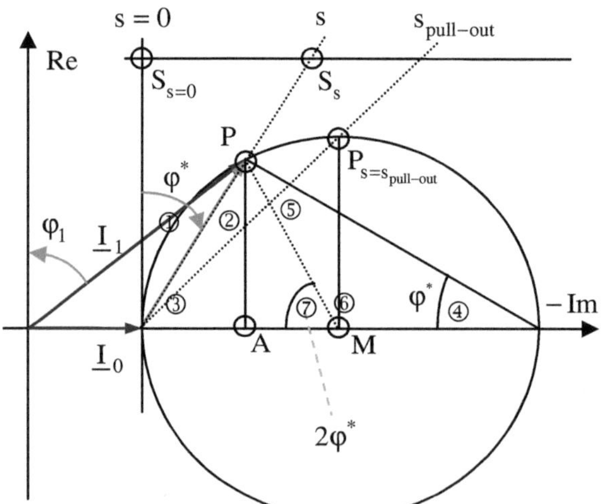

angle relations:

① definition: φ^*

② geometry: φ^*

③ triangle:

$$\pi - \frac{\pi}{2} - \varphi^* = \frac{\pi}{2} - \varphi^*$$

④ triangle:

$$\pi - \frac{\pi}{2} - \left(\frac{\pi}{2} - \varphi^*\right) = \varphi^*$$

⑤ equal-sided triangle: φ^*

⑥ triangle: $\pi - 2\varphi^*$

⑦ straight line:

$$\pi - \left(\pi - 2\varphi^*\right) = 2\varphi^*$$

Fig. 4.14. Heyland-diagram of the induction motor: relation of different angles.

The torque in the operation point is:

$$T = \frac{3p}{\omega_1} U_1 \; C \; \overline{AP} \tag{4.50}$$

The torque in the pull-out point is:

$$T_{pull-out} = \frac{3p}{\omega_1} U_1 \; C \; \overline{MP_{s=s_{pull-out}}} \tag{4.51}$$

Therefore:

$$\begin{aligned}\frac{T}{T_{pull-out}} &= \frac{\overline{AP}}{\overline{MP_{s=s_{pull-out}}}} = \frac{\sin\left(2\varphi^*\right)\frac{1}{2}I_{\emptyset}}{\frac{1}{2}I_{\emptyset}} = \sin\left(2\varphi^*\right) \\ &= 2\sin\left(\varphi^*\right)\cos\left(\varphi^*\right) = 2\frac{\sin\left(\varphi^*\right)}{\cos\left(\varphi^*\right)}\cos^2\left(\varphi^*\right) \\ &= 2\tan\left(\varphi^*\right)\frac{1}{1+\tan^2\left(\varphi^*\right)} \\ &= \frac{2}{\tan\left(\varphi^*\right)+\frac{1}{\tan\left(\varphi^*\right)}}\end{aligned} \tag{4.52}$$

From this, directly *Kloss's Law* is obtained:

$$\frac{T}{T_{pull-out}} = \frac{2}{\frac{s}{s_{pull-out}} + \frac{s_{pull-out}}{s}} \tag{4.53}$$

The pull-out torque can be calculated like follows:

$$\begin{aligned} T_{pull-out} &= \frac{3p}{\omega_1} U_1 \; I_{1,real,pull-out} = \frac{3p}{\omega_1} U_1 \frac{I_{\varnothing}}{2} \\ &= \frac{3p}{\omega_1} U_1 \frac{U_1}{2X_1 \dfrac{\sigma}{1-\sigma}} = \frac{3p}{2\omega_1} \frac{U_1^2}{X_1 \dfrac{\sigma}{1-\sigma}} \end{aligned} \tag{4.54}$$

Consequently, the pull-out torque is determined by the total leakage coefficient (high torque for small leakage).

As approximation the graph of the torque-slip-characteristic can be separated into the following two regions:

$$s \ll s_{pull-out} \quad \Rightarrow \quad \frac{T}{T_{pull-out}} = \frac{2s}{s_{pull-out}} \tag{4.55}$$

$$s \gg s_{pull-out} \quad \Rightarrow \quad \frac{T}{T_{pull-out}} = \frac{2s_{pull-out}}{s} \tag{4.56}$$

Figure 4.15 illustrates the evaluation of Kloss's Law and this approximation:

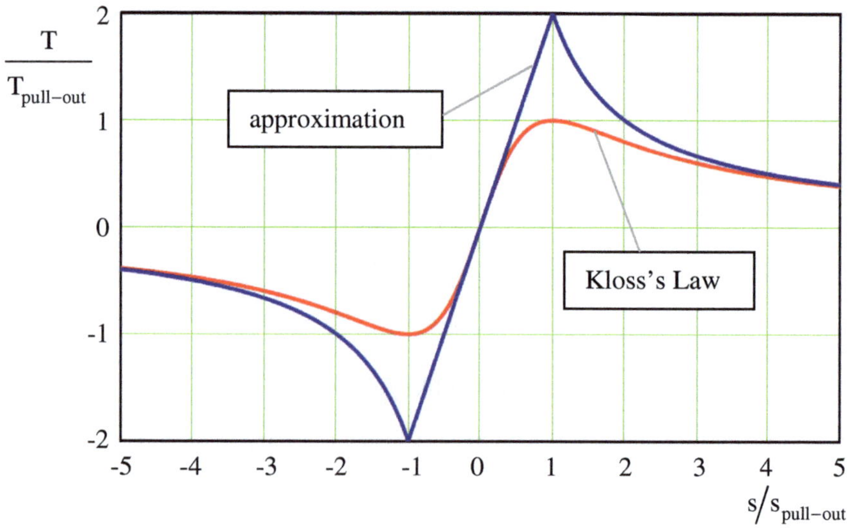

Fig. 4.15. Torque-slip-characteristic of the induction motor.

Kloss's Law and the pull-out torque are independent of the rotor resistance. Just the absolute value of the pull-out slip $s_{pull-out}$ is determined by the rotor resistance. The stable operating area of the induction motor is: $-1 < s/s_{pull-out} < 1$.

In Fig. 4.16 $T/T_{pull-out}$ is shown as a function of n/n_0 for $s_{pull-out} = 0.1$.

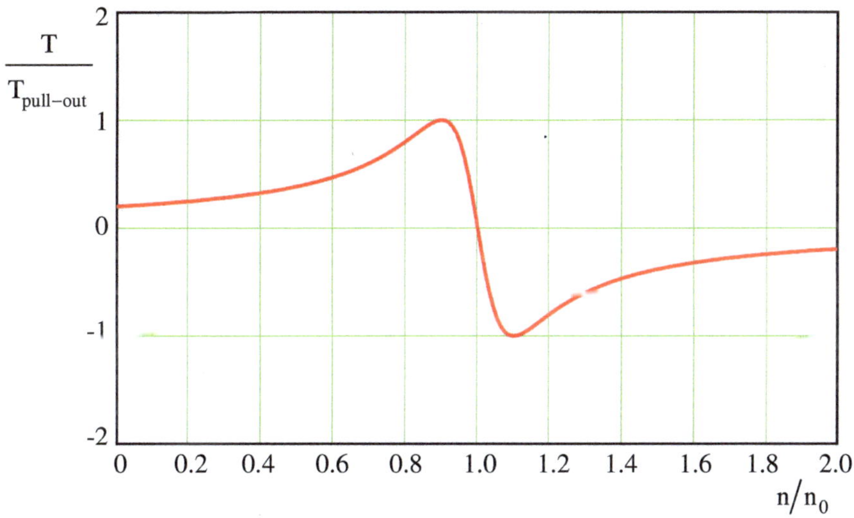

Fig. 4.16. Torque-speed-characteristic of the induction motor.

4.3.4 Series Resistance in the Rotor Circuit

Having machines equipped with slip rings on the rotor, series resistances can be added to the rotor circuit. Doing this the Heyland-diagram remains unchanged (it is independent from the rotor resistance, see above). But the parameterization is changed.

Transforming even the series resistance R_S to the stator system it follows:

$$R'_S = R_S \frac{m_1 (w_1 \xi_1)^2}{m_2 (w_2 \xi_2)^2} \tag{4.57}$$

The identical point on the circle is described, if for the new parameterization with s^* the following holds true:

$$\frac{R_2' + R_S'}{s^*} = \frac{R_2'}{s} \quad \Rightarrow \quad s^* = s\left(1 + \frac{R_S'}{R_2'}\right) \tag{4.58}$$

Analogously the pull-out slip is changed (see Fig. 4.17):

$$s^*_{pull-out} = s_{pull-out}\left(1 + \frac{R_S'}{R_2'}\right) \tag{4.59}$$

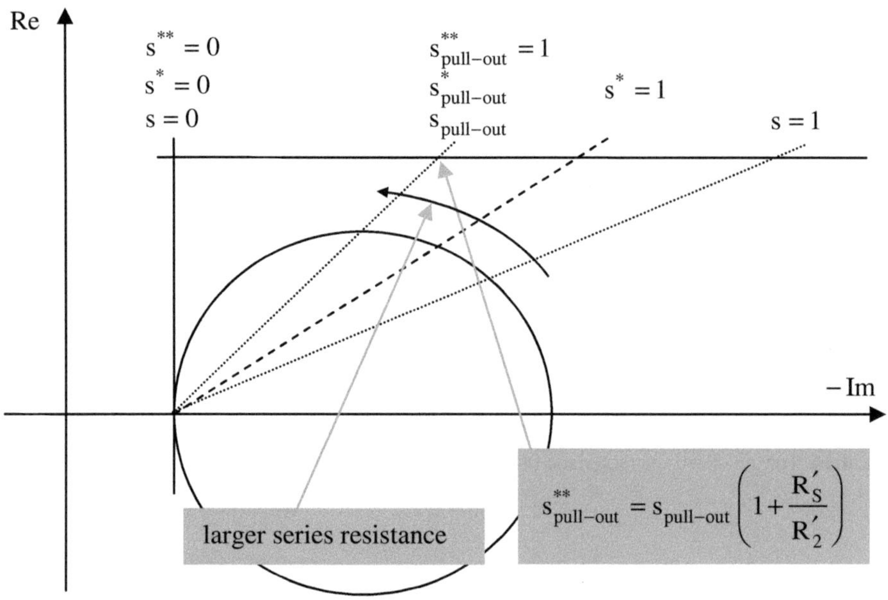

Fig. 4.17. Heyland-diagram of the induction motor: series resistance in the rotor circuit.

When introducing series resistances to the rotor circuit the absolute value of the pull-out torque remains unchanged; but the starting point is shifted into the direction of the pull-out torque. It is even possible to start-up the machine with the pull-out torque ($s^{**}_{pull-out} = 1$). Simultaneously, the phase current at stand-still is reduced. These advantages are opposed by higher rotor losses.

Moreover, with the series resistances an open-loop speed control under load is possible, but this generates losses in the series resistances and therefore it is used only occasionally.

In Fig. 4.18 $T/T_{pull-out}$ is shown as a function of n/n_0 for $s_{pull-out} = 0.1;\ 0.2$ and 1.0 (torque-speed-characteristics):

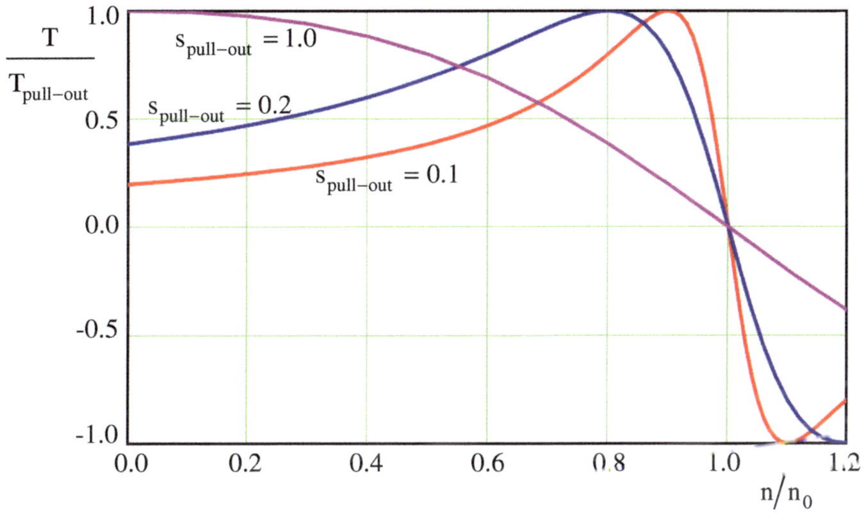

Fig. 4.18. Torque-speed-characteristics of the induction motor for different values of the pull-out slip.

4.3.5 Operation with Optimum Power Factor

During no-load of the induction machine the stator phase current is a pure reactive current $\underline{I}_0$. This corresponds to the magnetizing reactive (wattless) power required for the operation of the induction machine. This reactive power is supplied by the mains. There is:

$$\underline{I}_0 = \frac{U_1}{jX_1} = -j\frac{U_1}{\omega_1\left(1+\sigma_1\right)\frac{3}{2}\mu_0\left(\frac{w_1\xi_1}{p}\right)^2\frac{4}{\pi}\frac{\ell r}{\delta}} \sim \delta \tag{4.60}$$

To reduce the no-load current as far as possible, the air-gap should be as small as possible. But technical limits (bending of the axis, bearing tolerances) as well as economic limits (precision of the production) have to be considered. Usually it is reached

$$\frac{I_0}{I_N} \approx 0.2\left(\text{low pole number}\right)\cdots 0.5\left(\text{high pole number}\right) \tag{4.61}$$

The maximum reachable torque (pull-out torque) of the induction machine is determined by the diameter of the Heyland-diagram. Therefore, the current $I_\infty = I_0 + I_\varnothing$ should be as large as possible. The following is true:

$$\frac{I_\infty}{I_N} = \frac{I_0 + I_\varnothing}{I_N} = \frac{\dfrac{U_1}{X_1} + \dfrac{U_1}{X_1 \dfrac{\sigma}{1-\sigma}}}{I_N} = \frac{\dfrac{U_1}{\sigma X_1}}{I_N} = \frac{1}{\sigma}\frac{I_0}{I_N} \tag{4.62}$$

Generally reachable values for the total leakage coefficient are: $\sigma \approx 0.03 \cdots 0.1$. From this the ratio of the currents becomes:

$$\frac{I_\infty}{I_N} = 5 \cdots 7 \tag{4.63}$$

The nominal operating point of the induction machine often is chosen to maximize the power factor $\cos(\varphi)$ (then the load to the supplying mains because of the reactive current is smallest). The operating point with maximum $\cos(\varphi)$ is called "optimum point". It is obtained by a tangent of the stator current onto the Heyland-diagram (Fig. 4.19):

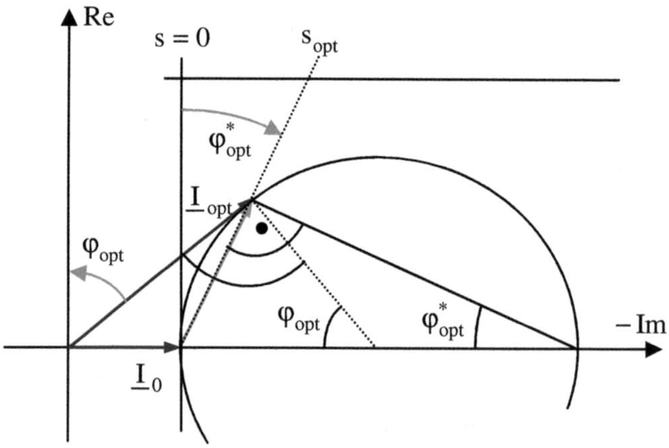

Fig. 4.19. Heyland-diagram of the induction motor: operation with optimum power factor.

Having the nominal operation point identical with the optimum point, the following is true:

$$\varphi_{opt} = 2\varphi^{*}_{opt} \tag{4.64}$$

$$\begin{aligned}\cos\left(\varphi_{opt}\right) &= \frac{\frac{1}{2}I_{\varnothing}}{\frac{1}{2}I_{\varnothing}+I_0} = \frac{\frac{1}{2}\left(I_{\infty}-I_0\right)}{\frac{1}{2}\left(I_{\infty}+I_0\right)} \\ &= \frac{\left(I_{\infty}-I_0\right)/I_N}{\left(I_{\infty}+I_0\right)/I_N} = \frac{\frac{1}{\sigma}\frac{I_0}{I_N}-\frac{I_0}{I_N}}{\frac{1}{\sigma}\frac{I_0}{I_N}+\frac{I_0}{I_N}} = \frac{1-\sigma}{1+\sigma}\end{aligned} \tag{4.65}$$

Usually reachable in practice is: $\cos\left(\varphi_{opt}\right) \approx 0.8\cdots0.9$.

$$\begin{aligned}I_{opt} &= \left(I_0+\frac{1}{2}I_{\varnothing}\right)\sin\left(\varphi_{opt}\right) = \frac{1}{2}\left(I_0+I_{\infty}\right)\sqrt{1-\cos^2\left(\varphi_{opt}\right)} \\ &= \frac{1}{2}\left(I_0+\frac{1}{\sigma}I_0\right)\sqrt{1-\left(\frac{1-\sigma}{1+\sigma}\right)^2} = \frac{I_0}{2}\frac{\sigma+1}{\sigma}\sqrt{\left(\frac{1+\sigma}{1+\sigma}\right)^2-\left(\frac{1-\sigma}{1+\sigma}\right)^2} \\ &= \frac{I_0}{2\sigma}\sqrt{\left(1+\sigma\right)^2-\left(1-\sigma\right)^2} = \frac{I_0}{2\sigma}\sqrt{4\sigma} \\ &= \frac{I_0}{\sqrt{\sigma}}\end{aligned} \tag{4.66}$$

For the slip in the optimum point the following is true:

$$\tan\left(\varphi_{opt}^{*}\right)=\frac{s_{opt}}{s_{pull-out}}$$

$$\begin{aligned}\Rightarrow \quad s_{opt} &= s_{pull-out}\tan\left(\frac{\varphi_{opt}}{2}\right)=s_{pull-out}\sqrt{\frac{1-\cos\left(\varphi_{opt}\right)}{1+\cos\left(\varphi_{opt}\right)}}\\ &= s_{pull-out}\sqrt{\frac{1-\dfrac{1-\sigma}{1+\sigma}}{1+\dfrac{1-\sigma}{1+\sigma}}}=s_{pull-out}\sqrt{\frac{(1+\sigma)-(1-\sigma)}{(1+\sigma)+(1-\sigma)}}\\ &= s_{pull-out}\sqrt{\sigma}\end{aligned} \tag{4.67}$$

In practice the pull-out slip is about $s_{pull-out} \approx 0.05\cdots0.2$ and the nominal slip is about $s_N \approx 0.01\cdots0.05$.

The torque ratio in the pull-out operating point and the optimum point is:

$$\frac{T_{pull-out}}{T_{opt}}=\frac{\dfrac{s_{pull-out}}{s_{opt}}+\dfrac{s_{opt}}{s_{pull-out}}}{2}=\frac{\dfrac{1}{\sqrt{\sigma}}+\sqrt{\sigma}}{2}=\frac{1+\sigma}{2\sqrt{\sigma}} \tag{4.68}$$

With the given values for the total leakage coefficient it follows for the overload capability: $T_{pull-out}/T_{opt} \approx 1.7\cdots3.0$.

Up to now the stator copper losses, the iron losses, and the friction losses were neglected; only the rotor copper losses were considered. Then the efficiency is:

$$\eta=\frac{P_{mech}}{P_1}=\frac{P_{mech}}{P_{rot}}=1-s \tag{4.69}$$

For the optimum point it follows:

$$\eta_{opt}=1-s_{opt}=1-s_{pull-out}\sqrt{\sigma} \tag{4.70}$$

For $s=0$ or $s\rightarrow\infty$ the equivalent circuit diagrams (see Fig. 4.20) can be deduced:

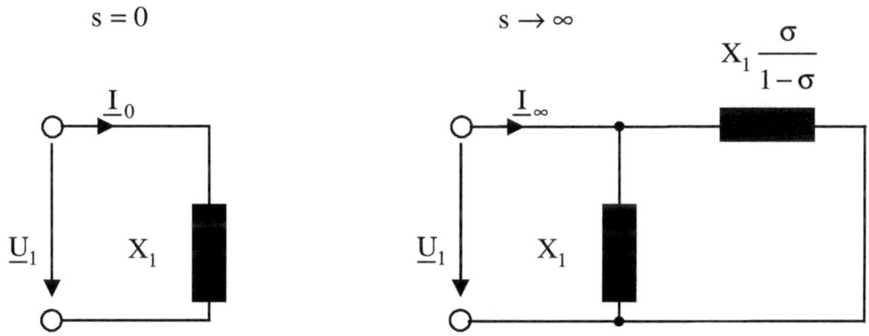

Fig. 4.20. Equivalent circuit diagrams of the induction machine.

Directly from Fig. 4.20 the stator currents for $s = 0$ and $s \to \infty$ can be calculated:

$$I_0 - \frac{U_1}{X_1} \tag{4.71}$$

$$\begin{aligned} I_\infty &= U_1 \left(\frac{X_1 \, X_1 \dfrac{\sigma}{1-\sigma}}{X_1 + X_1 \dfrac{\sigma}{1-\sigma}} \right)^{-1} = U_1 \left(\frac{X_1}{\dfrac{1-\sigma}{\sigma} + 1} \right)^{-1} \\ &= U_1 \left(\sigma X_1 \right)^{-1} = \frac{U_1}{\sigma X_1} \end{aligned} \tag{4.72}$$

Further there is:

$$\begin{aligned} \sigma X_1 &= \left(1 - \frac{1}{(1+\sigma_1)(1+\sigma_2)} \right) (1+\sigma_1) X_{1m} \\ &= \left((1+\sigma_1) - \frac{1}{(1+\sigma_2)} \right) X_{1m} \\ &\approx ((1+\sigma_1) - (1-\sigma_2)) X_{1m} = (\sigma_1 + \sigma_2) X_{1m} = X_{1\sigma} + X'_{2\sigma} \end{aligned} \tag{4.73}$$

and therefore:

$$I_\infty \approx \frac{U_1}{X_{1\sigma} + X'_{2\sigma}} \tag{4.74}$$

Remark: For the total leakage coefficient the following holds true:

$$\begin{aligned}\sigma &= 1 - \frac{1}{(1+\sigma_1)(1+\sigma_2)} = \frac{\sigma_1 + \sigma_2 + \sigma_1\sigma_2}{1+\sigma_1+\sigma_2+\sigma_1\sigma_2} \\ &\approx \frac{\sigma_1+\sigma_2}{1+\sigma_1+\sigma_2} \approx \sigma_1 + \sigma_2\end{aligned} \tag{4.75}$$

Changes of the machine parameters have the following influences onto the Heyland-diagram and therefore onto the operational characteristics:

- Increasing the air-gap the main reactance X_1 is decreased. This means that the magnetizing current I_0 is enlarged.
- By decreasing the leakage reactance (e.g. decreasing $X'_{2\sigma}$ by utilizing the skin-effect) the current I_∞ increases. By this the pull-out torque and the stand-still current are getting larger.
- Increasing the rotor resistance R'_2 (e.g. by utilizing the skin-effect) changes the parameterization of the slip line. It is possible to start-up the motor with higher torque and lower phase current, but this is opposed by higher rotor losses.

4.3.6 Further Equations for Calculating the Torque

In the following the equivalent circuit diagram with stator and rotor leakage reactance is used (Fig. 4.21).

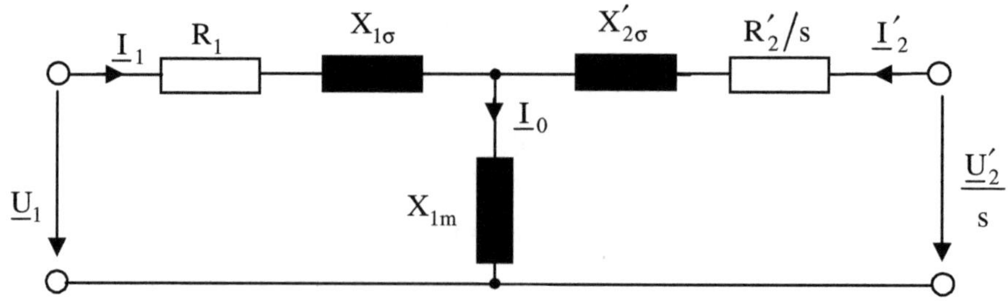

Fig. 4.21. Equivalent circuit diagram of the induction machine.

The appropriate phasor diagram is shown in Fig. 4.22 (if $R_1 = 0$ is true).

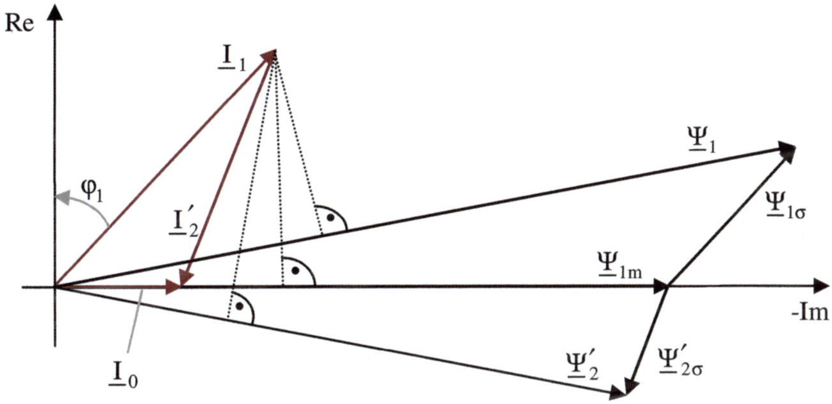

Fig. 4.22. Phasor diagram of the induction machine.

From the rotating field theory it follows:

$$\begin{aligned} T &= 3\ell r w_1 \xi_1 \sqrt{2}\, I_1 B_\delta \cos(\varphi_1) \\ &= \frac{3}{2} p \Psi_{1m} \sqrt{2}\, I_1 \cos(\varphi_1) = \frac{3}{2} p \Psi_{1m} \sqrt{2}\, I_1 \sin\left(\frac{\pi}{2} - \varphi_1\right) \end{aligned} \tag{4.76}$$

With $\underline{\hat{I}}_1 = \sqrt{2}\, I_1\, e^{-j\varphi_1}$ and $\underline{\Psi}_{1m} = \Psi_{1m}\, e^{-j\frac{\pi}{2}}$ it follows:

$$\underline{\Psi}_{1m} \times \underline{\hat{I}}_1 = \Psi_{1m} \sqrt{2}\, I_1 \sin\left(\frac{\pi}{2} - \varphi_1\right) \tag{4.77}$$

and consequently:

$T = \frac{3}{2} p \left(\underline{\Psi}_{1m} \times \underline{\hat{I}}_1\right)$ *main flux and stator current*

- With $\underline{\Psi}_1 = L_{1\sigma} \underline{\hat{I}}_1 + \underline{\Psi}_{1m}$ it follows: $\underline{\Psi}_{1m} \times \underline{\hat{I}}_1 = \underline{\Psi}_1 \times \underline{\hat{I}}_1$ and therefore:

$T = \frac{3}{2} p \left(\underline{\Psi}_1 \times \underline{\hat{I}}_1\right)$ *stator flux and stator current*

- With $\underline{\Psi}_{1m} = L_{1m} \left(\underline{\hat{I}}_1 + \underline{\hat{I}}'_2\right)$ it follows: $\underline{\Psi}_{1m} \times \underline{\hat{I}}_1 = L_{1m} \underline{\hat{I}}'_2 \times \underline{\hat{I}}_1$ and therefore:

$T = \frac{3}{2} p L_{1m} \left(\underline{\hat{I}}'_2 \times \underline{\hat{I}}_1\right)$ *rotor current and stator current*

- With $\underline{\Psi}_2' = L_2'\hat{\underline{I}}_2' + L_{1m}\hat{\underline{I}}_1$ it follows: $\underline{\Psi}_2' \times \hat{\underline{I}}_1 = L_2'\hat{\underline{I}}_2' \times \hat{\underline{I}}_1$ and therefore:

$$T = \frac{3}{2} p \frac{L_{1m}}{L_2'} \left(\underline{\Psi}_2' \times \hat{\underline{I}}_1 \right) \qquad \textit{rotor flux and stator current}$$

- With $\underline{\Psi}_1 = L_1\hat{\underline{I}}_1 + L_{1m}\hat{\underline{I}}_2'$ it follows: $\underline{\Psi}_1 \times \hat{\underline{I}}_2' = L_1\hat{\underline{I}}_1 \times \hat{\underline{I}}_2'$ and therefore:

$$T = -\frac{3}{2} p \frac{L_{1m}}{L_1} \left(\underline{\Psi}_1 \times \hat{\underline{I}}_2' \right) \qquad \textit{stator flux and rotor current}$$

- With $L_{1m}\left(\hat{\underline{I}}_1 \times \hat{\underline{I}}_2' \right) = L_{1m}\left(\left(\hat{\underline{I}}_1 + \hat{\underline{I}}_2' \right) \times \hat{\underline{I}}_2' \right) = \underline{\Psi}_{1m} \times \hat{\underline{I}}_2'$ it follows:

$$T = -\frac{3}{2} p \left(\underline{\Psi}_{1m} \times \hat{\underline{I}}_2' \right) \qquad \textit{main flux and rotor current}$$

- With $\underline{\Psi}_{1m} = \underline{\Psi}_2' - L_{2\sigma}'\hat{\underline{I}}_2'$ it follows: $\underline{\Psi}_{1m} \times \hat{\underline{I}}_2' = \underline{\Psi}_2' \times \hat{\underline{I}}_2'$ and therefore:

$$T = -\frac{3}{2} p \left(\underline{\Psi}_2' \times \hat{\underline{I}}_2' \right) \qquad \textit{rotor flux and rotor current}$$

- With $\underline{\Psi}_2' = L_2'\hat{\underline{I}}_2'$ it follows:

$$T = -\frac{3}{2} p \frac{L_{1m}}{L_1 L_2'} \left(\underline{\Psi}_1 \times \underline{\Psi}_2' \right) \qquad \textit{stator flux and rotor flux}$$

With $A \times B = -B \times A$ further equations can easily be deduced.

The above equations are valid for supplying the machine with sinusoidal voltages and currents (the capital letters symbolize rms-values or peak-values, depending if electrical or magnetic variables are indicated). The torque calculation for arbitrary time dependencies of the supplying voltages and currents is performed in Sect. 8.9. With arbitrary time dependencies even dynamic operation conditions can be calculated, generally the torque is then no longer constant but time-dependent (in contrast to the steady-state behavior regarded in this chapter).

4.4 Squirrel Cage Rotor

4.4.1 Fundamentals

The squirrel cage rotor (cage rotor) of an induction machine in its simplest form is composed of stacked iron laminations; in the slots of this stack bars (made from copper or aluminum) are inserted and these bars are connected with rings (made from copper or aluminum) in both axial end regions. For machines with small power the slots are filled by aluminum die-cast and the end rings are realized in the same production step. The induction machine with cage rotor is the most often used electrical machine. It is simple, robust, and cost-effective (see Fig. 4.23 for an example).

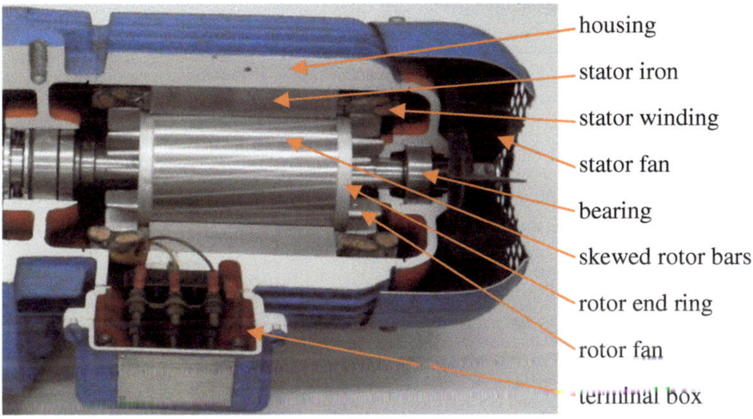

Fig. 4.23. Photograph of a squirrel cage induction machine.

Having N_2 rotor slots the cage winding can be interpreted as a multi-phase winding with N_2 phases. Here, each phase is composed of a single bar (better: each two neighboring bars, one half each, together with the connecting end ring segments, are composing the winding of a phase). The number of turns is then:

$$w_2 = \frac{N_2 z_N}{2 m_2 a} = \frac{1}{2}, \qquad z_N = 1, \quad a = 1 \tag{4.78}$$

The fundamental wave has the winding factor:

$$\xi_{2,1} = 1 \tag{4.79}$$

The number of pole pairs of the cage rotor is not determined by the winding, but because of the inductive effect the number of pole pairs of the stator is adopted.

The fundamental wave of the rotor magneto-motive force is (analogously to the three-phase winding):

$$\Theta_{2,1} = \frac{m_2}{2}\frac{w_2\xi_{2,1}}{p}\frac{4}{\pi}\sqrt{2}I_2 = \frac{N_2}{2}\frac{4}{\pi}\frac{\frac{1}{2}^{-1}}{p}\sqrt{2}I_{bar}$$
$$= \frac{\sqrt{2}}{\pi}\frac{N_2}{p}I_{bar} \tag{4.80}$$

Here $I_2 = I_{bar}$ is the current of a rotor bar.

Now the relation between the bar current and the ring current shall be calculated. Figure 4.24 illustrates the notation:

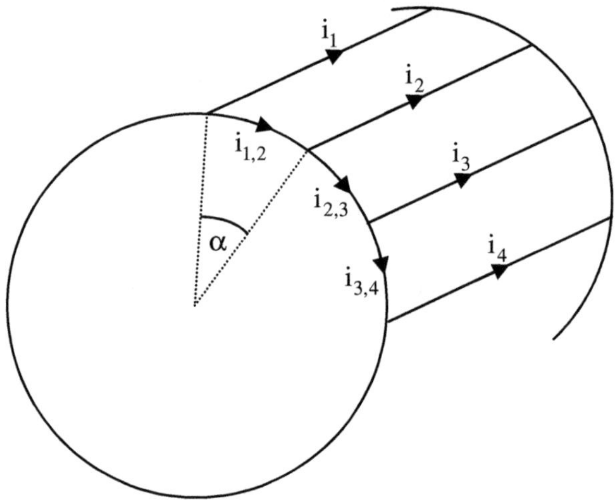

Fig. 4.24. Rotor currents of a squirrel cage induction machine.

There is:

$$\beta = p\alpha = \frac{p\ 2\pi}{N_2} \tag{4.81}$$

From

$$\sum_{k=1}^{N_2} i_k = 0 \tag{4.82}$$

(there are no current sources in the rotor) and

$$\begin{aligned} & i_{k-1,k} = i_k + i_{k,k+1} \\ \Rightarrow \quad & i_k + i_{k,k+1} - i_{k-1,k} = 0 \end{aligned} \tag{4.83}$$

the phasor diagram can be deduced (Fig. 4.25):

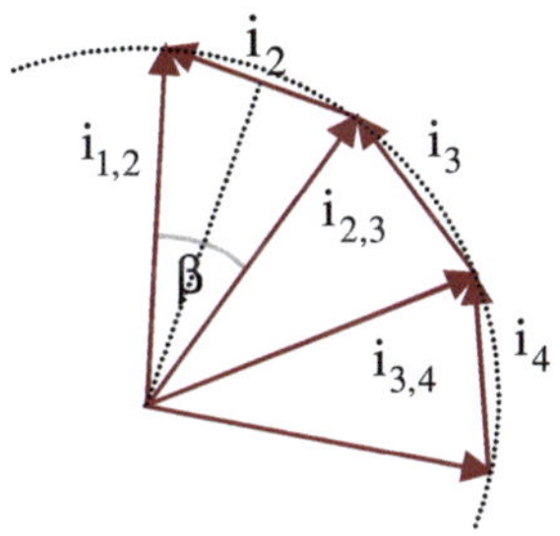

Fig. 4.25. Phasor diagram of the rotor currents of a squirrel cage induction machine.

The phase shift between two neighboring bar currents is equal to the phase shift of two neighboring ring currents. It follows:

$$I_{ring} = \frac{\frac{1}{2} I_{bar}}{\sin\left(\frac{\beta}{2}\right)} = \frac{I_{bar}}{2\sin\left(\frac{p\pi}{N_2}\right)} \tag{4.84}$$

Therefore, the cross section of the end rings have to be dimensioned accordingly larger than the cross section of the bars (because usually equal current densities should be achieved).

The voltage transformation ratio is (see Sect. 4.1):

$$r_u = \frac{U_2'}{U_{bar}} = \frac{w_1 \xi_1}{w_2 \xi_2} = \frac{w_1 \xi_1}{\frac{1}{2} 1} = 2 w_1 \xi_1 \tag{4.85}$$

The current transformation ratio is (see Sect. 4.1):

$$r_i = \frac{I_2'}{I_{bar}} = \frac{m_2}{m_1} \frac{w_2 \xi_2}{w_1 \xi_1} = \frac{N_2 \frac{1}{2} 1}{3 w_1 \xi_1} = \frac{N_2}{6 w_1 \xi_1} \tag{4.86}$$

The rotor resistance is composed of the bar resistance and the corresponding segmental ring resistance. The transformation of the rotor resistance onto the stator system is performed on the basis of equal losses:

$$\begin{aligned} m_1 I_2'^2 R_2' &= m_2 I_2^2 R_2 \\ &= N_2 I_{bar}^2 R_{bar} + 2N_2 I_{ring}^2 R_{ring} \\ &= N_2 I_{bar}^2 R_{bar} + 2N_2 R_{ring} \left(\frac{I_{bar}}{2\sin(p\pi/N_2)} \right)^2 \end{aligned} \tag{4.87}$$

Further it follows:

$$\begin{aligned} R_2' &= \frac{N_2 I_{bar}^2}{m_1 I_2'^2} \left(R_{bar} + 2R_{ring} \frac{1}{2\sin^2(p\pi/N_2)} \right) \\ &= \frac{N_2}{m_1 I_2'^2} \left(\frac{2m_1 w_1 \xi_1}{N_2} I_2' \right)^2 \left(R_{bar} + 2R_{ring} \frac{1}{2\sin^2(p\pi/N_2)} \right) \\ &= \frac{4m_1 (w_1 \xi_1)^2}{N_2} \left(R_{bar} + 2R_{ring} \frac{1}{2\sin^2(p\pi/N_2)} \right) \\ &= \frac{12 (w_1 \xi_1)^2}{N_2} \left(R_{bar} + 2R_{ring} \frac{1}{2\sin^2(p\pi/N_2)} \right) \\ &= r_R R_2, \qquad R_2 = R_{bar} + 2R_{ring} \frac{1}{2\sin^2(p\pi/N_2)} \end{aligned} \tag{4.88}$$

4.4.2 Skewed Rotor Slots

Apart from the fundamental wave there are even harmonic waves of the flux density distribution in the air-gap of the machine, e.g. because of the winding distribution and the slotting effect. These harmonic waves can be regarded as "leakage fields" (air-gap leakage, harmonic leakage). However, having a cage rotor these field waves can induce voltages and corresponding currents like the fundamental wave and consequently generate (disturbing) torque components. By skewing the rotor bars the effect of the harmonic waves can be reduced, because the flux penetrating a cage mesh is different:

- for skewed and unskewed bars the flux of the fundamental wave is nearly the same, whereas
- the flux of a higher harmonic wave is large for unskewed bars and it is reduced for skewed rotor bars.

Figure 4.26 schematically illustrates a rotor with skewed rotor bars.

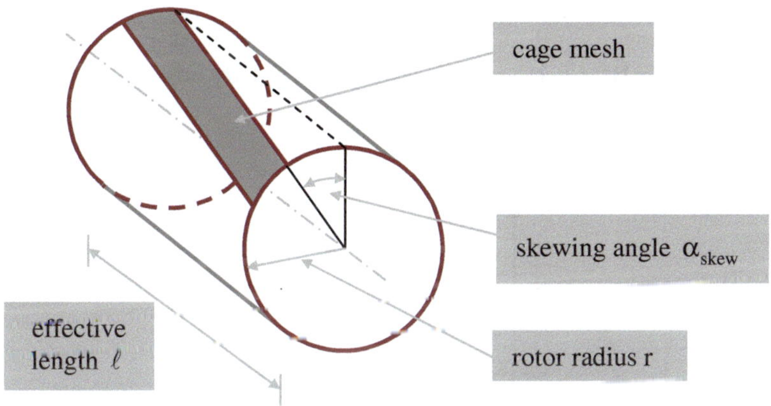

Fig. 4.26. Sketch of a squirrel cage rotor with skewed rotor slots.

The ordinal numbers of the rotating field generated by a symmetric three-phase winding are (please refer to Chap. 3 "Rotating Field Theory"):

$$\begin{aligned} \nu &= 2m_1 g + 1 \\ &= 6g + 1, \qquad g = 0, \pm 1, \pm 2, \ldots \end{aligned} \tag{4.89}$$

For $g = 0$, i.e. $\nu = 1$, the fundamental wave is obtained.

Regarding a skewed rotor mesh in a "wound-off" representation (x-coordinate: circumferential direction; y-coordinate: axial direction), Fig. 4.27 can be drawn.

For the fundamental wave and the harmonic waves of the air-gap field the following is true (for better distinction here the amplitude of the magnetic flux density is marked by "^"):

$$B_{\delta,\nu} = \hat{B}_{\delta,\nu} \cos(\nu\gamma_2) \tag{4.90}$$

Consequently, the flux in the (skewed) rotor mesh becomes:

$$\Phi_{\delta,\nu} = \int\limits_{-\frac{\ell}{2}}^{\frac{\ell}{2}} \int\limits_{-\frac{\pi r}{N_2} - \alpha_{skew} r \frac{y}{\ell}}^{+\frac{\pi r}{N_2} - \alpha_{skew} r \frac{y}{\ell}} B_{\delta,\nu} \; dx \; dy$$

$$= \hat{B}_{\delta,\nu} \frac{r}{p} \int\limits_{-\frac{\ell}{2}}^{\frac{\ell}{2}} \int\limits_{-\frac{p\pi}{N_2} - \alpha_{skew} p \frac{y}{\ell}}^{+\frac{p\pi}{N_2} - \alpha_{skew} p \frac{y}{\ell}} \cos(\nu\gamma_2) \; d\gamma_2 \; dy \tag{4.91}$$

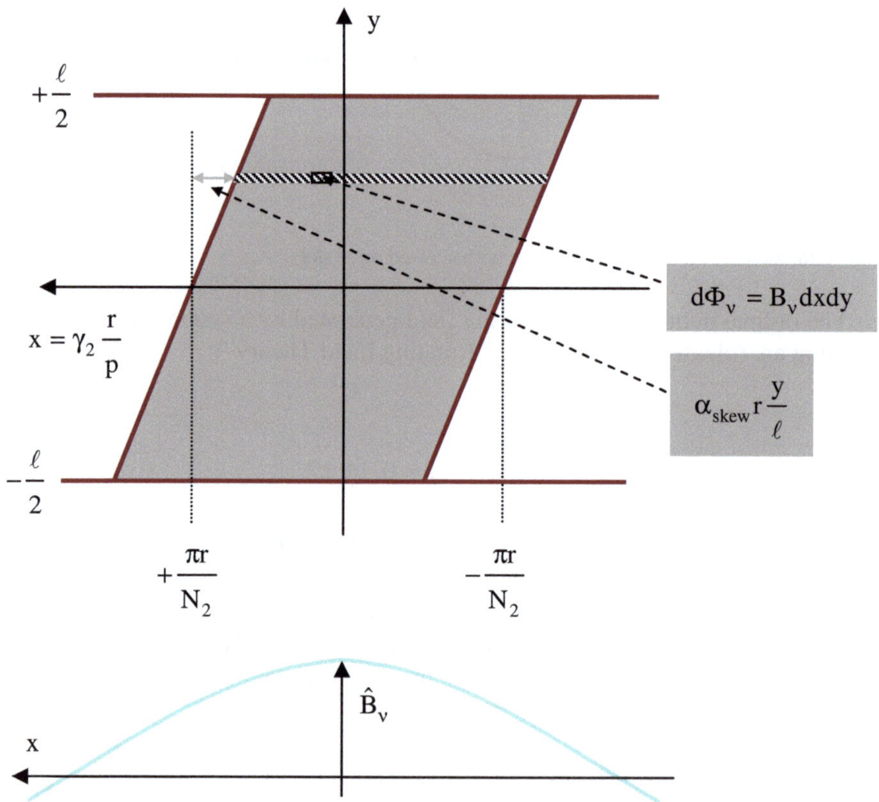

Fig. 4.27. Sketch of a squirrel cage rotor with skewed rotor slots.

Solving firstly the inner integral of Eq. (4.91), it follows:

$$\Phi_{\delta,\nu} = \hat{B}_{\delta,\nu} \frac{r}{p} \int_{-\frac{\ell}{2}}^{\frac{\ell}{2}} \left[\frac{1}{\nu} \sin(\nu\gamma_2) \Big|_{\gamma_2 = -\frac{p\pi}{N_2} - \alpha_{skew} p \frac{y}{\ell}}^{\gamma_2 = +\frac{p\pi}{N_2} - \alpha_{skew} p \frac{y}{\ell}} \right] dy \tag{4.92}$$

and further:

$$\begin{aligned} \Phi_{\delta,\nu} &= \hat{B}_{\delta,\nu} \frac{r}{p} \frac{1}{\nu} \int_{-\frac{\ell}{2}}^{\frac{\ell}{2}} \left[\sin\left(+\frac{\nu p \pi}{N_2} - \nu\alpha_{skew} p \frac{y}{\ell} \right) - \right. \\ &\qquad \left. \sin\left(-\frac{\nu p \pi}{N_2} - \nu\alpha_{skew} p \frac{y}{\ell} \right) \right] dy \\ &= \hat{B}_{\delta,\nu} \frac{2r}{\nu p} \int_{-\frac{\ell}{2}}^{\frac{\ell}{2}} \left[\cos\left(-\nu\alpha_{skew} p \frac{y}{\ell} \right) \sin\left(\frac{\nu p \pi}{N_2} \right) \right] dy \\ &= \hat{B}_{\delta,\nu} \frac{2r}{\nu p} \sin\left(\frac{\nu p \pi}{N_2} \right) \int_{-\frac{\ell}{2}}^{\frac{\ell}{2}} \cos\left(-\nu\alpha_{skew} p \frac{y}{\ell} \right) dy \end{aligned} \tag{4.93}$$

Now the remaining integral is solved:

$$\begin{aligned} \Phi_{\delta,\nu} &= \hat{B}_{\delta,\nu} \frac{2r}{\nu p} \sin\left(\frac{\nu p \pi}{N_2} \right) \frac{-\ell}{\nu p \alpha_{skew}} \sin\left(-\nu p \alpha_{skew} \frac{y}{\ell} \right) \Big|_{y=-\frac{\ell}{2}}^{y=\frac{\ell}{2}} \\ &= \hat{B}_{\delta,\nu} \frac{2r\ell}{\nu p} \sin\left(\frac{\nu p \pi}{N_2} \right) \frac{\sin\left(\frac{\nu p \alpha_{skew}}{2} \right) - \sin\left(\frac{-\nu p \alpha_{skew}}{2} \right)}{\nu p \alpha_{skew}} \\ &= \hat{B}_{\delta,\nu} \frac{2r\ell}{\nu p} \sin\left(\frac{\nu p \pi}{N_2} \right) \frac{2\cos(0) \sin\left(\frac{\nu p \alpha_{skew}}{2} \right)}{\nu p \alpha_{skew}} \end{aligned} \tag{4.94}$$

A transformation gives:

$$\begin{aligned}
\Phi_{\delta,\nu} &= \hat{B}_{\delta,\nu} \frac{2r\ell}{\nu p} \frac{\nu p\pi}{N_2} \frac{\sin\left(\frac{\nu p\pi}{N_2}\right)}{\frac{\nu p\pi}{N_2}} \frac{\sin\left(\frac{\nu p\alpha_{skew}}{2}\right)}{\frac{\nu p\alpha_{skew}}{2}} \\
&= \hat{B}_{\delta,\nu} \frac{2\pi r\ell}{N_2} \frac{\sin\left(\frac{\nu p\pi}{N_2}\right)}{\frac{\nu p\pi}{N_2}} \frac{\sin\left(\frac{\nu p\alpha_{skew}}{2}\right)}{\frac{\nu p\alpha_{skew}}{2}} \\
&= \hat{B}_{\delta,\nu} \frac{2\pi r\ell}{N_2} \mathrm{si}\left(\frac{\nu p\pi}{N_2}\right) \mathrm{si}\left(\frac{\nu p\alpha_{skew}}{2}\right)
\end{aligned} \tag{4.95}$$

The flux of a harmonic wave in a single rotor mesh therefore is composed of the maximum value of the harmonic flux density wave, the air-gap area of a rotor mesh, the coupling factor

$$\xi_{C,\nu} = \mathrm{si}\left(\frac{\nu p\pi}{N_2}\right) \tag{4.96}$$

and the skewing factor

$$\xi_{skew,\nu} = \mathrm{si}\left(\frac{\nu p\alpha_{skew}}{2}\right) \tag{4.97}$$

For the coupling factor it holds:

$$\xi_{C,\nu} \rightarrow 1 \text{ for } N_2 \rightarrow \infty \tag{4.98}$$

The effect of skewing is that the harmonic waves are damped to a large extent, whereas the fundamental wave is hardly influenced. The skewing factor becomes zero for:

$$\begin{aligned}
&\xi_{skew,\nu} = 0 \quad \Rightarrow \quad \frac{\nu p\alpha_{skew}}{2} = k\pi, \qquad k = 1, 2, 3, \ldots \\
&\Rightarrow \quad \frac{p\alpha_{skew}}{2\pi} = \frac{k}{\nu}, \qquad k = 1, 2, 3, \ldots
\end{aligned} \tag{4.99}$$

The skewing factor for the fundamental wave and the first four harmonic waves is shown in Fig. 4.28 as a function of the skewing angle.

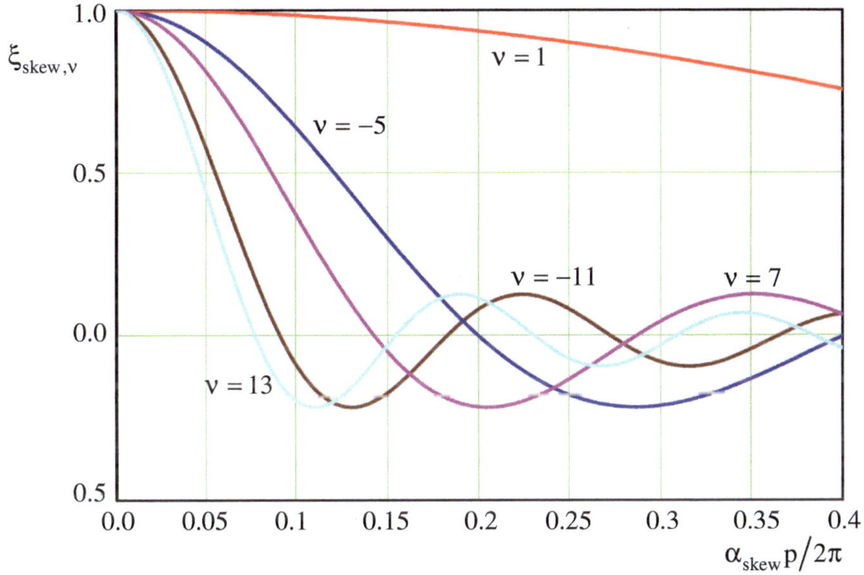

Fig. 4.28. Influence of skewed rotor slots onto the skewing factor for different harmonics.

4.4.3 Skin Effect

Because of the massive bars there is always a single-sided skin effect (towards the slot opening) with cage rotors: the slot leakage fields cross the bars and induce voltages, that are accompanied by eddy currents and consequently uneven current distribution. According to Lenz's Law these induced voltages are directed in such a way, that the hereby generated eddy currents act against their reason; please refer to Fig. 4.29 for the principle geometry, the current density distribution $J(x)$ and the flux density distribution $B(x)$.

The integral of the superposed eddy currents across the slot cross section is equal to zero, i.e. the total current is unchanged (only the current distribution is changed across the cross section).

Generally, the skin effect is unwanted in electrical machines, because this generates additional losses in the bars and therefore is the reason for higher heating and lower efficiency.

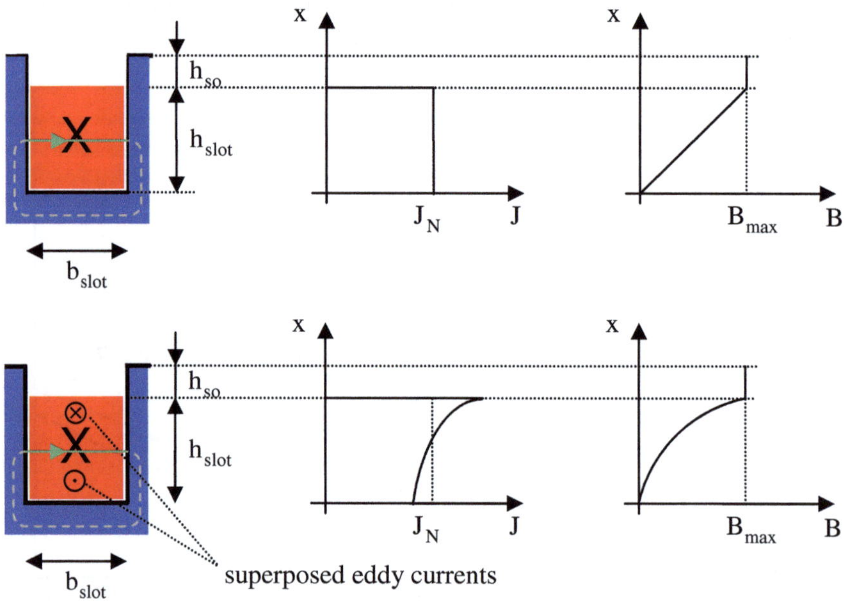

Fig. 4.29. Current density and flux density distribution in massive rotor bars (*above*: DC current; *below*: AC current).

For induction machines with cage rotor this skin effect however can be used to improve the starting behavior: Because of the skin effect the rotor resistance is higher at starting operation (i.e. at high frequency of the rotor currents) and therefore even the starting torque is higher. Simultaneously, the leakage reactance becomes smaller (the flux density in the rotor bar is smaller), which means an increase of the diameter of the Heyland-diagram.

With increasing speed of the motor the skin effect gets smaller (because the frequency of the rotor currents gets smaller), until it is hardly noticeable at nominal operation. The Heyland-diagram C can now be composed of a diagram C_{st} for starting and a diagram C_{op} for operation (the current $\underline{I}_0$ is identical in both cases, because X_1 is unchanged), see Fig. 4.30.

It is:

$s = 1$ (i.e. $f_2 = f_1$) $\quad$ X_1 unchanged; $X'_{2\sigma}$ small, R'_2 large

$s \to 0$ (i.e. $f_2 \to 0$) $\quad$ X_1 unchanged; $X'_{2\sigma}$ large, R'_2 small

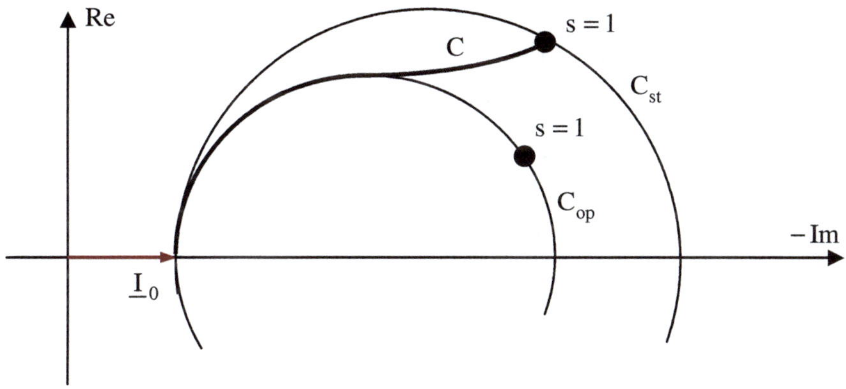

Fig. 4.30. Heyland-diagram of the induction motor: starting diagram and operating diagram.

This effect specifically is utilized in so-called "skin effect rotors" (Fig. 4.31).

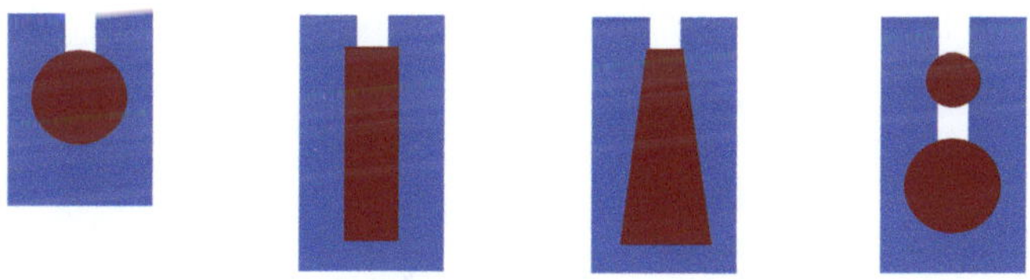

Fig. 4.31. Different kinds of skin effect rotors; from *left* to *right*: round bar (small skin effect), rectangular bar, keyed bar, double bar.

Changing the rotor resistance and the rotor leakage inductivity with frequency usually is described by the following factors:

$$\begin{aligned} R_2(f_2) &= K_R(f_2)\ R_2(f_2 = 0) \\ K_R(0) &= 1 \\ K_R(f_2 = f_1) &\approx 3 \ldots 5 \end{aligned} \tag{4.100}$$

$$\begin{aligned} L_{2\sigma,slot}(f_2) &= K_I(f_2)\ L_{2\sigma,slot}(f_2 = 0) \\ K_I(0) &= 1 \\ K_I(f_2 = f_1) &\approx 0.25 \ldots 0.4 \end{aligned} \tag{4.101}$$

For rectangular bar cages with the slot height h and the specific resistance ρ the following is true (here without deduction):

$$K_R(f_2) = \zeta \frac{\sinh(2\zeta) + \sin(2\zeta)}{\cosh(2\zeta) - \cos(2\zeta)}$$
$$K_I(f_2) = \frac{3}{2\zeta} \frac{\sinh(2\zeta) - \sin(2\zeta)}{\cosh(2\zeta) - \cos(2\zeta)} \quad , \quad \zeta(f_2) = h\sqrt{\frac{\pi\mu_0 f_2}{\rho}} \tag{4.102}$$

Figures 4.32 and 4.33 illustrate the factors $K_R(f_2)$ and $K_I(f_2)$ as function of $\zeta(f_2)$ for the rectangular bar cage.

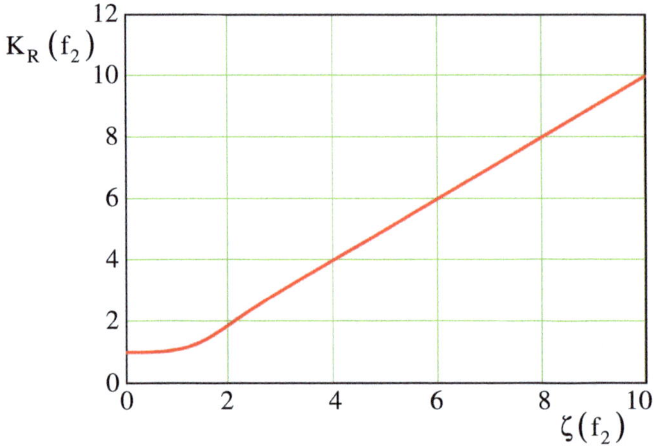

Fig. 4.32. Rotor resistance modifying factor for rectangular bars.

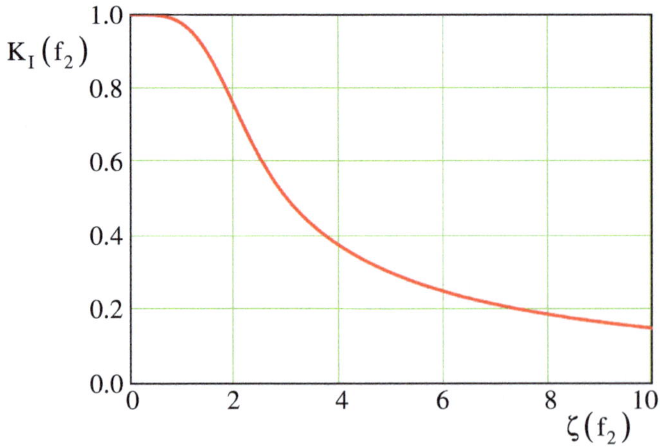

Fig. 4.33. Rotor leakage inductance modifying factor for rectangular bars.

4.5 Possibilities for Open-Loop Speed Control

The main possibilities for open-loop speed control of induction motors can be deduced from the following equation:

$$n = \frac{f_1}{p}(1-s) \tag{4.103}$$

4.5.1 Changing (Increasing) the Slip

Either by resistances in the rotor circuit of a slip ring rotor or by decreasing the terminal voltage ($T \sim U_1^2$) the slip is increased. In both cases the additional slip power is transformed to heat and the efficiency is decreased:

$$\eta = 1 - s \tag{4.104}$$

The no-load speed does not change.

4.5.2 Changing the Supply Frequency

The power is taken from the 50Hz three-phase mains, rectified and given to an inverter via an intermediate circuit (DC-current or DC-voltage) that supplies the induction machine with variable voltage (Fig. 4.34):

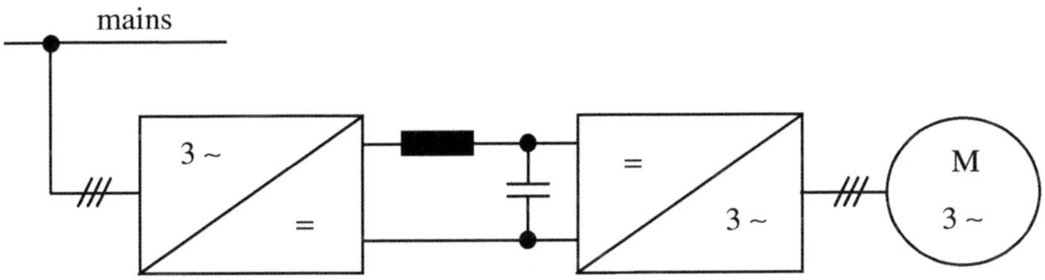

Fig. 4.34. Induction motor with inverter supply.

For the Heyland-diagram (stator current locus) it follows for variable frequency:

$$I_0 = \frac{U_1}{X_1} \sim \frac{U_1}{f_1} \tag{4.105}$$

$$I_\infty = \frac{I_0}{\sigma} \sim \frac{U_1}{f_1} \tag{4.106}$$

Therefore, the circle remains unchanged in its size, if the supplying voltage is changed proportional to the frequency. Then even the pull-out torque remains unchanged:

$$T_{pull-out} = \frac{3p}{\omega_1} \frac{U_1^2}{X_1 \frac{\sigma}{1-\sigma}} \sim \left(\frac{U_1}{f_1}\right)^2 \tag{4.107}$$

But the parameterization of the Heyland-diagram is changed:

$$\tan\left(\varphi^*\right) = \frac{X_1 \frac{\sigma}{1-\sigma}}{R_2'\left(1+\sigma_1\right)^2}\, s \sim s\, f_1 = f_2 \tag{4.108}$$

With decreasing supply frequency the stand-still operating point (starting, short-circuit) is shifted in the direction of the no-load operation point (Fig. 4.35).

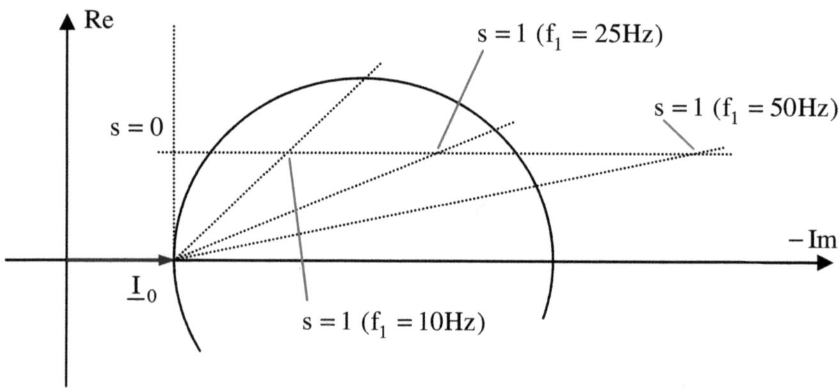

Fig. 4.35. Heyland-diagram of the induction motor: different supply frequencies.

The torque-speed-characteristics are illustrated in Fig. 4.36.

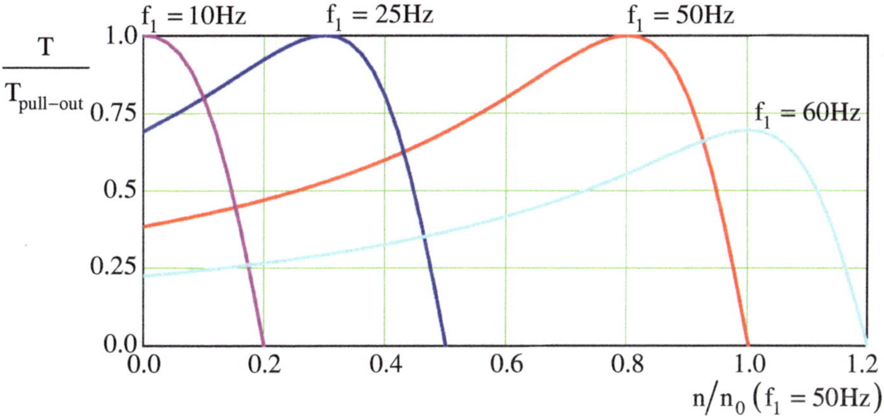

Fig. 4.36. Torque-speed-characteristics of the induction motor for different supply frequencies (with nominal supply frequency of 50Hz).

If the supply frequency f_1 gets smaller, the stator resistance R_1 has to be considered, because then R_1 cannot be neglected any more against X_1. Considering R_1, the Heyland-diagram gets smaller and the pull-out torque decreases (because of the voltage drop at R_1). For compensating the voltage drop at R_1 at low frequency the voltage is reduced less than proportional to the frequency (to reach a constant stator flux).

If the further increase of the voltage with frequency is not possible (this point is called "nominal operation point") the pull-out torque decreases inversely proportional to the squared frequency, see the characteristic for $f_1 = 60Hz$ in Fig. 4.36. This operating region is called "field weakening region".

4.5.3 Changing the Number of Pole Pairs

As the cage rotor adopts itself to any number of pole pairs, the synchronous speed (no-load speed) $n_0 = f_1/p$ can be varied in discrete steps by changing the number of pole pairs p. For this the stator

- can be equipped with several separated three-phase windings, placed in the same slots (high effort, small utilization) or
- contains a single winding with switchable winding segments (see Fig. 4.37).

The pole switching is used for example in the so-called "Dahlander circuit".

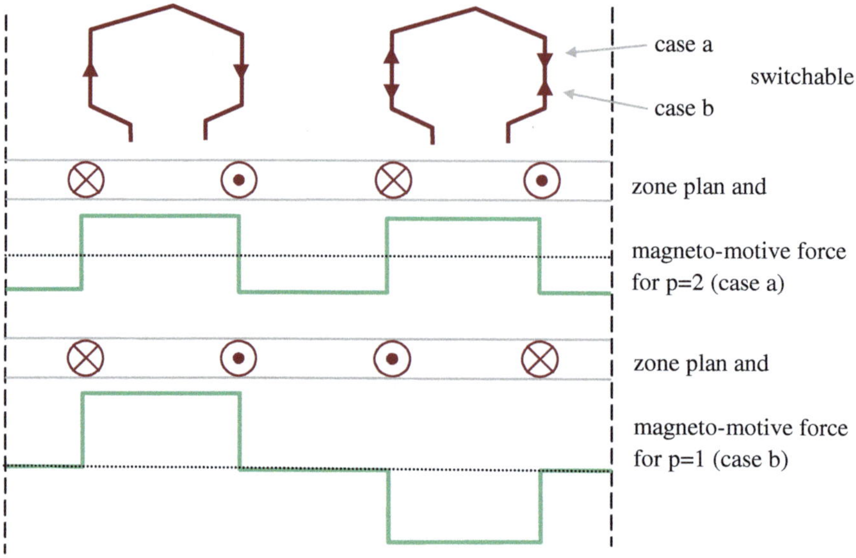

Fig. 4.37. Pole switching winding (with zone plan and magneto-motive force characteristics).

4.6 Star-Delta-Switching

When switching the induction machine with zero rotor speed to the mains the entire stall (short-circuit) current flows. This is a multiple of the nominal current. As this means a high load to the machine, the starting current has to be limited.

A possibility for limitation of this current is the introduction of series resistances to the rotor circuit, if a slip ring rotor is present (please refer to Sect. 4.3).

This is not possible with cage rotors. Here, the voltage is decreased by means of a starting transformer (at high power) or by means of star-delta-switching (low and medium power).

If the motor phases are connected in star there is:

$$U_{phase,Y} = \frac{1}{\sqrt{3}} U_{line}, \qquad I_{phase,Y} = I_{line,Y} \tag{4.109}$$

For delta connection it is true:

$$U_{phase,\Delta} = U_{line}, \qquad I_{phase,\Delta} = \frac{1}{\sqrt{3}} I_{line,\Delta} \tag{4.110}$$

As the torque is proportional to the squared phase voltage it follows:

$$\frac{T_Y}{T_\Delta} = \left(\frac{U_{phase,Y}}{U_{phase,\Delta}}\right)^2 = \frac{1}{3} \tag{4.111}$$

For the mains current it follows:

$$\frac{I_{line,Y}}{I_{line,\Delta}} = \frac{I_{phase,Y}}{\sqrt{3}\ I_{phase,\Delta}} = \frac{I_{phase,Y}}{\sqrt{3}\sqrt{3}\ I_{phase,Y}} = \frac{1}{3} \tag{4.112}$$

Consequently, torque and mains current are decreased by a factor of 3 in starting operation. The switching to the delta connection is realized near the nominal operating point.

4.7 Doubly-Fed Induction Machine

The slip-ring rotor of an induction machine can be supplied with voltages having slip frequency (formerly the slip power was supplied by another machine set-up, today generally this is done using an inverter) (Fig. 4.38).

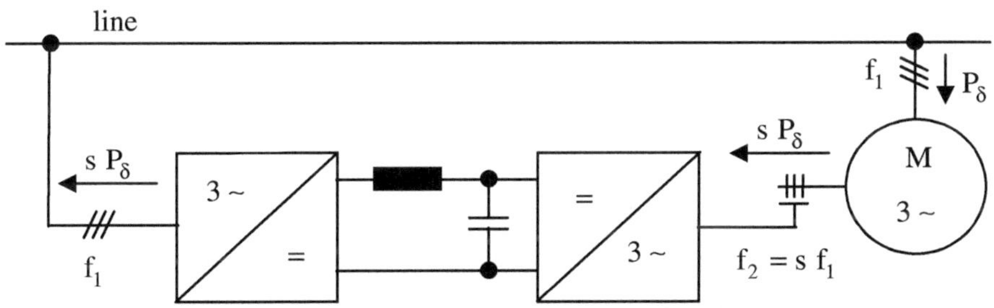

Fig. 4.38. Doubly-fed induction machine.

Via the inverter additional power can be fed to or drawn from the rotor. Feeding the power an operation above synchronous speed is reached, drawing the power an operation below synchronous speed is realized (for motor operation).

Apart from the inverter losses this kind of speed control (nearly) is lossless. The advantage of this alternative is that the inverter has to be dimensioned only for the slip power. A typical application of this kind of speed control e.g. is in wind power plants (generator operation).

To calculate the torque characteristic the equivalent circuit diagram in Fig. 4.39 is regarded.

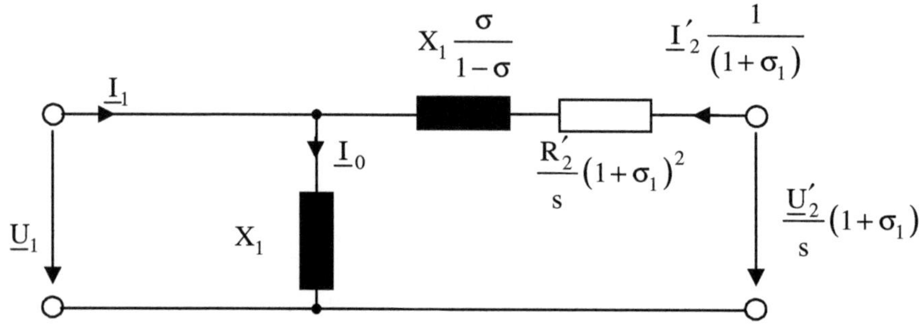

Fig. 4.39. Equivalent circuit diagram of the doubly-fed induction machine.

There is:

$$\underline{I}_1 = \underline{I}_0 - \underline{I}'_2 \frac{1}{1+\sigma_1}$$

$$\underline{I}_0 = \frac{\underline{U}_1}{jX_1}$$

$$\underline{I}'_2 \frac{1}{1+\sigma_1}\left(\frac{R'_2}{s}(1+\sigma_1)^2 + jX_1\frac{\sigma}{1-\sigma}\right) = \frac{\underline{U}'_2}{s}(1+\sigma_1) - \underline{U}_1 \qquad (4.113)$$

$$\Rightarrow \quad \underline{I}'_2 \frac{1}{1+\sigma_1} = \frac{\underline{U}'_2(1+\sigma_1) - s\underline{U}_1}{R'_2(1+\sigma_1)^2 + jsX_1\frac{\sigma}{1-\sigma}}$$

It can be deduced that the rotor current is composed from the superposition of two parts: The first part comes from the terminal voltage of the stator $\underline{U}_1$, the second part comes from the rotor voltage $\underline{U}'_2$ which is used as control variable.

In no-load operation $\underline{I}'_2 = 0$ is true. This is fulfilled for

$$s = s'_0 = \frac{\underline{U}'_2(1+\sigma_1)}{\underline{U}_1} \qquad (4.114)$$

The value s'_0 is called no-load slip. As the no-load slip has to be a real number, even $\underline{U}'_2/\underline{U}_1$ has to be a real number, i.e. these two voltages have to have the same phase angle (or opposite phase angle). Therefore, the no-load speed

$$n = n_0' = n_0 \left(1 - s_0'\right) = \frac{\omega_1}{2\pi p}\left(1 - s_0'\right) \tag{4.115}$$

can be adjusted by the voltage $\underline{U}_2'$; e.g. it follows

- for $0 < s_0' < 0.5$ a no-load speed of $n_0 > n_0' > 0.5n_0$ is obtained;
- for $-0.5 < s_0' < 0$ a no-load speed of $1.5n_0 > n_0' > n_0$ is obtained.

The torque can be calculated from the power of the rotating field. With

$$\mathrm{Re}\{\underline{I}_1\} = \mathrm{Re}\left\{-\underline{I}_2' \frac{1}{1+\sigma_1}\right\} \tag{4.116}$$

it follows:

$$\begin{aligned} T &= \frac{p}{\omega_1} P_\delta = \frac{3p}{\omega_1} U_1 \,\mathrm{Re}\left\{-\underline{I}_2' \frac{1}{1+\sigma_1}\right\} \\ &= \frac{3p}{\omega_1} \underline{U}_1 \,\mathrm{Re}\left\{\frac{s\underline{U}_1 - \underline{U}_2'\left(1+\sigma_1\right)}{R_2'\left(1+\sigma_1\right)^2 + jsX_1 \dfrac{\sigma}{1-\sigma}}\right\} \end{aligned} \tag{4.117}$$

With

$$s_{\text{pull-out}} = \frac{R_2'\left(1+\sigma_1\right)^2}{X_1 \dfrac{\sigma}{1-\sigma}} \tag{4.118}$$

it follows further, if $\underline{U}_1 = U_1$ is chosen being real:

$$T=\frac{3p}{\omega_1}\frac{U_1^2}{X_1\dfrac{\sigma}{1-\sigma}}\operatorname{Re}\left\{\frac{s-\dfrac{\underline{U}_2'}{U_1}(1+\sigma_1)}{s_{pull-out}+js}\right\}$$
$$=\frac{3p}{\omega_1}\frac{U_1^2(1-\sigma)}{\sigma X_1}\operatorname{Re}\left\{\frac{\left(s-\dfrac{\underline{U}_2'}{U_1}(1+\sigma_1)\right)\left(s_{pull-out}-js\right)}{s_{pull-out}^2+s^2}\right\} \tag{4.119}$$

Now just the rotor voltages that are in phase with the stator voltage are considered (it has been shown above that only for those voltages no-load of the machine is possible), then it follows:

$$T=\frac{3p}{\omega_1}\frac{U_1^2(1-\sigma)}{\sigma X_1}\operatorname{Re}\left\{\frac{\left(s-\dfrac{U_2'}{U_1}(1+\sigma_1)\right)\left(s_{pull-out}-js\right)}{s_{pull-out}^2+s^2}\right\}$$
$$=\frac{3p}{\omega_1}\frac{U_1^2(1-\sigma)}{\sigma X_1}\left(s-\frac{U_2'}{U_1}(1+\sigma_1)\right)\frac{s_{pull-out}}{s_{pull-out}^2+s^2} \tag{4.120}$$

Setting this torque in relation to the pull-out torque at zero rotor voltage (see Sect. 4.3), the result is:

$$T_{pull-out,U_2'=0}=\frac{3p}{2\omega_1}\frac{U_1^2}{X_1\dfrac{\sigma}{1-\sigma}}$$
$$\Rightarrow\quad\frac{T}{T_{pull-out,U_2'=0}}=2\left(s-\frac{U_2'}{U_1}(1+\sigma_1)\right)\frac{s_{pull-out}}{s_{pull-out}^2+s^2} \tag{4.121}$$

A further transformation finally gives:

$$T_{ratio} = \frac{T}{T_{pull-out,U_2'=0}} = \left(1 - \frac{\frac{U_2'}{s}(1+\sigma_1)}{U_1}\right) \frac{2}{\frac{s_{pull-out}}{s} + \frac{s}{s_{pull-out}}} \qquad (4.122)$$

Figure 4.40 exemplarily shows the torque (in relation to the torque at $U_2' = 0$) for a machine with $s_{pull-out} \approx 0.4$ as a function of speed (in relation to the synchronous speed).

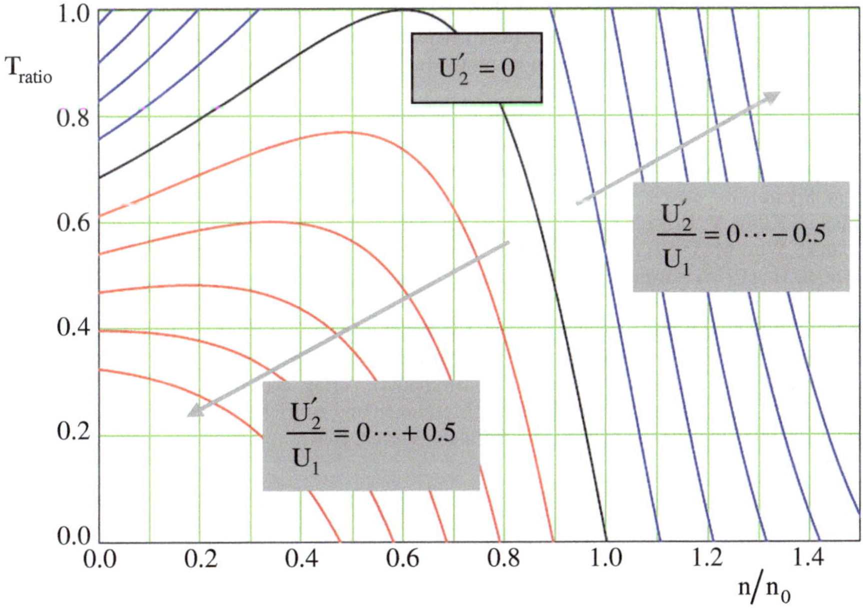

Fig. 4.40. Torque-speed-characteristics of the doubly-fed induction machine.

The advantage of this kind of speed control is that the inverter has to be dimensioned only for the slip power. Therefore it is smaller than an inverter supplying the machine on the stator side.

A disadvantage of this alternative of speed control is that only a limited speed variation can be reached (the larger the speed variation, the larger is the required power of the inverter) and that the machine has to be equipped with slip rings. This pays off only for large power machines (typically more than 500kW).

Designing the machine-side converter as a pure rectifier, it is only possible to draw power from the machine. Consequently, only the speed area below synchronous speed is reachable with the advantage that the power electronic complexity and costs is reduced considerably. Such a layout is called *sub-synchronous converter cascade.*

4.8 References for Chapter 4

Ban D, Zarko D, Mandic I (2003) Turbogenerator end winding leakage inductance calculation using a 3 D analytical approach based on the solution of Neumann integrals. In: IEEE International Electric Machines and Drives Conference (IEMDC), Madison, Wisconsin, USA

Gerling D (1999) Analytical calculation of leakage and inductivity for motors / actuators with concentrated windings. In: International Conference on Modeling and Simulation of Electric Machines, Converters and Systems (ELECTRIMACS), Lisbon, Portugal

Gerling D, Schramm A (2005) Analytical calculation of the end winding leakage inductance based on the solution of Neumann integrals. In: International Symposium on Industrial Electronics (ISIE), Dubrovnik, Croatia

Gerling D, Schramm A (2005) Calculation of the magnetic field in the end winding region of unsaturated electrical machines with small air-gap. In: International Symposium on Electromagnetic Fields in Mechatronics, Electrical and Electronic Engineering (ISEF), Baiona, Vigo, Spain

Jordan H, Klima V, Kovacs KP (1975) Asynchronmaschinen. Vieweg Verlag, Braunschweig

Jordan H, Weis M (1969) Asynchronmaschinen. Vieweg Verlag, Braunschweig

Kleinrath H (1975) Grundlagen elektrischer Maschinen. Akademische Verlagsgesellschaft, Wiesbaden

Müller G, Ponick B (2009) Theorie elektrischer Maschinen. Wiley-VCH Verlag, Weinheim

Nürnberg W (1976) Die Asynchronmaschine. Springer-Verlag, Berlin

Richter R (1954) Elektrische Maschinen IV. Birkhäuser Verlag, Basel

Seinsch HO (1993) Grundlagen elektrischer Maschinen und Antriebe. Teubner Verlag, Stuttgart

Spring E (1998) Elektrische Maschinen. Springer-Verlag, Berlin

5 Synchronous Machines

5.1 Equivalent Circuit and Phasor Diagram

Like the induction machine the synchronous machine contains a stator with three-phase winding (it is a rotating field machine), but the rotor winding is supplied with DC-current. In the following the voltage equations and the equivalent circuit of the synchronous machine will be derived from those of the induction machine.

The three-phase winding of the stator is supplied from a three-phase mains with constant voltage U_1 and constant frequency f_1. The rotor shall have a three-phase winding of the same pole number as well, which is connected to slip rings. Between two of these slip rings a DC-current is fed (excitation or field current I_F). Therefore, the frequency of the rotor currents is $f_2 = 0$.

According to the rotating field theory the synchronous machine is only able to produce a torque that is constant in time and different from zero, if the frequency condition is fulfilled:

$$f_2 = s\, f_1 \tag{5.1}$$

With $f_2 = 0$ and $f_1 = f_{line}$ it follows:

$$s = 0 \qquad \Rightarrow \qquad n = n_0 = \frac{f_1}{p} \tag{5.2}$$

Consequently, in stationary operating conditions the rotor always rotates with the synchronous speed n_0. At any different speed $n \neq n_0$ an oscillating torque with a mean value according to time equal to zero is generated.

In contrary to the induction machine, which does not generate a torque for $n = n_0$, the synchronous machine generates a torque only at $n = n_0$.

Now, coming from the general circuit diagram of the induction machine with slip ring rotor, the circuit diagram of the synchronous machine will be deduced. Here the “energy generation system” is applied, because synchronous machines mainly are used as generators (Fig. 5.1).

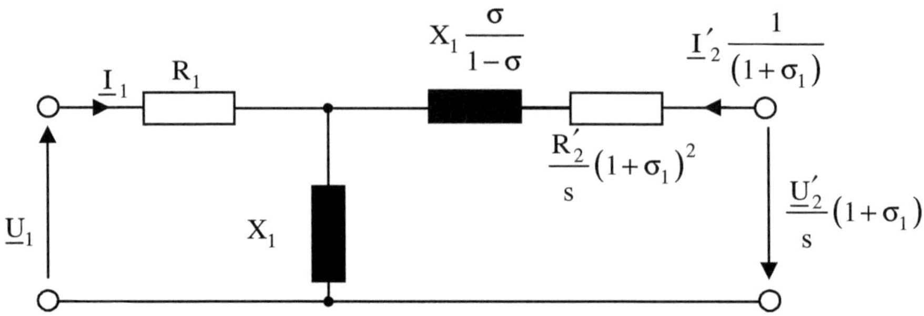

Fig. 5.1. Equivalent circuit diagram of the synchronous machine.

The voltage equations are:

$$\begin{aligned} &\underline{U}_1 + R_1\underline{I}_1 + jX_1\left(\underline{I}_1 + \frac{\underline{I}'_2}{1+\sigma_1}\right) = 0 \\ &\frac{\underline{U}'_2}{s}(1+\sigma_1) = \frac{\underline{I}'_2}{1+\sigma_1}\left(\frac{R'_2}{s}(1+\sigma_1)^2 + jX_1\frac{\sigma}{1-\sigma}\right) \\ &\qquad + jX_1\left(\underline{I}_1 + \frac{\underline{I}'_2}{1+\sigma_1}\right) \end{aligned} \tag{5.3}$$

Multiplying the voltage equation of the rotor with s and regarding that $s = 0$ holds true, it follows:

$$\begin{aligned} &\underline{U}_1 + R_1\underline{I}_1 + j\left(X_{1\sigma} + X_{1m}\right)\underline{I}_1 = -jX_{1m}\underline{I}'_2 \\ &\underline{U}'_2 = R'_2\underline{I}'_2 \end{aligned} \tag{5.4}$$

$\underline{I}'_2$ is the excitation current I_F transformed to the stator side. A current $\underline{I}'_2$ in the stator winding having the mains frequency will generate the same rotating magneto-motive force (MMF) like the DC-current I_F flowing in the rotor having synchronous speed. The relation between $\underline{I}'_2$ and I_F is as follows:

1. The number of turns of the excitation winding is:

$$w_F = 2w_2 \tag{5.5}$$

2. The winding factor of the excitation (field) winding (nominated in the following with the index "F") can be calculated by means of the rotating field theory from the distribution factor and the short-pitch factor. For the

fundamental wave and a distribution of the winding in a large number of slots ($q \rightarrow \infty$) it follows, please refer to Fig. 5.2:

$$\xi_F = \xi_Z\ \xi_S = \mathrm{si}\left(\frac{\pi}{6}\right)\sin\left(\frac{s}{\tau_p}\frac{\pi}{2}\right)$$
$$= \frac{\sin\left(\pi/6\right)}{\pi/6}\sin\left(\frac{2\pi/3}{\pi}\frac{\pi}{2}\right) = \frac{1/2}{\pi/6}\sin\left(\frac{\pi}{3}\right) \tag{5.6}$$
$$= \frac{3}{\pi}\frac{\sqrt{3}}{2}$$

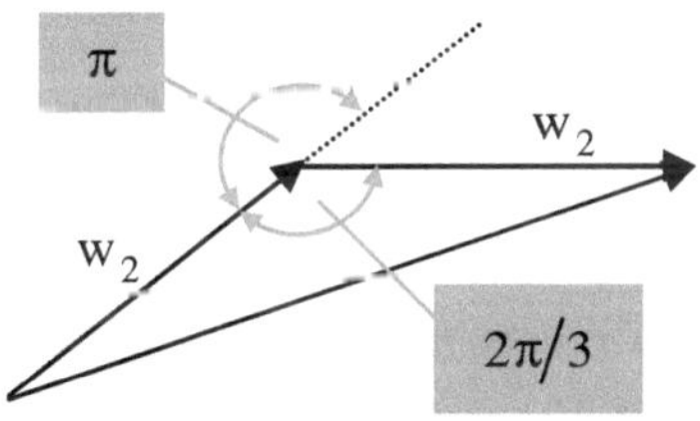

Fig. 5.2. Explanation for calculating the winding factor.

3. The fundamental waves of the MMF have to be the same in both cases:

$$\frac{3}{2}\frac{4}{\pi}\frac{w_1\xi_1}{p}\sqrt{2}I_2' = \frac{4}{\pi}\frac{w_F\xi_F}{p}I_F$$
$$\Rightarrow \quad I_2' = \frac{1}{\sqrt{2}}\frac{w_F\xi_F}{p}I_F\frac{2}{3}\frac{p}{w_1\xi_1} = \frac{2}{3}\frac{w_F\xi_F}{w_1\xi_1}\frac{I_F}{\sqrt{2}} \tag{5.7}$$
$$= \frac{2}{3}\frac{2w_2\frac{\sqrt{3}}{2}\frac{3}{\pi}}{w_1\xi_1}\frac{I_F}{\sqrt{2}} = \frac{2w_2}{w_1\xi_1}\frac{\sqrt{3}}{\pi}\frac{I_F}{\sqrt{2}}$$

4. The complex value can be obtained from the fact that the current $\underline{I}_2'$ in the stator winding having the mains frequency generates the same rotating magneto-motive force (MMF) in the air-gap like the DC-current I_F flowing in the rotor having synchronous speed. Against the voltage $\underline{U}_1$ the current $\underline{I}_2'$ has a phase shift of angle φ_2 (see the phasor diagram in Fig. 5.4):

$$\underline{I}'_2 = I'_2 \, e^{-j\varphi_2} \tag{5.8}$$

The excitation current I'_2 transformed to the stator side induces a voltage at the main reactance X_{1m}. This voltage is called *internal machine voltage* (or *open-circuit voltage* or *no-load voltage*):[6]

$$\underline{U}_P = -jX_{1m}\underline{I}'_2 \tag{5.9}$$

The stator voltage equation of the synchronous machine becomes:

$$\underline{U}_1 + R_1\underline{I}_1 + j\left(X_{1\sigma} + X_{1m}\right)\underline{I}_1 = \underline{U}_P \tag{5.10}$$

From the voltage equation of the stator the circuit diagram of the synchronous machine can be deduced. The rotor part must not be considered separately, because the voltage U_P induced from the excitation field into the stator winding is already included and in stationary operating points there is no reaction from the stator to the rotor. In Fig. 5.3 the directions of $\underline{U}_1$ and $\underline{I}_1$ are reversed against the beginning of this chapter. Therefore, even here the "energy generation system" is used.

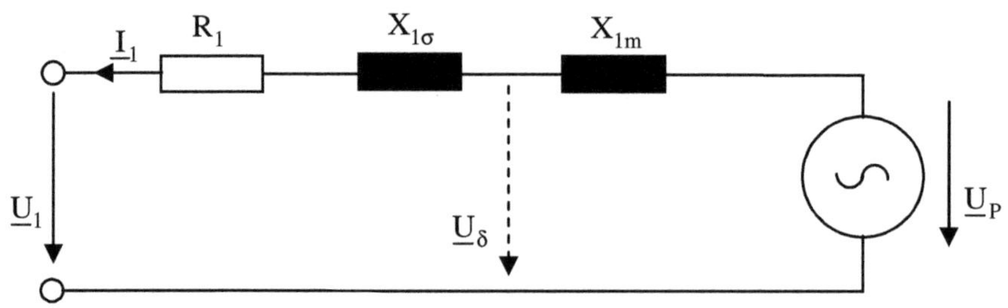

Fig. 5.3. Equivalent circuit diagram of the synchronous machine.

The internal machine voltage U_P can be measured directly at the terminals of the machine, if excitation with I_F, driving with synchronous speed n_0 and no-load operation ($I_1 = 0$) is used.

The entire phasor diagram of the synchronous machine in generator mode with resistive-inductive load is shown in Fig. 5.4.

[6] This voltage sometimes is also called back emf (electromotive force) and nominated with "e".

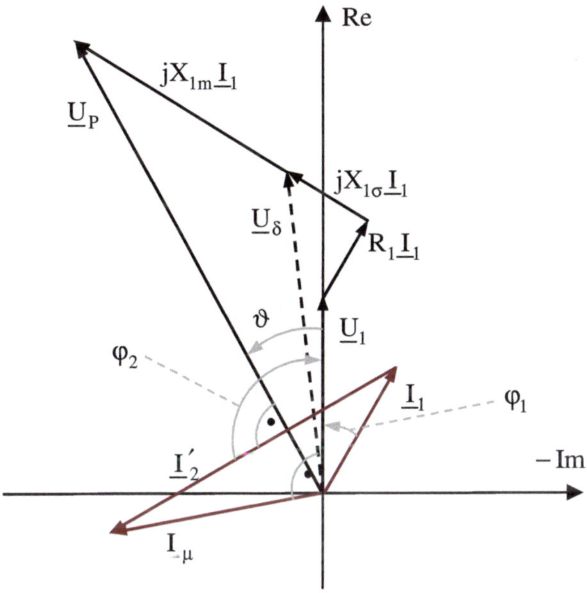

Fig. 5.4. Phasor diagram of the synchronous machine in generator mode.

The induced voltage $\underline{U}_\delta$ coming from the resulting rotating air-gap field corresponds to the magnetizing condition (saturation) of the machine:

$$\underline{U}_\delta = -jX_{1m}\underline{I}_\mu \tag{5.11}$$

The internal machine voltage U_P is:

$$\underline{U}_P = -jX_{1m}\underline{I}_2' \tag{5.12}$$

The armature reaction becomes:

$$-jX_{1m}\underline{I}_1 \tag{5.13}$$

The angle ϑ is called *rotor angle*. It shows the phase shift of the internal machine voltage $\underline{U}_P$ against the voltage at the terminals $\underline{U}_1$. In generator operation ϑ is positive, in motor operation ϑ is negative. In no-load operation ($\underline{I}_1 = 0$) and operating the synchronous machine purely with reactive power ($\varphi_1 = \pm\pi/2$ and $R_1 = 0$) the rotor angle is equal to zero ($\vartheta = 0$).

The angle

$$\delta_G = \vartheta + \varphi_1 \tag{5.14}$$

is called *load angle*. In generator operation the excitation MMF is leading the armature MMF by $\frac{\pi}{2} + \delta_G$, in motor operation it is lagging by $\frac{\pi}{2} - \delta_G$.

Having large synchronous machines generally the phase resistance R_1 can be neglected against the phase reactance X_1. Moreover, for the description of the operating performance only the stator voltage equation is required. Therefore, the indices may be omitted. Then there is the following equation:

$$\underline{U}_P = \underline{U} + jX\underline{I} \tag{5.15}$$

From this the equivalent circuit and phasor diagram (see Fig. 5.5) can be deduced (in the following the angle φ_2 is not needed any longer, therefore the phase angle φ_1 will be used without index: $\varphi = \varphi_1$).

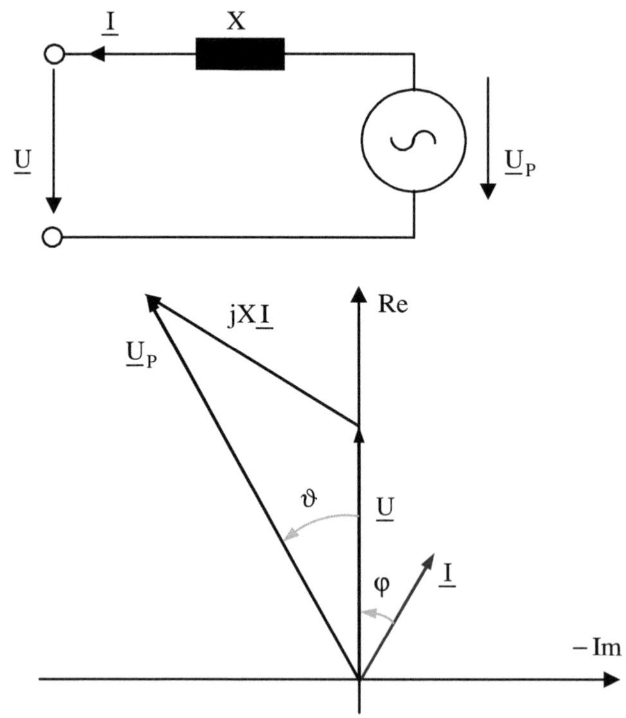

Fig. 5.5. Equivalent circuit (*above*) and phasor diagram (*below*) of the synchronous machine.

5.2 Types of Construction

5.2.1 Overview

Synchronous machines have the same stator construction like induction machines: A three-phase winding is placed in the slots of the lamination stack. For the rotor there are two different types of construction, see Figs. 5.6 and 5.7.

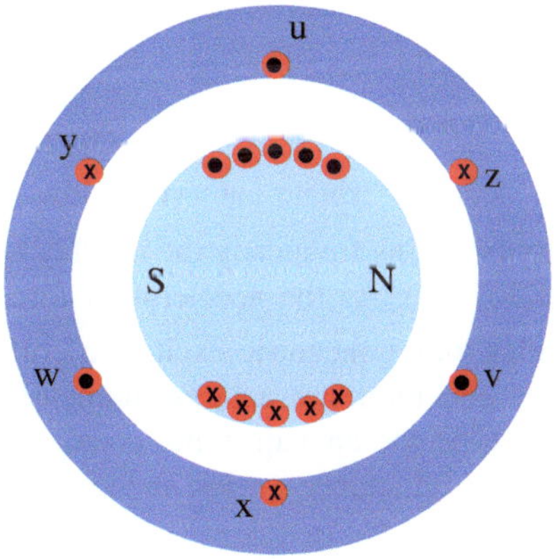

Fig. 5.6. Cylindrical rotor (non-salient pole) synchronous machine (example: p=1).

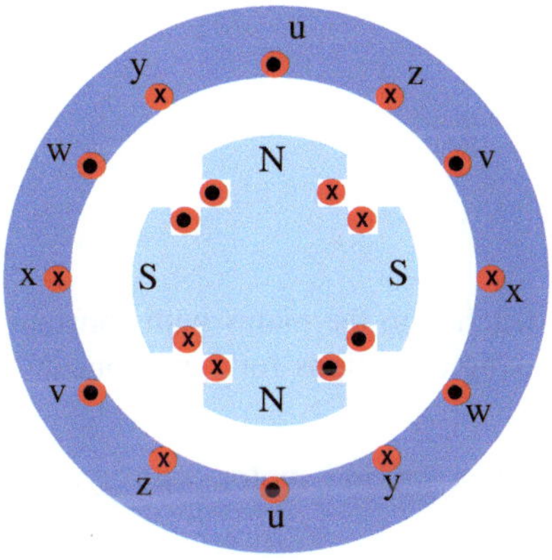

Fig. 5.7. Salient pole synchronous machine (example: p=2).

5.2.2 High-Speed Generator with Cylindrical Rotor

If synchronous generators are driven by steam or gas turbines (thermal power station) the speed is chosen as high as possible to reach an as good as possible turbine efficiency. For a mains frequency of 50Hz the maximum speed is $3000\,\text{min}^{-1}$ (2-pole construction, i.e. $p = 1$). The rotor diameter is limited by the accelerating forces. Because of the high mechanical stress the rotor construction is chosen being cylindrical. The volume required for the desired power is achieved by using a quite long rotor.

5.2.3 Salient-Pole Generator

The turbines of hydroelectric power stations rotate at very low speed ($100 \cdots 750\,\text{min}^{-1}$). To adapt this speed to the mains frequency the number of pole pairs have to be chosen very large ($p = 30 \cdots 4$). As the acceleration forces are low (because of the low speed) single poles with concentric excitation coils may be realized. The salient-pole synchronous generator has a large diameter and a short axial length. Using this construction type the air-gap is not constant at the circumference of the rotor, i.e. the magnetic reluctance varies at the circumference.

5.3 Operation at Fixed Mains Supply

5.3.1 Switching to the Mains

The synchronous machine may only be switched to the mains with constant voltage and frequency (Fig. 5.8), if the following conditions for synchronization are fulfilled:

- The synchronous machine is rotated by a driving motor with synchronous speed: $n = n_0$.
- The excitation current I_F of the synchronous machine is adjusted so that the generator voltage is equal to the mains voltage: $U_{gen} = U_{line}$.

- The phase sequence of the terminal voltages of generator and mains have to be the same: abc – uvw.
- The phase shift of the voltage systems of generator and mains must be identical, i.e. the voltage difference at the terminals that shall be connected must be zero: $\Delta U = 0$.

If these conditions for synchronization are not fulfilled, there are very high torque and current pulsations after switching the generator to the mains.

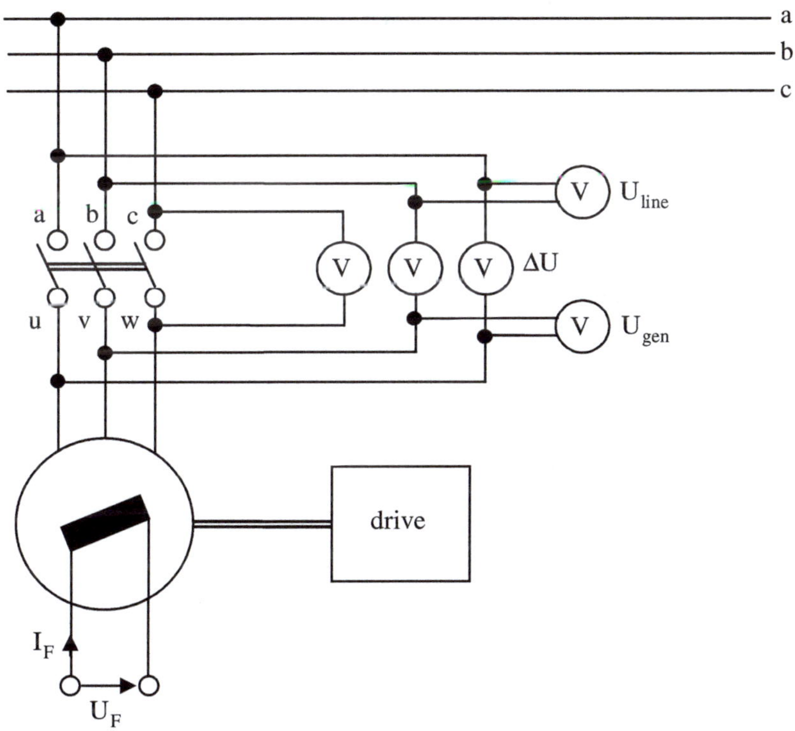

Fig. 5.8. Principle diagram of switching a synchronous machine to the mains

The phase shift of the voltage systems of generator and mains is illustrated in Fig. 5.9.

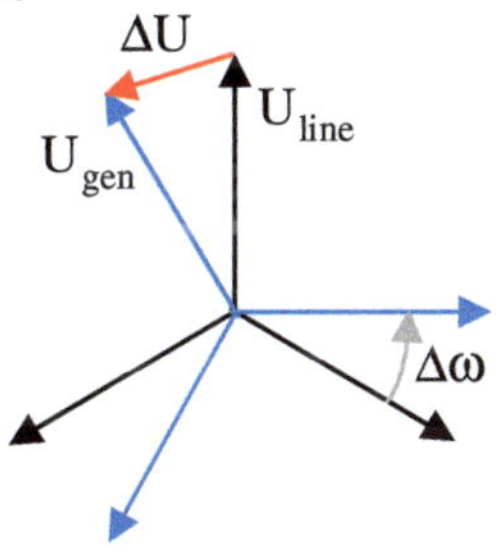

Fig. 5.9. Phase shift of the voltage systems of generator and mains.

5.3.2 Torque Generation

The torque can be calculated from the rotating field power divided by the synchronous angular frequency. Neglecting the stator losses ($R_1 = 0$) the input active power equals this rotating field power:

$$T = \frac{P_\delta}{\Omega_0} = \frac{3UI\cos(\varphi)}{\omega_1/p} \tag{5.16}$$

The phasor diagram of the synchronous machine is shown in Fig. 5.10:

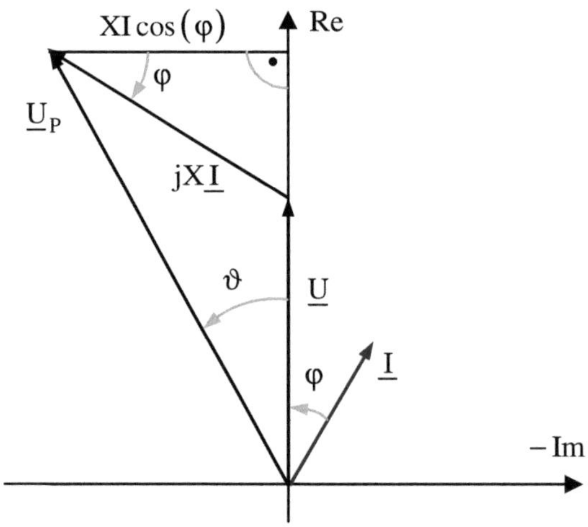

Fig. 5.10. Phasor diagram of the synchronous machine.

From this it can be deduced:

$$\begin{aligned} & XI\cos(\varphi) = U_P\sin(\vartheta) \\ \Rightarrow \quad & I\cos(\varphi) = \frac{U_P}{X}\sin(\vartheta) \end{aligned} \tag{5.17}$$

Therefore, the torque becomes:

$$T = \frac{3p}{\omega_1}\frac{UU_P}{X}\sin(\vartheta) = T_{pull-out}\sin(\vartheta) \tag{5.18}$$

This torque equation is true only for stationary operating points with $I_F = \text{const.}$ and $n = n_0$. Generator operation is given for $\vartheta > 0$, motor operation is given for $\vartheta < 0$. A stable operation is possible only for $-\pi/2 < \vartheta < \pi/2$: Increasing the load slowly, the torque and the rotor angle ϑ are increased as well, until the synchronous machines reaches the pull-out torque at $\vartheta = \pm\pi/2$ and the machine falls out of synchronism. As a motor the machine stops, as a generator it runs away. High oscillating torque components do occur, combined with high currents pulses. In this case the synchronous machine has to be disconnected from the mains immediately.

These characteristics are illustrated in Fig. 5.11.

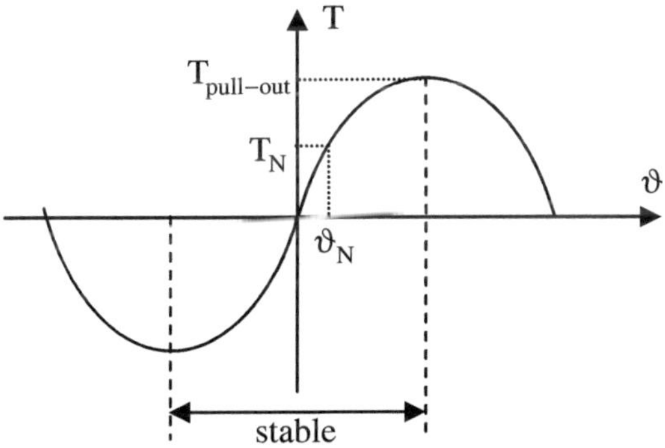

Fig. 5.11. Torque versus rotor angle characteristic of the synchronous machine.

The overload capability of a synchronous machine is:

$$\frac{T_{pull-out}}{T_N} = \frac{\dfrac{3p}{\omega_1}\dfrac{U_N U_P}{X}}{\dfrac{3p}{\omega_1} U_N I_N \cos(\varphi_N)} = \frac{U_P}{X I_N \cos(\varphi_N)} \tag{5.19}$$

In practice, often an overload margin of $\dfrac{T_{pull-out}}{T_N} > 1{,}6$ is called for. A measure for the stability in stationary operation is the synchronizing torque (also see Fig. 5.12):

$$T_{syn} = \frac{dT}{d\vartheta} = T_{pull-out} \cos(\vartheta) > 0 \tag{5.20}$$

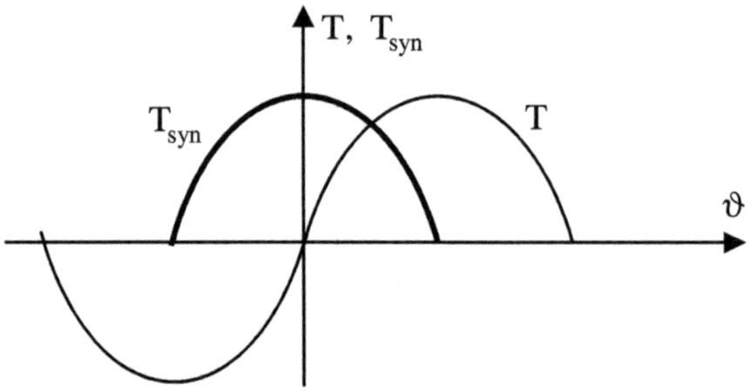

Fig. 5.12. Synchronizing torque of the synchronous machine.

The larger $dT/d\vartheta$, the larger is the restoring torque T_{syn} after a load step. The smaller the absolute value of ϑ, the more stable is the operating point.

5.3.3 Operating Areas

There is:

$$\begin{aligned} \underline{U}_P &= U_P e^{j\vartheta} = U_P\left(\cos(\vartheta) + j\sin(\vartheta)\right) \\ \underline{I} &= Ie^{-j\varphi} = I\left(\cos(\varphi) - j\sin(\varphi)\right) \end{aligned} \tag{5.21}$$

Consequently (if the terminal voltage $\underline{U}$ is defined being in the real axis):

$$\begin{aligned} &\underline{U}_P = \underline{U} + jX\underline{I} \\ \Rightarrow\quad &U_P\left(\cos(\vartheta) + j\sin(\vartheta)\right) = U + jXI\left(\cos(\varphi) - j\sin(\varphi)\right) \end{aligned} \tag{5.22}$$

Separated into real and imaginary parts it follows:

$$\begin{aligned} U_P\cos(\vartheta) &= U + XI\sin(\varphi) \quad \Rightarrow \quad I\sin(\varphi) = \frac{U_P\cos(\vartheta) - U}{X} \\ U_P\sin(\vartheta) &= XI\cos(\varphi) \quad \Rightarrow \quad I\cos(\varphi) = \frac{U_P}{X}\sin(\vartheta) \end{aligned} \tag{5.23}$$

From this four operating areas can be deduced. In the “power generation model” they look like follows:

- $I\cos(\varphi) > 0$ ($\vartheta > 0$): production of active power (generator)
- $I\cos(\varphi) < 0$ ($\vartheta < 0$): consumption of active power (motor)
- $I\sin(\varphi) > 0$ ($U_P\cos(\vartheta) > U$): delivering reactive power (over-excited) machine operates as a capacitor
- $I\sin(\varphi) < 0$ ($U_P\cos(\vartheta) < U$): consumption of reactive power (under-excited) machine operates as an inductor

The characteristic phasor diagrams are illustrated in Fig. 5.14.

The active power is determined only by the driving turbine (generator operation) or by the load torque (motor operation).

The reactive power can be adjusted *independently* just by the excitation (delivering reactive power when being over-excited, consuming reactive power when being under-excited). This is illustrated in Fig. 5.13; the phasors belonging to different excitations are shown in different colors (the phasors of voltage and current are presented in the same color).

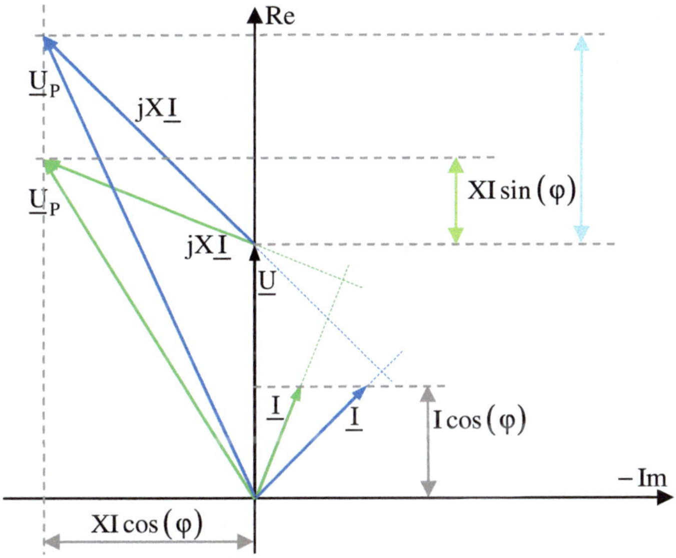

Fig. 5.13. Phasor diagram of the synchronous machine and different excitations.

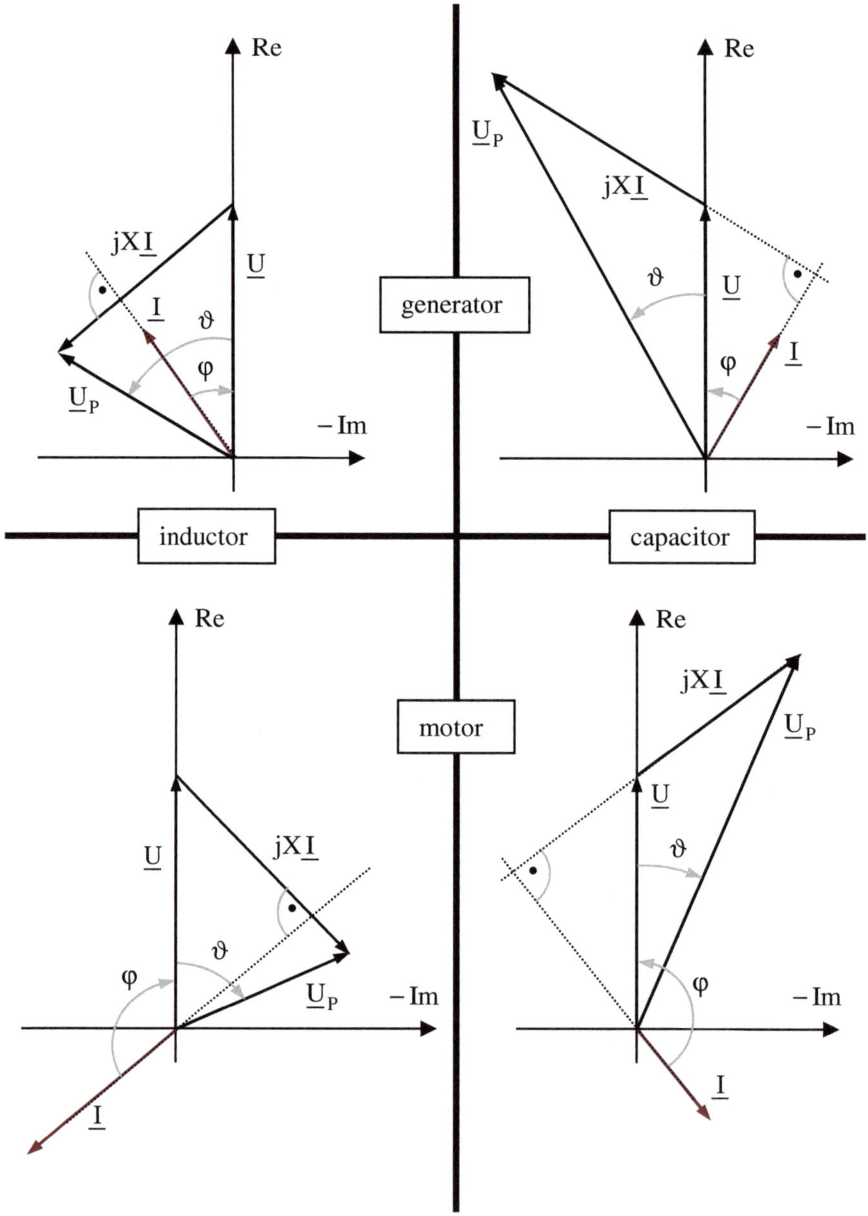

Fig. 5.14. Operating areas of the synchronous machine.

Sometimes synchronous machines are used as reactive power generators without producing active power for *phase shift operation*. With this the inductive reactive power of transformers or induction machines can be compensated and therefore the load of the mains can be reduced. Then the internal machine voltage $\underline{U}_P$ is in phase with the terminal voltage $\underline{U}$; the current $\underline{I}$ is a pure reactive current (in phase shift operation the machine may be over-excited or under-excited). In no-load operation the current is $\underline{I} = 0$ and $\underline{U}_P$ and $\underline{U}$ are identical (same amplitude and phase).

5.3.4 Operating Limits

From

$$\underline{U}_P = \underline{U} + jX\underline{I} \tag{5.24}$$

and

$$\underline{U}_P = U_P e^{j\vartheta}, \qquad \underline{U} = U_{N,phase} \tag{5.25}$$

it follows for the current:

$$\begin{aligned} \underline{I} &= \frac{U_P e^{j\vartheta} - U_{N,phase}}{jX} = j\frac{U_{N,phase}}{X} - je^{j\vartheta}\frac{U_P}{X} \\ \Rightarrow \quad \frac{\underline{I}}{I_N} &= e^{j\frac{\pi}{2}}\frac{U_{N,phase}}{I_N X} + e^{j\left(-\frac{\pi}{2}+\vartheta\right)}\frac{U_P}{U_{N,phase}}\frac{U_{N,phase}}{I_N X} \end{aligned} \tag{5.26}$$

With $U_P \sim I_F$ and $U_P\left(I_F = I_{F,0}\right) = U_{N,phase}$ (no-load) it follows:

$$\frac{U_P}{U_{N,phase}} = \frac{I_F}{I_{F,0}} \tag{5.27}$$

The reactance X related to the nominal impedance is:

$$x = \frac{X}{U_{N,phase}/I_N} = \frac{I_N X}{U_{N,phase}} \tag{5.28}$$

Consequently the current becomes:

$$\frac{\underline{I}}{I_N} = e^{j\frac{\pi}{2}} \frac{1}{x} + e^{j\left(-\frac{\pi}{2}+\vartheta\right)} \frac{I_F}{I_{F,0}} \frac{1}{x} \tag{5.29}$$

From this equation the current diagram of the synchronous machine together with the operation limits can be deduced (Fig. 5.15).

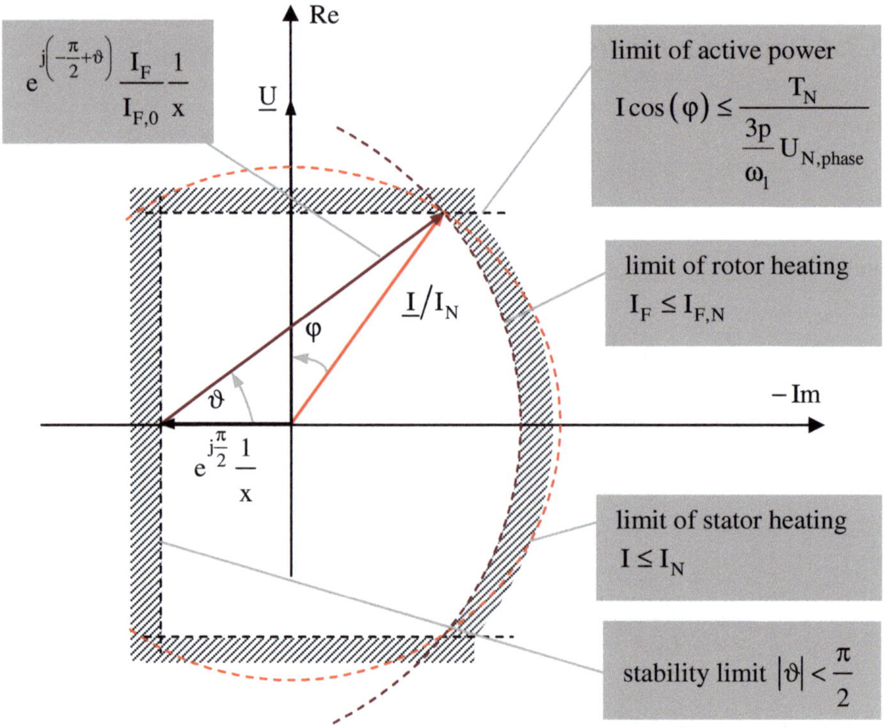

Fig. 5.15. Operating limits of the synchronous machine.

5.4 Isolated Operation

5.4.1 Load Characteristics

Having a single synchronous machine as a generator connected with some loads, this is called "isolated operation". Typically, such an operation is used where a connection to the grid would require very long distances and therefore would imply unreasonable high costs. Frequently, synchronous machines in isolated operation are driven by wind or hydroelectric power stations.

Contrary to the previous considerations, no stable grid can be assumed any more, but the voltage at the terminals of the generator changes with the load (even with constant excitation current).

At first, the well-known phasor diagram of the synchronous machine (Fig. 5.16) can be taken as a basis:

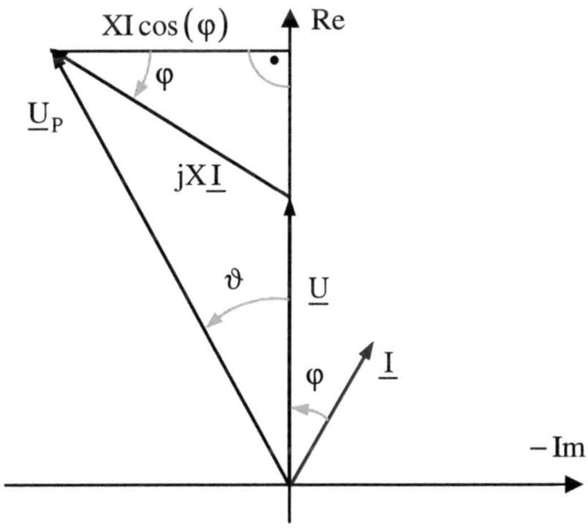

Fig. 5.16. Phasor diagram of the synchronous machine.

From this phasor diagram it follows:

$$
\begin{aligned}
&(U + XI\sin(\varphi))^2 + (XI\cos(\varphi))^2 = U_P^2 \\
\Rightarrow\quad & U^2 + 2UXI\sin(\varphi) + (XI)^2 = U_P^2
\end{aligned}
\tag{5.30}
$$

For $n = n_0$ and $I_F = I_{F,0}$ there is: $U_P = U_{N,phase}$ (this voltage can be measured at the terminals of the synchronous machine at no-load operation). Further:

$$U^2 + 2UXI\sin(\varphi) + (XI)^2 = U_{N,phase}^2$$

$$\Rightarrow \left(\frac{U}{U_{N,phase}}\right)^2 + 2\frac{U}{U_{N,phase}}\frac{XI}{U_{N,phase}}\sin(\varphi) + \left(\frac{XI}{U_{N,phase}}\right)^2 = 1 \qquad (5.31)$$

$$\Rightarrow \left(\frac{U}{U_{N,phase}}\right)^2 + 2\frac{U}{U_{N,phase}}x\frac{I}{I_N}\sin(\varphi) + \left(x\frac{I}{I_N}\right)^2 = 1$$

Herewith, the so-called *load characteristics* (terminal voltage U of the generator depending on the load current I) of the synchronous machine in isolated operation are given. Figure 5.17 shows the terminal voltage U as a function of the load current I for different loads $\cos(\varphi)$.

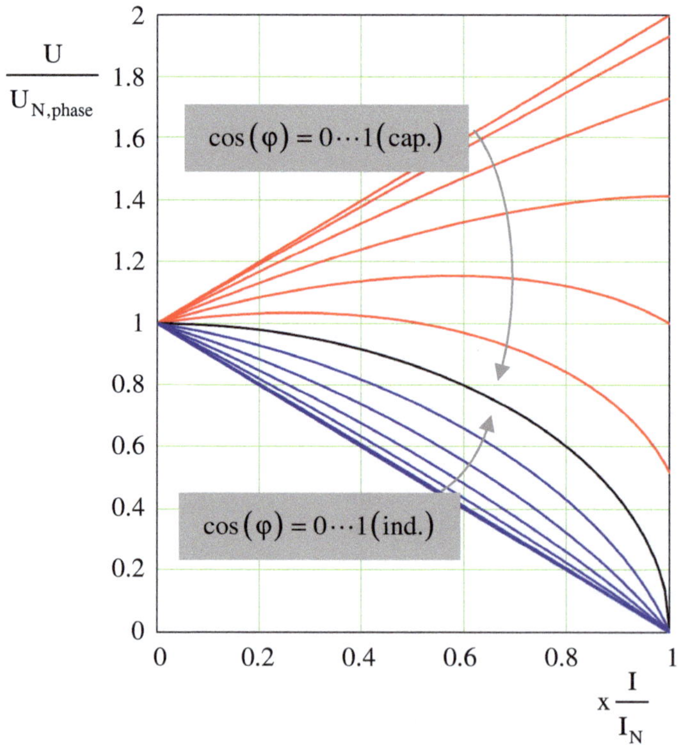

Fig. 5.17. Load characteristics of the synchronous machine in isolated operation.

For resistive-inductive loads the voltage decreases with increasing load, for purely capacitive loads the voltage increases with increasing load (for resistive-capacitive loads with phase angles less than $30° \hat{=} \pi/6$ the voltage may even decrease, depending on the value of the load current).

5.4.2 Control Characteristics

Even in isolated operation the loads should be connected to a constant voltage source, independent from the load current. Therefore, the excitation current has to be controlled in dependency of the load (amplitude and phase shift of the load current). From the phasor diagram (Fig. 5.16) it follows with the requirement $U = U_{N,phase}$:

$$U_{N,phase}^2 + 2U_{N,phase} XI \sin(\varphi) + (XI)^2 = U_P^2 \tag{5.32}$$

With $\frac{U_P}{U_{N,phase}} = \frac{I_F}{I_{F,0}}$ it follows further:

$$\begin{aligned} \frac{I_F}{I_{F,0}} &= \sqrt{1 + \frac{2XI\sin(\varphi)}{U_{N,phase}} + \left(\frac{XI}{U_{N,phase}}\right)^2} \\ &= \sqrt{1 + 2x\frac{I}{I_N}\sin(\varphi) + \left(x\frac{I}{I_N}\right)^2} \end{aligned} \tag{5.33}$$

With this the so-called *control characteristics* (excitation current I_F of the generator in dependency of the load current I) of the synchronous machine in isolated operation are given, to fix the terminal voltage to the constant nominal voltage at synchronous speed $n = n_0$. Figure 5.18 shows the excitation current I_F as a function of the load current I for different loads $\cos(\varphi)$.

For resistive-inductive loads the excitation current has to be increased to fix the terminal voltage to the nominal voltage at increasing load; for purely capacitive loads the excitation current has to be decreased (for resistive-capacitive loads with phase angles less than $30° \hat{=} \pi/6$ the excitation current maybe has to be increased, depending on the value of the load current).

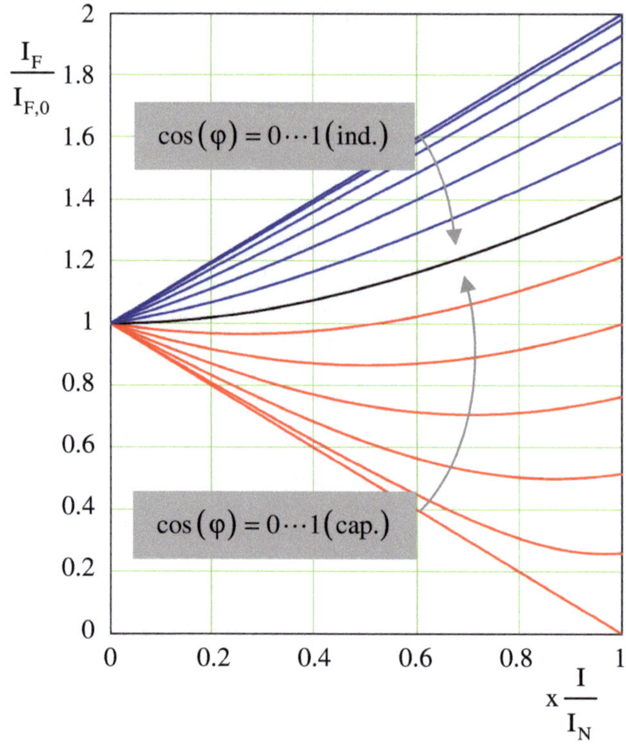

Fig. 5.18. Control characteristics of the synchronous machine in isolated operation.

In Fig. 5.19 the main load and control characteristics are summarized.

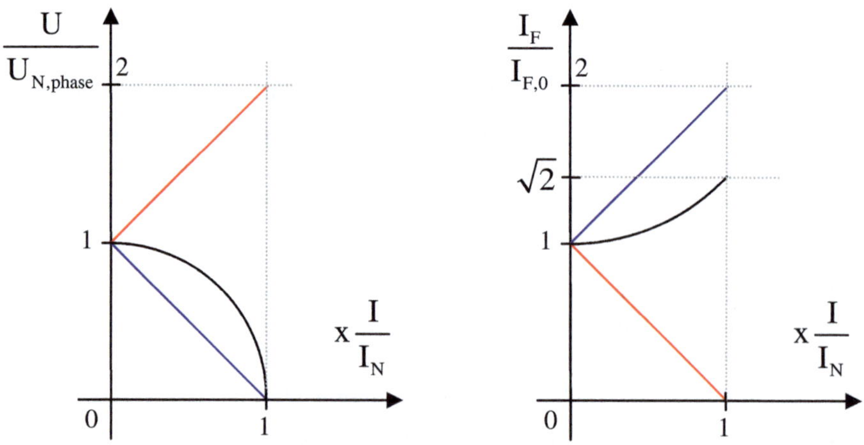

Fig. 5.19. Main load (*left*) and control (*right*) characteristics of the synchronous machine in isolated operation for zero (capacitive and inductive) and unity power factor in *red*, *blue*, and *black*, respectively.

5.5 Salient-Pole Synchronous Machines

Because of the distinct single poles the air-gap on the circumference of the salient-pole machine is not constant (contrary to the induction machine or the non salient-pole synchronous machine). Therefore, a simple summation of the rotating magneto-motive forces of stator and rotor is not allowed to get the resulting air-gap field. In fact, the magneto-motive force of the stator has to be decomposed in two components, one in parallel to the rotor pole axis (d-axis, direct axis), and the other perpendicular to the rotor pole axis (q-axis, quadrature axis):

$$\begin{aligned} \Theta_d &= \Theta \sin\left(\delta_G\right) \\ \Theta_q &= \Theta \cos\left(\delta_G\right) \end{aligned} \tag{5.34}$$

This is illustrated in Fig. 5.20.

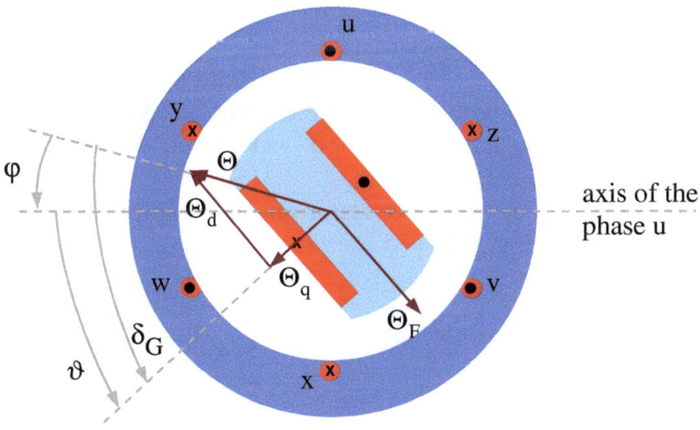

Fig. 5.20. Magneto-motive forces of the salient-pole synchronous machine.

Similarly, the main reactance has to be decomposed according to the d- and q-axis (because of the different air-gap widths in d- and q-axis):

$$\begin{aligned} X_d &= X_{md} + X_{1\sigma} \\ X_q &= X_{mq} + X_{1\sigma} \end{aligned}, \quad X_{mq} < X_{md} \tag{5.35}$$

Now, the phasor diagram (Fig. 5.21) can be drawn (the stator resistance R_1 further is neglected).

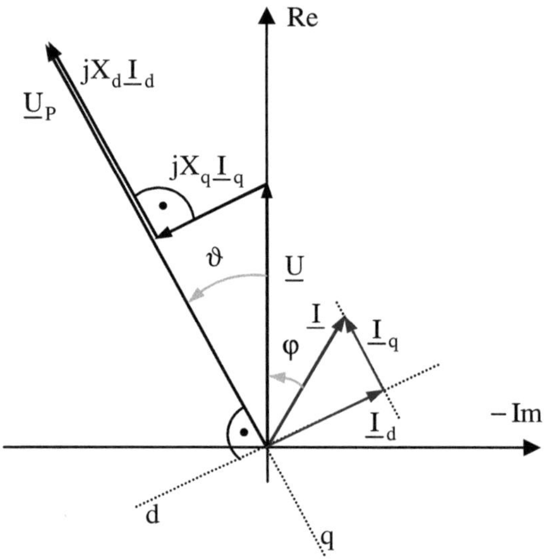

Fig. 5.21. Phasor diagram of the salient-pole synchronous machine.

By decomposing the magneto-motive force of the stator into the components, the armature reaction is determined separately for d- and q-direction and the result is superposed.

Thus, from the original system with three stator phases a two-phase system is generated as a replacement, see Fig. 5.22.

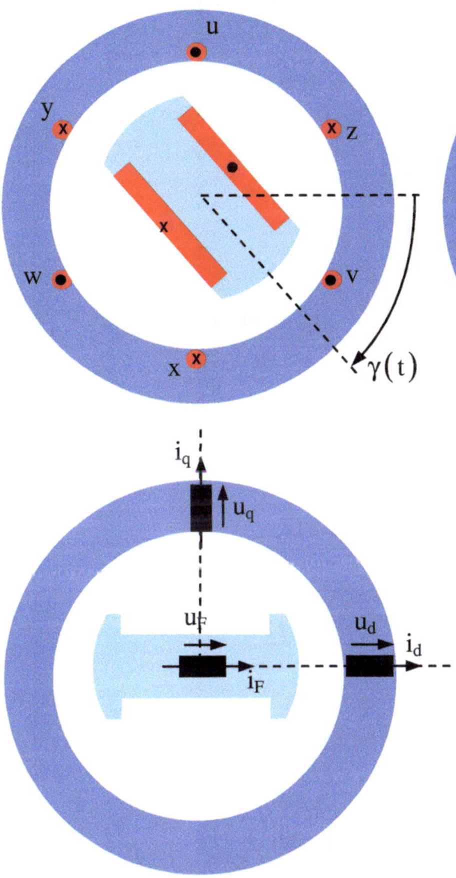

Fig. 5.22. Sketch of the salient-pole synchronous machine: original system (*above left*), replacement (*above right*) and two-phase replacement (*below*).

The main reactances (synchronous reactances) X_d and X_q can be measured: For this the stator winding is energized and the rotor – with open excitation winding – is driven with nearly synchronous speed. Because of the small slip between rotating field of the stator and the rotor the respective axes are alternately coincident or perpendicular. From the ratio of the oscillographically measured phase voltage and phase current the reactance can be calculated; this reactance is oscillating between the extreme values X_d and X_q.

Performing this measurement with a single phase at stand-still, the result will be falsified because of the short-circuited damper winding and the eddy currents in the massive iron parts. The measured values will be too small.

Decomposing the voltage $\underline{U}$ into the components in d- and q-direction, there is (please refer to the phasor diagram in Fig. 5.21):

$$\begin{aligned} \underline{U}\cos(\vartheta)e^{j\vartheta} &= \underline{U}_P - jX_d\underline{I}_d \\ j\underline{U}\sin(\vartheta)e^{j\vartheta} &= jX_q\underline{I}_q \end{aligned} \tag{5.36}$$

From this it follows for the stator current:

$$\underline{I} = \underline{I}_d + \underline{I}_q = \frac{\underline{U}\cos(\vartheta)e^{j\vartheta} - \underline{U}_P}{-jX_d} + \frac{j\underline{U}\sin(\vartheta)e^{j\vartheta}}{jX_q} \tag{5.37}$$

With $e^{j\vartheta} = \cos(\vartheta) + j\cdot\sin(\vartheta)$ it follows further:

$$\begin{aligned} \underline{I} &= j\frac{\underline{U}\cos(\vartheta)(\cos(\vartheta)+j\ \sin(\vartheta)) - \underline{U}_P}{X_d} \\ &\quad + \frac{\underline{U}\sin(\vartheta)(\cos(\vartheta)+j\ \sin(\vartheta))}{X_q} \\ &= \frac{1}{X_d}\left(j\underline{U}\cos^2(\vartheta) - \underline{U}\cos(\vartheta)\sin(\vartheta) - j\underline{U}_P\right) \\ &\quad + \frac{1}{X_q}\left(\underline{U}\cos(\vartheta)\sin(\vartheta) + j\underline{U}\sin^2(\vartheta)\right) \end{aligned} \tag{5.38}$$

With $\cos(x)\sin(x) = \frac{1}{2}\sin(2x)$ the following is true:

$$\begin{aligned} \underline{I} &= \frac{1}{X_d}\left(j\underline{U}\cos^2(\vartheta) - \frac{1}{2}\underline{U}\sin(2\vartheta) - j\underline{U}_P\right) \\ &\quad + \frac{1}{X_q}\left(\frac{1}{2}\underline{U}\sin(2\vartheta) + j\underline{U}\sin^2(\vartheta)\right) \end{aligned} \tag{5.39}$$

and further

$$\underline{I} = \frac{1}{2}\underline{U}\sin(2\vartheta)\left(\frac{1}{X_q} - \frac{1}{X_d}\right) + j\left(\underline{U}\cos^2(\vartheta)\frac{1}{X_d} + \underline{U}\sin^2(\vartheta)\frac{1}{X_q} - \underline{U}_P\frac{1}{X_d}\right) \tag{5.40}$$

An additional transformation with $\sin^2(x) = \frac{1}{2}(1-\cos(2x))$ and $\cos^2(x) = \frac{1}{2}(1+\cos(2x))$ leads to:

$$\begin{aligned}\underline{I} &= \frac{1}{2}\underline{U}\sin(2\vartheta)\left(\frac{1}{X_q} - \frac{1}{X_d}\right) \\ &\quad + j\left(\frac{1}{2}\underline{U}(1+\cos(2\vartheta))\frac{1}{X_d} + \frac{1}{2}\underline{U}(1-\cos(2\vartheta))\frac{1}{X_q} - \underline{U}_P\frac{1}{X_d}\right) \\ &= \frac{1}{2}\underline{U}\sin(2\vartheta)\left(\frac{1}{X_q} - \frac{1}{X_d}\right) \\ &\quad + j\left(\frac{1}{2}\underline{U}\left(\frac{1}{X_d} + \frac{1}{X_q}\right) + \frac{1}{2}\underline{U}\cos(2\vartheta)\left(\frac{1}{X_d} - \frac{1}{X_q}\right) - \underline{U}_P\frac{1}{X_d}\right)\end{aligned} \tag{5.41}$$

Choosing the terminal voltage as real value ($\underline{U} = U$, please refer to the phasor diagram in Fig. 5.21), it follows with $\underline{U}_P = U_P\cos(\vartheta) + jU_P\sin(\vartheta)$ and $\underline{I} = Ie^{-j\varphi} = I[\cos(\varphi) - j\sin(\varphi)]$:

$$\begin{aligned}\mathrm{Im}\{\underline{I}\} &= -I\sin(\varphi) \\ &= \frac{1}{2}U\left(\frac{1}{X_q} + \frac{1}{X_d}\right) - \frac{1}{2}U\cos(2\vartheta)\left(\frac{1}{X_q} - \frac{1}{X_d}\right) \\ &\quad - U_P\cos(\vartheta)\frac{1}{X_d}\end{aligned} \tag{5.42}$$

and

$$\begin{aligned}\mathrm{Re}\{\underline{I}\} &= I\cos(\varphi)\\ &= \frac{1}{2}U\sin(2\vartheta)\left(\frac{1}{X_q}-\frac{1}{X_d}\right)+U_P\sin(\vartheta)\frac{1}{X_d}\end{aligned} \tag{5.43}$$

Neglecting the losses the torque can be calculated from the power of the rotating field:

$$\begin{aligned}T &= \frac{P_\delta}{2\pi n_0}=\frac{P_1}{\omega_1/p}=\frac{3UI\cos(\varphi)}{\omega_1/p}\\ &= \frac{3p}{\omega_1}\left(\frac{UU_P}{X_d}\sin(\vartheta)+\frac{U^2}{2}\left(\frac{1}{X_q}-\frac{1}{X_d}\right)\sin(2\vartheta)\right)\end{aligned} \tag{5.44}$$

The first summand corresponds to the torque of the non salient-pole synchronous machine which is *excitation-dependent*, the second summand is the so-called reaction torque which is *excitation-independent* (caused by the difference of the magnetic reluctances in d- and q-axis). Because of this reaction torque the pull-out torque of the salient-pole synchronous machine is reached at smaller rotor angles than $\pi/2$, please refer to Fig. 5.23. Moreover, it can be deduced from that figure, that the salient-pole machine – because of the additional reaction torque – delivers a higher pull-out torque than the non salient-pole machine, compared at the same excitation current (assuming $X_{d,\text{salient-pole}} = X_{\text{non salient-pole}}$). Additionally, it becomes obvious that for the assumed ratio $X_d = 2X_q$ the pull-out torque without excitation (i.e. the reaction pull-out torque) is only half of the pull-out torque of the non salient-pole synchronous machine.

From the equations for the real part and the imaginary part of the stator current the ratio of reactive power to active power (as a function of rotor angle and excitation) can be deduced:

$$\begin{aligned}\frac{Q}{P} &= \frac{UI\sin(\varphi)}{UI\cos(\varphi)}\\ &= \frac{-\left(\frac{1}{X_q}+\frac{1}{X_d}\right)+\cos(2\vartheta)\left(\frac{1}{X_q}-\frac{1}{X_d}\right)+2\frac{U_P}{U}\cos(\vartheta)\frac{1}{X_d}}{\sin(2\vartheta)\left(\frac{1}{X_q}-\frac{1}{X_d}\right)+2\frac{U_P}{U}\sin(\vartheta)\frac{1}{X_d}}\end{aligned} \tag{5.45}$$

From active and reactive power the power factor can be calculated like follows:

$$\cos(\varphi) = \frac{P}{\sqrt{P^2 + Q^2}} \tag{5.46}$$

In Figs. 5.23 to 5.25 the characteristics are presented as a function of the rotor angle ϑ and for different excitations U_P/U.

- T_{ratio} : the torque of the salient-pole machine relative to the pull-out torque of the non salient-pole machine with nominal excitation ($U_P/U = 1.0$) and comparable machine data ($X_{d,salient-pole} = X_{non\ salient-pole}$) is shown in Fig. 5.23,
- P_{ratio} : the ratio of reactive power to active power is presented in Fig. 5.24,
- $\cos(\varphi)$: Figure 5.25 depicts the power factor.

Especially for non-excited rotors (i.e. synchronous reluctance machines, red characteristic in Figs. 5.23 to 5.25) it is decisive to realize a high ratio of X_d / X_q to improve the torque capability as well as the power factor.

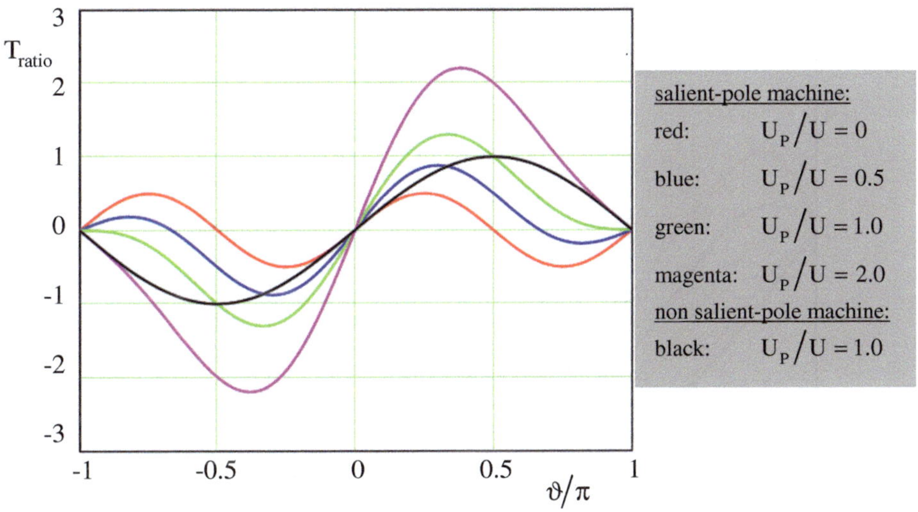

Fig. 5.23. Torque versus rotor angle with the parameter excitation.

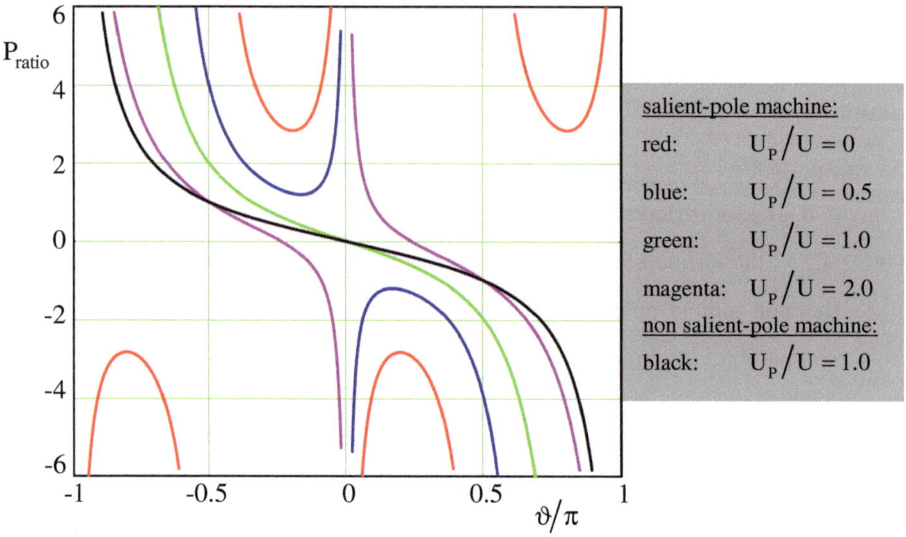

Fig. 5.24. Ratio of reactive power to active power with the parameter excitation.

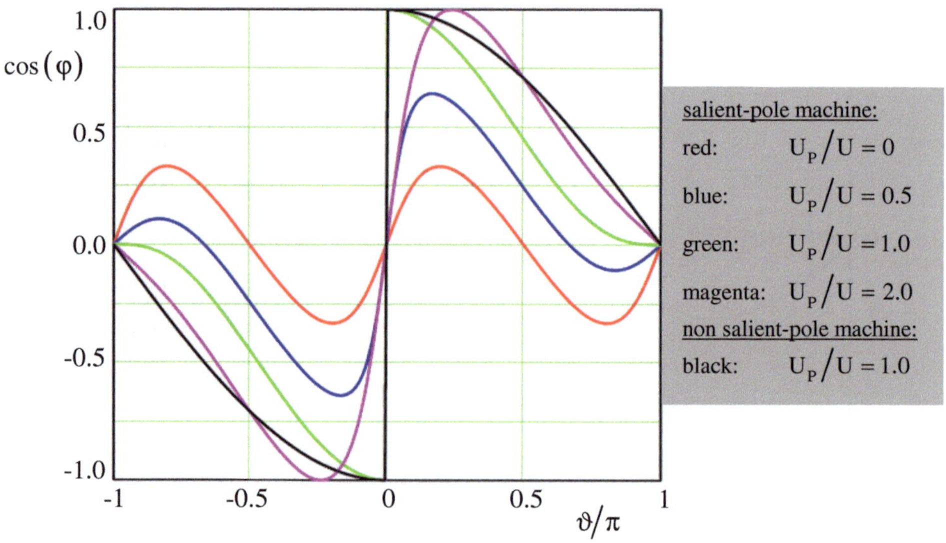

Fig. 5.25. Power factor versus rotor angle with the parameter excitation.

5.6 References for Chapter 5

Kleinrath H (1975) Grundlagen elektrischer Maschinen. Akademische Verlagsgesellschaft, Wiesbaden

Müller G, Ponick B (2009) Theorie elektrischer Maschinen. Wiley-VCH Verlag, Weinheim

Richter R (1963) Elektrische Maschinen II. Birkhäuser Verlag, Basel

Seinsch HO (1993) Grundlagen elektrischer Maschinen und Antriebe. Teubner Verlag, Stuttgart

Spring E (1998) Elektrische Maschinen. Springer-Verlag, Berlin

6 Permanent Magnet Excited Rotating Field Machines

6.1 Rotor Construction

Having a synchronous machine and substituting the DC excitation current (which generates a constant magnetic field according to time) by an excitation with permanent magnets, the following is saved

- voltage source for the excitation current,
- excitation winding and
- excitation current supply via slip rings and brushes.

However, the excitation field is no longer controllable. Figure 6.1 illustrates different alternatives for the positioning of the permanent magnets (surface mounted permanent magnet machines, also called SPM or SMPM machines and interior permanent magnet machines, also called IPM machines, which carry the permanent magnets inside the rotor iron).

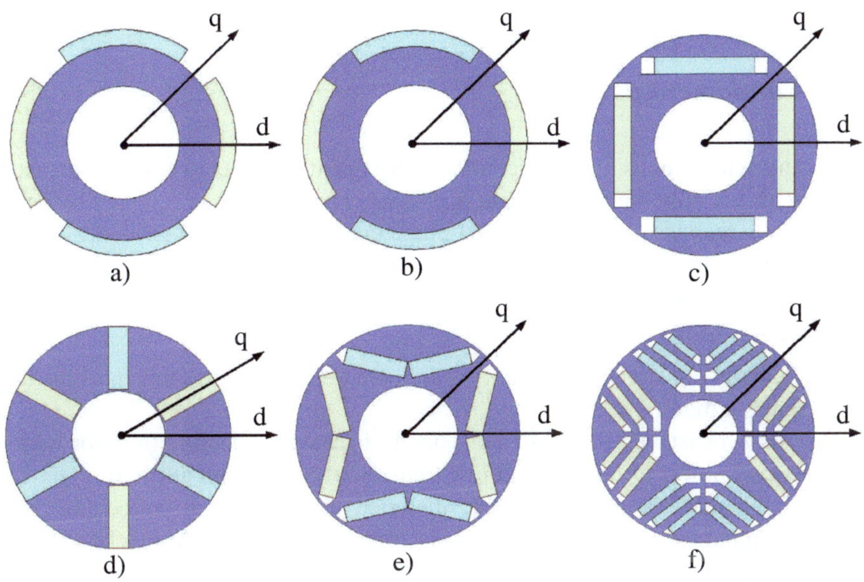

Fig. 6.1. Different alternatives of positioning the permanent magnets in the rotor: a) surface magnets; b) inset magnets; c) to f) buried magnets; the main magnetic axes are described with d and q.

6.2 Linestart-Motor

Incorporating a starting cage into such a permanent magnet synchronous machine, it is called "linestart-motor": The motor is supplied directly from the line voltage, the speed-up is realized as an induction motor. Near to the synchronous speed the rotor is synchronized to the rotating field. Then the machine operates as synchronous motor at the line.

The advantages of self-starting, good power factor, and high efficiency during operation are opposed by a small utilization (because of the combination of two different machine types in the rotor). In addition, during starting operation the permanent magnets produce considerable torque pulsations as the frequencies of the magneto-motive force waves are not identical (please refer to Sect. 3.7). Applications for such linestart-motors are drives with very long operation duration and small power (e.g. small pumps and blowers, etc.).

6.3 Electronically Commutated Rotating Field Machine with Surface Mounted Magnets

6.3.1 Fundamentals

Again, the synchronous machine contains permanent magnets to generate the excitation field, but now no starting cage is present. The machine is supplied by an inverter which realizes a sinusoidal three-phase current system.

For explaining the operational behavior it is recommended to start with the simplified equivalent circuit of the synchronous machine (Sect. 5.1), but with the following changes:

- The "energy consumption system" is applied, because mostly this machine type is used as a motor.
- The stator resistance R will be considered, because neglecting (e.g. for motors with small power or operation at low frequency) often is not allowed.
- In this chapter rotors with surface mounted magnets are regarded (so-called surface-mounted permanent magnet machines, SPM or SMPM machines). This means that the magnetically effective air-gap length is constant along the circumference, as permanent magnets have a relative permeability near to 1 (i.e. near to the value of air). Therefore, the inductivities in d- and q-direction are the same and they do not have to be distinguished.

The equivalent circuit diagram is shown in Fig. 6.2.[7]

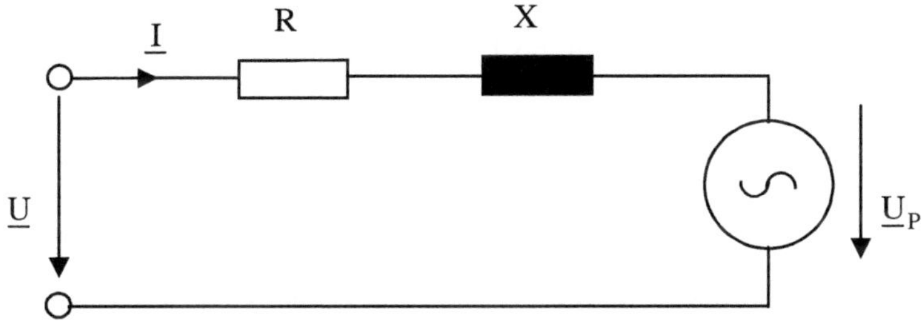

Fig. 6.2. Equivalent circuit diagram of the permanent magnet excited rotating field machine.

The fundamental frequency of the supplying three-phase system determines the frequency of the rotating magneto-motive force and therefore even the rotor speed. The rotating magneto-motive force together with the field of the permanent magnet rotor generates the torque. Mostly, this torque shall be as smooth as possible. The rotation of the rotating stator field is realized depending on the rotor position by means of the inverter in such a way, that the electrical angle between rotating magneto-motive force of the stator and the rotor field is $\pi/2$ (i.e. $\vartheta = -\varphi$). With this the load angle in the "energy consumption system" becomes $\delta_M = -\delta_G = -\vartheta - \varphi = 0$ (please compare to the load angle δ_G in the "energy generation system" in Sect. 5.1). The rotor position can be measured by using sensors or it can be deduced from the terminal voltages and/or terminal currents. The explained operation is described by the phasor diagram (see Fig. 6.3).

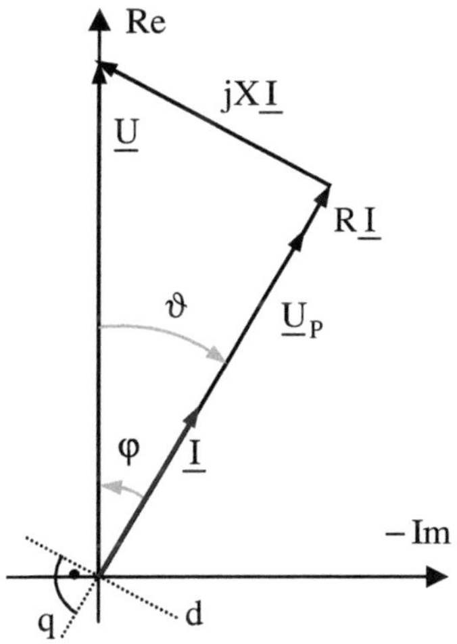

Fig. 6.3. Phasor diagram of the permanent magnet excited rotating field machine.

---

[7] The internal machine voltage (open-circuit voltage, no-load voltage) U_P sometimes is also called "back emf" (electromotive force) and nominated with "e".

As this machine type mostly is used as a motor, the relations are given here for the "energy consumption system".

An operation is obtained that does no longer correspond to the synchronous machine, but to the DC-machine:

- DC-machine: magneto-motive force of the armature and excitation field build an electrical angle of $\pi/2$; this adjustment is realized mechanically by means of the commutator.
- Synchronous machine: the rotor angle ϑ and the phase angle φ are adjusted depending on the operation point; there is no active influence on the phase shift between magneto-motive force of the stator and excitation field.
- Electronically commutated permanent magnet excited rotating field machine: magneto-motive force of the stator and rotor field build an electrical angle of $\pi/2$; this adjustment is realized electronically by means of the inverter.

This machine topology shows very good dynamics and the control is quite simple. The brushless technology is wear-resistant and maintenance-free. This kind of motor often is used for machine tool drives and for robot drives.

From the phasor diagram there is:

$$\underline{U} = \underline{U}_P + R\underline{I} + jX\underline{I} \tag{6.1}$$

The input active power is:

$$P_1 = 3\left(U_P I + RI^2\right) = P_\delta + P_{loss,1} \tag{6.2}$$

Now, the torque can be calculated from the rotating field power:

$$T = \frac{P_\delta}{\omega_1/p} = \frac{3p}{\omega_1} U_P I \tag{6.3}$$

To achieve good operational characteristics as a motor (generally the torque should be as smooth as possible), the *permanent magnet field*, the *stator winding* and the *motor supply* have to be adjusted to each other very carefully. Some examples are given in the next sections.

6.3.2 Brushless DC-Motor

Firstly, a motor without short-pitch winding (i.e. a motor with diameter winding) and a two-pole ($p = 1$), radially magnetized permanent magnet ring is regarded.

The three phases of the motor (usually in star connection) are supplied from a three-phase inverter (illustrated by the six switches in Fig. 6.4). The electrical power is drawn from the DC-voltage U_{DC}.

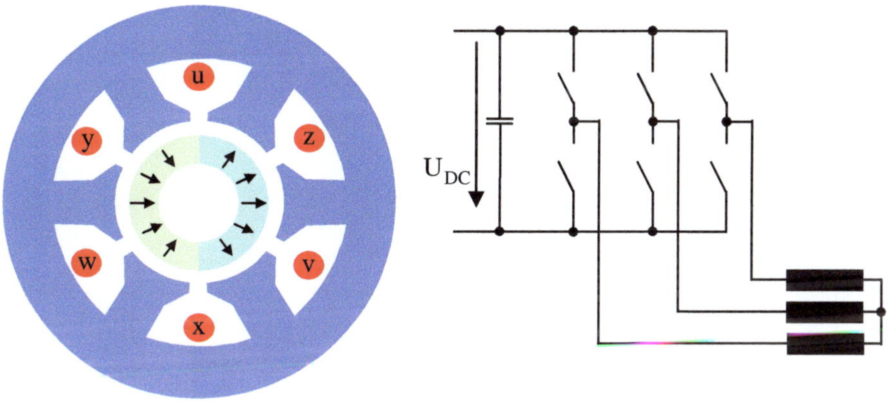

Fig. 6.4. Two-pole permanent magnet excited rotating field machine (*left*) and power electronic supply (*right*).

In the following the torque will be calculated from the change of the flux linkage with respect to the rotational angle (in the linear case it follows for a single phase of the motor $T = \frac{\partial}{\partial \alpha} W_{mag}$ and $W_{mag} = \int_0^{\Psi} i d\tilde{\Psi} = \frac{1}{2} i \Psi$):

$$T = \sum_{k=1}^{m} T_k = \frac{1}{2} \sum_{k=1}^{m} \frac{d\Psi_k}{d\alpha} i_k \tag{6.4}$$

Here $\alpha = \beta/p$ is the mechanical angle (β is the electrical angle) and m is the number of phases. In this special case there is $\alpha = \beta$ (because of $p = 1$).

The characteristics of flux linkage, change of flux linkage, current, and torque are shown in Fig. 6.5. It can be deduced that the chosen current characteristics lead to an ideally smooth torque characteristic.

Because of the kind of operation, which is similar to the operation of a DC-motor, and because of the sectional DC-characteristic of the currents (which is the same as for DC-motors), the names of this kind of motor (together with the kind of operation) are derived:

- electronically commutated DC-motor (sometimes referred to as EC-motor);
- brushless DC-motor, BLDC-motor (which is most common).

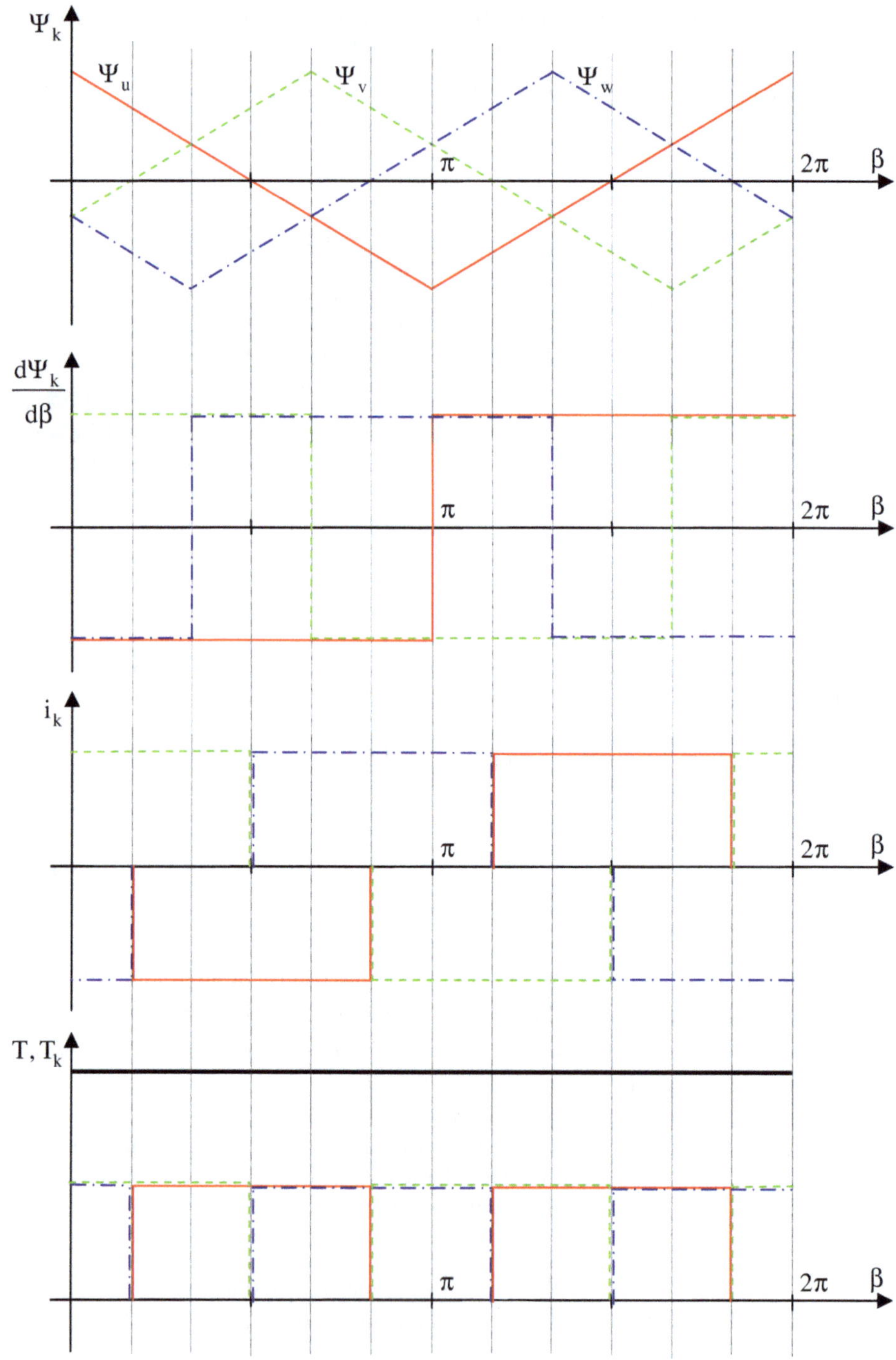

Fig. 6.5. Flux linkage, its derivative, phase current and torque versus electrical angle.

It can be deduced that for square-wave currents only two phases are energized at the same time. This current waveform is optimally adapted to the radial magnetization of the permanent magnets to achieve the smoothest possible torque characteristic (when neglecting the slotting effect).

It has to be emphasized here that the Eqs. (6.1) to (6.3) and the phasor diagram (Fig. 6.3) are only valid for sinusoidal voltages and currents. Nevertheless, the torque calculation according to Eq. (6.4) is valid for any time-dependency, please refer to Chap. 1. Therefore the shown characteristics in Fig. 6.5 are valid.

However, to achieve a torque characteristic as smooth as possible the flux per phase characteristics of the brushless DC-motor may be different from the above shown functions *outside* of the power-on time of the respective phase currents. From this additional degree of freedom the possibility to choose the magnetization or the winding layout differently arises.

For example the torque contribution of each phase and the entire torque remain unchanged, if the value of the derivative of the flux linkage of each phase is changed at the discontinuity of the flux linkage characteristic in a range of $\pm\pi/6$ (e.g. because of a gap between two poles or because of magnetic leakage between two poles). Therefore, trapezoidal flux linkage characteristics because of leakage together with ideal square-wave currents generate a constant torque, as long as the constant part of this trapezoidal characteristic is not less than the critical power-on time period of $2\pi/3$.

However, the practical realization of steep current slopes (square-wave currents) require especially in the middle and upper speed region with high induced back emf voltages a high voltage reserve of the inverter. This means an overdimensioning of the inverter and consequently high costs. In addition, high acoustic noise is generated, which is not acceptable for many applications. Therefore, in many cases a deviation from the idealized square-wave current is accepted, which is accompanied by a higher torque oscillation, but even lower costs and lower acoustic noise.

Another often used winding topology can be deduced from the above described diameter winding by incorporating an extreme short-pitch: the winding of a phase is concentrated onto one stator tooth; please refer to Fig. 6.6 for a four-pole motor.

This winding topology contains two main advantages:

- The conductor length in the end winding is extremely short; this means a low stator resistance as well as low copper weight. Consequently, cost and losses are reduced.
- With a respective geometry of the lamination (e.g. stator teeth with parallel shoulders or separated lamination) the coils can be wound before and then mounted to the teeth. This reduces the manufacturing costs.

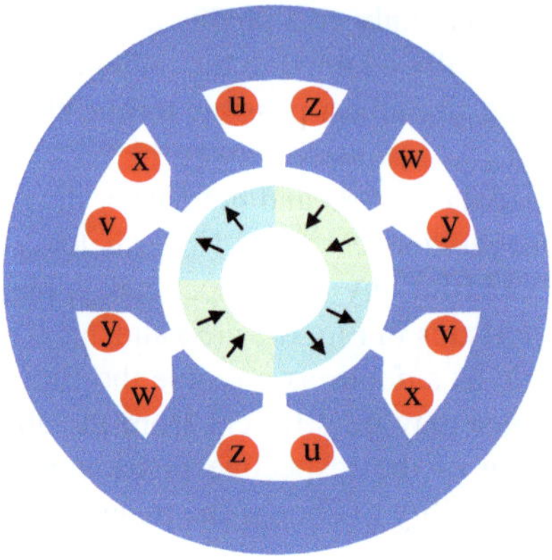

Fig. 6.6. Four-pole permanent magnet excited rotating field machine.

Generally, using this concentrated winding means that the width of a rotor pole is larger than the width of a stator tooth. Therefore, the flux linkage characteristic is trapezoidal: for a part of the rotor rotation, always parts of $2\pi/2 - 2\pi/3 = \pi/3$ electrical degrees, the flux in the stator coil is unchanged; this is different to the above example of the two-pole machine with diameter winding.

In practice this extreme short-pitch winding is characterized by torque dips, because the parts with constant $d\Psi/d\alpha$ are shortened and the magnetic leakage at the pole edges (transition between north and south pole) becomes more prominent.

Flux linkage, change of flux linkage, current, and torque for this case are shown in Fig. 6.7 for idealized assumptions:

To achieve a torque as smooth as possible even in practice the following means can be implemented:

- skewing of the stator slots or the rotor poles (disadvantages: high costs, flux leakage);
- distributed, short-pitch two-layer winding (disadvantage: high costs);
- choosing a fractional number for the ratio of stator slots and rotor poles (disadvantage: for non-symmetric winding topologies there are radial forces onto the rotor);
- sinusoidal flux linkage combined with sinusoidal currents (disadvantage: complex current shaping).

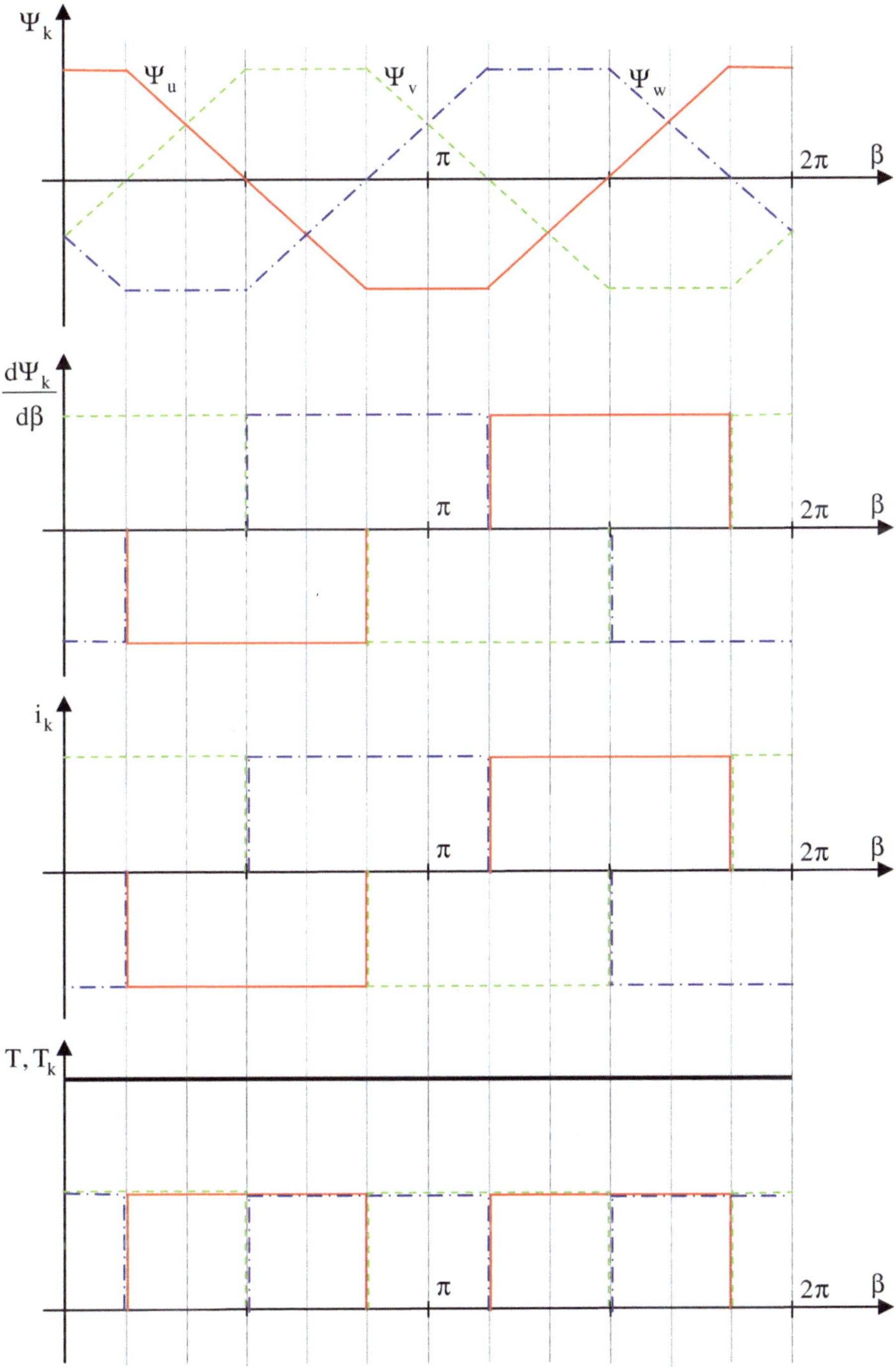

Fig. 6.7. Flux linkage, its derivative, phase current and torque versus electrical angle.

6.3.3 Electronically Commutated Permanent Magnet Excited Synchronous Machine

Operating the supplying inverter appropriately with high switching frequency (at low power the switching frequency for the voltage waveform generally is above the human threshold of hearing of about 20kHz), the phase currents of the motor can be nearly sinusoidal because of the low-pass effect of the phase impedances. With this an artificial three-phase system with variable voltage and variable frequency is realized. The permanent magnet excited synchronous machine supplied by such a frequency variable sinusoidal current system in literature often is referred to as "permanent magnet synchronous machine (PMSM)" or "brushless AC operation".

A motor construction that is adapted concerning winding topology and magnet design to this sinusoidal operation (to reach a smooth torque) is illustrated in Fig. 6.8.

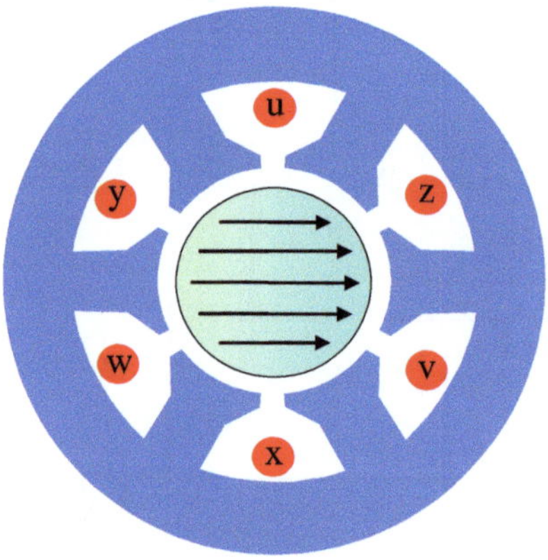

Fig. 6.8. Two-pole (diametrically magnetized) permanent magnet excited rotating field machine.

The two-pole rotor is magnetized diametrically and it generates a sinusoidal air-gap field. The stator contains a diameter winding.

The idealized characteristics of flux linkage, change of flux linkage, currents, and torque are shown in Fig. 6.9.

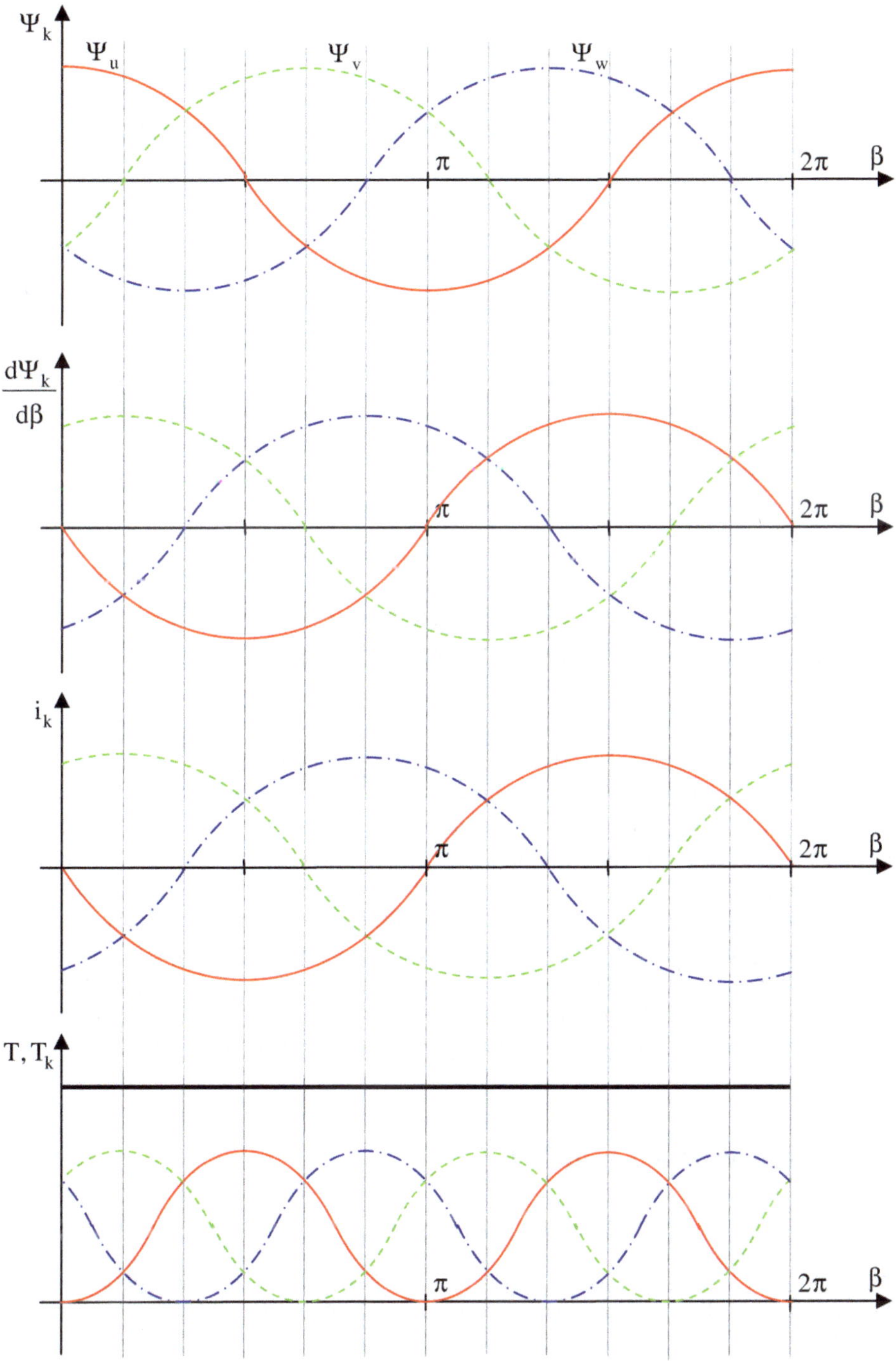

Fig. 6.9. Flux linkage, its derivative, phase current and torque versus electrical angle.

From

$$T = \frac{1}{2}\sum_{k=1}^{m}\frac{d\Psi_k}{d\alpha}\, i_k, \qquad \alpha = \frac{\beta}{p} \tag{6.5}$$

it follows:

$$T = \frac{p}{2}\sum_{k=1}^{m}\frac{d\Psi_k}{d\beta}\, i_k \tag{6.6}$$

However, considering the magnetic leakage at the pole edges (transition between north and south poles with not perfectly shaped poles) the torque is not increasing linearly with the number of pole pairs. With increasing number of pole pairs the leakage percentage (depending on the geometry) gets increasing relevance and consequently there is an optimum at moderate numbers of pole pairs.

6.4 Calculation of the Operational Characteristics; Permanent Magnet Excited Machines with Buried Magnets

To calculate the operation conditions of the above described machines, it is possible using the "Rotating Field Theory" (Chap. 3) for the permanent magnet excited synchronous machine (PMSM), because this machine shows sinusoidal voltages and currents. However, this is not possible for the brushless DC-machine (BLDC), because this kind of operation makes step-by-step DC-currents necessary. As the machine topology of both alternatives can be described with the same equations (just the supply is different), the calculation of the operational characteristics shall be done using a theory where arbitrary current waveforms can be used. This theory is the "Space Vector Theory" dealt with in Chap. 11. Therefore, the operational characteristics of PMSM and BLDC will be described in the later Chap. 14. In this chapter even the possibility of buried magnets (see Sect. 6.1) is described.

6.5 References for Chapter 6

Dajaku G (2006) Electromagnetic and Thermal Modeling of Highly Utilized PM Machines. Shaker-Verlag, Aachen

Gieras JF, Wing M (2002) Permanent Magnet Motor Technology. Marcel Dekker, New York

Krishnan R (2010) Permanent magnet synchronous and brushless DC motor drives. CRC Press, Boca Raton

7 Reluctance Machines

7.1 Synchronous Reluctance Machines

The torque of the salient-pole synchronous machine is composed of two parts; the first one is generated by the excitation current and the second one by the different reluctance in d- and q-axis:

$$T = \frac{3p}{\omega_1}\left(\frac{UU_P}{X_d}\sin(\vartheta) + \frac{U^2}{2}\left(\frac{1}{X_q} - \frac{1}{X_d}\right)\sin(2\vartheta)\right) \tag{7.1}$$

Omitting the excitation winding, the slip rings, and the brushes the torque because of the different reluctance is remaining:

$$T = \frac{3p}{\omega_1}\frac{U^2}{2}\left(\frac{1}{X_q} - \frac{1}{X_d}\right)\sin(2\vartheta) \tag{7.2}$$

The advantages of such a machine are:

- simple construction (no excitation winding, no slip rings, no brushes);
- no (excitation) losses in the rotor.

A challenge is the fact that the reachable torque is depending on the reactances in d- and q-axis: E.g. for $X_d = 2X_q$ it had been calculated for the salient-pole synchronous machine (Sect. 5.5) that the reluctance torque was just half of the torque generated by the nominal excitation current. Another disadvantage is the poor power factor $\cos(\varphi)$ (see Sect. 5.5), so that an inverter with large power rating has to be used.

Because of this synchronous reluctance machines have only relevance, if the difference of the reactances in d- and q-axis can be increased considerably (e.g. by proper rotor design with multiple flux barriers).

7.2 Switched Reluctance Machines

7.2.1 Construction and Operation

The construction of the switched reluctance machine (SR-machine, SRM) is simple, robust and cost-effective. The stator is composed of teeth (poles) with concentrated coils. Generally, opposite coils are representing a winding phase. The rotor contains teeth without windings, the number of rotor teeth is lower than the number of stator teeth. Figure 7.1 illustrates a switched reluctance motor with six stator poles (stator teeth) and four rotor teeth (6/4-motor).

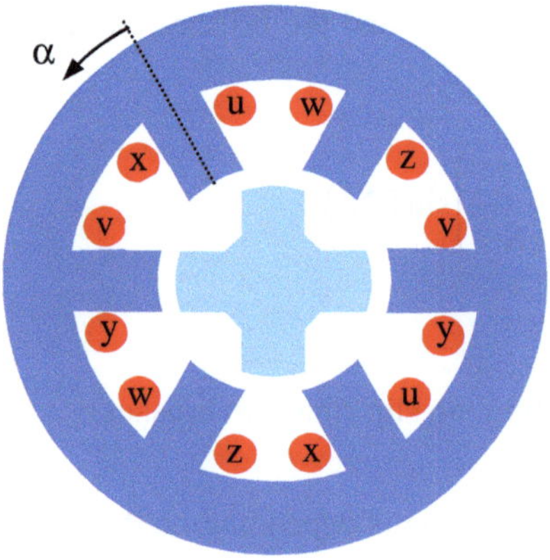

Fig. 7.1. Principle construction of a 6/4 switched reluctance machine.

The shown SRM has three phases. To rotate the rotor the stator phases have to be switched on and off cyclically depending on the actual rotor position.

For moving the rotor clockwise (relative to the mechanical angle α in negative direction), the phase with coils w-z has to be switched on next at the above presented motor. If in the following the rotor teeth are aligned to the stator teeth of this phase („aligned position"), it is not possible to generate further torque (in the desired direction) with this phase. It has to be switched off and the phase with coils u-x will be switched on. After accordant rotation of the rotor, this phase will be switched off and the phase with coils v-y will be switched on. Then the rotor moves further until it reaches the above presented position; here this phase will be switched off again and the phase with coils w-z will be switched on.

For knowing the actual rotor position it must be measured: directly via sensors or indirectly via the terminal values (current and voltage).

From the described operation of the SRM it becomes obvious that the rotor moves in opposite direction to the rotating field (energizing the stator phases in anti-clockwise direction a clockwise movement of the rotor is created).

Having m stator phases and 2p stator poles per phase, the number of stator teeth is:

$$N_S = 2pm \tag{7.3}$$

With every impulse the stator field rotates by the angle:

$$\alpha_S = \frac{2\pi}{N_S} = \frac{2\pi}{2pm} \tag{7.4}$$

The number of rotor teeth is:

$$N_R = 2p(m-1) < N_S \tag{7.5}$$

With every impulse the rotor field rotates by the angle:

$$\alpha_R = \frac{2\pi}{N_R m} \tag{7.6}$$

Therefore, the rotor moves by the factor [8]

$$a = \frac{\alpha_R}{\alpha_S} = \frac{2p}{N_R} \tag{7.7}$$

slowlier than the rotating stator field. Therefore, depending on the application a speed reduction gear-set can be omitted because of the motor design. However, the switching frequency has to be accordingly larger, if the speed of the SRM should be as high as for a motor rotating in synchronism with the rotating field.

[8] There are even SRM alternatives with $N_R = 2p(m+1) > N_S$. For this number of rotor teeth the rotor moves in the same direction like the rotating stator field. Generally, such variants are not used, because the mechanical construction is less robust and the frequency is higher for a certain speed. Therefore, in the following these alternatives are not regarded further.

7.2.2 Torque

The voltage equation for a single phase is:

$$\begin{aligned} u &= Ri + \frac{d\psi(i,\alpha)}{dt} \\ &= Ri + \frac{\partial\psi(i,\alpha)}{\partial i}\frac{di}{dt} + \frac{\partial\psi(i,\alpha)}{\partial\alpha}\frac{d\alpha}{dt} \end{aligned} \tag{7.8}$$

The sum on the right-hand side is composed of three parts:

- voltage drop at the Ohmic resistance;
- induced voltage because of current change (transformer effect);
- induced voltage because of rotational movement of the rotor.

The torque can be calculated from a power balance (please compare the force calculation of the lifting magnet in Chap. 1). Multiplying Eq. (7.8) with the phase current, it follows:

$$\begin{aligned} ui &= Ri^2 + i\frac{\partial\psi(i,\alpha)}{\partial i}\frac{di}{dt} + i\frac{\partial\psi(i,\alpha)}{\partial\alpha}\frac{d\alpha}{dt} \\ \Rightarrow \quad uidt &= Ri^2dt + i\frac{\partial\psi(i,\alpha)}{\partial i}di + i\frac{\partial\psi(i,\alpha)}{\partial\alpha}d\alpha \end{aligned} \tag{7.9}$$

The electrical input energy at the terminals is:

$$W_{el} = \int uidt \tag{7.10}$$

The electrical losses are:

$$W_{loss} = \int i^2Rdt \tag{7.11}$$

The magnetic energy is:

$$W_{mag} = \int id\psi \tag{7.12}$$

The magnetic co-energy is:

$$W'_{mag} = \int \psi di \tag{7.13}$$

The mechanical energy is:

$$W_{mech} = \int T d\alpha \tag{7.14}$$

As the change of electrical energy at the terminals has to cover the change of electrical losses, the change of magnetic energy and the change of mechanical energy, it follows:

$$\begin{aligned} & dW_{el} = dW_{loss} + dW_{mag} + dW_{mech} \\ \Rightarrow \quad & uidt = Ri^2 dt + dW_{mag} + dW_{mech} \end{aligned} \tag{7.15}$$

By comparing with the above equation it is:

$$dW_{mag} + dW_{mech} = i \frac{\partial \psi(i,\alpha)}{\partial i} di + i \frac{\partial \psi(i,\alpha)}{\partial \alpha} d\alpha \tag{7.16}$$

In an intermediate step the change of magnetic energy will be calculated:

As the magnetic energy is dependent on the phase current as well as on the rotor position, it follows:

$$dW_{mag} = \frac{\partial W_{mag}}{\partial i} di + \frac{\partial W_{mag}}{\partial \alpha} d\alpha \tag{7.17}$$

For a fixed rotor position angle α there is (the tilde is introduced to differentiate between integration limit and integration variable) (see Fig. 7.2):

$$W_{mag} = \int_0^{\psi} i d\tilde{\psi} = i\psi - W'_{mag} = i\psi - \int_0^{i} \psi d\tilde{i} \tag{7.18}$$

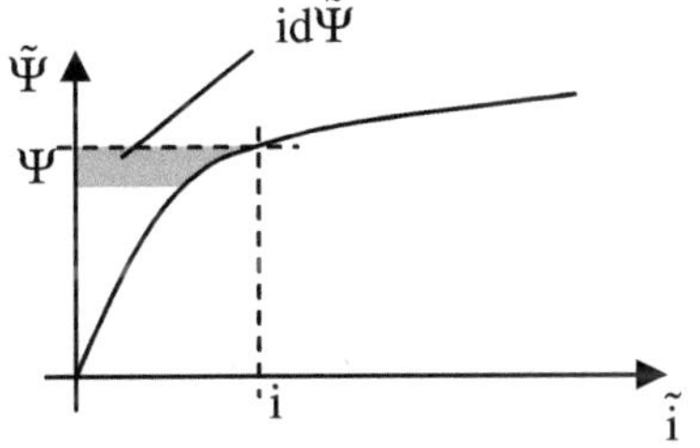

Fig. 7.2. Flux linkage versus current diagram of the switched reluctance machine.

Further it follows:

$$\frac{\partial W_{mag}}{\partial i} = \psi + i\frac{\partial \psi}{\partial i} - \int_0^i \frac{\partial \psi}{\partial i} d\tilde{i} = i\frac{\partial \psi}{\partial i} \tag{7.19}$$

and:

$$\frac{\partial W_{mag}}{\partial \alpha} = \frac{\partial (i\psi)}{\partial \alpha} - \frac{\partial W'_{mag}}{\partial \alpha} = i\frac{\partial \psi}{\partial \alpha} - \frac{\partial W'_{mag}}{\partial \alpha} \tag{7.20}$$

In total it follows from this intermediate step:

$$dW_{mag} = i\frac{\partial \psi}{\partial i} di + \left(i\frac{\partial \psi}{\partial \alpha} - \frac{\partial W'_{mag}}{\partial \alpha} \right) d\alpha \tag{7.21}$$

Inserting the result from this intermediate step (Eq. (7.21)) into Eq. (7.16), there is:

$$\begin{aligned}
dW_{mag} + dW_{mech} &= i\frac{\partial \psi(i,\alpha)}{\partial i} di + i\frac{\partial \psi(i,\alpha)}{\partial \alpha} d\alpha \\
\Rightarrow \quad i\frac{\partial \psi}{\partial i} di + \left(i\frac{\partial \psi}{\partial \alpha} - \frac{\partial W'_{mag}}{\partial \alpha} \right) d\alpha + dW_{mech} & \\
&= i\frac{\partial \psi(i,\alpha)}{\partial i} di + i\frac{\partial \psi(i,\alpha)}{\partial \alpha} d\alpha \\
\Rightarrow \quad dW_{mech} &= \frac{\partial W'_{mag}}{\partial \alpha} d\alpha \\
\Rightarrow \quad T d\alpha &= \frac{\partial W'_{mag}}{\partial \alpha} d\alpha \\
\Rightarrow \quad T &= \frac{\partial W'_{mag}}{\partial \alpha}
\end{aligned} \tag{7.22}$$

Consequently, the torque of a single phase is obtained from the partial differentiation of the magnetic co-energy of the regarded phase with respect to the rotor position angle. The torque of the entire machine is calculated by summation of all phase torques.

The magnetic co-energy has a very important relevance:

$$\psi = \frac{\partial W'_{mag}}{\partial i} \tag{7.23}$$

$$T = \frac{\partial W'_{mag}}{\partial \alpha} \tag{7.24}$$

Consequently it follows:

$$\frac{\partial \psi}{\partial \alpha} = \frac{\partial T}{\partial i} \tag{7.25}$$

Torque and flux linkage of the switched reluctance machine are directly linked to each other via the rotor position and the phase current.

If there is no saturation during machine operation, the following is true: $\psi(i,\alpha) = L(\alpha)\ i$. Then it follows:

$$\begin{aligned} T &= \frac{\partial W'_{mag}}{\partial \alpha} = \frac{\partial}{\partial \alpha}\left(\int_0^i \psi(\tilde{i},\alpha)\,d\tilde{i}\right) = \frac{\partial}{\partial \alpha}\left(\int_0^i L(\alpha)\,\tilde{i}\,d\tilde{i}\right) \\ &= \frac{\partial}{\partial \alpha}\left(L(\alpha)\frac{1}{2}i^2\right) = \frac{1}{2}i^2\frac{\partial L(\alpha)}{\partial \alpha} \end{aligned} \tag{7.26}$$

Because in the linear case (i.e. if there is no saturation) $W_{mag} = W'_{mag}$ and $W_{mag} = \frac{1}{2}Li^2$ are true, the same solution for the torque even with a calculation by means of the magnetic energy is obtained. From the equation for the linear operation

$$T = \frac{1}{2}i^2\frac{\partial L(\alpha)}{\partial \alpha} \tag{7.27}$$

the following can be deduced:

- The torque is proportional to the squared current. This means that the current direction (sign "+" or "-") has no influence on the direction of the torque (this corresponds to the general experience: an iron object is attracted from an (electro-) magnet independent from the polarity of the magnet).

- The torque increases the higher the difference in inductivity between "aligned position" and "unaligned position" is (i.e. dependent on the rotor position).

Therefore, the switched reluctance machine in contrary to induction machines, synchronous machines or brushless DC-machines can be operated with unipolar currents.

The torque as change of the magnetic co-energy with respect to the rotational angle can be deduced from the $\psi - i -$ diagram. Because of the different inductivities in the "aligned position" and the "unaligned position" different characteristics are obtained (Fig. 7.3).

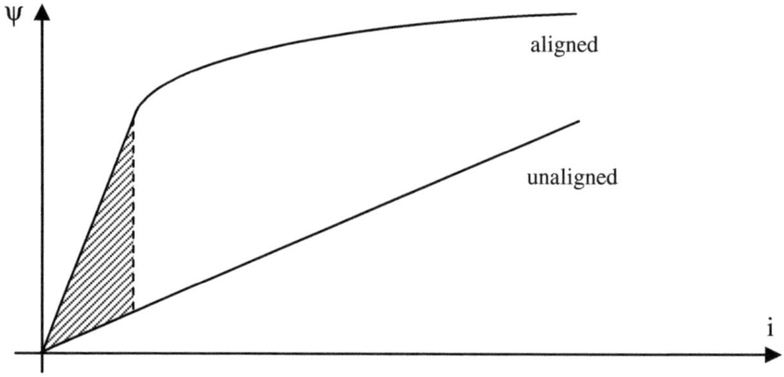

Fig. 7.3 Torque of the switched reluctance machine without saturation.

Operating the machine in the unsaturated region a torque per switch-on period of a single stator tooth is generated that corresponds to the shaded co-energy area in Fig. 7.3.

In contrast, operating the machine far in saturation (in the "aligned position"), a far higher torque is achievable (this can be deduced from the considerably larger co-energy area per switch-on period, see Fig. 7.4):

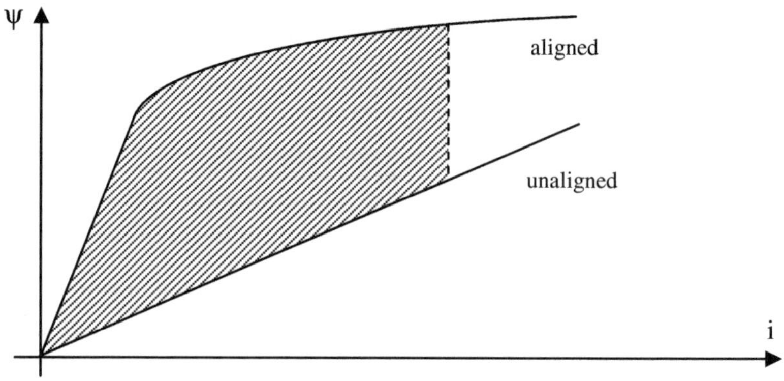

Fig. 7.4 Torque of the switched reluctance machine with saturation.

In this idealized presentation the co-energy area is limited on the right by a vertical line, because generally the phase current is limited (either thermally or by the supplying inverter).

Only in the operation mode with high saturation the switched reluctance motor is comparable to the induction motor in terms of torque density. For the exact calculation of the torque numerical methods (e.g. finite element method, FEM) are required because of the non-linearity coming from high saturation.

7.2.3 Modes of Operation

The torque-speed-plane of the switched reluctance machine can be separated into two main areas: pulsed operation and block-mode operation.

- During *pulsed operation* a speed-independent maximum torque can be generated; this area corresponds to the armature control range (base speed range) of the induction machine. Here the inverter has to be pulsed with high frequency, so that the phase current remains within its limits. Because of this active control an as far as possible square-wave current will be adjusted.
- In *block-mode operation* the phase current will only be switched on and off once per period; the phase current is not adjusted actively. With this the maximum possible torque is about proportional to $1/n$. Increasing the speed further – and therefore it is necessary to switch off the respective phase during current increase – the maximum possible torque decreases about $1/n^2$.

Figure 7.5 illustrates these operation modes schematically; especially the transition from pulsed operation to block-mode operation is not at a certain point, but it is variable and depends on many parameters (e.g. torque level, resistance, voltage source, etc.).

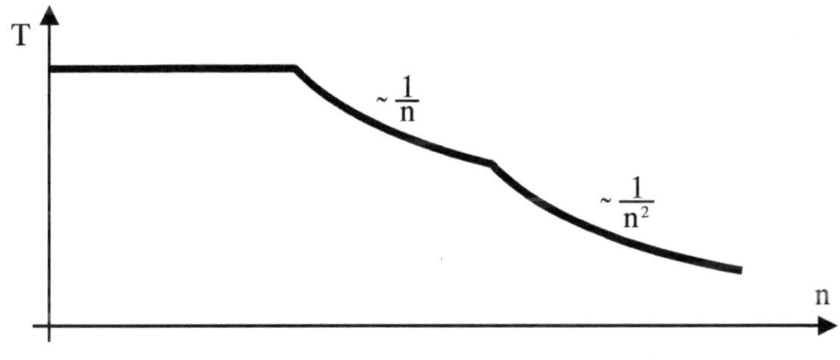

Fig. 7.5 Torque versus speed of the switched reluctance machine.

The pulsed operation will be explained in the following by means of a single phase. Each phase of the machine is connected via two power transistors (shown as simple switches in Fig. 7.6) and two power diodes as a half bridge. The different switching states are illustrated in Fig. 7.6, where the machine phase is symbolized as an inductivity.

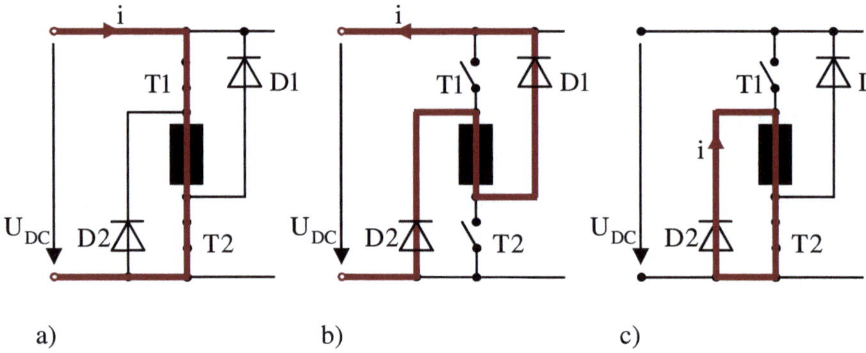

Fig. 7.6 Different switching states of a half bridge of the switched reluctance machine inverter.

Neglecting the voltage drops across the transistors and diodes, in case a) of Fig. 7.6 the voltage U_{DC} supplies the phase winding; the current increases in the shown direction.

If the phase current exceeds the desired current I_{Hi} by a specific amount, both transistors T1 and T2 are opened and the voltage U_{DC} supplies the phase winding in opposite direction to case a): In this case b) of Fig. 7.6 the current decreases until a certain value below the desired current is reached. At this point in time it is switched back to case a).

The current and voltage characteristics (Fig. 7.7) show these time-dependent functions (so-called "hard chopping") including the magnetizing and demagnetizing of the phase.

If the current does not reach the desired value, the voltage will be switched on at the rotor position angle α_{on} (case a)) and switched off at the rotor position angle α_{off} (case b)). Then block-mode operation is active. At the rotor position angle α_{end} the phase current is zero again. The time function of the phase current is shown in Fig. 7.8.

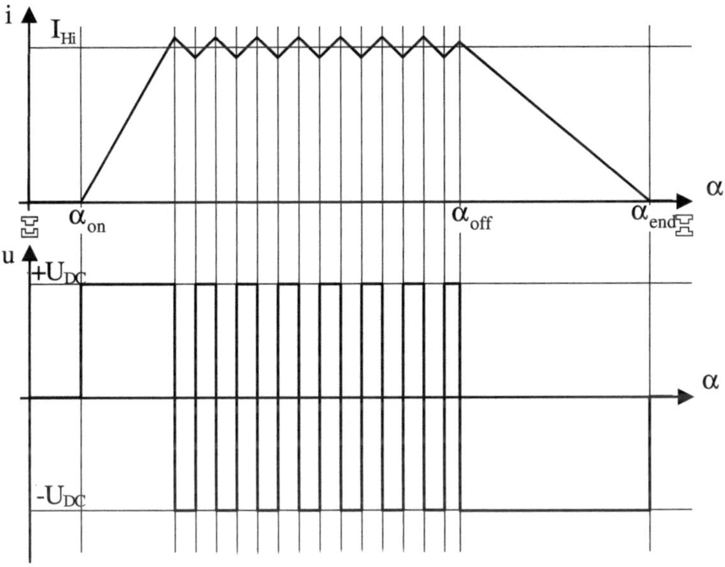

Fig. 7.7 Current and voltage during hard chopping operation.

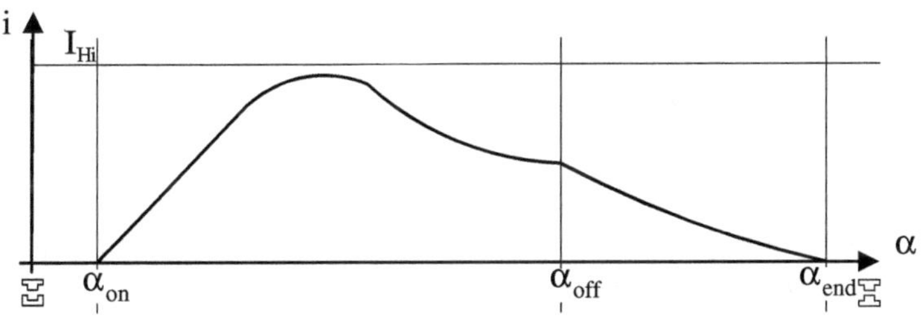

Fig. 7.8 Current during block-mode operation.

During the pulsed operation there are high switching losses because of the high switching frequencies of the power transistors. To avoid this, per switching period only one of both power transistors is switched and for the following switching period the other power transistor is switched (in Fig. 7.6 in case c) the switching of the power transistor T1 is shown as an example). The current and voltage characteristics for this so-called "soft chopping" look like it is shown in Fig. 7.9.

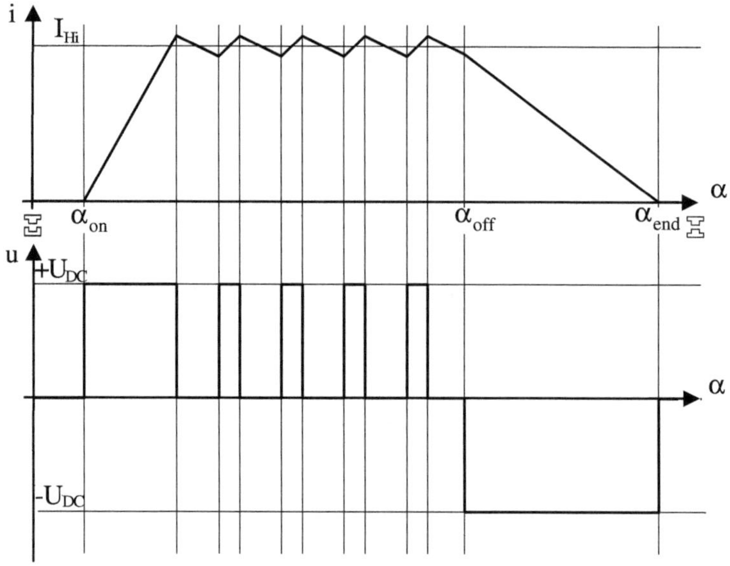

Fig. 7.9 Current and voltage during soft chopping operation.

7.2.4 Alternative Power Electronic Circuits

The inverter already shown in Fig. 7.6 for one phase is called 2n inverter, because it contains 2n switchable power electronic devices, if the number of phases is called n (see Fig. 7.10 for a four-phase machine).

This configuration has the advantage of being most flexible concerning current waveforms, but there is even the disadvantage of a large number of power electronic devices, resulting in high costs.

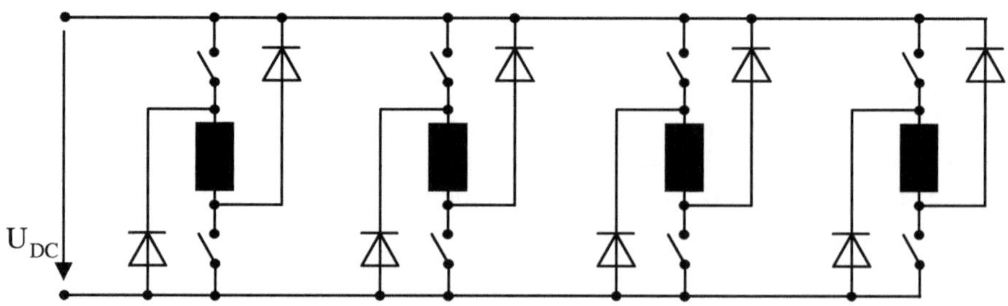

Fig. 7.10 Four-phase switched reluctance machine with 2n inverter.

This circuit can be simplified by far, if the so-called n+1 inverter is used. This inverter (see Fig. 7.11) is characterized by one low-side power electronic switch that may switch at low frequency and just selects the phase of the machine that shall be energized. In addition, there is one single high-side switch for all phases together, being responsible for high-frequency PWM switching. This alternative reduces the effort concerning number of power electronic switches and their accompanied driving circuit and concerning the quality of most of the switches (just one power electronic device has to be capable to switch at high PWM frequency).

On the other hand, this circuit has the disadvantage that the phases cannot be switched independently, which decreases the degree of freedom for machine control.

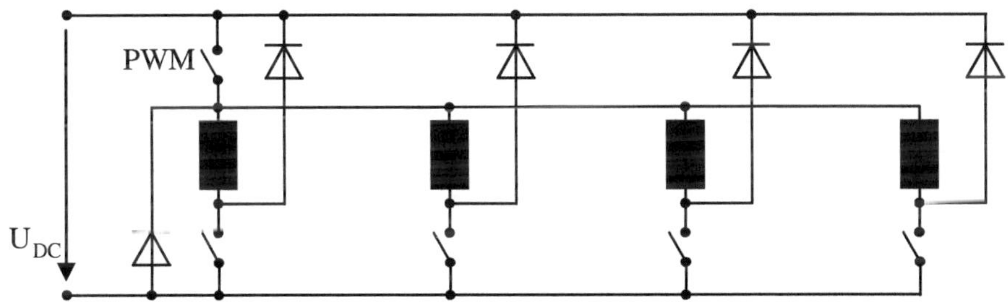

Fig. 7.11 Four-phase switched reluctance machine with n+1 inverter.

A compromise between low effort and high degree of freedom is the so-called n+2 inverter. Here, the phases of the machine are organized in two groups, so that two phases may be energized simultaneously, see Fig. 7.12.

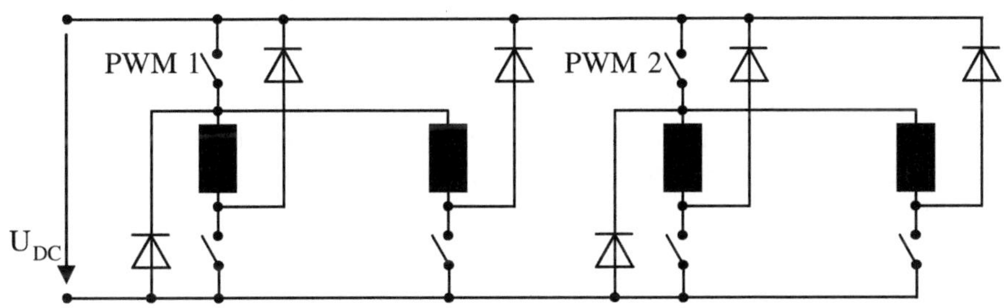

Fig. 7.12 Four-phase switched reluctance machine with n+2 inverter.

7.2.5 Main Characteristics

The main positive characteristics of switched reluctance machines are:

- The construction is simple, robust, and cost-effective.
- Heating because of Ohmic losses only does occur in the stator (and there good possibilities for cooling exist).
- Short-term overload is no problem for the SRM.
- The rotor has low inertia and it is robust; therefore it is qualified for high-speed applications.
- The torque is independent from the current direction; consequently simple electronic circuits for the inverter may be used.

The main disadvantages are:

- The actually energized coil has to be switched off at high phase current and maximum stored energy: Therefore, the inverter usage is relatively low.
- To achieve a high difference in inductivity between the "aligned position" and the "unaligned position" the air-gap width in the "aligned position" has to be very small. This increases the costs and it makes the SRM sensitive to production tolerances.
- The torque is pulsating. To smooth it, e.g. the number of phases can be increased or the teeth (reasonably the rotor teeth, as these do not carry windings) can be skewed. Both measures increase the costs and decrease the utilization.
- Because of the high difference in inductivity between the "aligned position" and the "unaligned position" high pulsating radial forces occur. These forces are the reason for serious acoustic noise.
- By influencing the current waveform the torque and/or the acoustic noise can be influenced. To do so, generally a voltage reserve is necessary and the utilization is reduced.
- In analogy to induction machines, a reactive current component is necessary for the magnetization of the machine. With this reactive current component the apparent power of the inverter is increased.

7.3 References for Chapter 7

DeDoncker RW, Pulle DWJ, Veltman A (2011) Advanced electrical drives. Springer-Verlag, Berlin

Barnes M, Pollock C (1998) Power electronic converters for switched reluctance drives, IEEE Transactions on Power Electronics, 13:1100-1111

Krishnan R (2001) Switched reluctance motor drives. CRC Press, Boca Raton

Miller TJE (1993) Switched reluctance motors and their control. Magna Physics Publishing, Oxford

Miller TJE, McGilp M (1990) Nonlinear theory of the switched reluctance motor for rapid computer-aided design, Proc. Inst. Elect. Eng. B, 137:337-347

Schinnerl B (2009) Analytische Berechnung geschalteter Reluktanzmaschinen, Shaker-Verlag, Aachen

Schramm A (2006) Redundanzkonzepte für geschaltete Reluktanzmaschinen. Shaker-Verlag, Aachen

8 Small Machines for Single-Phase Operation

8.1 Fundamentals

The generation of electrical energy, its distribution and its industrial application for high-power drives is done by means of three-phase systems with high voltage. For this, three-phase machines are used that produce a torque that is constant in time.

At the low-voltage level for low power in households mostly there is just a single-phase system available. For such applications, single-phase machines have to be used. The construction of such single-phase machines often deviates significantly from high-power machines, because:

- inherent to their functional principle they are non-symmetric,
- often they have to be integrated into the application and
- concessions to manufacturing needs of the (usually existing) high-volume production have to be made.

There is a steadily growing market for these kinds of machines and their economic relevance is increasing. Main importance for the direct connection to the single-phase mains have the universal motor and the single-phase induction machine.

In addition, more and more BLDC-machines with inverter supply are used with the single-phase mains. Because of the electronic supply these machine are able to draw a constant power from the mains and to deliver a constant power (torque) to the application.

8.2 Universal Motor

Fundamentally, the construction of the universal motor is like the DC-motor, but it has a laminated stator. It can be operated by DC currents and AC currents, hence the name comes from. For operation with AC currents of the frequency f the main equations of the DC-machine in their time-dependent formulation are valid.

Induced voltage:

$$U_i(t) = k\phi(t)n, \qquad \phi(t) = \phi\sin(\omega t) \tag{8.1}$$

Torque:

$$T(t) = \frac{k}{2\pi} I(t)\phi(t), \qquad I(t) = \sqrt{2}\, I \sin(\omega t - \rho) \tag{8.2}$$

It follows:

$$\begin{aligned} T(t) &= \frac{k}{2\pi} \sqrt{2} I\phi \sin(\omega t - \rho) \sin(\omega t) \\ &= \frac{k}{2\pi} \sqrt{2} I\phi \frac{1}{2} \left(\cos(-\rho) - \cos(2\omega t - \rho)\right) \\ &= \frac{k}{2\pi} \frac{I\phi}{\sqrt{2}} \left(\cos(\rho) - \cos(2\omega t - \rho)\right) \end{aligned} \tag{8.3}$$

Consequently, the torque is composed of two components (see Fig. 8.1):

- a constant component, which is proportional to the cosine of the angle ρ (the phase shift between flux and current) and
- an alternating component, that oscillates with the double mains frequency.

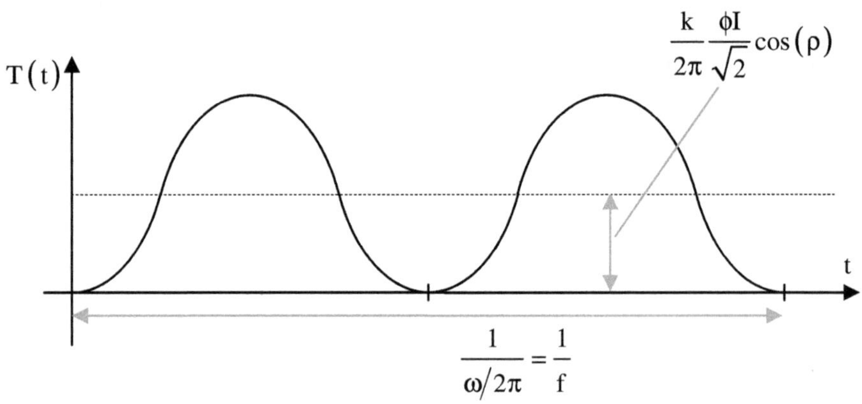

Fig. 8.1. Torque-time characteristic of a universal motor.

To maximize the constant component, it follows:

$$\cos(\rho) \to 1 \quad \Rightarrow \quad \rho \to 0 \tag{8.4}$$

This means that flux and armature current have to be in phase. Figure 8.2 compares shunt-wound and series-wound motors with respect to the requirements of Eq. (8.4).

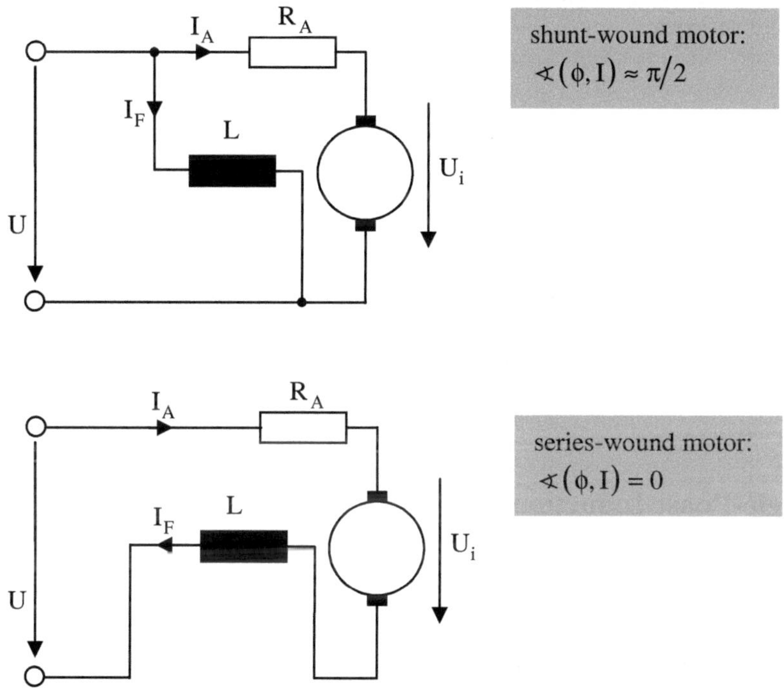

Fig. 8.2. Equivalent circuit diagrams of shunt-wound and series-wound motors.

The requirement mentioned above ($\sphericalangle(\phi, I) = 0$) is fulfilled only for the series-wound machine. The ripple torque is damped by the inertias of the rotor and the load, so that in steady-state operation there are only minimal speed variations. The useful torque is the mean value of the torque (constant component).

Compared to the DC-motor the universal motor delivers a lower torque when operated by AC voltage (if the DC voltage is equal to the rms-value of the AC voltage), because there is an additional voltage drop at the reactance $X = \omega L$ (see the qualitative speed-torque-characteristics in Fig. 8.3).

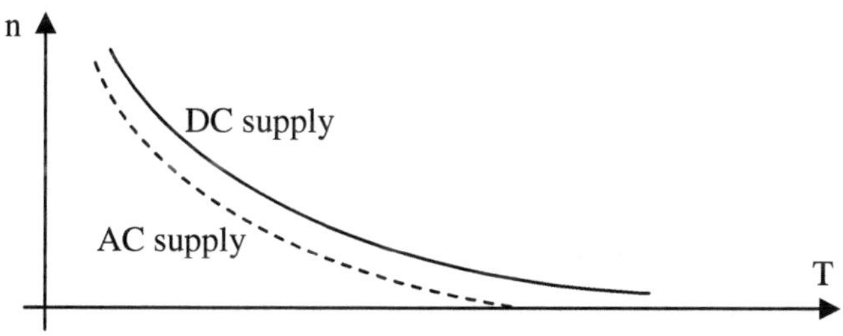

Fig. 8.3. Speed-torque-characteristics of the universal motor (for different supply).

For the sake of completeness Fig. 8.4 shows the same qualitative fact as torque-speed-characteristics:

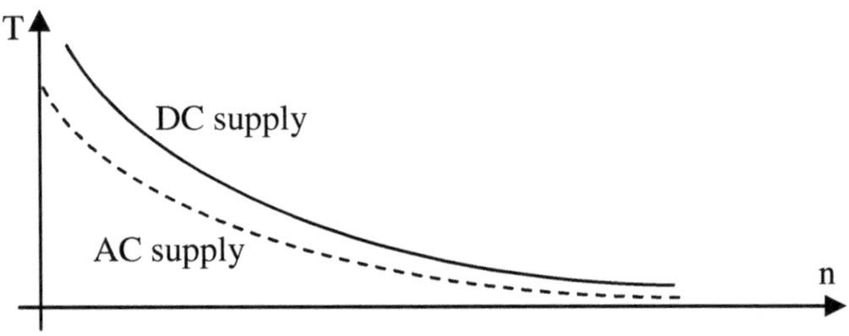

Fig. 8.4. Torque-speed-characteristic of the universal motor (for different supply).

8.3 Single-Phase Induction Machine

8.3.1 Single-Phase Operation of Three-Phase Induction Machine

Disconnecting one phase of the three-phase induction machine from the symmetric three-phase mains, a single-phase supply at two phases of the machine is obtained. The stator MMF produces an alternating air-gap field, which – according to the "Rotating Field Theory" (Chap. 3) – can be regarded as being composed of two rotating fields travelling in opposite direction and having half the amplitude.

That field component travelling in the same direction like the rotor (defined as positive direction, index "p") induces voltages into the rotor windings, that produce the currents $I_{2,p}$ with the frequency $f_{2,p} = s_p f_1$. Here, $s_p = s$ is the same slip like for the three-phase machine.

The slip of the rotor relative to the field component rotating in negative direction (index "n") is:

$$s_n = \frac{-n_0 - n}{-n_0} = 1 + \frac{n}{n_0} = 1 + \frac{n_0}{n_0} - \left(\frac{n_0}{n_0} - \frac{n}{n_0}\right) = 2 - \frac{n_0 - n}{n_0} = 2 - s \qquad (8.5)$$

Therefore, additional rotor currents $I_{2,n}$ occur, having the frequency $f_{2,n} = s_n f_1$.

Both rotating components of the stator field together with both rotor currents produce some torque, i.e. there are four torque components. The torque component coming from the positive rotating stator field together with the negative rotating rotor MMF and the torque component coming from the negative rotating stator field together with the positive rotating rotor MMF are oscillating torques with the mean value equal to zero. They do not deliver any useful torque. In addition there are the torque components of the positive and negative rotating stator fields, produced with their "own" (i.e. the self-induced) rotor currents. These two components deliver a mean torque different from zero.

At stand-still the rotor has the slip $s = 1$ to the positive rotating stator field as well as to the negative rotating stator field. The torque components from both rotating fields have the same amplitude, but opposite direction. The motor does not accelerate.

If the rotor rotates in one direction, there are different slip values for s_p and s_n. Consequently, the reaction of the squirrel-cage rotor onto both rotating stator MMFs is different. The negative rotating field is strongly damped at $s_n \approx 2$, the positive rotating field at $s_p \approx 0$ is exposed only to a small reaction. Therefore, an elliptical rotating air-gap field is generated, which results in producing a useful torque in the direction of rotor rotation. Consequently, this means: The three-phase induction machine rotates further, if one phase is disconnected from the mains and the load torque is not too high (but the slip will increase and therefore the efficiency will decrease, see the torque-speed-characteristics in Fig. 8.5). By mechanical starting the rotor from zero speed, the motor can accelerate further.

A detailed calculation can be performed by means of the symmetric components (see Sect. 1.6). The resulting torque-speed-characteristic is identical to that of two three-phase induction motors connected in series, which rotors are mechanically coupled and which stators are supplied with opposite phase sequence at $\sqrt{3}$ times the phase voltage. The qualitative torque-speed-characteristics look like it is shown in Fig. 8.5.

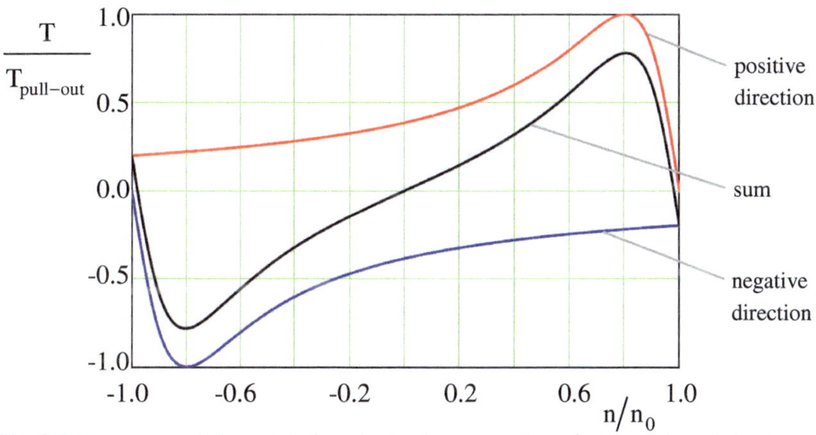

Fig. 8.5. Torque-speed-characteristics: single-phase operation of a three-phase induction motor.

8.3.2 Single-Phase Induction Motor with Auxiliary Phase

If the induction motor in single-phase operation shall deliver a torque even at zero speed, at least an elliptical rotating air-gap field has to be present at stand-still. This can be realized by means of an auxiliary winding "a", that is shifted against the main winding "m" by a spatial angle ε and that is supplied by a current with a phase shift by an (electrical) angle φ.

If $p\varepsilon = \varphi = 0$ holds true, a purely oscillating air-gap field is obtained. If $p\varepsilon = \varphi = \pi/2$ holds, the amplitude of the negative rotating field is minimal and the amplitude of the positive rotating field is maximal. The phase shift between the current $\underline{I}_m$ of the main winding and the current $\underline{I}_a$ of the auxiliary winding is realized by an additional impedance in the auxiliary winding, like it is shown in the equivalent circuit diagram in Fig. 8.6.

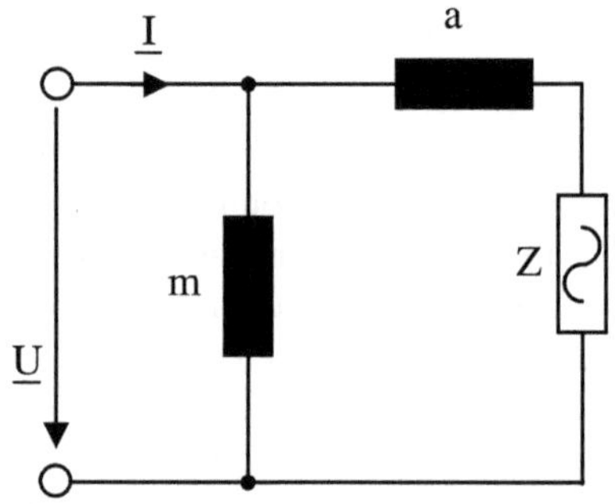

Fig. 8.6. Equivalent circuit diagram of the single-phase induction motor with auxiliary phase.

There are three possibilities for this impedance in the auxiliary winding:

- Resistance: A resistance is very cost-effective, but only a small starting torque is generated; because of the losses it has to be switched off after starting the motor.
- Inductivity: An inductivity delivers only a small starting torque (a pure inductivity produces no phase shift of the current; only because of the unavoidable resistance there is a generally small phase shift); in addition the inductivity is costly and heavy.
- Capacity: Starting capacity (switching off by centrifugal switch) or operating capacity (improvement of power factor and efficiency); the use of a capacity results in a high starting torque, from technical point of view this is the best solution.

8.3.3 Shaded-Pole (Split-Pole) Motor

The shaded-pole motor is a special case of the single-phase induction motor with auxiliary winding: The main winding is located on two salient poles and contains concentrated coils. It is supplied from the single-phase mains. The auxiliary winding is realized as a short-circuited ring that encloses only a part of the salient pole. It is supplied by means of induction from the main winding. The rotor is a squirrel-cage rotor. A principle sketch is shown in Fig. 8.7.

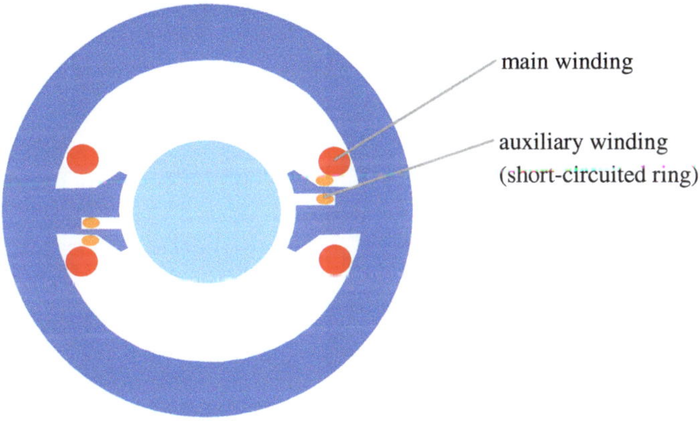

Fig. 8.7. Principle sketch of a shaded-pole motor.

Only by means of the resistance R_a of the auxiliary winding the phase shift of the currents of main and auxiliary winding is realized. As there is even a spatial shift of the windings, an elliptical rotating air-gap field is generated. Therefore, this motor can start from stand-still. As the field of the main winding is preceding the field of the auxiliary winding (the short-circuited ring delays the change of the field) the rotor always moves from the main pole to the auxiliary pole.

In many cases the short-circuited ring is made from bronze and not from copper to increase the resistance R_a.

Because of the losses in the short-circuited ring and because of the opposite rotating field, shaded-pole machines have a quite low efficiency of about 20 to 40%. The starting torque is lower and the starting current is higher compared to the motor with capacity in the auxiliary phase. Therefore, in spite of being simple and very cost-effective the shaded-pole motor is used only for small power applications up to about 100W.

8.4 References for Chapter 8

Stepina J (1982) Die Einphasenasynchronmotoren. Springer-Verlag, Wien

Stölting HD, Beisse A (1987) Elektrische Kleinmaschinen. Teubner-Verlag, Stuttgart

Stölting HD, Kallenbach E (2001) Handbuch Elektrische Kleinantriebe. Hanser Verlag, München

Veinott CG (1959) Theory and design of small induction motors. McGraw Hill Book Company, New York

Veinott CG (1970) Fractional- and subfractional-horsepower electric motors. McGraw Hill Book Company, New York

9 Fundamentals of Dynamic Operation

9.1 Fundamental Dynamic Law, Equation of Motion

9.1.1 Translatory Motion

A mass m with the velocity $\vec{v}$ has the impulse $m\vec{v}$ (kinetic quantity). The sum of all from outside acting forces leads to a time-dependent change of the kinetic quantity.

$$\sum_{\nu} \vec{F}_{\nu} = \vec{F}_{a} = \frac{d}{dt}(m\vec{v}) = m\frac{d}{dt}(\vec{v}) + \vec{v}\frac{d}{dt}(m) \tag{9.1}$$

For $m = \text{const.}$ it follows (Newton's equation of motion):

$$\vec{F}_{a} = m\frac{d\vec{v}}{dt} = m\vec{a} \tag{9.2}$$

If all outside forces can be described by a driving force $\vec{F}$ and a load force $\vec{F}_{load}$, where both are acting in the same direction (x-axis), it follows:

$$F - F_{load} = F_{a} = ma = m\frac{dv}{dt} = m\frac{d^2x}{dt^2} \tag{9.3}$$

9.1.2 Translatory / Rotatory Motion

In the following a combined motion of two bodies will be regarded, translatory (body 1) and rotatory (body 2) – see Fig. 9.1. The bodies are closely coupled to each other. The following forces F and torques T exist (the rotating body has the mass $m = 0$, ω is the angular frequency of this rotating body):

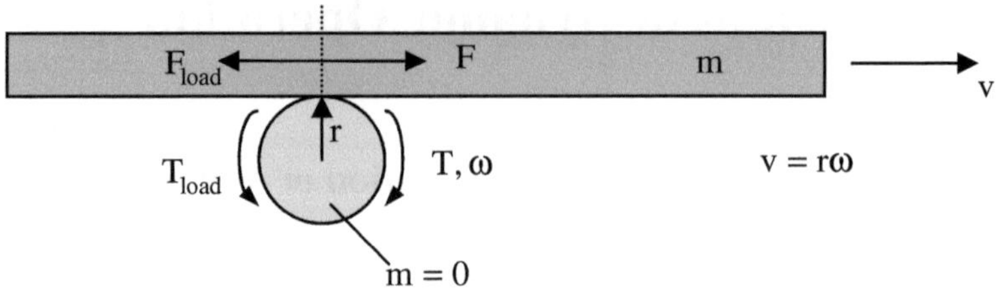

Fig. 9.1. Principle of combined translatory and rotatory motion.

It follows:

$$\begin{aligned} F - F_{load} &= m\frac{dv}{dt} \\ \Rightarrow r(F - F_{load}) &= rm\frac{d(r\omega)}{dt} \\ \Rightarrow T - T_{load} &= mr^2\frac{d\omega}{dt} = \Theta^*\frac{d\omega}{dt} \end{aligned} \tag{9.4}$$

The inertia of the rotating body Θ is zero, because the mass was assumed being zero. $\Theta^* = mr^2$ is the translatory moved mass, transformed to the rotatory movement.

9.1.3 Rotatory Motion

A rotating mass Θ with the angular frequency ω has the rotating impulse $\Theta\omega$ (kinetic quantity). The sum of all from outside acting torques leads to a time-dependent change of the kinetic quantity.

$$\sum_{\nu} T_{\nu} = T_a = \frac{d}{dt}(\Theta\omega) = \Theta\frac{d}{dt}(\omega) + \omega\frac{d}{dt}(\Theta) \tag{9.5}$$

For $\Theta = \text{const.}$ and T_{load} as sum of all load torques it follows:

$$T - T_{load} = T_a = \Theta\frac{d\omega}{dt} = \Theta\frac{d^2\gamma}{dt^2} \tag{9.6}$$

9.1.4 Stability

From the equation of motion

$$T - T_{load} = \Theta \frac{d\omega}{dt} = 2\pi\Theta \frac{dn}{dt} \tag{9.7}$$

it follows (see Fig. 9.2):

$\frac{dn}{dt} > 0$	if	$T > T_{load}$	acceleration, starting
$\frac{dn}{dt} < 0$	if	$T < T_{load}$	deceleration, braking
$\frac{dn}{dt} = 0$	if	$T = T_{load}$	constant speed, static balance (stable or unstable)

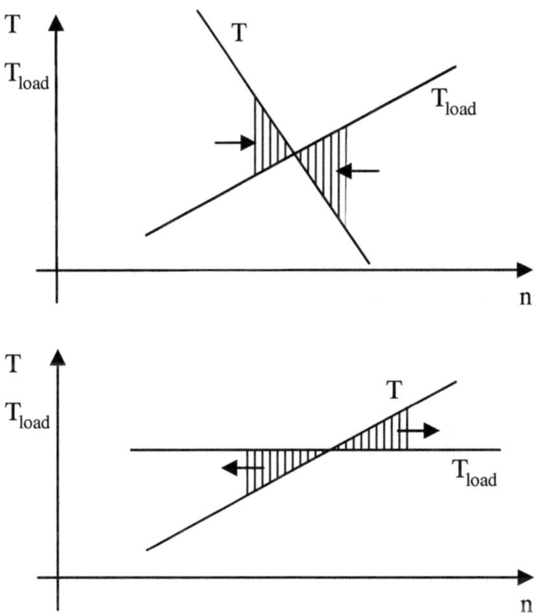

Fig. 9.2. Static stability (*above*) and static instability (*below*).

Static stability is given for $\frac{\partial T_{load}}{\partial n} > \frac{\partial T}{\partial n}$, static instability for $\frac{\partial T_{load}}{\partial n} < \frac{\partial T}{\partial n}$.

9.2 Mass Moment of Inertia

9.2.1 Inertia of an Arbitrary Body

An arbitrarily sized, inelastic body shall rotate around an arbitrary axis (see Fig. 9.3). The equation of motion for a mass element dm is:

$$dT_a = r\ dF_a = r\ dm\ \frac{dv}{dt} = r^2\ \frac{d\omega}{dt}\ dm \tag{9.8}$$

The total acceleration torque is:[9]

$$T_a = \int_0^{T_a} d\tilde{T}_a = \frac{d\omega}{dt}\int_0^m r^2 d\tilde{m} = \Theta\frac{d\omega}{dt} \tag{9.9}$$

with the inertia:

$$\Theta = \int_0^m r^2 d\tilde{m} \tag{9.10}$$

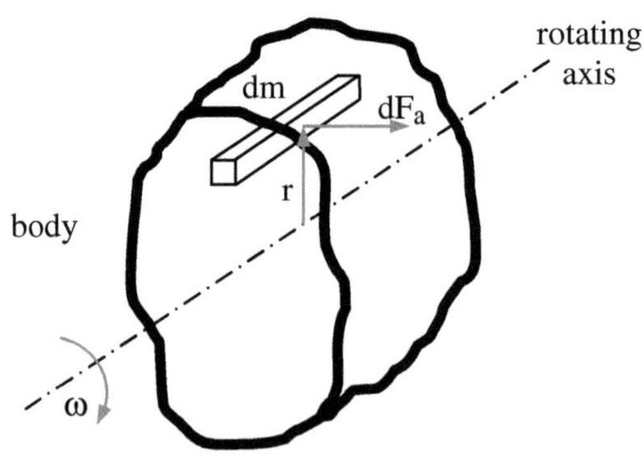

Fig. 9.3. Calculating the inertia of an arbitrary body.

[9] The tilde is introduced to distinguish between integration limit and integration variable.

9.2.2 Inertia of a Hollow Cylinder

From Sect. 9.2.1 it follows for a hollow cylinder (see Fig. 9.4):

$$\Theta = \int_0^m r^2 d\tilde{m} = \int_0^V r^2 \rho d\tilde{V}$$
$$= \int_{r_i}^{r_a} r^2 \rho \ell 2\pi r dr = 2\pi\rho\ell \int_{r_i}^{r_a} r^3 dr \tag{9.11}$$
$$= \frac{\pi}{2}\rho\ell\left(r_a^4 - r_i^4\right)$$

Here ρ is the specific weight and ℓ is the axial length.

Introducing the mass $m = \rho\pi\ell\left(r_a^2 - r_i^2\right)$ the following is obtained:

$$\Theta = m\frac{r_a^2 + r_i^2}{2} = mr^{*2} \tag{9.12}$$

The quadratic mean value of the radii $r^* = \sqrt{\frac{r_a^2 + r_i^2}{2}}$ is called “inertia radius”.

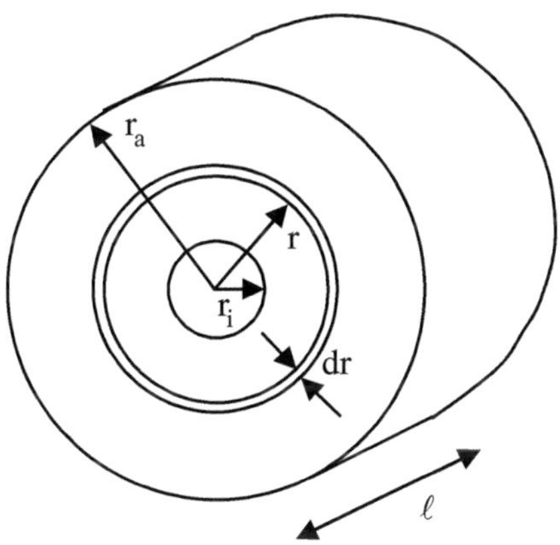

Fig. 9.4. Calculating the inertia of a hollow cylinder.

9.3 Simple Gear-Sets

9.3.1 Assumptions

The transmission (gear-set) is assumed being lossless and form-fit: Then there is no slip, no backlash, no hysteresis and no elasticity.

9.3.2 Rotation / Rotation (e.g. Gear Transmission)

In Fig. 9.5 the principle rotation / rotation gear set is shown:

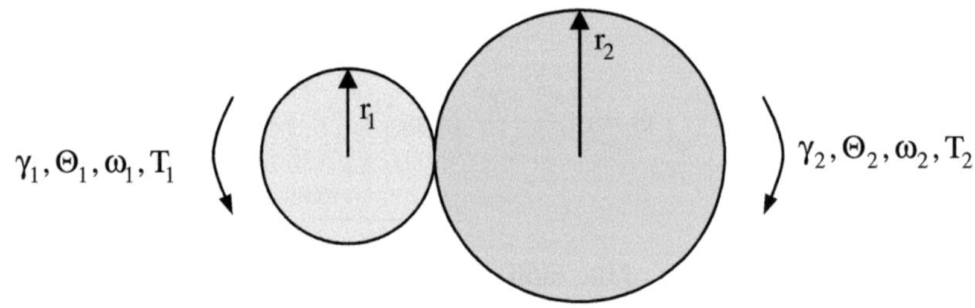

Fig. 9.5. Rotation / rotation gear-set.

Defining $\frac{r_1}{r_2} = \frac{1}{u}$, the condition of identical displacement at the point of transmission delivers:

$$\gamma_1 r_1 = \gamma_2 r_2 \qquad \Rightarrow \qquad \frac{\gamma_1}{\gamma_2} = u \tag{9.13}$$

From identical speed ($v = \omega r$) at the point of transmission it follows:

$$\frac{\omega_1}{\omega_2} = u \tag{9.14}$$

From "actio = reactio" at the point of transmission the following can be deduced:

$$\frac{T_1}{r_1} = \frac{T_2}{r_2} \qquad \Rightarrow \qquad \frac{T_1}{T_2} = \frac{1}{u} \tag{9.15}$$

The acceleration torque of the rotating bodies is (if they are regarded as single, independent bodies):

$$T_{a,n} = \Theta_n \frac{d}{dt}\omega_n, \qquad n = 1, 2 \tag{9.16}$$

Because of the inelastic coupling of these bodies, body 2 has to be accelerated as well, if body 1 is accelerated. Then it follows from the law "actio = reactio":

$$\begin{aligned} T_{a,1} &= \Theta_1 \frac{d}{dt}\omega_1 + \frac{r_1}{r_2}\Theta_2 \frac{d}{dt}\omega_2 \\ &= \Theta_1 \frac{d}{dt}\omega_1 + \frac{1}{u}\Theta_2 \frac{d}{dt}\left(\frac{\omega_1}{u}\right) \\ &= \left(\Theta_1 + \frac{1}{u^2}\Theta_2\right)\frac{d}{dt}\omega_1 \end{aligned} \tag{9.17}$$

Consequently, the transformation of inertia Θ_2 onto axis 1 gives:

$$\Theta_{1,tot} = \Theta_1 + \frac{1}{u^2}\Theta_2 \tag{9.18}$$

9.3.3 Rotation / Translation (e.g. Lift Application)

In Fig. 9.6 the principle rotation / translation gear set is shown.

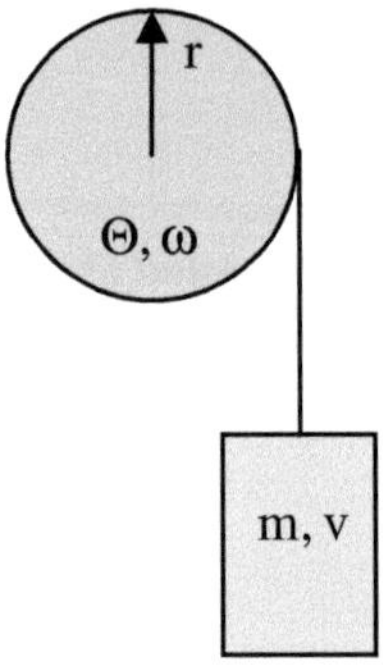

Fig. 9.6. Rotation / translation gear-set.

The acceleration torque is:

$$\begin{aligned} T &= \Theta \frac{d}{dt}\omega + r\ m \frac{d}{dt} v = \Theta \frac{d}{dt}\omega + r\ m \frac{d}{dt}(\omega r) \\ &= \Theta \frac{d}{dt}\omega + r^2 m \frac{d}{dt}\omega \\ &= \left(\Theta + r^2 m\right)\frac{d}{dt}\omega \end{aligned} \tag{9.19}$$

Consequently, the transformation of the mass m onto the rotational axis is:

$$\Theta_{tot} = \Theta + r^2 m \tag{9.20}$$

9.4 Power and Energy

From the equation of motion

$$T - T_{load} = \Theta \frac{d\omega}{dt} \tag{9.21}$$

the power balance is obtained after multiplication with ω:

$$T\omega = T_{load}\omega + \Theta\omega \frac{d\omega}{dt} \tag{9.22}$$

This equation means: The input power ($T\omega$) is equal to the sum of power of the load ($T_{load}\omega$) and change of kinetic energy ($\Theta\omega\frac{d\omega}{dt}$).

The energy balance can be calculated as follows: In the time period $\Delta t = t_2 - t_1$, in which the speed of the drive is changed as $\Delta\omega = \omega_2 - \omega_1$, the work

$$\int_{t_1}^{t_2} T\omega dt = \int_{t_1}^{t_2} T_{load}\omega dt + \int_{t_1}^{t_2} \Theta\omega\frac{d\omega}{dt}dt \tag{9.23}$$

is supplied. The increase of kinetic energy of the rotating masses is:

$$\int_{t_1}^{t_2} \Theta\omega\frac{d\omega}{dt}dt = \int_{\omega_1}^{\omega_2} \Theta\omega d\omega = \frac{1}{2}\Theta\left(\omega_2^2 - \omega_1^2\right) \tag{9.24}$$

After an acceleration from $\omega_1 = 0$ to $\omega_2 = \omega_0$ the kinetic energy of

$$E_{kin} = \frac{1}{2}\Theta\omega_0^2 \tag{9.25}$$

is stored in the rotating masses of the drive.

If there is a complex drive with different speeds of the different rotating masses, their effect can be concentrated virtually into a single rotating body. In most cases the rotating mass is calculated relative to the motor axis (motor speed). During transformation the kinetic energy is unchanged. Therefore, the following holds true (transformed values are marked with " ′ "):

$$\begin{aligned} &\frac{1}{2}\Theta'\omega'^2 = \frac{1}{2}\Theta\omega^2 \\ &\Rightarrow \quad \Theta' = \Theta\left(\frac{\omega}{\omega'}\right)^2 = \Theta\left(\frac{n}{n'}\right)^2 \end{aligned} \tag{9.26}$$

9.5 Slow Speed Change

9.5.1 Fundamentals

From the equation of motion (Eq. (9.21)) the mechanical transient operation of a drive can be calculated, if the torque characteristics as function of time or angular frequency are known.

The torque of the electric machine can be calculated by means of the system equations of the electromagnetic circuits. The phase currents of the machine are determined even by the equation of motion. Therefore, generally there is a coupling between the electrical and mechanical transient behavior (dynamic operation).

This coupling can be neglected, if the operation is a quasi steady-state one. Such an operation is characterized by slow speed changes (compared to the electrical time constants). Then the steady-state characteristics $T(\omega)$ and $T_{load}(\omega)$ can be used.

9.5.2 First Example

A constant acceleration torque $T_a = \text{const.}$ is assumed (i.e. the electrical machine always produces a torque that is constantly higher than the load torque). It follows:

$$\frac{d\omega}{dt} = \frac{T_a}{\Theta} \tag{9.27}$$

For the acceleration the following holds true:

$$\omega = \frac{T_a}{\Theta} t \tag{9.28}$$

Consequently, there is a linear speed increase with time. Dividing this equation by the final angular frequency ω_0 it follows:

$$\frac{\omega}{\omega_0} = \frac{t}{\tau}, \qquad \tau = \frac{\omega_0 \Theta}{T_a} \tag{9.29}$$

If the acceleration torque is equal to the nominal torque ($T_a = T_N$) and the final angular frequency is equal to its nominal value ($\omega_0 = \omega_{0,N}$), the mechanical time constant becomes:

$$\tau_{mech} = \frac{\omega_{0,N}\Theta}{T_N} \tag{9.30}$$

9.5.3 Second Example

If the acceleration torque decreases linearly with increasing speed $T_a = T_{a,0}\dfrac{\omega_0 - \omega}{\omega_0}$ (i.e. the difference between torque of the electrical machine and load torque gets smaller and smaller with increasing speed), it follows:

$$\frac{d\omega}{dt} = \frac{T_{a,0}}{\omega_0\Theta}(\omega_0 - \omega) \tag{9.31}$$

Consequently:

$$\frac{d\omega}{dt} + \frac{1}{\tau'_{mech}}\omega = \frac{T_{a,0}}{\Theta} \qquad \text{with} \qquad \tau'_{mech} = \frac{\omega_0\Theta}{T_{a,0}} \tag{9.32}$$

For the run-up characteristic it follows (Fig. 9.7):

$$\frac{\omega}{\omega_0} = 1 - e^{-\frac{t}{\tau'_{mech}}} \tag{9.33}$$

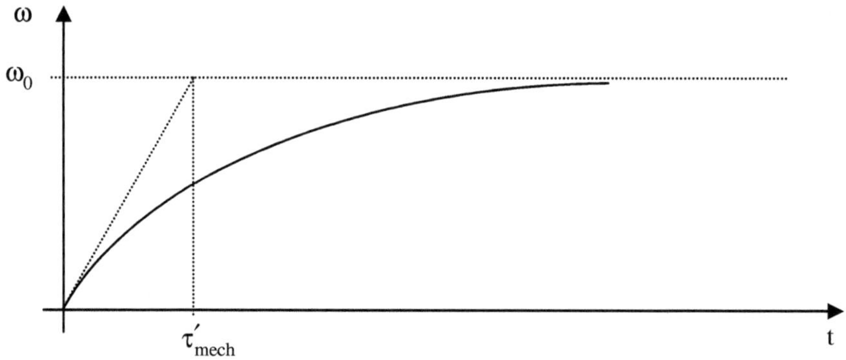

Fig. 9.7. Run-up characteristic.

Often the acceleration torque T_a is not known explicitly as a function, but as a characteristic $T_a(\omega)$ that is complex to describe analytically. In these cases a numerical integration is strongly recommended. The differential equation is then substituted by an equation of differences

$$T_a = \Theta \frac{\Delta\omega}{\Delta t} \tag{9.34}$$

The acceleration torque T_a is approximated by a constant value in every time interval (but generally different for separate time intervals). Then for every time interval $\Delta t_\nu = t_{\nu+1} - t_\nu$ the change of angular frequency is calculated as

$$\Delta\omega_\nu = \frac{T_{a,\nu}}{\Theta} \Delta t_\nu \tag{9.35}$$

and added to the value of the preceding time interval:

$$\omega(t_{\nu+1}) = \omega(t_\nu) + \Delta\omega_\nu \tag{9.36}$$

9.6 Losses during Starting and Braking

9.6.1 Operation without Load Torque

The mechanical energy during acceleration or braking of drives follows from the fundamental equation of dynamic operation:

$$E_{mech} = \int_{t_1}^{t_2} T\omega dt = \int_{t_1}^{t_2} T_{load}\omega dt + \Theta \int_{\omega_1}^{\omega_2} \omega d\omega \tag{9.37}$$

For the operation without load torque ($T_{load} = 0$) it follows:

$$E_{mech} = \int_{t_1}^{t_2} T\omega dt = \Theta \int_{\omega_1}^{\omega_2} \omega d\omega \tag{9.38}$$

As an example a three-phase induction machine will be regarded in the following. The inertia Θ is the inertia of the rotor and that of the coupled rotating masses ($\Theta = \Theta_{rotor} + \Theta_{load}$). The schematic energy distribution is shown in Fig. 9.8.

The angular frequency of an induction machine is (with s being the slip):

$$\begin{aligned} &\omega = \omega_0 (1 - s) \\ \Rightarrow \quad &\frac{d\omega}{dt} = -\omega_0 \frac{ds}{dt} \\ \Rightarrow \quad &d\omega = -\omega_0 ds \end{aligned} \tag{9.39}$$

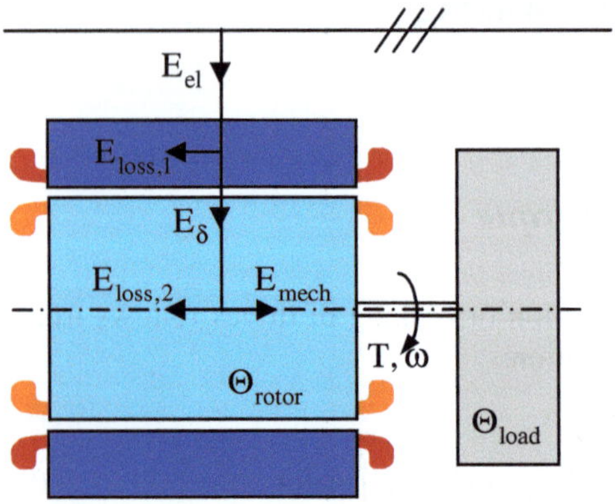

Fig. 9.8. Schematic energy distribution.

By introducing Eq. (9.39) into Eq. (9.38) it follows:

$$E_{mech} = \int_{t_1}^{t_2} T\omega dt = \Theta \int_{\omega_1}^{\omega_2} \omega d\omega$$

$$\Rightarrow \quad E_{mech} = \int_{t_1}^{t_2} T\omega_0 (1-s) dt = \Theta \int_{s_1}^{s_2} \omega_0 (1-s)(-\omega_0) ds$$

$$\Rightarrow \quad E_{mech} = \int_{t_1}^{t_2} P_\delta (1-s) dt = -\Theta\omega_0^2 \int_{s_1}^{s_2} (1-s) ds \tag{9.40}$$

$$\Rightarrow \quad E_{mech} = -\Theta\omega_0^2 \left[s - \frac{1}{2} s^2 \right]_{s_1}^{s_2}$$

$$\Rightarrow \quad E_{mech} = 2E_{kin} \left[(s_1 - s_2) - \frac{1}{2}(s_1^2 - s_2^2) \right] \quad \text{with} \quad E_{kin} = \frac{1}{2}\Theta\omega_0^2$$

In this equation P_δ is the air-gap power and E_{kin} the kinetic energy stored in the drive at synchronous speed. For the run-up from zero speed to (nearly) synchronous speed ($s_1 = 1, s_2 \approx 0$) it follows:

$$E_{mech} = E_{kin} \tag{9.41}$$

The rotor heat losses occurring during this run-up are (with $P_\delta = T\omega_0$):

$$\begin{aligned} E_{loss,2} &= \int_{t_1}^{t_2} sP_\delta dt = \int_{t_1}^{t_2} s\omega_0 T dt = -\int_{s_1}^{s_2} s\Theta\omega_0^2 ds \\ &= -\Theta\omega_0^2 \frac{1}{2}\left[s^2\right]_{s_1}^{s_2} = E_{kin}\left(s_1^2 - s_2^2\right) \end{aligned} \tag{9.42}$$

With $s_1 = 1$, $s_2 \approx 0$ (run-up from zero speed) it can be further deduced:

$$E_{loss,2} = E_{kin} \tag{9.43}$$

Consequently, the rotor heat losses during run-up are identical to the kinetic energy stored in the rotating masses after this run-up.

The heat losses in the stator will be calculated by using a quite simple approximation: Neglecting the magnetizing current means $I_1 \approx I_2'$. Then the heat values in stator and rotor are proportional to the respective Ohmic resistances:

$$\frac{E_{loss,1}}{E_{loss,2}} \approx \frac{R_1}{R_2'} \approx 1 \tag{9.44}$$

Consequently during this run-up there is:

$$E_{loss,1} = E_{kin} \tag{9.45}$$

From the mains the following energy has to be delivered for this run-up (neglecting the iron losses and friction losses):

$$\begin{aligned} E_{el} &= E_{loss,1} + E_{loss,2} + E_{mech} \\ &= 3E_{kin} \end{aligned} \tag{9.46}$$

For electrical braking ($s_1 = 2$, $s_2 = 1$) the energy values become:

$$E_{mech} = -E_{kin} \tag{9.47}$$

$$E_{loss,2} = 3E_{kin} \tag{9.48}$$

$$E_{loss,1} = 3E_{kin} \tag{9.49}$$

$$\begin{aligned} E_{el} &= E_{loss,1} + E_{loss,2} + E_{mech} \\ &= 5E_{kin} \end{aligned} \tag{9.50}$$

Starting the motor by changing the number of poles (from p' to p'') the following rotor losses are obtained:

$$E_{loss,2} = E'_{kin}\left(s_1'^2 - s_2'^2\right) + E''_{kin}\left(s_1''^2 - s_2''^2\right) \tag{9.51}$$

Selecting $p' = 2$ and $p'' = 1$, it follows:

$$\begin{aligned} &\omega_0' = \frac{1}{2}\omega_0'' = \frac{1}{2}\omega_0 \\ &s_1' = 1, \quad s_2' = 0, \quad s_1'' = 0.5, \quad s_2'' = 0 \\ &E'_{kin} = \frac{1}{4}E''_{kin} = \frac{1}{4}E_{kin} \end{aligned} \tag{9.52}$$

Consequently:

$$\begin{aligned} E_{loss,2} &= \frac{1}{4}E_{kin}\left(1^2 - 0^2\right) + E_{kin}\left(0.5^2 - 0^2\right) \\ &= \frac{1}{2}E_{kin} \end{aligned} \tag{9.53}$$

Having the same final kinetic energy like for the run-up without changing the number of poles (because the final speed is not changed), there are only half the losses in the rotor (and therefore there is considerably reduced rotor heating).

9.6.2 Operation with Load Torque

From

$$T - T_{load} = \Theta \frac{d\omega}{dt} \quad \Rightarrow \quad \frac{T}{T - T_{load}} \Theta \frac{d\omega}{dt} = T \tag{9.54}$$

it follows:

$$\begin{aligned} E_{loss,2} &= \int_{t_1}^{t_2} sP_\delta dt = \int_{t_1}^{t_2} s\omega_0 T dt = \int_{t_1}^{t_2} s\omega_0 \left(\frac{T}{T - T_{load}} \Theta \frac{d\omega}{dt} \right) dt \\ &= \int_{s_1}^{s_2} s\omega_0 \frac{T}{T - T_{load}} \Theta(-\omega_0) ds = -2\frac{1}{2}\Theta\omega_0^2 \int_{s_1}^{s_2} \frac{T}{T - T_{load}} sds \\ &= -2E_{kin} \int_{s_1}^{s_2} \frac{T}{T - T_{load}} sds \end{aligned} \tag{9.55}$$

The factor $\frac{T}{T - T_{load}} > 1$ (usually depending on ω or s) increases the losses (heating) during speed change against the case without load torque ($T_{load} = 0$).

For example this is important for $Y - \Delta -$ starting: in Y-connection the torque is reduced to $1/3$, but the load torque remains unchanged (this means $(T - T_{load})$ may become very small). The mains loading is reduced by smaller phase currents, but the driving machine has to withstand increased heating for $T_{load} > 0$. The same holds true for the run-up with reduced terminal voltage.

For the special case $T_a = T - T_{load} = 0$ (no acceleration torque, e.g. locked rotor) the loss energy $E_{loss,2}$ increases to infinity.

Only for the case $T_{load} = 0$ the loss energy $E_{loss,2}$ is independent from the kind of run-up.

The loss energy (heating) calculated above is present if the machine is operated at mains supply (constant voltage and constant frequency). With inverter supply at changing voltage and frequency the losses are much lower.

9.7 References for Chapter 9

Leonhard W (1973) Regelung in der elektrischen Antriebstechnik. Teubner-Verlag, Stuttgart

White DC, Woodson HH (1958) Electromechanical energy conversion. John Wiley & Sons, New York

10 Dynamic Operation and Control of DC-Machines

10.1 Set of Equations for Dynamic Operation

In comparison to the steady-state operation of the DC-machine the energy storage elements (inductivities L_A, L_F and inertia Θ) have to be considered additionally for the dynamic operation.

For a detailed analysis it is even necessary to consider the voltage drop across the brushes ΔU_B, the frictional torque ΔT_{fric} and the nonlinearity of the magnetic circuit because of saturation of the iron. In total the following set of equations and equivalent circuit diagram (Fig. 10.1) can be deduced for the dynamic operation:

$$\begin{aligned}
&U_A = I_A R_A + L_A \frac{dI_A}{dt} + U_i + \Delta U_B \\
&U_i = c\Omega\phi, \qquad c = \frac{k}{2\pi} = \frac{1}{2\pi} 4pw_A, \qquad \Omega = 2\pi n \\
&U_F = I_F R_F + w_F \frac{d\phi}{dt} \\
&T_i - T_{load} - \Delta T_{fric} = \Theta \frac{d\Omega}{dt} \\
&T_i = c\phi I_A
\end{aligned} \tag{10.1}$$

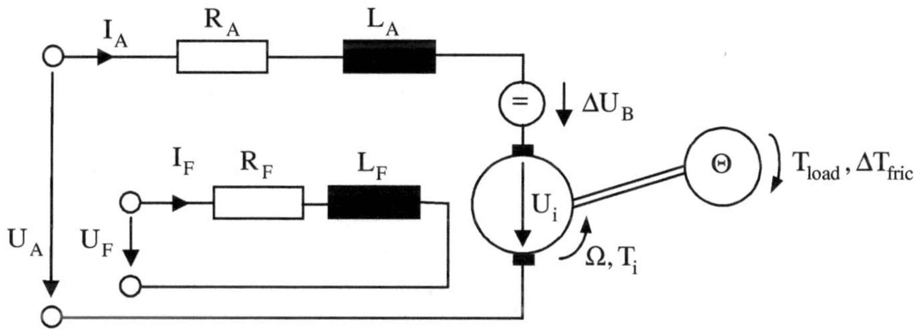

Fig. 10.1. Equivalent circuit diagram of the DC-machine for dynamic operation.

The nonlinear dependency of flux and exciting current (field current) is shown in Fig. 10.2 (nominal values are denoted with the index "N" in the following).

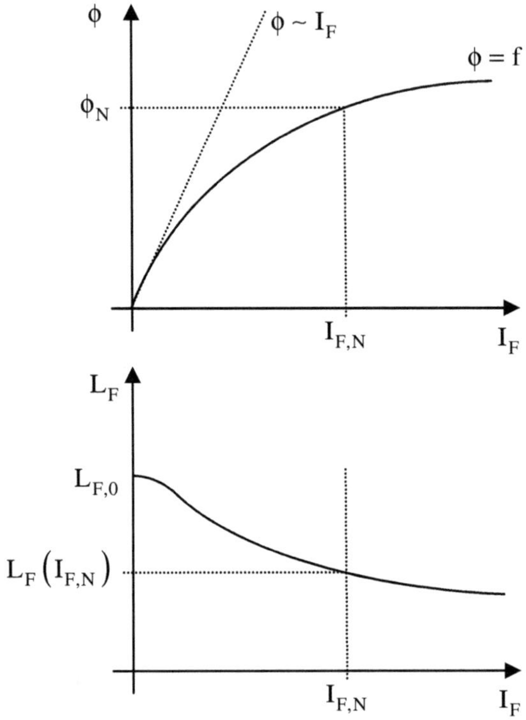

Fig. 10.2. Nonlinear characteristics of the DC-machine: flux versus exciting current (*above*) and field winding inductivity versus exciting current (*below*).

The inductivity of the field winding depends on the current: $L_F(I_F) = \dfrac{w_F \phi}{I_F}$.

At nominal operation the field exciting circuit usually is saturated: $L_{F,N} = \dfrac{w_F \phi_N}{I_{F,N}}$.

From the above relations a system of three coupled differential equations follows. With this system (armature circuit equation, field circuit equation, torque equation) all operational conditions of the DC-machine can be described:

$$U_A - \Delta U_B = U_A' = I_A R_A + L_A \frac{dI_A}{dt} + c\Omega\phi \tag{10.2}$$

$$U_F = I_F R_F + w_F \frac{d\phi}{dt}, \qquad \phi = f\left(I_F\right) \tag{10.3}$$

$$c\phi I_A = \Theta \frac{d\Omega}{dt} + \left(T_{load} + \Delta T_{fric}\right) = \Theta \frac{d\Omega}{dt} + T'_{load} \tag{10.4}$$

The differential equations of the armature circuit and the torque are coupled by the speed $\Omega = 2\pi n$ and the armature current I_A.

The inductivity of the field winding is assumed being dependent on the current, whereas the inductivity of the armature winding is assumed being constant. This assumption of constant armature inductivity is valid, because the armature circuit proceeds perpendicular to the main pole axis and here either a large air-gap is dominating or just a small leakage inductivity is present if commutation poles and / or compensation windings are used.

To get a universally valid solution a normalization to the nominal values is advantageous:

$$\begin{aligned}
&U_A^{\otimes} = \frac{U'_A}{U_N}, \qquad I_A^{\otimes} = \frac{I_A}{I_{A,N}}, \qquad R_A^{\otimes} = \frac{R_A I_{A,N}}{U_N} \\
&U_F^{\otimes} = \frac{U_F}{U_N}, \qquad I_F^{\otimes} = \frac{I_F}{I_{F,N}}, \qquad R_F^{\otimes} = \frac{R_F I_{F,N}}{U_N} \\
&\phi^{\otimes} = \frac{\phi}{\phi_N}, \qquad n^{\otimes} = \frac{\Omega}{\Omega_0}, \qquad T_{load}^{\otimes} = \frac{T_{load}}{T_N} \\
&L_{F,N} = \frac{w_F \phi_N}{I_{F,N}}, \quad U_N = c\phi_N \Omega_0, \quad T_N = c\phi_N I_{A,N}
\end{aligned} \tag{10.5}$$

In addition the armature time constant, the field time constant, and the nominal-starting time constant, respectively, are introduced as follows:

$$\tau_A = \frac{L_A}{R_A}, \qquad \tau_F = \frac{L_{F,N}}{R_F}, \qquad \tau_\Theta = \frac{\Theta\Omega_0}{T_N} \tag{10.6}$$

By means of transformations and substitutions the following set of equations is deduced:

$$\tau_A \frac{dI_A^{\otimes}}{dt} = \frac{1}{R_A^{\otimes}}\left(U_A^{\otimes} - n^{\otimes}\phi^{\otimes}\right) - I_A^{\otimes}$$

$$\tau_F \frac{d\phi^{\otimes}}{dt} = \frac{U_F^{\otimes}}{R_F^{\otimes}} - I_F^{\otimes} \tag{10.7}$$

$$\tau_{\Theta} \frac{dn^{\otimes}}{dt} = \phi^{\otimes} I_A^{\otimes} - T_g^{\otimes}$$

These equations lead to the block diagram that usually is used in control engineering (Fig. 10.3).

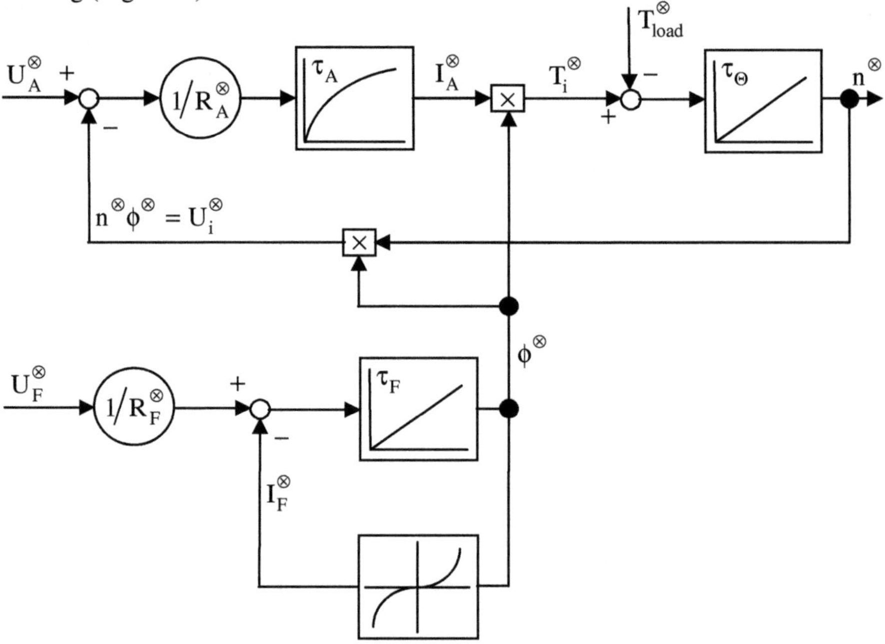

Fig. 10.3. Block diagram of the DC-machine in dynamic operation.

The input values are $U_A^{\otimes}$, $U_F^{\otimes}$ and $T_{load}^{\otimes}$. The system composed of three differential equations is nonlinear because of the multiplications $n^{\otimes}\ \phi^{\otimes}$ and $I_A^{\otimes}\ \phi^{\otimes}$ and because of the magnetizing characteristic $\phi^{\otimes} = f\left(I_F^{\otimes}\right)$. The coupling of these differential equations is performed via $n^{\otimes}$ and $I_A^{\otimes}$.

This coupled, nonlinear set of differential equations can be solved completely only by means of numerical methods. In the following some typical applications are discussed, which can be calculated analytically because of simplifications.

10.2 Separately Excited DC-Machines

10.2.1 General Structure

Speed-variable DC-machines are often operated with constant excitation (constant field e.g. by means of permanent magnets). Then torque and speed are adjusted by varying the armature voltage. In this case $\phi = \phi_N = \text{const.}$ is valid and the block diagram (Fig. 10.4) gets quite simple by $\phi^{\otimes} = 1$.

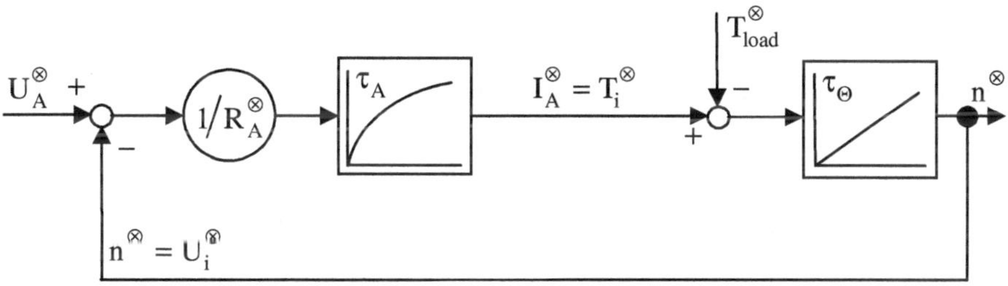

Fig. 10.4. Block diagram of the separately excited DC-machine in dynamic operation with $\phi = \phi_N = \text{const.}$.

By means of the Laplace transformation the representation as block diagram usually used in control engineering is deduced (Fig. 10.5).

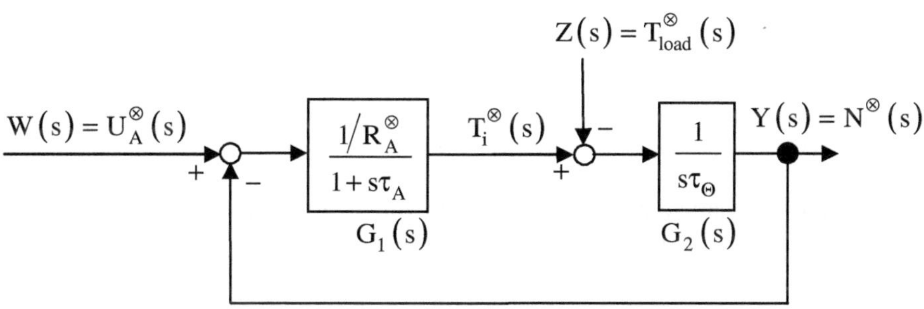

Fig. 10.5. Block diagram of the separately excited DC-machine in Laplace notation with $\phi = \phi_N = \text{const.}$.

Output value is the speed of the DC-machine, which is controlled by the set-point (armature voltage). The disturbance quantity is the load torque.

The following relations are valid:

$$\begin{aligned} Y &= G_2\left(T_i^{\otimes} - Z\right) \\ T_i^{\otimes} &= G_1\left(W - Y\right) \end{aligned} \tag{10.8}$$

By substitutions and transformations it is obtained:

$$Y = \frac{G_1G_2}{1+G_1G_2}W - \frac{G_2}{1+G_1G_2}Z \tag{10.9}$$

10.2.2 Response to Setpoint Changes

"Response to setpoint changes" is called the change of the output value because of a variation of the input (setpoint), when the disturbance is equal to zero.

Therefore, the response to setpoint changes of the DC-machine is the speed change when changing the armature voltage for $Z(s) = T_{load}^{\otimes}(s) = 0$. This situation is described with the block diagram in Fig. 10.6.

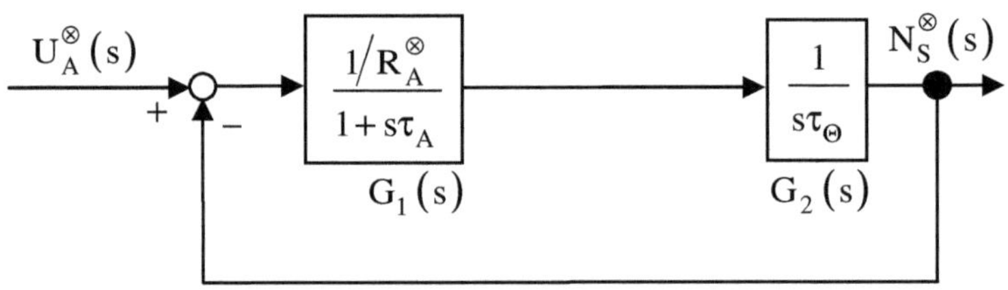

Fig. 10.6. Block diagram of the separately excited DC-machine in Laplace notation for the operation "response to setpoint changes".

For the mathematical description the following is obtained (the additional index "S" at speed and armature current indicates the characteristic functions for response to setpoint changes in the following):

$$\begin{aligned} &N_S^{\otimes} = G_1G_2\left(U_A^{\otimes} - N_S^{\otimes}\right) \\ \Rightarrow\quad &N_S^{\otimes} = \frac{G_1G_2}{1+G_1G_2}U_A^{\otimes} \end{aligned} \tag{10.10}$$

With

$$\frac{Y(s)}{W(s)} = \frac{N_S^{\otimes}(s)}{U_A^{\otimes}(s)} = \frac{G_1(s)G_2(s)}{1+G_1(s)G_2(s)} = \frac{\frac{1/R_A^{\otimes}}{1+s\tau_A}\frac{1}{s\tau_\Theta}}{1+\frac{1/R_A^{\otimes}}{1+s\tau_A}\frac{1}{s\tau_\Theta}} \tag{10.11}$$

it follows further

$$\frac{Y(s)}{W(s)} = \frac{1}{s\tau_\Theta R_A^{\otimes}(1+s\tau_A)+1} \tag{10.12}$$

Introducing the mechanical time constant

$$\tau_{mech} = \tau_\Theta R_A^{\otimes} = \frac{\Theta\Omega_0}{T_N}\frac{R_A I_{A,N}}{U_N} = \frac{\Theta\Omega_0}{T_N\frac{U_N}{R_A}\frac{1}{I_{A,N}}} = \frac{\Theta\Omega_0}{T_N\frac{I_{stall}}{I_{A,N}}} = \frac{\Theta\Omega_0}{T_{stall}} \tag{10.13}$$

it follows:

$$\frac{N_S^{\otimes}(s)}{U_A^{\otimes}(s)} = \frac{1}{1+(1+s\tau_A)s\tau_{mech}} = \frac{1}{\tau_{mech}\tau_A\left(s^2+\frac{s}{\tau_A}+\frac{1}{\tau_{mech}\tau_A}\right)} \tag{10.14}$$

Now at time $t = 0$ a step function of the setpoint shall happen, e.g. switching the nominal voltage to the DC-machine at zero speed. Consequently:

$$U_A^{\otimes}(s) = \frac{1}{s} \tag{10.15}$$

and therefore:

$$N_S^{\otimes}(s) = \frac{1}{s}\frac{1}{\tau_{mech}\tau_A\left(s^2+\frac{s}{\tau_A}+\frac{1}{\tau_{mech}\tau_A}\right)} \tag{10.16}$$

The time-dependent speed variation is obtained by reverse Laplace transformation:

$$n_S^{\otimes}(t) = L^{-1}\left\{\frac{1}{s}\frac{\omega_0^2}{s^2 + 2sD\omega_0 + \omega_0^2}\right\} \tag{10.17}$$

with

$$\begin{aligned} &\omega_0^2 = \frac{1}{\tau_{mech}\tau_A}, \qquad 2D\omega_0 = \frac{1}{\tau_A} \\ &\Rightarrow \quad D = \sqrt{\frac{\tau_{mech}}{4\tau_A}} \end{aligned} \tag{10.18}$$

The solution is (taken from literature concerning Laplace transformation):

$$n_S^{\otimes}(t) = 1 - \frac{e^{-D\omega_0 t}}{\sqrt{1-D^2}}\sin\left(\omega_0\sqrt{1-D^2}\,t + \arcsin\left(\sqrt{1-D^2}\right)\right) \tag{10.19}$$

For $D = 1$ the time-dependent speed is obtained by considering small x-values: $\sin(x) \approx x$, $\arcsin(x) \approx x$. It follows:

$$\begin{aligned} n_S^{\otimes}(t, D=1) &= 1 - \frac{e^{-D\omega_0 t}}{\sqrt{1-D^2}}\left(\omega_0\sqrt{1-D^2}\,t + \sqrt{1-D^2}\right) \\ &= 1 - e^{-D\omega_0 t}\left(\omega_0 t + 1\right) \end{aligned} \tag{10.20}$$

Figure 10.7 shows the results of these equations for different values of τ_{mech}/τ_A.

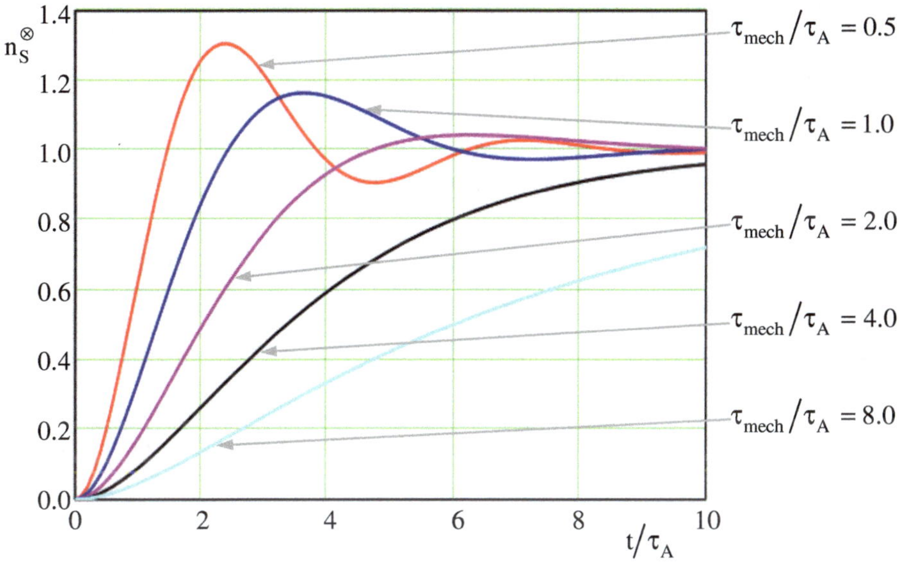

Fig. 10.7. Normalized speed versus time of the separately excited DC-machine for the operation "response to setpoint changes".

From

$$N_S^{\otimes}(s) = G_2(s) I_{A,S}^{\otimes}(s) \tag{10.21}$$

the armature current becomes:

$$I_{A,S}^{\otimes}(s) = s\tau_{\Theta} N_S^{\otimes}(s) \tag{10.22}$$

This means a differentiation for the reverse Laplace transformation:

$$I_{A,S}^{\otimes}(t) = L^{-1}\left\{s\tau_{\Theta} N_S^{\otimes}(s)\right\} = \tau_{\Theta} \frac{dn_S^{\otimes}}{dt} = \frac{1}{R_A^{\otimes}} \tau_{mech} \frac{dn_S^{\otimes}}{dt} \tag{10.23}$$

In total it follows:

$$
\begin{aligned}
I_{A,S}^{\otimes}(t) &= \frac{\tau_{mech}}{R_A^{\otimes}}\left[\frac{D\omega_0}{\sqrt{1-D^2}} e^{-D\omega_0 t} \sin\left(\omega_0\sqrt{1-D^2}\,t + \arcsin\left(\sqrt{1-D^2}\right)\right)\right. \\
&\qquad \left. -\omega_0 e^{-D\omega_0 t} \cos\left(\omega_0\sqrt{1-D^2}\,t + \arcsin\left(\sqrt{1-D^2}\right)\right)\right] \\
&= \frac{\tau_{mech}}{R_A^{\otimes}}\omega_0 e^{-D\omega_0 t}\left[\frac{D}{\sqrt{1-D^2}} \sin\left(\omega_0\sqrt{1-D^2}\,t + \arcsin\left(\sqrt{1-D^2}\right)\right)\right. \\
&\qquad \left. -\cos\left(\omega_0\sqrt{1-D^2}\,t + \arcsin\left(\sqrt{1-D^2}\right)\right)\right]
\end{aligned}
\tag{10.24}
$$

Even here the result for $D = 1$ is obtained by considering small arguments:

$$
\begin{aligned}
I_{A,S}^{\otimes}(t) &= \frac{\tau_{mech}}{R_A^{\otimes}}\omega_0 e^{-D\omega_0 t}\left[\frac{D}{\sqrt{1-D^2}} \sin\left(\omega_0\sqrt{1-D^2}\,t + \arcsin\left(\sqrt{1-D^2}\right)\right)\right. \\
&\qquad \left. -\cos\left(\omega_0\sqrt{1-D^2}\,t + \arcsin\left(\sqrt{1-D^2}\right)\right)\right] \\
&= \frac{\tau_{mech}}{R_A^{\otimes}}\omega_0 e^{-D\omega_0 t}\left[\frac{D}{\sqrt{1-D^2}} \left(\omega_0\sqrt{1-D^2}\,t + \sqrt{1-D^2}\right)\right. \\
&\qquad \left. -\cos\left(\omega_0\sqrt{1-D^2}\,t + \arcsin\left(\sqrt{1-D^2}\right)\right)\right]
\end{aligned}
\tag{10.25}
$$

Further:

$$
\begin{aligned}
I_{A,S}^{\otimes}(t) &= \frac{\tau_{mech}}{R_A^{\otimes}}\omega_0 e^{-D\omega_0 t}\left[D(\omega_0 t + 1) - \cos(0)\right] \\
&= \frac{\tau_{mech}}{R_A^{\otimes}}\omega_0 e^{-D\omega_0 t}\left[D(\omega_0 t + 1) - 1\right]
\end{aligned}
\tag{10.26}
$$

These functions are shown in Fig. 10.8.

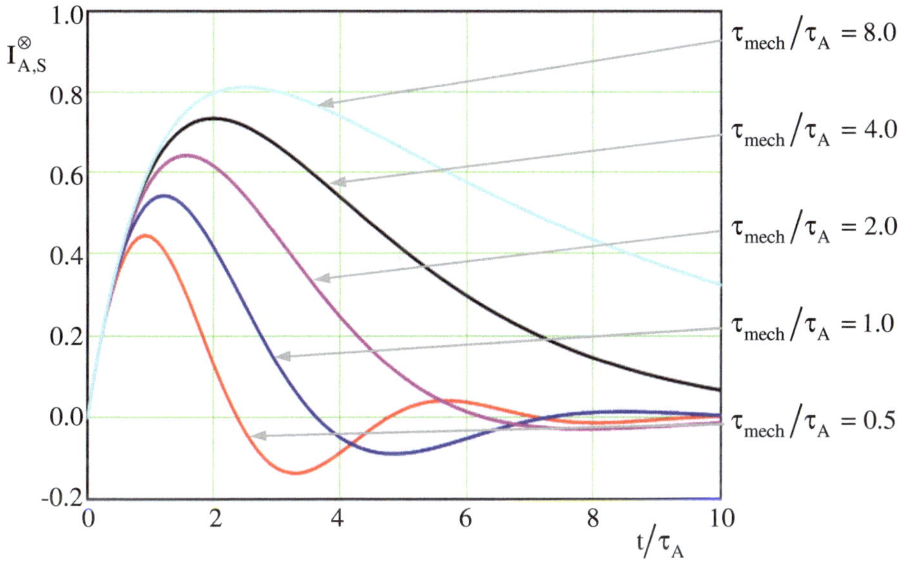

Fig. 10.8. Normalized armature current versus time of the separately excited DC machine for the operation "response to setpoint changes".

10.2.3 Response to Disturbance Changes

"Response to disturbance changes" is called the change of the output value because of a variation of the disturbance, when the input (setpoint) is equal to zero.

Therefore, the response to disturbance changes of the DC-machine is the speed change when changing the load torque for $W(s) = U_A^{\otimes}(s) = 0$. The resulting block diagram is shown in Fig. 10.9 (the additional index "D" at the speed indicates the characteristic functions for response to disturbance changes in the following).

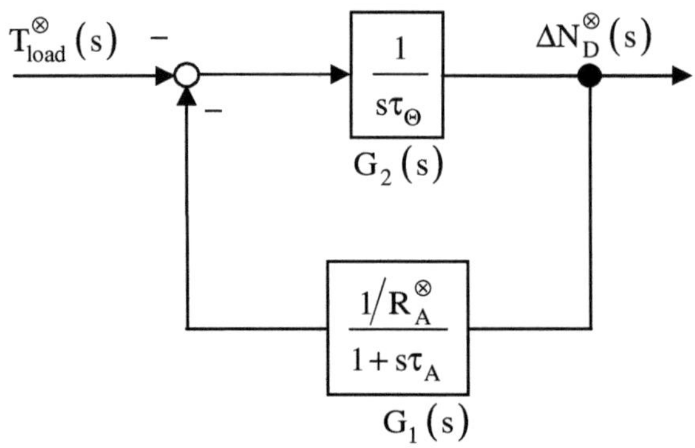

Fig. 10.9. Block diagram of the separately excited DC-machine in Laplace notation for the operation "response to disturbance changes".

Therefore:

$$\Delta N_D^{\otimes}(s) = G_2(s)\left(-T_{load}^{\otimes}(s) - G_1(s)\Delta N_D^{\otimes}(s)\right)$$
$$\Rightarrow \quad \Delta N_D^{\otimes}(s) = \frac{-G_2(s)}{1+G_1(s)G_2(s)} T_{load}^{\otimes}(s) \tag{10.27}$$

and further:

$$\frac{\Delta N_D^{\otimes}(s)}{T_{load}^{\otimes}(s)} = \frac{-\dfrac{1}{s\tau_\Theta}}{1+\dfrac{1/R_A^{\otimes}}{1+s\tau_A}\dfrac{1}{s\tau_\Theta}} = \frac{-R_A^{\otimes}(1+s\tau_A)}{1+R_A^{\otimes}s\tau_\Theta(1+s\tau_A)}$$
$$= \frac{-R_A^{\otimes}(1+s\tau_A)}{1+s\tau_{mech}+s^2\tau_{mech}\tau_A} \tag{10.28}$$

Switching on the nominal torque $T_{load}^{\otimes}(s) = 1/s$, the speed change is:

$$\Delta n_D^{\otimes}(t) = L^{-1}\left\{\frac{-R_A^{\otimes}(1+s\tau_A)}{s\left(1+s\tau_{mech}+s^2\tau_{mech}\tau_A\right)}\right\}$$
$$= -R_A^{\otimes}L^{-1}\left\{N_S^{\otimes}(s) + s\tau_A N_S^{\otimes}(s)\right\} \tag{10.29}$$

With the solution of the preceding section it follows:

$$\begin{aligned}\Delta n_D^{\otimes}(t) &= -R_A^{\otimes}\left(n_S^{\otimes}(t)+\tau_A \frac{dn_S^{\otimes}(t)}{dt}\right)\\ &= -R_A^{\otimes}\left(n_S^{\otimes}(t)+\tau_A \frac{R_A^{\otimes}}{\tau_{mech}} I_{A,S}^{\otimes}(t)\right)\\ &= -R_A^{\otimes} n_S^{\otimes}(t)-\left(R_A^{\otimes}\right)^2 \frac{\tau_A}{\tau_{mech}} I_{A,S}^{\otimes}(t)\end{aligned} \tag{10.30}$$

Now, the speed-time-characteristic shall be calculated, when just before the load change the DC-machine is in no-load operation ($n^{\otimes}=1$, $I_A^{\otimes}=0$) at nominal excitation ($\phi^{\otimes}=1$, see Sect. 10.1). Then it follows:

$$n_D^{\otimes}(t) = 1+\Delta n_D^{\otimes}(t) \tag{10.31}$$

For different values of the parameter $\frac{\tau_{mech}}{\tau_A}$ the speed change is shown in Fig. 10.10.

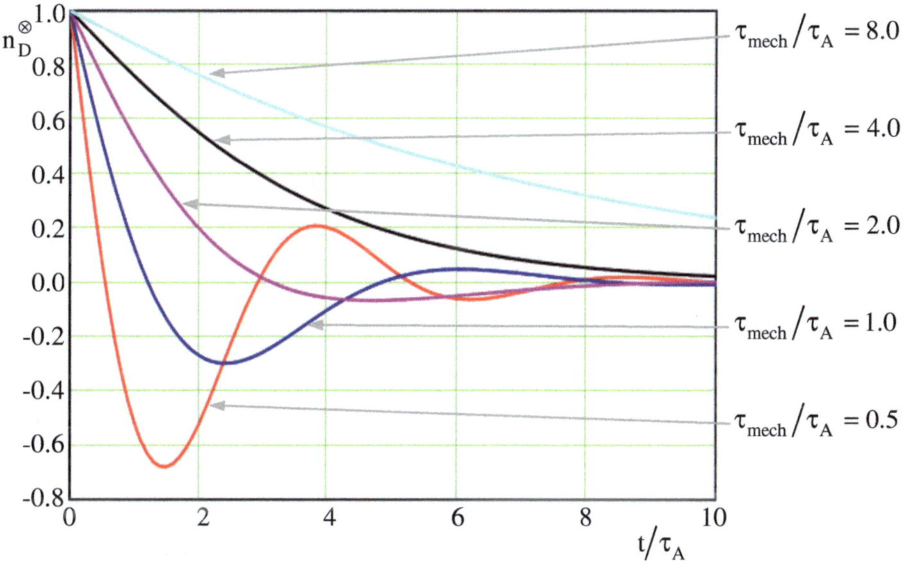

Fig. 10.10. Normalized speed versus time of the separately excited DC-machine for the operation "response to disturbance changes".

The separately excited DC-machine with its energy storages L_A and Θ is an oscillatory system for $D = \sqrt{\frac{\tau_{mech}}{4\tau_A}} < 1$.

For $D = \sqrt{\frac{\tau_{mech}}{4\tau_A}} > 1$ aperiodic characteristics are obtained when changing the armature voltage or the load torque. The case $D = 1$ is called "critical damping".

10.3 Shunt-Wound DC-Machines

For DC-machines with variable exciting flux ϕ the investigation of dynamic operation has to be performed by numerical calculation, because the set of differential equations is nonlinear and cannot be solved analytically. Such a case is present e.g. for starting the shunt-wound DC-machine.

At the time $t = 0$ the machine at standstill shall be switched to the mains. During the entire run-up the machine is unloaded ($T_{load} = 0$). The time-dependent characteristics of armature current, speed, torque, and field exciting current are calculated in the following by means of normalized quantities.

With $U_A^{\otimes} = U_F^{\otimes} = 1$ it follows:

$$\begin{aligned}
\tau_A \frac{dI_A^{\otimes}}{dt} &= \frac{1}{R_A^{\otimes}}\left(1 - n^{\otimes}\phi^{\otimes}\right) - I_A^{\otimes} \\
\tau_F \frac{d\phi^{\otimes}}{dt} &= \frac{1}{R_F^{\otimes}} - I_F^{\otimes}, \qquad \phi^{\otimes} = f\left(I_F^{\otimes}\right) \\
\tau_\Theta \frac{dn^{\otimes}}{dt} &= \phi^{\otimes} I_A^{\otimes}
\end{aligned} \tag{10.32}$$

The iron saturation is considered by using the function $\phi^{\otimes} = f\left(I_F^{\otimes}\right)$.

This set of equations will be solved step by step by numerical integration using a digital computer. The values at time t_{k+1} are calculated from the values at time t_k and the respective changes in the time interval $\Delta t = t_{k+1} - t_k$. This is done by transforming the differential equations into equations of differences:

$$\tau_A \frac{I^{\otimes}_{A,k+1} - I^{\otimes}_{A,k}}{\Delta t} = \frac{1}{R^{\otimes}_A}\left(1 - n^{\otimes}_k \phi^{\otimes}_k\right) - I^{\otimes}_{A,k}$$

$$\tau_F \frac{\phi^{\otimes}_{k+1} - \phi^{\otimes}_k}{\Delta t} = \frac{1}{R^{\otimes}_F} - I^{\otimes}_{F,k}, \qquad \phi^{\otimes}_k = f\left(I^{\otimes}_{F,k}\right) \tag{10.33}$$

$$\tau_\Theta \frac{n^{\otimes}_{k+1} - n^{\otimes}_k}{\Delta t} = \phi^{\otimes}_k I^{\otimes}_{A,k}$$

An additional transformation of the equations gives:

$$I^{\otimes}_{A,k+1} = \left(1 - \frac{\Delta t}{\tau_A}\right) I^{\otimes}_{A,k} + \frac{\Delta t}{\tau_A}\left(\frac{1 - n^{\otimes}_k \phi^{\otimes}_k}{R^{\otimes}_A}\right)$$

$$\phi^{\otimes}_{k+1} = \phi^{\otimes}_k + \frac{\Delta t}{\tau_F}\left(\frac{1}{R^{\otimes}_F} - I^{\otimes}_{F,k}\right), \qquad \phi^{\otimes}_k = f\left(I^{\otimes}_{F,k}\right) \tag{10.34}$$

$$n^{\otimes}_{k+1} = n^{\otimes}_k + \frac{\Delta t}{\tau_\Theta} \phi^{\otimes}_k I^{\otimes}_{A,k}$$

An evaluation of this set of equations during run-up of the DC-machine is shown in Fig. 10.11 for typical time constants and typical normalized resistances (all quantities are shown as normalized values; armature current in red, speed in blue, torque in black, and field exciting current in magenta).

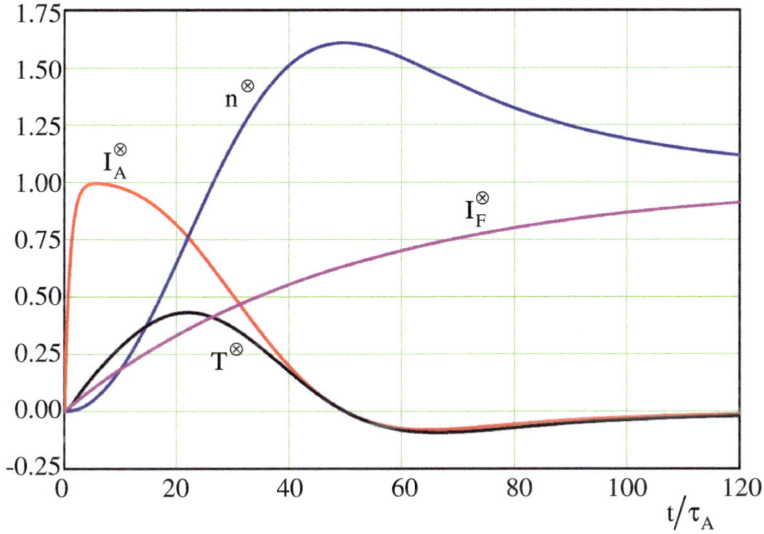

Fig. 10.11. Normalized characteristics of the shunt-wound DC-machine in dynamic operation.

The steep increase of the armature current and the slow increase of the field exciting current lead to a reduced torque during acceleration. Because of $D = \sqrt{\frac{\tau_{mech}}{4\tau_A}} > 1$ (see Sect. 10.2, "separately excited DC-machine") aperiodic characteristics would be expected initially, but the speed and the armature current show overshooting characteristics. The reason for this difference between the separately excited DC-machine and the shunt-wound DC-machine is:

- For the separately excited DC-machine the flux always is constant at its nominal value, i.e. for all times $\phi^{\otimes} = 1 = \text{const.}$
- The field of the shunt-wound DC-machine is increased during starting, i.e. the nominal value is reached delayed.

10.4 Cascaded Control of DC-Machines

For control purposes in the electrical drive engineering often PI- (proportional-integral-) controllers are used. In addition to a simple structure they have the advantage of stationary preciseness (i.e. after disturbances the initial value is reached again, after change of setpoint the new value is reached, both without any stationary difference).

Block diagram and transfer function of a PI-controller are shown in Fig. 10.12:

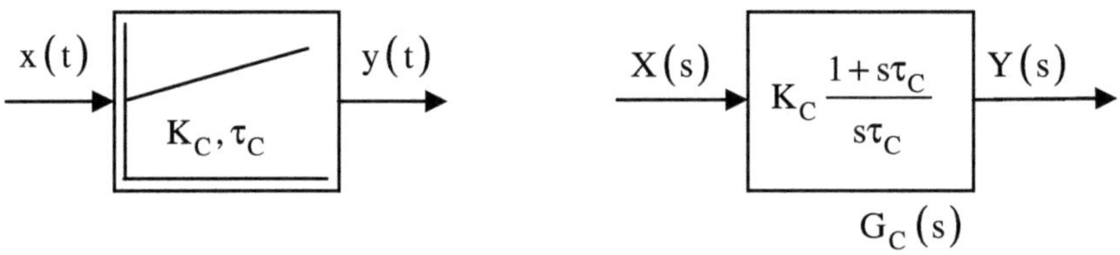

Fig. 10.12. Block diagram and transfer function of a PI-controller.

Figure 10.13 shows a commonly used control circuit composed of permanent magnet excited DC-machine, power electronic converter, and cascaded control. The cascaded control consists of a speed control circuit (realized by a PI-controller: $G_{C,n}(s)$) and a subordinate current control circuit (realized by a PI-controller as well: $G_{C,I}(s)$). In addition, a limitation for the armature current is introduced.

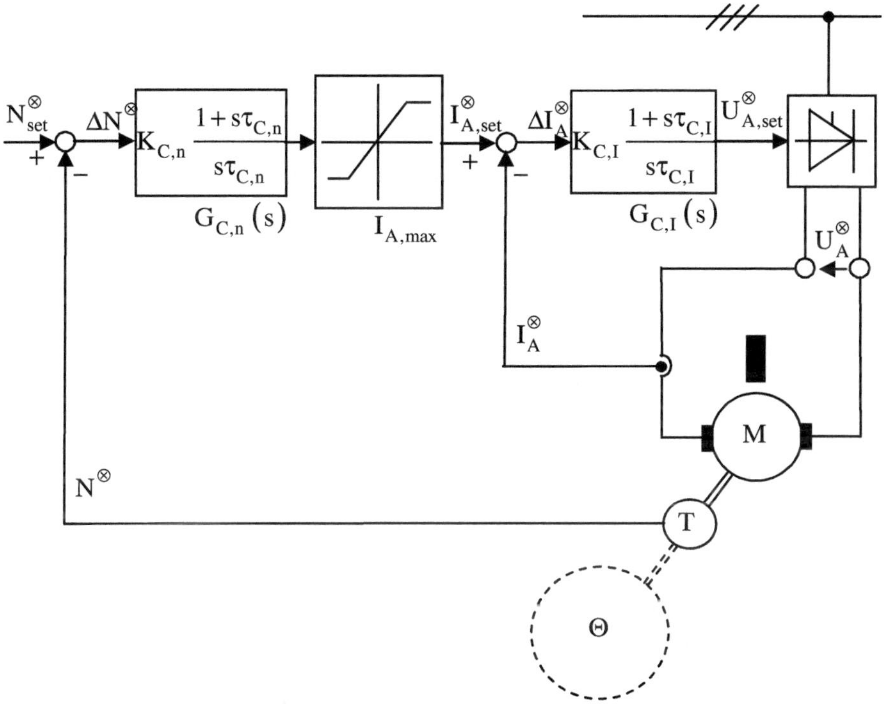

Fig. 10.13. Block diagram of the cascaded control circuit of DC-machines.

For the step by step solution of the set of differential equations numerical methods are used. Therefore, the PI-controllers have to be discretized. There is:

$$\begin{aligned} G_C(s) &= \frac{Y(s)}{X(s)} = K_C \frac{1+s\tau_C}{s\tau_C} \\ \Rightarrow \quad s\tau_C Y(s) &= K_C(1+s\tau_C)X(s) \end{aligned} \tag{10.35}$$

In the time domain this gives:

$$\begin{aligned} \tau_C \frac{dy}{dt} &= K_C\left(x + \tau_C \frac{dx}{dt}\right) \\ \Rightarrow \quad y(t) &= K_C\left(\frac{1}{\tau_C}\int x(t)\,dt + x(t)\right) \end{aligned} \tag{10.36}$$

In discretized description this equation is:

$$y_k = K_C\left(\frac{1}{\tau_C}\sum_{i=1}^{k-1} x_i \ \Delta t_i + x_k\right) \tag{10.37}$$

For the following calculation of the dynamic operation the dead time (time constant) of the power electronic converter will be neglected, i.e. there is $U_A^{\otimes} \equiv U_{A,set}^{\otimes}$. For the permanent magnet excited DC-machine ($\phi^{\otimes} = 1$) the equations in normalized form are:

$$\begin{aligned} \tau_A \frac{dI_A^{\otimes}}{dt} &= \frac{1}{R_A^{\otimes}}\left(U_A^{\otimes} - n^{\otimes}\right) - I_A^{\otimes} \\ \tau_{\Theta} \frac{dn^{\otimes}}{dt} &= I_A^{\otimes} - T_{load}^{\otimes} \end{aligned} \tag{10.38}$$

Figure 10.14 shows the time-dependent characteristics of speed (solid blue line) and armature current (solid red line) together with their respective set values (dotted lines) for a step function of the speed set value from 0 to 1 and later to -1 (starting and reversing) without any load. In this case the armature current is limited to the double nominal value. Because of the constant excitation the armature current (in normalized representation) is identical to the torque. From the overshootings and oscillations can be deduced that the parameters of the controllers are not adjusted optimally (concerning oscillations, speed adjustment, and preciseness).

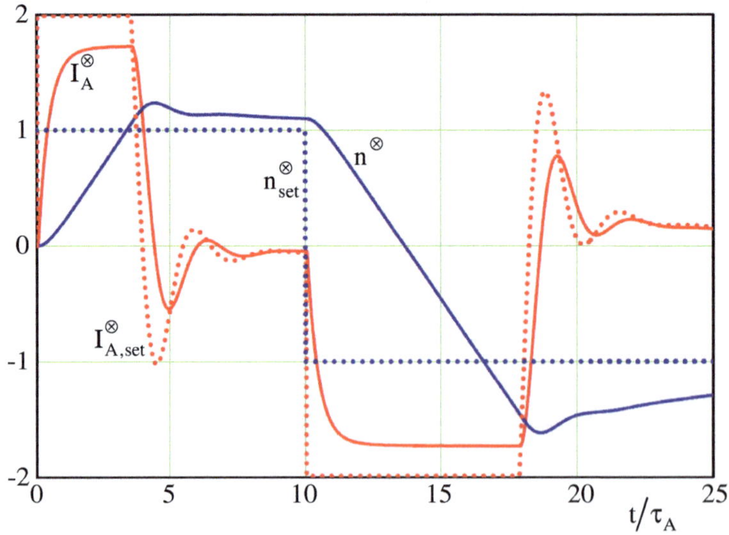

Fig. 10.14. Characteristics of the separately excited DC-machine in cascaded control operation.

10.5 Adjusting Rules for PI-Controllers

10.5.1 Overview

In electrical drive engineering controlled systems with PI-controllers often can be reduced (at least approximately) to the fundamental structure depicted in Fig. 10.15.

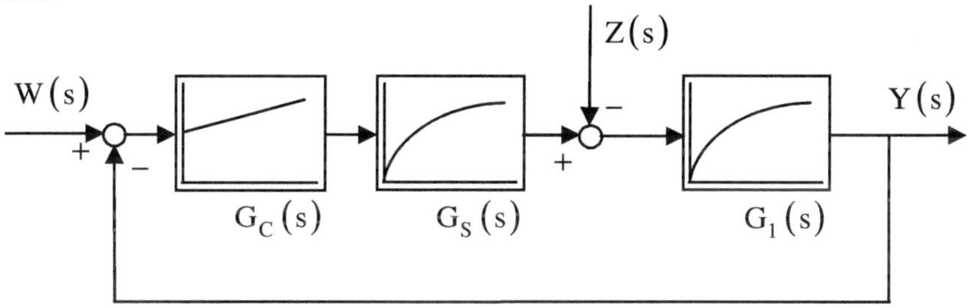

Fig. 10.15. General block diagram of a controlled system.

The transfer functions of the PI-controller and the controlled system in the Laplace domain are:

$$G_C(s) = K_C \frac{1 + s\tau_C}{s\tau_C} \tag{10.39}$$

$$G_S(s) = K_S \frac{1}{1 + s\tau_S}, \qquad G_1(s) = \frac{1}{1 + s\tau_1} \tag{10.40}$$

The parameters are:

- K_C : the gain of the PI-controller
- τ_C : the time constant of the PI-controller
- K_S : the entire gain of the controlled system
- τ_S : the sum of all small time constants of the controlled system
- τ_1 : the large time constant of the controlled system

A practical rule means that the distinction between "sum of all small time constants" and "large time constant" is valid, if

$$\tau_1 \geq 4\tau_S \tag{10.41}$$

is true. Without detailed derivation adjusting rules for the parameters of the PI-controller will be given for two special cases.

10.5.2 Adjusting to Optimal Response to Setpoint Changes (Rule "Optimum of Magnitude")

For the optimum of magnitude the controller parameters have to be chosen as follows; the results are illustrated in Fig. 10.16.

$$\tau_C = \tau_1 \qquad K_C = \frac{1}{2}\frac{1}{K_S}\frac{\tau_1}{\tau_S} \tag{10.42}$$

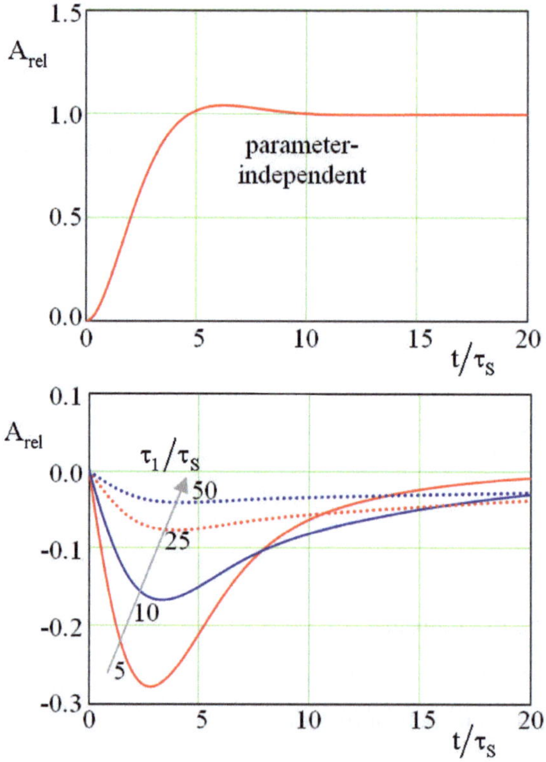

Fig. 10.16. Rule "optimum of magnitude": relative output amplitude versus time for response to setpoint changes (*above*) and response to disturbances (*below*).

10.5.3 Adjusting to Optimal Response to Disturbances (Rule "Symmetrical Optimum")

For the symmetrical optimum the controller parameters have to be chosen as follows:

$$\tau_C = 4\tau_S \qquad K_C = \frac{1}{2}\frac{1}{K_S}\frac{\tau_1}{\tau_S} \tag{10.43}$$

Testing by means of step functions (of the setpoint or the disturbance) the characteristics shown in Fig. 10.17 are obtained.

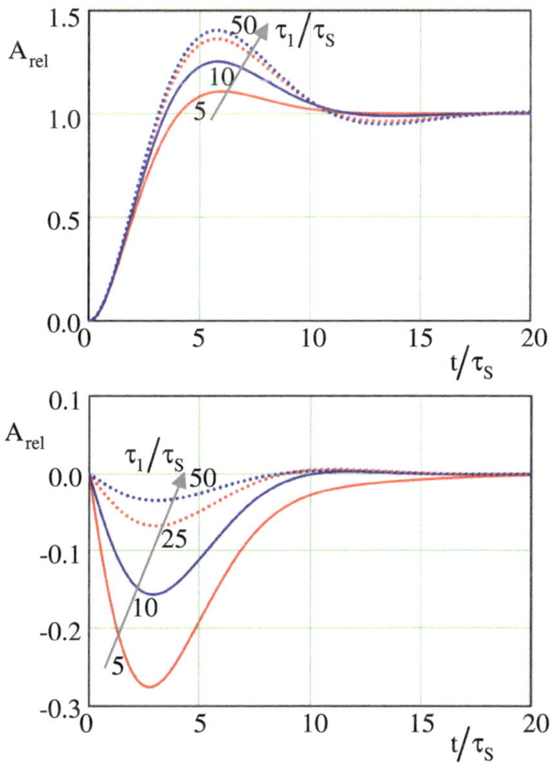

Fig. 10.17. Rule "symmetrical optimum": relative output amplitude versus time for response to setpoint changes (*above*) and response to disturbances (*below*).

10.5.4 Application of the Adjusting Rules to the Cascaded Control of DC-Machines

Applying the rules of optimal response to setpoint changes ("optimum of magnitude") to the current controller and to the speed controller of the cascaded control (see Sect. 10.4) the following results are obtained (Fig. 10.18).

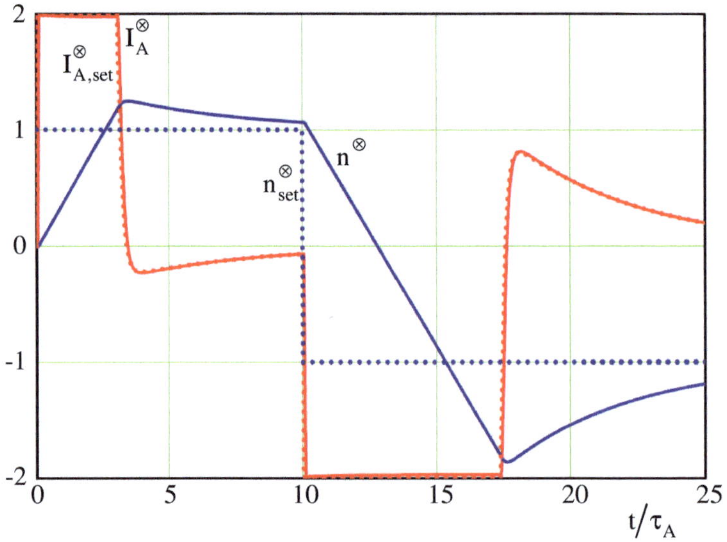

Fig. 10.18. Characteristics of the separately excited DC-machine in cascaded control operation when applying the rules of optimal response to setpoint changes.

In this case, the "optimum of magnitude" is applied to both controllers, because speed and current are exposed to changing setpoints. For the current controller this is the usual case, for the speed controller sometimes even the "symmetrical optimum" rule is applied. This is the case if not changing setpoints, but changing load torques have to be controlled.

From Fig. 10.18 it can be deduced that the armature current follows its set value quite precisely, i.e. the adjusting rule "optimum of magnitude" works very good. Between the speed and its set value there is quite a large deviation. This comes from the fact, that the structure analyzed for the adjusting rules is just an approximation to the speed control loop. Nevertheless, compared to the results shown in Fig. 10.14, the control behavior is improved by far.

By varying the control parameters even a better performance will be achievable.

10.6 References for Chapter 10

Krishnan R (2001) Electric motor drives. Prentice Hall, London
Nasar SA (1970) Electromagnetic energy conversion devices and systems. Prentice Hall, London
Pfaff G, Meier C (1992) Regelung elektrischer Antriebe II. Oldenbourg Verlag, München
Schröder D (1995) Elektrische Antriebe 2. Springer-Verlag, Berlin
White DC, Woodson HH (1958) Electromechanical energy conversion. John Wiley & Sons, New York

11 Space Vector Theory

11.1 Methods for Field Calculation

For the description and calculation of electro-magnetic fields in electrical machines mainly the following four methods are used:

1. The *wave-description* of the electro-magnetic fields (see Sects. 3.3 to 3.5) has proven its value if stationary characteristics are to be calculated. This method may be used for the fundamental-wave characteristics of electrical machines, but it may be even used if additionally to the fundamental wave higher harmonics are to be considered. E.g. this is the case if harmonic torque components (torque oscillations) or acoustic noise shall be computed.
2. The *symmetric components* (please refer to Sect. 1.6) are mainly used for the examination of asymmetric events with constant frequency (this may be even transient reactions).
3. The usage of *complex space vectors* is advantageous if the transient characteristics of (controlled or uncontrolled) electrical drives are regarded. This method will be explained in the following sections.
4. When using the *Finite Element Method (FEM)* the electrical machine is divided in many small parts ("finite elements") and the electro-magnetic behavior of the machine is calculated numerically for each of these elements and for each operating point: The Maxwell's equations are solved in each element and the solutions are adapted to each other at the element borders. The disadvantages of this method are on the one hand a quite high computation time, on the other hand it is more suited for the analysis of a known machine than for the design of a new one (the relevance of different influencing factors on the machine characteristics cannot be directly observed). The main advantage of this method is that virtually all relevant attributes can be considered simultaneously, whereas for the other three mentioned (analytical) methods always limiting constraints have to be regarded.

As can be seen from the above description of the different methods the choice of the suitable alternative depends on the task that has to be solved.

11.2 Requirements for the Application of the Space Vector Theory

For the following considerations some limiting assumptions are made:

1. Each phase winding of stator and rotor produce a sinusoidal magneto-motive force in space, all of the same wave length. This means a limitation to the fundamental wave of the magneto-motive forces (and therefore even a limitation to the fundamental waves of current loading and air-gap flux density); the winding factors of all harmonic waves are assumed being zero. Each single wave can be represented by a vector; the location of the vector shows the instantaneous position of the maximum of the wave, the length of the vector represents this maximum value.
2. Magnetically the machine is completely symmetric (i.e. constant air-gap along the circumference) and the influence of the slots is neglected. Within one machine part (stator or rotor) the self- and mutual-inductivities are independent from the rotor position.
3. Partly, these requirements can be disclaimed: If stator or rotor contains two magnetically or electrically perpendicular preferred orientations, the space vector theory still can be applied if the coordinate system is fixed to the asymmetric machine part.
4. Saturation is neglected, i.e. the magnetic permeances are independent from the magneto-motive forces, and the magnetic voltage drop in iron is neglected ($\mu_{Fe} \rightarrow \infty$); linear relationships do exist. Now (as a main advantage of the representation of the waves by vectors) the common effect of the single waves can be calculated by vector addition.[10]

It is important for understanding the space vector theory, that there is no limiting requirement for the time-dependency of the single currents: The currents may have arbitrary time-dependency which even may be asymmetric (this is crucial for the calculation of transient characteristics). In spite of the *arbitrary time-dependency* of the currents the magneto-motive force, the current loading, and the air-gap flux density generated by each single current are *sinusoidal in space* at every moment;[11] this is realized by a clever winding distribution in the slots of the machine.

In the following only the very important three-phase system is regarded. Nevertheless, the method of the complex space vector is applicable for arbitrary number of phases.

---

[10] This requirement can be attenuated: It is sufficient, that there is a constant saturation condition inside the machine. By linearizing at the operation point the vector addition is still applicable.

[11] This gives a hint for the origin of the name "space vector theory": main importance have the *spatially* sinusoidal waves inside the machine, there are no requirements to special time-dependencies.

11.3 Definition of the Complex Space Vector

The representation of the waves by means of vectors will now be transformed to a representation by means of complex numbers. For this, a complex plain is defined, where the real axis and the axis of the phase "u" enclose a (time-dependent) angle $\alpha(t)$, please refer to Fig. 11.1.

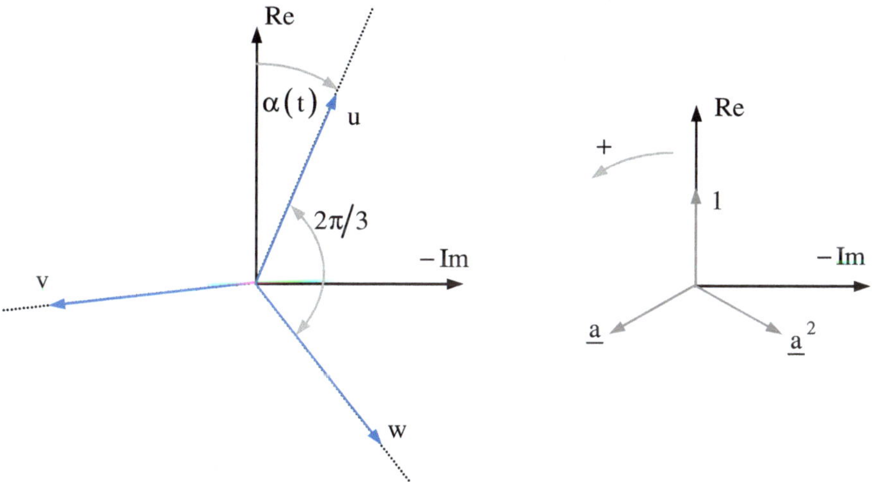

Fig. 11.1. Machine axes in a complex plain (*left*) and definition of the complex operator $\underline{a}$ (*right*).

Now a complex number (called "space vector" in the following) is defined as follows (here shown for the example of the phase currents of a three-phase system):[12]

$$\underset{\rightarrow}{i}(t) = \frac{2}{3}\left(i_u(t) + \underline{a}\, i_v(t) + \underline{a}^2\, i_w(t)\right) e^{-j\alpha(t)} \tag{11.1}$$

[12] In the following complex space vectors, for distinguishing them from other complex numbers, are identified by arrows under the variable. The calculation rules for complex numbers are valid even here; the arrows under the variable just label the special definition of these complex numbers. In literature even a simple underscore can be found to label complex space vectors. For systems with m phases the space vector is: $\underset{\rightarrow}{i}(t) = \frac{2}{m}\sum_{k=1}^{m} \underline{\ell}^{k-1}\, i_k(t)\, e^{-j\alpha}$, with $\underline{\ell} = e^{j\frac{2\pi}{m}}$.
By means of this definition even the rotor of a squirrel cage induction machine can be described with space vectors.

The multiplication with the operator $\underline{a} = e^{j\frac{2\pi}{3}}$ means a rotation by 120° in positive direction. The factor $2/3$ secures a scaling of the absolute value of the space vector in a way, that this absolute value is equal to the amplitude of a phase current when having symmetric supply of all phases. To simplify the notation, in the following the explicit statement of the time-dependency of the angle α is omitted.

The projection of the complex space vector onto the respective phase axis gives the instantaneous value of the phase current.

Regarding a symmetric three-phase current system at the time $t = 0$, at which the current $i_u(t)$ shall be maximum (i.e. the real axis coincides with the axis of the phase “u“, this means $\alpha = 0$), it follows in normalized description (please refer to Fig. 11.2):

$$i_u(t=0) = 1$$
$$i_v(t=0) = i_w(t=0) = -\frac{1}{2} \tag{11.2}$$

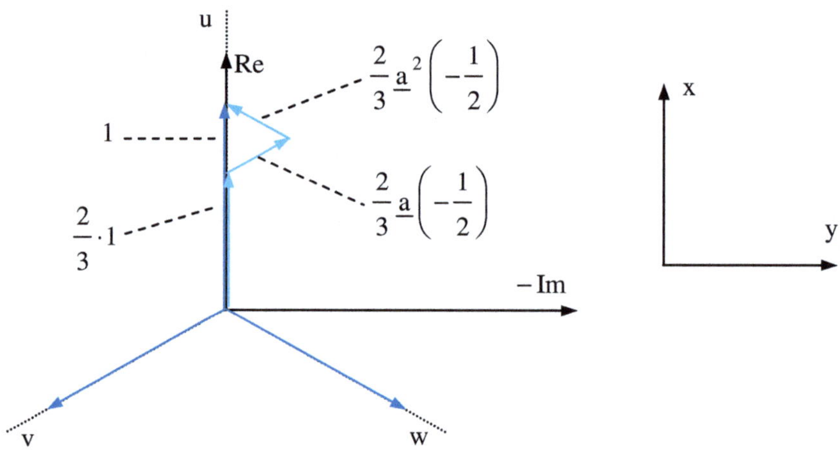

Fig. 11.2. Symmetric three-phase current system in a complex plain (*left*) and definition of the x-y-coordinate system (*right*).

Separating the complex space vector into real and imaginary part, there is:

$$\begin{aligned} \underline{i}(t) &= \mathrm{Re}\{\underline{i}(t)\} + j\,\mathrm{Im}\{\underline{i}(t)\} \\ &= i_x(t) - j\,i_y(t) \end{aligned} \tag{11.3}$$

From

$$\underset{\rightarrow}{\underline{i}}(t) = \frac{2}{3}\left(i_u(t) + \underline{a}\, i_v(t) + \underline{a}^2\, i_w(t)\right) e^{-j\alpha} = i_x(t) - j\, i_y(t) \tag{11.4}$$

both components of the space vector can be computed:

$$\begin{aligned} i_x(t) &= \frac{2}{3}\Bigg[\cos(-\alpha)\, i_u(t) + \cos\left(-\alpha + \frac{2\pi}{3}\right) i_v(t) \\ &\qquad + \cos\left(-\alpha + \frac{4\pi}{3}\right) i_w(t)\Bigg] \\ &= \frac{2}{3}\Bigg[\cos(\alpha)\, i_u(t) + \cos\left(\alpha - \frac{2\pi}{3}\right) i_v(t) \\ &\qquad + \cos\left(\alpha + \frac{2\pi}{3}\right) i_w(t)\Bigg] \end{aligned} \tag{11.5}$$

and

$$\begin{aligned} i_y(t) &= -\frac{2}{3}\Bigg[\sin(-\alpha)\, i_u(t) + \sin\left(-\alpha + \frac{2\pi}{3}\right) i_v(t) \\ &\qquad + \sin\left(-\alpha + \frac{4\pi}{3}\right) i_w(t)\Bigg] \\ &= \frac{2}{3}\Bigg[\sin(\alpha)\, i_u(t) + \sin\left(\alpha - \frac{2\pi}{3}\right) i_v(t) \\ &\qquad + \sin\left(\alpha + \frac{2\pi}{3}\right) i_w(t)\Bigg] \end{aligned} \tag{11.6}$$

Together with the equation for the zero component of the current $i_0(t)$:[13]

$$i_u(t) + i_v(t) + i_w(t) = 3\, i_0(t) \tag{11.7}$$

the following matrix equation can be set-up:

---

[13] The factor 3 in the equation for the zero component of the current is chosen arbitrarily.

$$\begin{bmatrix} i_x(t) \\ i_y(t) \\ i_0(t) \end{bmatrix} = \left[T_{xy0}\right] \begin{bmatrix} i_u(t) \\ i_v(t) \\ i_w(t) \end{bmatrix}$$

$$\left[T_{xy0}\right] = \frac{2}{3} \begin{bmatrix} \cos(\alpha) & \cos\left(\alpha - \frac{2\pi}{3}\right) & \cos\left(\alpha + \frac{2\pi}{3}\right) \\ \sin(\alpha) & \sin\left(\alpha - \frac{2\pi}{3}\right) & \sin\left(\alpha + \frac{2\pi}{3}\right) \\ \frac{1}{2} & \frac{1}{2} & \frac{1}{2} \end{bmatrix} \tag{11.8}$$

Likewise, the inversion can be easily calculated:

$$\begin{bmatrix} i_u(t) \\ i_v(t) \\ i_w(t) \end{bmatrix} = \left[T_{xy0}\right]^{-1} \begin{bmatrix} i_x(t) \\ i_y(t) \\ i_0(t) \end{bmatrix}$$

$$\left[T_{xy0}\right]^{-1} = \begin{bmatrix} \cos(\alpha) & \sin(\alpha) & 1 \\ \cos\left(\alpha - \frac{2\pi}{3}\right) & \sin\left(\alpha - \frac{2\pi}{3}\right) & 1 \\ \cos\left(\alpha + \frac{2\pi}{3}\right) & \sin\left(\alpha + \frac{2\pi}{3}\right) & 1 \end{bmatrix} \tag{11.9}$$

If, as a special condition, the zero component of the current is (time-independent) always equal to zero, as it is the case for star-connected three-phase systems without neutral line, the above equations can be simplified to:

$$i_0(t) = 0 \quad \Rightarrow \quad i_w(t) = -i_u(t) - i_v(t) \tag{11.10}$$

$$\begin{bmatrix} i_x(t) \\ i_y(t) \end{bmatrix} = \left[T_{xy}\right] \begin{bmatrix} i_u(t) \\ i_v(t) \end{bmatrix}$$

$$\left[T_{xy}\right] = \frac{2}{3} \begin{bmatrix} \cos(\alpha) & \cos\left(\alpha - \frac{2\pi}{3}\right) \\ \sin(\alpha) & \sin\left(\alpha - \frac{2\pi}{3}\right) \end{bmatrix} \tag{11.11}$$

$$\begin{bmatrix} i_u(t) \\ i_v(t) \end{bmatrix} = \left[T_{xy}\right]^{-1} \begin{bmatrix} i_x(t) \\ i_y(t) \end{bmatrix}$$

$$\left[T_{xy}\right]^{-1} = \begin{bmatrix} \cos(\alpha) & \sin(\alpha) \\ \cos\left(\alpha - \frac{2\pi}{3}\right) & \sin\left(\alpha - \frac{2\pi}{3}\right) \end{bmatrix} \tag{11.12}$$

In the same way like the space vector of the current at the beginning of this section, even the space vectors of voltage and flux linkage are defined:

$$\underset{\rightarrow}{u}(t) = \frac{2}{3}\left(u_u(t) + \underline{a}\; u_v(t) + \underline{a}^2\; u_w(t)\right) e^{-j\alpha} \tag{11.13}$$

$$\underset{\rightarrow}{\psi}(t) = \frac{2}{3}\left(\psi_u(t) + \underline{a}\; \psi_v(t) + \underline{a}^2\; \psi_w(t)\right) e^{-j\alpha} \tag{11.14}$$

11.4 Voltage Equation in Space Vector Notation

In the following, the space vector theory will be developed in the energy consumption system, which becomes obvious from the signs in the used voltage equations (please refer to Sect. 1.2).

Firstly, the three voltage equations of the (symmetric) system

$$
\begin{aligned}
u_u(t) &= R\, i_u(t) + \frac{d\psi_u(t)}{dt},\\
u_v(t) &= R\, i_v(t) + \frac{d\psi_v(t)}{dt},\\
u_w(t) &= R\, i_w(t) + \frac{d\psi_w(t)}{dt}
\end{aligned}
\tag{11.15}
$$

are multiplied with $\frac{2}{3}e^{-j\alpha}$, $\frac{2}{3}\underline{a}\, e^{-j\alpha}$ and $\frac{2}{3}\underline{a}^2\, e^{-j\alpha}$, respectively. Afterwards these equations are summed up. By means of the relation for the differential of the flux linkage with respect to time

$$
\begin{aligned}
\frac{d\vec{\psi}(t)}{dt} &= \frac{2}{3}\left(\frac{d}{dt}\psi_u(t) + \underline{a}\frac{d}{dt}\psi_v(t) + \underline{a}^2\frac{d}{dt}\psi_w(t)\right)e^{-j\alpha} + \\
&\quad \frac{2}{3}\left(\psi_u(t) + \underline{a}\,\psi_v(t) + \underline{a}^2\,\psi_w(t)\right)(-j)\frac{d\alpha}{dt}e^{-j\alpha}\\
&= \frac{2}{3}\left(\frac{d}{dt}\psi_u(t) + \underline{a}\frac{d}{dt}\psi_v(t) + \underline{a}^2\frac{d}{dt}\psi_w(t)\right)e^{-j\alpha}\\
&\quad - j\frac{d\alpha}{dt}\vec{\psi}(t)\\
\Rightarrow\quad & \frac{2}{3}\left(\frac{d}{dt}\psi_u(t) + \underline{a}\frac{d}{dt}\psi_v(t) + \underline{a}^2\frac{d}{dt}\psi_w(t)\right)e^{-j\alpha}\\
&= \frac{d\vec{\psi}(t)}{dt} + j\frac{d\alpha}{dt}\vec{\psi}(t)
\end{aligned}
\tag{11.16}
$$

the voltage equation in space vector notation is found:[14]

$$
\vec{u}(t) = R\,\vec{i}(t) + \frac{d\vec{\psi}(t)}{dt} + j\frac{d\alpha}{dt}\vec{\psi}(t)
\tag{11.17}
$$

[14] For the special case $\alpha(t) = \text{const.}$ (i.e. the real axis of the coordinate system has a time-independent angle to the axis of the phase "u") it is true: $\vec{u}(t) = R\,\vec{i}(t) + \frac{d\vec{\psi}(t)}{dt}$. This is even the case, if the real axis coincides with the axis of the phase "u", i.e. for $\alpha(t) = 0$.

For this derivation it is important that the linear combination of the three generally valid voltage equations requires no limitation concerning the spatial distribution of the fields or concerning the time-dependent functions of the currents.

11.5 Interpretation of the Space Vector Description

Like shown in the preceding sections, the space vector can be calculated from the values of the three phases or from the sum of real and imaginary part. With other words this means that the three-phase system can be transformed into a two-phase system, like it is schematically shown in Fig. 11.3.

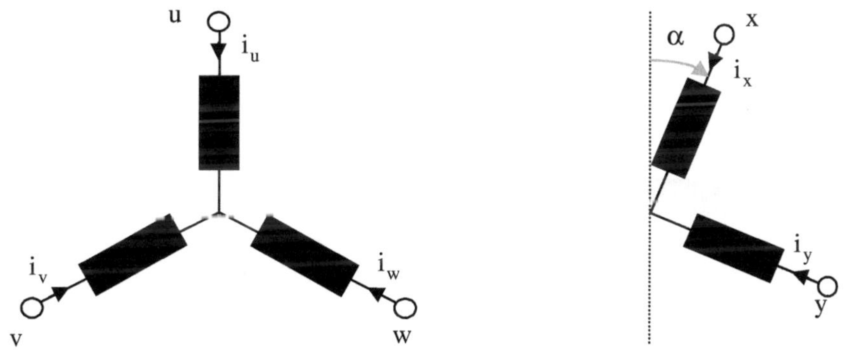

Fig. 11.3. Interpretation of the space vector description: three-phase system (*left*), two-phase system (*right*).

Additionally, by means of an arbitrary angle α a transformation into a coordinate system rotating with arbitrary angular frequency is successful: The time-dependency of the angle α is not limited.

As the currents (and the voltages and flux linkages) of an arbitrary m -phase system can be described in two (perpendicular) coordinates, the space vector theory is applicable to arbitrary phase numbers. The phase numbers of stator and rotor may be even different (an example for this is the squirrel-cage induction machine).

By applying the space vector defined in Sect. 11.3 in total the following transformation is achieved:

- from the m-phase system in stationary coordinates
- into a two-phase system in (arbitrarily) rotating coordinates.

11.6 Coupled Systems

Regarding three-phase systems of stator (index "1" and "I") and rotor (index "2" and "II"), respectively, the spatial position of the rotor system against the stator system has to be considered by the time-dependent angle $\gamma(t)$. It is assumed here that the rotor values are already transformed to the stator system (a special notation for this is omitted to simplify the description).

The coupling of both systems is realized via the flux linkages. As it is obvious from Fig. 11.4 (example for $\alpha = 0$), the coupling via the flux linkages is dependent on the rotor position because of the time-dependent angle $\gamma(t)$.

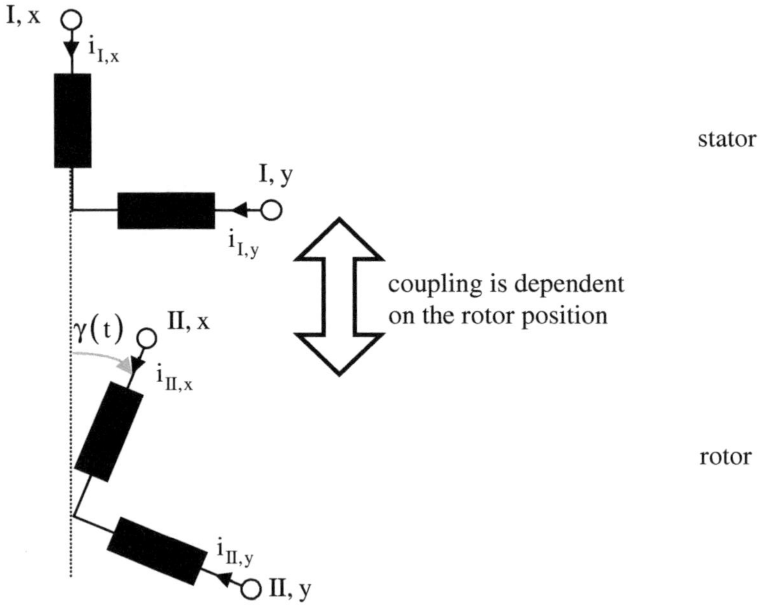

Fig. 11.4. Coupling between stator and rotor system.

Introducing an additional rotation with $e^{j\gamma}$ for the space vectors of the rotor (Fig. 11.5), both systems are transformed to a commonly rotating coordinate system.

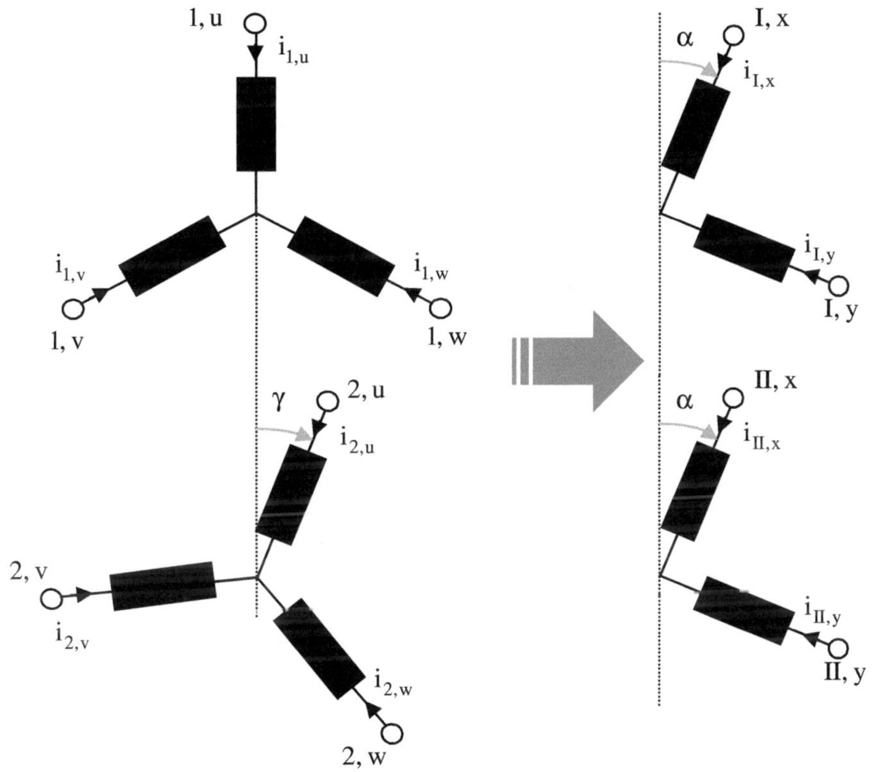

Fig. 11.5. Transforming three-phase systems (*left*) to two-phase systems (*right*); stator: *above*, rotor: *below*.

Then the space vectors of the rotor system are:

$$\underset{\rightarrow}{i}_{II}(t) = \frac{2}{3}\left(i_{2,u}(t) + \underline{a}\, i_{2,v}(t) + \underline{a}^2\, i_{2,w}(t)\right) e^{-j(\alpha-\gamma)} \tag{11.18}$$

$$\underset{\rightarrow}{u}_{II}(t) = \frac{2}{3}\left(u_{2,u}(t) + \underline{a}\, u_{2,v}(t) + \underline{a}^2\, u_{2,w}(t)\right) e^{-j(\alpha-\gamma)} \tag{11.19}$$

$$\underset{\rightarrow}{\psi}_{II}(t) = \frac{2}{3}\left(\psi_{2,u}(t) + \underline{a}\, \psi_{2,v}(t) + \underline{a}^2\, \psi_{2,w}(t)\right) e^{-j(\alpha-\gamma)} \tag{11.20}$$

whereas the space vectors of the stator system against the definition in Sect. 11.3 simply get the respective indices:

$$\underline{i}_I(t) = \frac{2}{3}\left(i_{1,u}(t) + \underline{a}\, i_{1,v}(t) + \underline{a}^2\, i_{1,w}(t)\right) e^{-j\alpha} \tag{11.21}$$

$$\underline{u}_I(t) = \frac{2}{3}\left(u_{1,u}(t) + \underline{a}\, u_{1,v}(t) + \underline{a}^2\, u_{1,w}(t)\right) e^{-j\alpha} \tag{11.22}$$

$$\underline{\psi}_I(t) = \frac{2}{3}\left(\psi_{1,u}(t) + \underline{a}\, \psi_{1,v}(t) + \underline{a}^2\, \psi_{1,w}(t)\right) e^{-j\alpha} \tag{11.23}$$

By means of the deduction from Sect. 11.4 the voltage equations of stator and rotor are in space vector notation:

$$\underline{u}_I(t) = R_I\, \underline{i}_I(t) + \frac{d\underline{\psi}_I(t)}{dt} + j\frac{d\alpha}{dt}\underline{\psi}_I(t) \tag{11.24}$$

and

$$\underline{u}_{II}(t) = R_{II}\, \underline{i}_{II}(t) + \frac{d\underline{\psi}_{II}(t)}{dt} + j\frac{d(\alpha-\gamma)}{dt}\underline{\psi}_{II}(t) \tag{11.25}$$

With the angular frequency of the coordinate system $\omega_{CS} = d\alpha/dt$ and the mechanical angular frequency of the rotor $\omega_{mech} = d\gamma/dt$ the voltage equations of stator and rotor become in space vector notation:[15]

$$\underline{u}_I(t) = R_I\, \underline{i}_I(t) + \frac{d\underline{\psi}_I(t)}{dt} + j\,\omega_{CS}\,\underline{\psi}_I(t) \tag{11.26}$$

---

[15] It has to be considered that the angle γ describes the relative movement between stator and rotor in electrical degrees (relative movement of the magnetic fluxes). The relation between the mechanical angular frequency and the rotor speed is obtained by means of the number of pole pairs p : $\omega_{mech} = p\Omega = 2\pi pn$, with n being the mechanical speed.

$$\vec{u}_{II}(t) = R_{II}\,\vec{i}_{II}(t) + \frac{d\vec{\psi}_{II}(t)}{dt} + j(\omega_{CS} - \omega_{mech})\,\vec{\psi}_{II}(t) \tag{11.27}$$

It is to emphasize here that there are no limitations concerning the time-dependent angles $\alpha(t)$ and $\gamma(t)$. Therefore, even the angular frequencies ω_{CS} and ω_{mech} may have arbitrary time-dependencies. This is crucial for the calculation of dynamic or transient operation conditions. In the following the explicit description of the time-dependency of the angular frequencies (analogously to the angles α and γ) will be omitted to simplify the writing.

11.7 Power in Space Vector Notation

The instantaneous electrical power of the machine can be calculated from the sum of the instantaneous power of all three phases:

$$\begin{aligned} p(t) = {} & u_{1,u}(t)i_{1,u}(t) + u_{1,v}(t)i_{1,v}(t) + u_{1,w}(t)i_{1,w}(t) + \\ & u_{2,u}(t)i_{2,u}(t) + u_{2,v}(t)i_{2,v}(t) + u_{2,w}(t)i_{2,w}(t) \end{aligned} \tag{11.28}$$

The following is true:

$$\begin{aligned} & \mathrm{Re}\left\{\vec{u}_I(t)\,\vec{i}_I^*(t)\right\} \\ & = \mathrm{Re}\left\{\left[\frac{2}{3}\left(u_{1,u}(t) + \underline{a}\,u_{1,v}(t) + \underline{a}^2\,u_{1,w}(t)\right)e^{-j\alpha}\right]\cdot\right. \\ & \qquad \left.\left[\frac{2}{3}\left(i_{1,u}(t) + \underline{a}\,i_{1,v}(t) + \underline{a}^2\,i_{1,w}(t)\right)e^{-j\alpha}\right]^*\right\} \\ & = \left(\frac{2}{3}\right)^2 \mathrm{Re}\left\{u_{1,u}(t)i_{1,u}(t) + u_{1,v}(t)i_{1,v}(t) + u_{1,w}(t)i_{1,w}(t) + \right. \\ & \qquad u_{1,u}(t)\left[\underline{a}^*\,i_{1,v}(t) + \left(\underline{a}^2\right)^* i_{1,w}(t)\right] + \\ & \qquad u_{1,v}(t)\left[\underline{a}\,i_{1,u}(t) + \underline{a}\left(\underline{a}^2\right)^* i_{1,w}(t)\right] + \\ & \qquad \left. u_{1,w}(t)\left[\underline{a}^2\,i_{1,u}(t) + \underline{a}^2\,\underline{a}^*\,i_{1,v}(t)\right]\right\} \end{aligned} \tag{11.29}$$

With

$$\mathrm{Re}\{\underline{a}\} = \mathrm{Re}\left\{e^{j\frac{2\pi}{3}}\right\} = \mathrm{Re}\{\underline{a}^2\} = \mathrm{Re}\{\underline{a}^*\} = \mathrm{Re}\left\{\left(\underline{a}^2\right)^*\right\} = -\frac{1}{2} \tag{11.30}$$

it follows further

$$\begin{aligned}
&\mathrm{Re}\{\underline{u}_I(t)\,\underline{i}_I^*(t)\} \\
&= \left(\frac{2}{3}\right)^2 \Big[u_{1,u}(t)i_{1,u}(t) + u_{1,v}(t)i_{1,v}(t) + u_{1,w}(t)i_{1,w}(t) + \\
&\qquad \left(-\frac{1}{2}\right)u_{1,u}(t)\left(i_{1,v}(t)+i_{1,w}(t)\right) + \\
&\qquad \left(-\frac{1}{2}\right)u_{1,v}(t)\left(i_{1,u}(t)+i_{1,w}(t)\right) + \\
&\qquad \left(-\frac{1}{2}\right)u_{1,w}(t)\left(i_{1,u}(t)+i_{1,v}(t)\right)\Big]
\end{aligned} \tag{11.31}$$

If the zero component of the current $i_0(t) = 0$ holds, there is further[16]

$$\begin{aligned}
&\mathrm{Re}\{\underline{u}_I(t)\,\underline{i}_I^*(t)\} \\
&= \left(\frac{2}{3}\right)^2 \Big[u_{1,u}(t)i_{1,u}(t) + u_{1,v}(t)i_{1,v}(t) + \\
&\qquad u_{1,w}(t)i_{1,w}(t) + \frac{1}{2}\Big(u_{1,u}(t)i_{1,u}(t) + \\
&\qquad u_{1,v}(t)i_{1,v}(t) + u_{1,w}(t)i_{1,w}(t)\Big)\Big] \\
&= \left(\frac{2}{3}\right)\left(u_{1,u}(t)i_{1,u}(t) + u_{1,v}(t)i_{1,v}(t) + u_{1,w}(t)i_{1,w}(t)\right)
\end{aligned} \tag{11.32}$$

[16] It can be shown that the following deduction is true even for the general case $i_0(t) \neq 0$. Because of simplification the general derivation is omitted here. As a hint may be taken that the zero component of the current does not contribute to the space vector because of $i_0 + \underline{a}\,i_0 + \underline{a}^2\,i_0 = 0$.

Performing an analogous calculation for the rotor values and comparing this to Eq. (11.28) the electrical power in space vector notation becomes:

$$p(t) = \frac{3}{2} \operatorname{Re}\left\{ \underline{u}_I(t)\, \underline{i}_I^*(t) + \underline{u}_{II}(t)\, \underline{i}_{II}^*(t) \right\} \tag{11.33}$$

Now the voltage equations Eq. (11.26) and Eq. (11.27) are introduced to this equation:

$$\begin{aligned} p(t) = \frac{3}{2} \operatorname{Re}\Bigg\{ & \left[R_I\, \underline{i}_I(t) + \frac{d\underline{\psi}_I(t)}{dt} + j\,\omega_{CS}\, \underline{\psi}_I(t) \right] \underline{i}_I^*(t) + \\ & \left[R_{II}\, \underline{i}_{II}(t) + \frac{d\underline{\psi}_{II}(t)}{dt} + j(\omega_{CS} - \omega_{mech})\, \underline{\psi}_{II}(t) \right] \underline{i}_{II}^*(t) \Bigg\} \\ = \frac{3}{2} \operatorname{Re}\Bigg\{ & \left[R_I\, \underline{i}_I(t)\, \underline{i}_I^*(t) + R_{II}\, \underline{i}_{II}(t)\, \underline{i}_{II}^*(t) \right] + \\ & \left[\frac{d\underline{\psi}_I(t)}{dt} \underline{i}_I^*(t) + \frac{d\underline{\psi}_{II}(t)}{dt} \underline{i}_{II}^*(t) \right] + \\ & j\,\omega_{CS} \left[\underline{\psi}_I(t)\, \underline{i}_I^*(t) + \underline{\psi}_{II}(t)\, \underline{i}_{II}^*(t) \right] - \\ & j\,\omega_{mech}\, \underline{\psi}_{II}(t)\, \underline{i}_{II}^*(t) \Bigg\} \end{aligned} \tag{11.34}$$

Writing the flux linkages by means of self- and mutual inductivities (with $L_{I,II} = L_{II,I}$) it follows

$$\begin{aligned} & \underline{\psi}_I(t)\, \underline{i}_I^*(t) + \underline{\psi}_{II}(t)\, \underline{i}_{II}^*(t) \\ & \quad = \left[L_I\, \underline{i}_I(t) + L_{I,II}\, \underline{i}_{II}(t) \right] \underline{i}_I^*(t) + \\ & \qquad \left[L_{II}\, \underline{i}_{II}(t) + L_{I,II}\, \underline{i}_I(t) \right] \underline{i}_{II}^*(t) \\ & \quad = L_I\, \underline{i}_I(t)\, \underline{i}_I^*(t) + L_{II}\, \underline{i}_{II}(t)\, \underline{i}_{II}^*(t) + \\ & \qquad L_{I,II} \left[\underline{i}_{II}(t)\, \underline{i}_I^*(t) + \underline{i}_I(t)\, \underline{i}_{II}^*(t) \right] \\ & \quad = L_I\, \underline{i}_I(t)\, \underline{i}_I^*(t) + L_{II}\, \underline{i}_{II}(t)\, \underline{i}_{II}^*(t) + \\ & \qquad L_{I,II} \left[\underline{i}_{II}(t)\, \underline{i}_I^*(t) + \left(\underline{i}_{II}(t)\, \underline{i}_I^*(t) \right)^* \right] \end{aligned} \tag{11.35}$$

Consequently, this expression always is real. It follows

$$\operatorname{Re}\left\{j\,\omega_{CS}\left[\underline{\psi}_{I}(t)\,\underline{i}_{I}^{*}(t)+\underline{\psi}_{II}(t)\,\underline{i}_{II}^{*}(t)\right]\right\}=0 \tag{11.36}$$

The power in space vector notation is therefore simplified to

$$\begin{aligned}p(t)=\frac{3}{2}\operatorname{Re}\Bigg\{&\left[R_{I}\,\underline{i}_{I}(t)\,\underline{i}_{I}^{*}(t)+R_{II}\,\underline{i}_{II}(t)\,\underline{i}_{II}^{*}(t)\right]+\\&\left[\frac{d\underline{\psi}_{I}(t)}{dt}\underline{i}_{I}^{*}(t)+\frac{d\underline{\psi}_{II}(t)}{dt}\underline{i}_{II}^{*}(t)\right]-\\&j\,\omega_{mech}\,\underline{\psi}_{II}(t)\,\underline{i}_{II}^{*}(t)\Bigg\}\end{aligned} \tag{11.37}$$

In this equation three parts of the power can be separated:

1. The losses (in the resistances of stator and rotor)

$$p_{loss}(t)=\frac{3}{2}\operatorname{Re}\left\{R_{I}\,\underline{i}_{I}(t)\,\underline{i}_{I}^{*}(t)+R_{II}\,\underline{i}_{II}(t)\,\underline{i}_{II}^{*}(t)\right\} \tag{11.38}$$

2. The change of the stored magnetic energy

$$p_{\mu}(t)=\frac{3}{2}\operatorname{Re}\left\{\frac{d\underline{\psi}_{I}(t)}{dt}\underline{i}_{I}^{*}(t)+\frac{d\underline{\psi}_{II}(t)}{dt}\underline{i}_{II}^{*}(t)\right\} \tag{11.39}$$

3. The mechanical power

$$p_{mech}(t)=-\frac{3}{2}\operatorname{Re}\left\{j\,\omega_{mech}\,\underline{\psi}_{II}(t)\,\underline{i}_{II}^{*}(t)\right\} \tag{11.40}$$

11.8 Elements of the Equivalent Circuit

11.8.1 Resistances

For the Ohmic losses the following is true:

$$\begin{aligned} p_{loss}(t) &= R_1\, i_{1,u}^2(t) + R_1\, i_{1,v}^2(t) + R_1\, i_{1,w}^2(t) + \\ &\quad R_2'\, i_{2,u}^2(t) + R_2'\, i_{2,v}^2(t) + R_2'\, i_{2,w}^2(t) \\ &= \frac{3}{2} \mathrm{Re}\left\{R_I\, \underline{i}_I(t)\, \underline{i}_I^*(t) + R_{II}\, \underline{i}_{II}(t)\, \underline{i}_{II}^*(t)\right\} \end{aligned} \tag{11.41}$$

Considering the part of the stator losses it follows:

$$R_1\left[i_{1,u}^2(t) + i_{1,v}^2(t) + i_{1,w}^2(t)\right] = \frac{3}{2} \mathrm{Re}\left\{R_I\, \underline{i}_I(t)\, \underline{i}_I^*(t)\right\} \tag{11.42}$$

Inserting the stator current space vector leads to:

$$\begin{aligned} &R_1\left[i_{1,u}^2(t) + i_{1,v}^2(t) + i_{1,w}^2(t)\right] \\ &= \frac{3}{2} R_I\, \mathrm{Re}\left\{\frac{2}{3}\left[i_{1,u}(t) + \underline{a}\, i_{1,v}(t) + \underline{a}^2\, i_{1,w}(t)\right] e^{-j\alpha} \cdot \right. \\ &\qquad \left. \frac{2}{3}\left[i_{1,u}(t) + \underline{a}\, i_{1,v}(t) + \underline{a}^2\, i_{1,w}(t)\right]^* e^{j\alpha}\right\} \end{aligned} \tag{11.43}$$

With an analogous calculation like in Sect. 11.7 it follows:

$$R_1\left[i_{1,u}^2(t) + i_{1,v}^2(t) + i_{1,w}^2(t)\right] = \frac{3}{2} \mathrm{Re}\left\{R_I\, \underline{i}_I(t)\, \underline{i}_I^*(t)\right\} \tag{11.44}$$

By comparison of the coefficients it follows finally

$$R_1 = R_I \tag{11.45}$$

By means of an analogous calculation it is obvious[17]

$$R_2 = R_{II} \tag{11.46}$$

This calculation shows that the transformation in space vector notation is resistance-invariant and concerning the Ohmic losses it is power-invariant.

11.8.2 Inductivities

Considering firstly only the flux linkage of phase “1,u“ (please refer to Fig. 11.5), it follows with L_{11} as self-inductivity of the stator and $L_{1\sigma}$ as leakage inductivity of the stator:[18]

$$\begin{aligned}
\psi_{1,u}(t) = {} & [L_{11} + L_{1\sigma}]\, i_{1,u}(t) + \\
& L_{11}\, i_{1,v}(t)\cos\left(\frac{2\pi}{3}\right) + L_{11}\, i_{1,w}(t)\cos\left(\frac{4\pi}{3}\right) + \\
& \left[L_{11}\, i_{2,u}(t) + L_{11}\, i_{2,v}(t)\cos\left(\frac{2\pi}{3}\right) + \right. \\
& \left. L_{11}\, i_{2,w}(t)\cos\left(\frac{4\pi}{3}\right)\right] e^{j\gamma}
\end{aligned} \tag{11.47}$$

This expression can be transformed like follows:

$$\begin{aligned}
\psi_{1,u}(t) = {} & L_{1\sigma}\, i_{1,u}(t) + \\
& L_{11}\left(i_{1,u}(t) - \frac{1}{2} i_{1,v}(t) - \frac{1}{2} i_{1,w}(t)\right) + \\
& L_{11}\left(i_{2,u}(t) - \frac{1}{2} i_{2,v}(t) - \frac{1}{2} i_{2,w}(t)\right) e^{j\gamma}
\end{aligned} \tag{11.48}$$

---

[17] Like usually done the rotor parameters of the machine are transformed to the stator system (the representation with primed variables is omitted here to achieve a more simple writing). This transformation is described e.g. in Sect. 4.1 for the induction machine.

[18] The transformation of the rotor values to the stator system will be done analogously to Sect. 4.1; this transformation is assumed here (please see the beginning of Sect. 11.6). The difference to the deduction in Sect. 4.1. is that here not the stationary operation condition is calculated by means of the single-phase complex phasors, but all time-dependencies in all phases of stator and rotor are considered explicitly.

With an analogous derivation for the flux linkages of the other phases and the definition of the flux linkage space vector (see Eq. (11.23)) it follows:

$$
\begin{aligned}
\vec{\underline{\psi}}_{I}(t) &= \frac{2}{3}\left(\psi_{1,u}(t)+\underline{a}\,\psi_{1,v}(t)+\underline{a}^{2}\,\psi_{1,w}(t)\right)e^{-j\alpha} \\
&= \frac{2}{3}L_{1\sigma}\left(i_{1,u}(t)+\underline{a}\,i_{1,v}(t)+\underline{a}^{2}\,i_{1,w}(t)\right)e^{-j\alpha} + \\
&\quad \frac{2}{3}L_{11}\Big[\left(i_{1,u}(t)+\underline{a}\,i_{1,v}(t)+\underline{a}^{2}\,i_{1,w}(t)\right) - \\
&\qquad \frac{1}{2}\left(i_{1,v}(t)+\underline{a}\,i_{1,w}(t)+\underline{a}^{2}\,i_{1,u}(t)\right) - \\
&\qquad \frac{1}{2}\left(i_{1,w}(t)+\underline{a}\,i_{1,u}(t)+\underline{a}^{2}\,i_{1,v}(t)\right)\Big]e^{-j\alpha} + \\
&\quad \frac{2}{3}L_{11}\Big[\left(i_{2,u}(t)+\underline{a}\,i_{2,v}(t)+\underline{a}^{2}\,i_{2,w}(t)\right) - \\
&\qquad \frac{1}{2}\left(i_{2,v}(t)+\underline{a}\,i_{2,w}(t)+\underline{a}^{2}\,i_{2,u}(t)\right) - \\
&\qquad \frac{1}{2}\left(i_{2,w}(t)+\underline{a}\,i_{2,u}(t)+\underline{a}^{2}\,i_{2,v}(t)\right)\Big]e^{-j\alpha}e^{j\gamma}
\end{aligned}
\tag{11.49}
$$

Inserting the stator and rotor current space vectors (see Eq. (11.21) and Eq. (11.18)) gives:

$$
\begin{aligned}
\vec{\underline{\psi}}_{I}(t) = L_{1\sigma}\,\vec{\underline{i}}_{I}(t) + L_{11}\left[\vec{\underline{i}}_{I}(t) - \frac{1}{2}\underline{a}^{2}\,\vec{\underline{i}}_{I}(t) - \frac{1}{2}\underline{a}\,\vec{\underline{i}}_{I}(t)\right] + \\
L_{11}\left[\vec{\underline{i}}_{II}(t) - \frac{1}{2}\underline{a}^{2}\,\vec{\underline{i}}_{II}(t) - \frac{1}{2}\underline{a}\,\vec{\underline{i}}_{II}(t)\right]
\end{aligned}
\tag{11.50}
$$

With $\underline{a}+\underline{a}^{2}=-1$ and the stator main inductivity $L_{1m}=\frac{3}{2}L_{11}$ it follows further:[19]

[19] For calculation of the stator main inductivity, which is also called rotating field inductivity, please refer to Sect. 4.1.

$$\underline{\psi}_I(t) = L_{1\sigma}\ \underline{i}_I(t) + \frac{3}{2} L_{11}\ \underline{i}_I(t) + \frac{3}{2} L_{11}\ \underline{i}_{II}(t)$$
$$= L_{1\sigma}\ \underline{i}_I(t) + L_{1m}\ \underline{i}_I(t) + L_{1m}\ \underline{i}_{II}(t) \qquad (11.51)$$

Introducing the stator inductivity $L_1 = L_{1m} + L_{1\sigma}$ it follows:

$$\underline{\psi}_I(t) = L_1\ \underline{i}_I(t) + L_{1m}\ \underline{i}_{II}(t) \qquad (11.52)$$

Analogously it follows for the space vector of the rotor flux linkage (after transformation of the rotor values to the stator system and with $L_2 = L_{1m} + L_{2\sigma}$):

$$\underline{\psi}_{II}(t) = L_2\ \underline{i}_{II}(t) + L_{1m}\ \underline{i}_I(t) \qquad (11.53)$$

11.8.3 Summary of Results

The transformation of the machine parameters from the existing (three-phase) machine to the space vector notation is done like explained in Table 11.1. Consequently, the components of the machine in space vector notation are identical to the components, which are already deduced for the stationary operation of the machine. Therefore, the transformation is resistance- and inductivity-invariant.

Table 11.1. Parameters of existing machines and in space vector notation.

parameter	**existing machine (rotor values transformed to the stator system)**	**space vector notation (rotor values transformed to the stator system)**
stator resistance	R_1	$R_I = R_1$
stator leakage inductivity	$L_{1\sigma}$	$L_{1\sigma}$
stator main inductivity	L_{1m}	L_{1m}
stator inductivity	$L_1 = L_{1m} + L_{1\sigma}$	$L_1 = L_{1m} + L_{1\sigma}$
rotor resistance	R_2	$R_{II} = R_2$
rotor leakage inductivity	$L_{2\sigma}$	$L_{2\sigma}$
rotor inductivity	$L_2 = L_{1m} + L_{2\sigma}$	$L_2 = L_{1m} + L_{2\sigma}$

11.9 Torque in Space Vector Notation

11.9.1 General Torque Calculation

For a machine having an arbitrary number of pole pairs p the mechanical power can be calculated from torque and mechanical angular frequency like follows:

$$p_{mech}(t) = T(t)\,\Omega = T(t)\frac{\omega_{mech}}{p} = T(t)\frac{d\gamma/dt}{p}$$
$$\Omega(t) = 2\pi\, n(t) \tag{11.54}$$

Together with Eq. (11.38) for the mechanical power the torque in space vector notation becomes:

$$T(t) = -\frac{3}{2}p\ \mathrm{Re}\left\{j\,\underline{\psi}_{II}(t)\,\underline{i}_{II}^{*}(t)\right\} \tag{11.55}$$

This equation will be slightly transformed to calculate the torque from stator values. With

$$\underline{\psi}_{II}(t) = L_{2\sigma}\ \underline{i}_{II}(t) + L_{1m}\ \underline{i}_{II}(t) + L_{1m}\ \underline{i}_{I}(t) \tag{11.56}$$

and

$$\mathrm{Re}\left\{j\,\underline{i}_{II}(t)\,\underline{i}_{II}^{*}(t)\right\} = 0 \tag{11.57}$$

it follows

$$T(t) = -\frac{3}{2}p\ \mathrm{Re}\left\{j\,L_{1m}\ \underline{i}_{I}(t)\,\underline{i}_{II}^{*}(t)\right\} \tag{11.58}$$

Because $\underline{i}_{I}(t)\,\underline{i}_{I}^{*}(t)$ always is real, it follows further

$$
\begin{aligned}
T(t) &= -\frac{3}{2} p \ \mathrm{Re}\left\{ j L_{1m} \ \underline{i}_{I}(t) \underline{i}_{II}^{*}(t) + j\left(L_{1m} + L_{1\sigma}\right) \underline{i}_{I}(t) \underline{i}_{I}^{*}(t) \right\} \\
&= -\frac{3}{2} p \ \mathrm{Re}\left\{ j\left[L_{1m} \ \underline{i}_{II}^{*}(t) + \left(L_{1m} + L_{1\sigma}\right) \underline{i}_{I}^{*}(t) \right] \underline{i}_{I}(t) \right\} \\
&= -\frac{3}{2} p \ \mathrm{Re}\left\{ j \ \underline{i}_{I}(t) \underline{\psi}_{I}^{*}(t) \right\} \\
&= \frac{3}{2} p \ \mathrm{Im}\left\{ \underline{i}_{I}(t) \underline{\psi}_{I}^{*}(t) \right\}
\end{aligned}
\tag{11.59}
$$

11.9.2 Torque Calculation by Means of Cross Product from Stator Flux Linkage and Stator Current

The torque can be calculated even as cross product from flux linkage and current. This will be shown in the following. The space vectors of current and flux linkage can be written as (φ and ξ are the phase angles of current and flux linkage, respectively):

$$
\begin{aligned}
\underline{i}_{I}(t) &= i_{I}(t) e^{-j\varphi(t)} \\
\underline{\psi}_{I}(t) &= \psi_{I}(t) e^{-j\xi(t)}
\end{aligned}
\tag{11.60}
$$

where all amplitudes and phase angles may have arbitrary time-dependencies.

Then:

$$
\begin{aligned}
T(t) &= \frac{3}{2} p \ \mathrm{Im}\left\{ i_{I}(t) e^{-j\varphi(t)} \left[\psi_{I}(t) e^{-j\xi(t)} \right]^{*} \right\} \\
&= \frac{3}{2} p \, i_{I}(t) \psi_{I}(t) \, \mathrm{Im}\left\{ e^{-j\left[\varphi(t) - \xi(t)\right]} \right\} \\
&= \frac{3}{2} p \, i_{I}(t) \psi_{I}(t) \sin\left\{ \xi(t) - \varphi(t) \right\} \\
&= \frac{3}{2} p \ \underline{\psi}_{I}(t) \times \underline{i}_{I}(t)
\end{aligned}
\tag{11.61}
$$

11.9.3 Torque Calculation by Means of Cross Product from Stator and Rotor Current

The above equation

$$T(t) = -\frac{3}{2} p \ \mathrm{Re}\left\{ j\, L_{1m} \ \underline{i}_{I}(t)\, \underline{i}_{II}^{*}(t) \right\} \qquad \text{(see Eq. (11.58))}$$

can be transformed to

$$\begin{aligned} T(t) &= \frac{3}{2} p \ \mathrm{Im}\left\{ L_{1m} \ \underline{i}_{I}(t)\, \underline{i}_{II}^{*}(t) \right\} \\ &= \frac{3}{2} p \, L_{1m} \, \mathrm{Im}\left\{ \underline{i}_{I}(t)\, \underline{i}_{II}^{*}(t) \right\} \end{aligned} \tag{11.62}$$

Analogously to the above calculation it follows:

$$T(t) = \frac{3}{2} p \, L_{1m} \ \underline{i}_{II}(t) \times \underline{i}_{I}(t) \tag{11.63}$$

11.9.4 Torque Calculation by Means of Cross Product from Rotor Flux Linkage and Rotor Current

According to Sect. 11.9.1 it is true:

$$T(t) = -\frac{3}{2} p \ \mathrm{Re}\left\{ j\, \underline{\psi}_{II}(t)\, \underline{i}_{II}^{*}(t) \right\} \qquad \text{(see Eq. (11.55))}$$

Analogously to the above calculation it follows:

$$\begin{aligned} T(t) &= \frac{3}{2} p \, \mathrm{Im}\left\{ \underline{\psi}_{II}(t)\, \underline{i}_{II}^{*}(t) \right\} \\ &= \frac{3}{2} p \ \underline{i}_{II}(t) \times \underline{\psi}_{II}(t) \end{aligned} \tag{11.64}$$

11.9.5 Torque Calculation by Means of Cross Product from Stator and Rotor Flux Linkage

From the torque equation in Sect. 11.9.1

$$T(t) = \frac{3}{2} p \,\mathrm{Im}\left\{ \underset{\rightarrow}{i}_I(t) \underset{\rightarrow}{\psi}^*_I(t) \right\} \qquad \text{(see Eq. (11.59))}$$

and

$$\begin{aligned}
\underset{\rightarrow}{\psi}_I &= L_1 \underset{\rightarrow}{i}_I + L_{1m} \underset{\rightarrow}{i}_{II} \\
&= L_1 \underset{\rightarrow}{i}_I + L_{1m} \frac{L_2}{L_2} \underset{\rightarrow}{i}_{II} \\
&= \frac{L_{1m}}{L_2}\left(L_2 \underset{\rightarrow}{i}_{II} + L_{1m} \underset{\rightarrow}{i}_I\right) + L_1 \underset{\rightarrow}{i}_I - \frac{L_{1m}}{L_2} L_{1m} \underset{\rightarrow}{i}_I \\
&= \frac{L_{1m}}{L_2} \underset{\rightarrow}{\psi}_{II} + \left((1+\sigma_1) - \frac{1}{1+\sigma_2} \right) L_{1m} \underset{\rightarrow}{i}_I \\
&= \frac{L_{1m}}{L_2} \underset{\rightarrow}{\psi}_{II} + \left(1 - \frac{1}{(1+\sigma_2)(1+\sigma_1)} \right) L_1 \underset{\rightarrow}{i}_I \\
&= \frac{L_{1m}}{L_2} \underset{\rightarrow}{\psi}_{II} + \sigma L_1 \underset{\rightarrow}{i}_I
\end{aligned} \tag{11.65}$$

it follows

$$T = \frac{3}{2} \frac{p}{\sigma L_1} \mathrm{Im}\left\{ \left(\underset{\rightarrow}{\psi}_I - \frac{L_{1m}}{L_2} \underset{\rightarrow}{\psi}_{II} \right) \underset{\rightarrow}{\psi}^*_I \right\} \tag{11.66}$$

With

$$\mathrm{Im}\left\{ \underset{\rightarrow}{\psi}_I \underset{\rightarrow}{\psi}^*_I \right\} = 0 \tag{11.67}$$

it follows further

$$T = -\frac{3}{2} \frac{p}{\sigma L_1} \frac{L_{1m}}{L_2} \mathrm{Im}\left\{ \underset{\rightarrow}{\psi}_{II} \underset{\rightarrow}{\psi}^*_I \right\} \tag{11.68}$$

Together with

$$\begin{aligned}\frac{L_{1m}}{L_1 L_2} &= \frac{L_{1m}}{(1+\sigma_1)L_{1m}(1+\sigma_2)L_{1m}} \\ &= \frac{1}{L_{1m}}\frac{1}{(1+\sigma_1)(1+\sigma_2)} \\ &= \frac{1}{L_{1m}}(1-\sigma)\end{aligned} \tag{11.69}$$

it follows

$$T = -\frac{3}{2}p\frac{1-\sigma}{\sigma L_{1m}}\operatorname{Im}\left\{\underset{\rightarrow}{\psi}_{II}\ \underset{\rightarrow}{\psi}_{I}^{*}\right\} \tag{11.70}$$

Analogously to the above calculation this results in:

$$\begin{aligned}T &= -\frac{3}{2}p\frac{1-\sigma}{\sigma L_{1m}}\underset{\rightarrow}{\psi}_{I}\times\underset{\rightarrow}{\psi}_{II} \\ &= \frac{3}{2}p\frac{1-\sigma}{\sigma L_{1m}}\underset{\rightarrow}{\psi}_{II}\times\underset{\rightarrow}{\psi}_{I}\end{aligned} \tag{11.71}$$

11.10 Special Coordinate Systems

For rotating field machines it is often necessary to perform the calculations in different coordinate systems. Examples are:

- Realizing constant mutual inductivities for salient-pole synchronous machines → here the calculation in coordinate system fixed to the rotor is advantageous;
- Field-oriented control of induction machines → here the usage of a field-oriented coordinate system is advantageous.

Because of this reason the general transformation to a coordinate system rotating with arbitrary angular frequency $\omega_{CS} = \frac{d\alpha}{dt}$ is very beneficial. The set of equations is generally applicable and depending on the machine topology or usefulness the angle $\alpha(t)$ can be chosen arbitrarily, e.g.:

- $\alpha(t)=0$ stationary coordinate system (fixed to the stator), in literature the axes " x " and " y " of the coordinate system are often called in this case " α " and " β ";
- $\omega_{CS}=\frac{d\gamma}{dt}=\omega_{mech}$ coordinate system rotating with the rotor speed (fixed to the rotor), in literature the axes " x " and " y " of the coordinate system are then often called " q " and " d ";
- $\omega_{CS}=\omega_1$ coordinate system rotating with the synchronous speed;
- $\omega_{CS}=\omega_\mu$ coordinate system rotating with the air-gap flux, even for this coordinate system in literature the axes " x " and " y "are often called "q-axis " and "d-axis ".

11.11 Relation between Space Vector Theory and Two-Axis-Theory

In addition to the space vector theory described in the preceding sections even the two-axis-theory is known to calculate dynamic operating conditions in electrical machines.

Both theories are strongly linked to each other, especially both theories require the same assumptions that have to be fulfilled for their application (please refer to Sect. 11.2).

The main difference can be found in the definition of the vectors. In the two-axis-theory the vectors are defined as follows (here exemplarily shown for the currents):

$$\vec{i}(t)=\sqrt{\frac{2}{3}}\left(i_u(t)+\underline{a}\,i_v(t)+\underline{a}^2\,i_w(t)\right)e^{-j\alpha(t)} \tag{11.72}$$

Similar definitions hold true even for the voltages and the flux linkages. As a result for the torque of the machine it is obtained:

$$T(t)=p\,\mathrm{Im}\left\{\vec{i}_I(t)\,\vec{\psi}_I^*(t)\right\} \tag{11.73}$$

Apart from the factor $3/2$ this equation is identical to the torque equation in space vector notation.

Therefore, the space vector notation has the advantage that voltages and currents can be interpreted quite clearly: Having a stationary, symmetric operation the amplitudes of current, voltage, and flux linkage space vectors are identical to the amplitudes of the phase values. But this transformation is not power-invariant, which can be deduced from the factor $3/2$ in the torque equation.

In contrary, it can be shown that the two-axis-theory is power-invariant. However, the interpretation of the voltages and currents are not so clear (i.e. for calculation of the really flowing currents a respective factor has to be introduced).

11.12 Relation between Space Vectors and Phasors

Between the space vectors (e.g. Sect. 11.3) and the phasors (e.g. Sect. 1.6) there is a *formal similarity*. Regarding the current phasor of the positive system when having symmetric components (see Sect. 1.6), the following is true:

$$\underline{I}_{p} = \frac{1}{3}\left(\underline{I}_{u} + \underline{a}\;\underline{I}_{v} + \underline{a}^{2}\;\underline{I}_{w}\right) \tag{11.74}$$

Here $\underline{I}_{p}$ is the current phasor of the positive system and $\underline{I}_{u}$, $\underline{I}_{v}$ and $\underline{I}_{w}$ are the current phasors of the three phase currents.

A main requirement for the application of the symmetric components was that the three phase currents are sinusoidal with the same frequency (then it can be calculated with rms-values, which is indicated by the capital letters in the above equation). With other words: Steady-state (but asymmetric) operation conditions can be calculated smartly by means of complex phasors.

The space vectors were defined in the preceding Sect. 11.3. For the currents and the special case $\alpha = 0$ there is:

$$\underline{i}(t) = \frac{2}{3}\left(i_{u}(t) + \underline{a}\;i_{v}(t) + \underline{a}^{2}\;i_{w}(t)\right) \tag{11.75}$$

When introducing the space vectors it was explicitly emphasized that there are *no restrictions* for the time-dependency of the currents.[20] This is the *main difference with regards to content* to the phasors, and only because of this difference dynamic operation conditions are able to be calculated by space vectors, but not by phasors.

11.13 References for Chapter 11

Dajaku G (2006) Electromagnetic and thermal modeling of highly utilized PM machines. Shaker-Verlag, Aachen

Kleinrath H (1980) Stromrichtergespeiste Drehfeldmaschinen. Springer Verlag, Wien

Krause PC (1986) Analysis of electric machinery. McGraw Hill Book Company, New York

Schröder D (1994) Elektrische Antriebe 1. Springer-Verlag, Berlin

Vas P (1992) Electrical machines and drives - a space vector theory approach. Oxford University Press, Oxford

[20] Reminder: When introducing the space vectors it was required that magneto-motive force, current loading, and air-gap flux density are *spatially* sinusoidal for every point in time and that there is the same wavelength for all phases. This is realized by a clever winding distribution in the machine and a limitation to the fundamental waves (which, looked at precisely, is an approximation).

12 Dynamic Operation and Control of Induction Machines

12.1 Steady-State Operation of Induction Machines in Space Vector Notation at No-Load

12.1.1 Set of Equations

For the calculation of the dynamic operation of the induction machine the general set of equations for rotating field machines (space vector theory, please refer to Chap. 11) can be used. Because of the constant air-gap when neglecting the slotting effect any choice of $\alpha(t)$ is possible.

Initially, an arbitrary coordinate system is chosen, the angular frequency ω_{CS} and the initial value α_0 will be chosen later:

$$\alpha(t) = \omega_{CS} t + \alpha_0 \tag{12.1}$$

The mechanical speed is (please refer to the footnote in Sect. 11.6):

$$\frac{d\gamma}{dt} = \omega_{mech} = p\Omega = 2\pi p n \tag{12.2}$$

The angular synchronous speed is:

$$\omega_1 = 2\pi f_1 \tag{12.3}$$

After transforming the rotor quantities to the stator winding, the voltage equations of the induction machine with short-circuited rotor winding are (see Sect. 11.6):

$$\underline{u}_I(t) = R_1 \, \underline{i}_I(t) + \frac{d\underline{\psi}_I(t)}{dt} + j\,\omega_{CS} \, \underline{\psi}_I(t) \tag{12.4}$$

$$0 = R_2 \; \underset{\rightarrow}{i}_{II}(t) + \frac{d\underset{\rightarrow}{\psi}_{II}(t)}{dt} + j(\omega_{CS} - \omega_{mech})\underset{\rightarrow}{\psi}_{II}(t) \tag{12.5}$$

with the flux linkages (see Sect. 11.8):

$$\underset{\rightarrow}{\psi}_{I}(t) = L_1 \; \underset{\rightarrow}{i}_{I}(t) + L_{1m} \; \underset{\rightarrow}{i}_{II}(t) \tag{12.6}$$

$$\underset{\rightarrow}{\psi}_{II}(t) = L_2 \; \underset{\rightarrow}{i}_{II}(t) + L_{1m} \; \underset{\rightarrow}{i}_{I}(t) \tag{12.7}$$

and the torque equation (see Sect. 11.9):

$$T(t) = \frac{3}{2} p \, \mathrm{Im}\left\{\underset{\rightarrow}{i}_{I}(t) \underset{\rightarrow}{\psi}^{*}_{I}(t)\right\} \tag{12.8}$$

12.1.2 Steady-State Operation at No-Load

Now the stationary operation at no-load is to be regarded. It is:

$$\omega = \text{const.}, \qquad \frac{d\underset{\rightarrow}{\psi}}{dt} = 0, \qquad \underset{\rightarrow}{i}_{II}(t) = 0 \tag{12.9}$$

If in addition the stator resistance is neglected ($R_1 = 0$), the set of equations becomes:

$$\begin{aligned} \underset{\rightarrow}{u}_{I}(t) &= j\,\omega_{CS} \; L_1 \; \underset{\rightarrow}{i}_{I}(t) \\ 0 &= j(\omega_{CS} - \omega_{mech}) L_{1m} \; \underset{\rightarrow}{i}_{I}(t) \\ T(t) &= \frac{3}{2} p \; L_1 \; \mathrm{Im}\left\{\underset{\rightarrow}{i}_{I}(t) \, \underset{\rightarrow}{i}^{*}_{I}(t)\right\} \end{aligned} \tag{12.10}$$

These three equations will now be regarded closely. From the defining equation of the stator voltage space vector

$$\underset{\rightarrow}{u}_{I}(t) = \frac{2}{3}\left(u_{1,u}(t) + \underline{a} \; u_{1,v}(t) + \underline{a}^{2} \; u_{1,w}(t)\right) e^{-j\alpha} \tag{12.11}$$

and supplying the machine with a symmetrical voltage system

$$\begin{aligned}
u_{1,u}(t) &= \sqrt{2}\,U_1 \cos(\omega_1 t) = \sqrt{2}\,U_1 \frac{1}{2}\left[e^{j\omega_1 t} + e^{-j\omega_1 t}\right] \\
u_{1,v}(t) &= \sqrt{2}\,U_1 \cos\left(\omega_1 t - \frac{2\pi}{3}\right) = \sqrt{2}\,U_1 \frac{1}{2}\left[\underline{a}^2\, e^{j\omega_1 t} + \underline{a}\, e^{-j\omega_1 t}\right] \\
u_{1,w}(t) &= \sqrt{2}\,U_1 \cos\left(\omega_1 t - \frac{4\pi}{3}\right) = \sqrt{2}\,U_1 \frac{1}{2}\left[\underline{a}\, e^{j\omega_1 t} + \underline{a}^2\, e^{-j\omega_1 t}\right]
\end{aligned} \tag{12.12}$$

the stator voltage equation becomes:

$$\begin{aligned}
\underrightarrow{u}_I(t) &= \frac{2}{3}\sqrt{2}\,U_1 \frac{1}{2}\Big[\left(1 + \underline{a}\,\underline{a}^2 + \underline{a}^2\,\underline{a}\right) e^{j\omega_1 t} + \\
&\qquad \left(1 + \underline{a}\,\underline{a} + \underline{a}^2\,\underline{a}^2\right) e^{-j\omega_1 t}\Big] e^{-j\alpha} \\
&= \sqrt{2}\,U_1\, e^{j\omega_1 t}\, e^{-j\alpha}
\end{aligned} \tag{12.13}$$

Thus the stator voltage space vector is performing a circular movement (in space) with the angular frequency ω_1.

Now a coordinate system is chosen that rotates in synchronism with the rotating stator field ($\omega_{CS} = \omega_1$). In addition the initial value is set to $\alpha_0 = 0$. Consequently there is

$$\alpha(t) = \omega_{CS} t + \alpha_0 = \omega_1 t \tag{12.14}$$

and further

$$\underrightarrow{u}_I(t) = \sqrt{2}\,U_1\, e^{j\omega_1 t}\, e^{-j\omega_1 t} = \sqrt{2}\,U_1 \tag{12.15}$$

The real component of the stator voltage space vector is identical to the peak value of the phase voltage, the imaginary component is zero. As the coordinate system rotates in synchronism with the stator frequency ($\omega_{CS} = \omega_1$ had been chosen), this is valid for any point in time. With other words: In stationary operation and having this choice of ω_{CS} all stator quantities become DC values (for the stator voltages this is shown in the last equation: right of the equal sign there is no time-dependency any more).

Further evaluating the stator voltage equation gives:

$$\sqrt{2}\, U_1 = j\, \omega_{CS}\, L_1\, \underline{i}_I(t) \tag{12.16}$$

and therefore

$$\underline{i}_I(t) = -j\frac{\sqrt{2}\, U_1}{\omega_{CS}\, L_1} \tag{12.17}$$

This expression is totally imaginary; it is the no-load current of the induction machine in stationary operation. This no-load current has a phase shift of 90° against the stator voltage, which is already known from Chap. 4.

The rotor voltage equation is

$$0 = j(\omega_{CS} - \omega_{mech})\, L_{1m}\, \underline{i}_I(t) \tag{12.18}$$

and can be fulfilled only for $\omega_{CS} = \omega_{mech}$. As the coordinate system rotates in synchronism with the rotating stator field (please refer to the choice of $\alpha(t)$ above), it follows that $\omega_1 = \omega_{mech}$ is true. With other words the machine rotates with synchronous speed. As it is well-known, this characterizes the no-load operation (when losses are neglected).

The torque equation gives:

$$\begin{aligned} T(t) &= \frac{3}{2}\, p\, L_I\, \mathrm{Im}\{\underline{i}_I(t)\, \underline{i}_I^*(t)\} \\ &= 0 \end{aligned} \tag{12.19}$$

because the multiplication of a complex number with its conjugate-complex value always gives a real number. Even this result is in accordance to the well-known torque of the induction machine in stationary operation at no-load.

12.2 Fast Acceleration and Sudden Load Change

In the following the fast acceleration of the induction machine will be calculated, if at the time $t = 0$ the machine at zero speed is switched to the nominal voltage. It is assumed that the supplying mains is fixed (concerning the rms-value U_1 and the angular frequency of the voltage $\omega_1 = 2\pi f_1$) and that the machine is loaded just by its inertia. After reaching a (nearly) steady-state condition the machine abruptly is loaded by its nominal torque.

As angular frequency and initial value the following is chosen

$$\alpha(t) = \omega_1 t \tag{12.20}$$

For numerical solution the set of equations is transformed into the following shape (here Θ is the inertia and T_{load} is the load torque; the equation of torque balance is: $\Theta \frac{d\Omega}{dt} = T - T_{load}$):

$$\begin{aligned}
\frac{d\underset{\rightarrow}{\psi}_I(t)}{dt} &= \underset{\rightarrow}{u}_I(t) - R_1\ \underset{\rightarrow}{i}_I(t) - j\,\omega_{CS}\ \underset{\rightarrow}{\psi}_I(t) \\
\frac{d\underset{\rightarrow}{\psi}_{II}(t)}{dt} &= -R_2\ \underset{\rightarrow}{i}_{II}(t) - j(\omega_{CS} - \omega_{mech})\underset{\rightarrow}{\psi}_{II}(t) \\
\frac{d\omega_{mech}}{dt} &= \frac{p}{\Theta}\left[\frac{3}{2} p \operatorname{Im}\left\{\underset{\rightarrow}{i}_I(t)\underset{\rightarrow}{\psi}_I^*(t)\right\} - T_{load}\right]
\end{aligned} \tag{12.21}$$

From the equations of the flux linkages (see Sect. 12.1) it follows

$$\begin{aligned}
\begin{bmatrix} \underset{\rightarrow}{i}_I(t) \\ \underset{\rightarrow}{i}_{II}(t) \end{bmatrix} &= \begin{bmatrix} L_1 & L_{1m} \\ L_{1m} & L_2 \end{bmatrix}^{-1} \begin{bmatrix} \underset{\rightarrow}{\psi}_I(t) \\ \underset{\rightarrow}{\psi}_{II}(t) \end{bmatrix} \\
&= \frac{1-\sigma}{\sigma L_{1m}} \begin{bmatrix} 1+\sigma_2 & -1 \\ -1 & 1+\sigma_1 \end{bmatrix} \begin{bmatrix} \underset{\rightarrow}{\psi}_I(t) \\ \underset{\rightarrow}{\psi}_{II}(t) \end{bmatrix}
\end{aligned} \tag{12.22}$$

with

$$\sigma = 1 - \frac{1}{(1+\sigma_1)(1+\sigma_2)} \tag{12.23}$$

The initial conditions for $t = 0$ are that all currents and voltages in this set of equations as well as the angular frequency and the load torque are zero. For $t > 0$ the excitation is:

$$\underset{\rightarrow}{u}_I(t) = \sqrt{2}\ U_1, \qquad T_{load} = 0 \tag{12.24}$$

After finishing the fast acceleration of the unloaded induction machine the following values are reached (please refer to the preceding section):

$$\underline{i}_I(t) = -j\frac{\sqrt{2}\,U_1}{\omega_1\,L_1}$$
$$\underline{i}_{II}(t) = 0 \qquad (12.25)$$

At the time $t = t_1$ the induction machine suddenly is loaded with its nominal torque $T_{load} = T_N$. The excitation quantities are then:

$$\underline{u}_I(t) = \sqrt{2}\,U_1, \qquad T_{load} = T_N \qquad (12.26)$$

In both cases (fast acceleration of the unloaded induction machine at fixed mains and sudden load change) there are transient responses. The data of the simulated machine are:

- stator resistance: $R_1 = 1\Omega$,
- rotor resistance: $R_2' = 1\Omega$
- stator main inductivity: $L_{1m} = 260\text{mH}$
- stator leakage coefficient: $\sigma_1 = 0.1$
- rotor leakage coefficient: $\sigma_2 = 0.1$
- number of pole pairs: $p = 2$
- inertia: $\Theta = 5 \cdot 10^{-3}\,\text{kgm}^2$

The machine is supplied with $U_1 = 230\text{V}$ and $f_1 = 50\text{Hz}$. With this supply the induction machine generates a pull-out torque of $T_{pull-out} \approx 26.8\text{Nm}$ at a pull-out slip of $s_{pull-out} \approx 0.058$. The nominal torque is $T_N = 15.0\text{Nm}$, therefore the overload capability is $T_{pull-out}/T_N \approx 1.8$.

The time-dependent characteristics of such an induction motor operation are shown in Figs. 12.1 to 12.4.

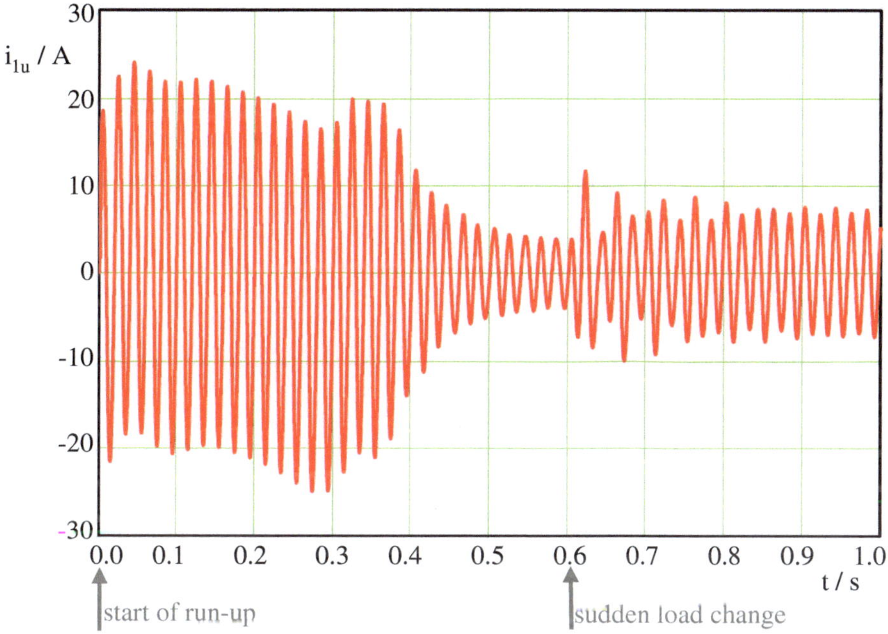

Fig. 12.1. Time dependent current in phase u.

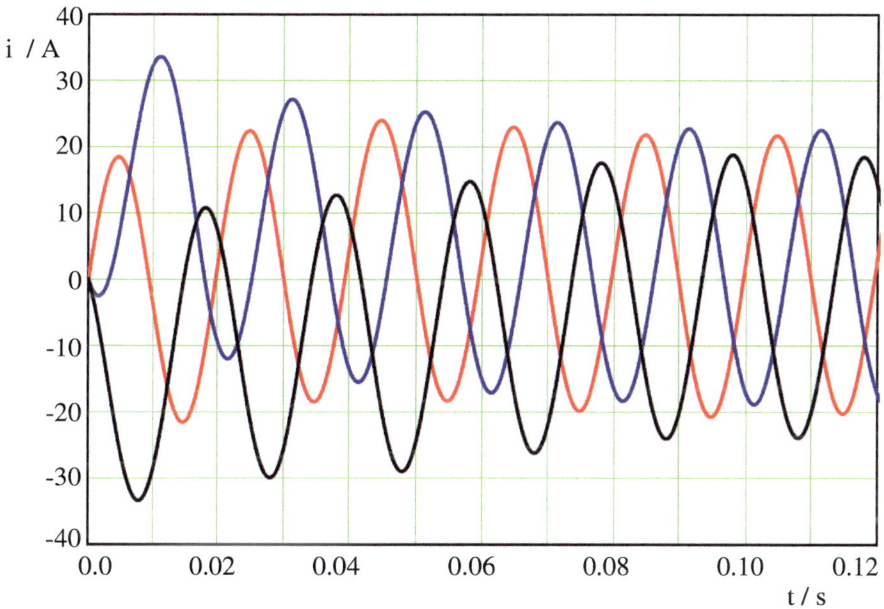

Fig. 12.2. Time dependent currents in phase u (*red*), phase v (*blue*), and phase w (*black*).

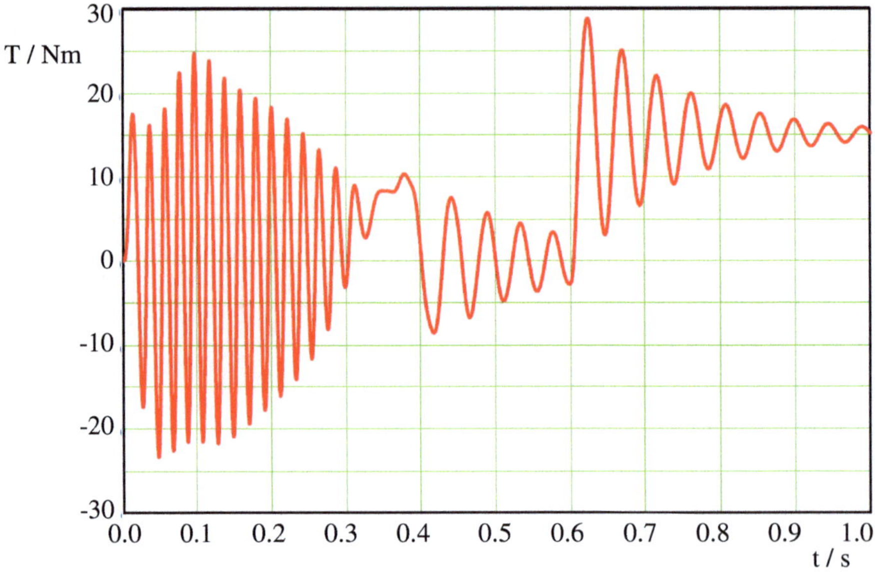

Fig. 12.3. Time dependent torque.

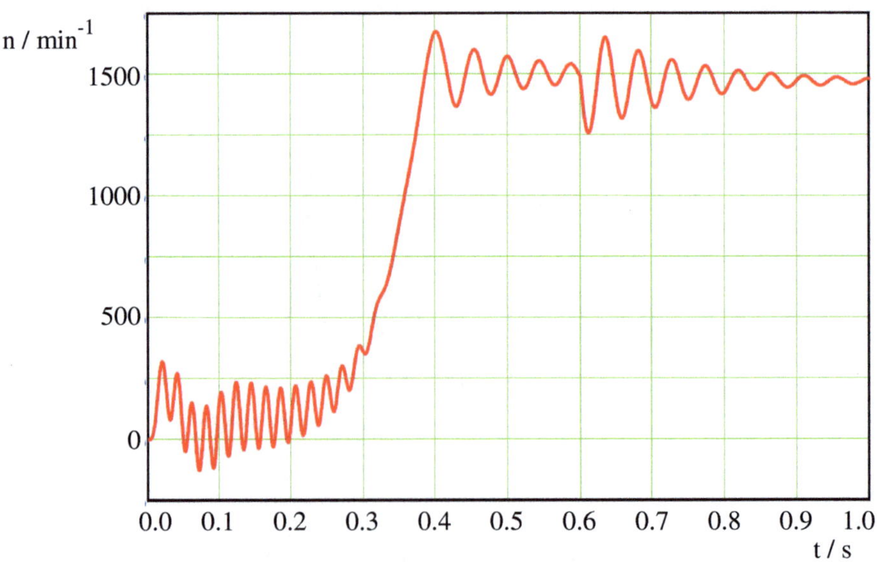

Fig. 12.4. Time dependent speed.

Figure 12.5 shows the torque-speed-characteristics of the regarded induction machine. In blue color the steady-state characteristic is shown, in red color the dynamic characteristic.

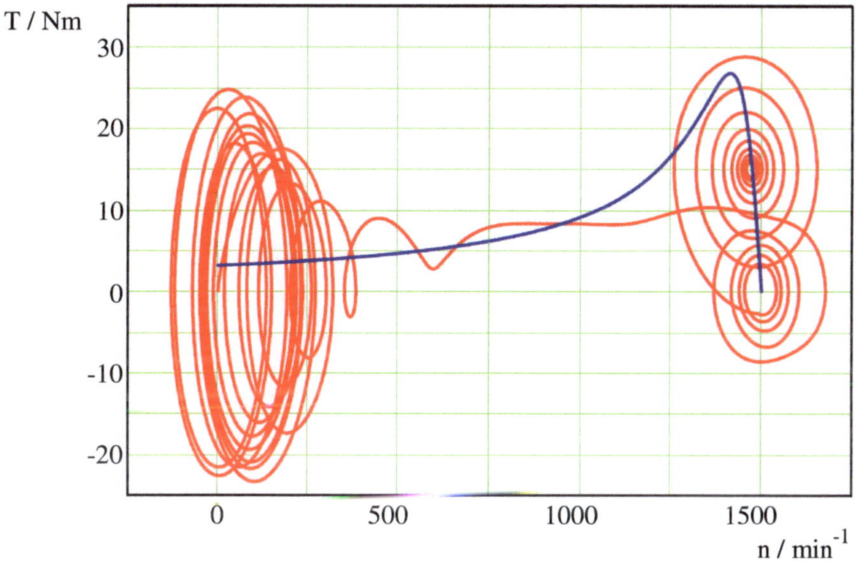

Fig. 12.5. Torque-speed-characteristics of the induction machine: steady-state (*blue*) and dynamic (*red*).

Switching the induction machine at zero speed to the mains at first there are high oscillating torque components (because of the DC current components) accompanied by high short-circuit AC currents. After decaying of these oscillating torque components the machine accelerates depending on the size of the coupled masses and it swings into the no-load operation (again with some oscillations).

The sudden load change initially decelerates the machine, until the electromagnetic torque is generated. Then an additional transient operation into the next steady-state operation point occurs. The steady-state speed is a little less than the synchronous speed.

The deviations from the steady-state characteristic are remarkable, see the comparison in Fig. 12.5. After decaying of all transient effects all operation points calculated by means of the equations for the dynamic operation are lying on the steady-state characteristic.

Figure 12.6 shows the stator current space vector for these transient operations in the complex plane (red characteristic). In addition, the blue characteristic illustrates the steady-state current circle diagram of the induction machine (circle diagram of the current amplitude, not the circle diagram of the current rms-value).

Even from these characteristics features of transient operations become obvious. High oscillating currents with large deviation from the steady-state characteristic do occur. For fast acceleration as well as for sudden load change the final steady-state values are lying on the circle diagram calculated for pure steady-state operation.

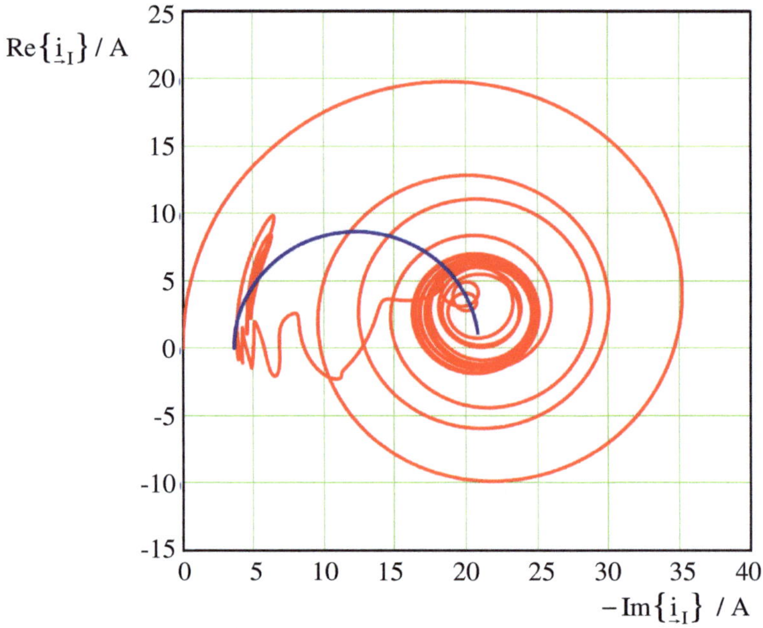

Fig. 12.6. Circle diagram of the induction machine: steady-state (*blue, current amplitude*) and dynamic (*red, current space vector*).

The effects of transient operation can be avoided if a slow acceleration according to the steady-state $U_1 - f_1$ – characteristic is realized; then the acceleration practically is quasi steady-state (acceleration on the steady-state characteristic).

In both cases (dynamic acceleration with oscillations or quasi steady-state acceleration) the induction machine is not suitable being used as a dynamic control unit in a drive system. However, in the following a solution will be developed how to employ the induction machine as such a dynamic control unit (similar to the DC-machine).

12.3 Field-Oriented Coordinate System for Induction Machines

Regarding the separately excited DC-machine the excitation flux Φ_F and the armature MMF (magneto-motive force) Θ_A always are perpendicular because of the impact of the commutator; their location in space is fixed.

The armature quadrature-axis field is assumed being completely compensated by the commutation poles and the compensation windings (in Fig. 12.7 this is not shown for the sake of clarity); because of this the exciting flux is not influenced by the armature current (the armature flux linkage in the quadrature-axis is zero: $\Psi_{A,q} = 0$). This means that the armature flux linkage in the direct-axis just depends on the field excitation current ($\Psi_{A,d} \sim I_F$); the torque is then produced by: $T \sim I_A \Phi_F$

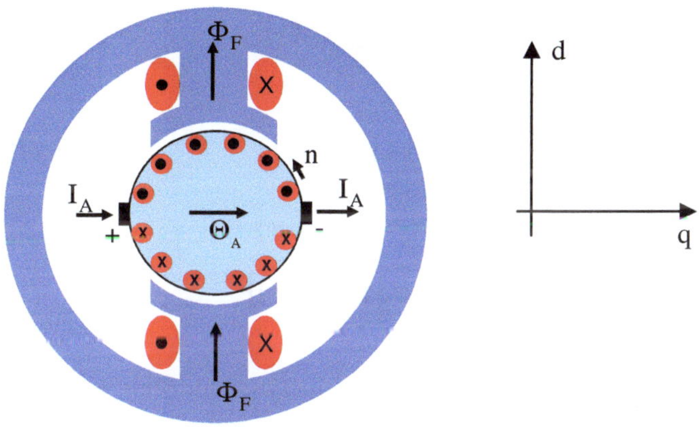

Fig. 12.7. Field excitation, armature MMF, and d-q-coordinate system of a DC-machine.

These features of the DC-machine can be transmitted even to the induction machine, if a coordinate system oriented to the rotor flux is chosen that rotates with the angular frequency of the rotor flux:

$$\alpha(t) = \omega_\mu t + \alpha_0 \tag{12.27}$$

The instantaneous value of the angular frequency of the rotor flux

$$\omega_\mu = \omega_{mech} + \omega_R \tag{12.28}$$

must not necessarily be identical to the steady-state value ω_1 of the angular frequency of the rotating stator field (with $\omega_{mech} = d\gamma/dt = p\Omega$ and ω_R being the angular frequency of the rotor currents).

To better show this analogy to the DC-machine, in the following the space vectors are decomposed into their components and these components are allocated to the respective axes. Firstly, the common set of equations follows from Eqs. (11.26), (11.27), (11.45), (11.46), (11.52), (11.53), and (11.55):

$$\underline{u}_I(t) = R_1\ \underline{i}_I(t) + \frac{d\underline{\psi}_I(t)}{dt} + j\,\omega_{CS}\ \underline{\psi}_I(t) \tag{12.29}$$

$$\underline{u}_{II}(t) = R_2\ \underline{i}_{II}(t) + \frac{d\underline{\psi}_{II}(t)}{dt} + j(\omega_{CS} - \omega_{mech})\,\underline{\psi}_{II}(t) \tag{12.30}$$

$$\underline{\psi}_I(t) = L_1\ \underline{i}_I(t) + L_{1m}\ \underline{i}_{II}(t) \tag{12.31}$$

$$\underline{\psi}_{II}(t) = L_2\ \underline{i}_{II}(t) + L_{1m}\ \underline{i}_I(t) \tag{12.32}$$

$$T(t) = -\frac{3}{2}p\ \mathrm{Re}\left\{j\,\underline{\psi}_{II}(t)\,\underline{i}_{II}^*(t)\right\} \tag{12.33}$$

The separation into real components and imaginary components is already introduced for the general space vector in Sect. 11.3:

$$\begin{aligned}\underline{i}(t) &= \mathrm{Re}\{\underline{i}(t)\} + j\ \mathrm{Im}\{\underline{i}(t)\}\\ &= i_x(t) - j\,i_y(t)\end{aligned} \qquad \text{(see Eq. (11.3))}$$

Using this separation of the components the following equations are obtained from the initial voltage and flux linkage equations:

$$\begin{aligned}u_{I,x}(t) &= R_1\ i_{I,x}(t) + \frac{d\psi_{I,x}(t)}{dt} + \omega_{CS}\ \psi_{I,y}(t)\\ u_{I,y}(t) &= R_1\ i_{I,y}(t) + \frac{d\psi_{I,y}(t)}{dt} - \omega_{CS}\ \psi_{I,x}(t)\end{aligned} \tag{12.34}$$

$$\begin{aligned}u_{II,x}(t) &= R_2\ i_{II,x}(t) + \frac{d\psi_{II,x}(t)}{dt} + (\omega_{CS} - \omega_{mech})\,\psi_{II,y}(t)\\ u_{II,y}(t) &= R_2\ i_{II,y}(t) + \frac{d\psi_{II,y}(t)}{dt} - (\omega_{CS} - \omega_{mech})\,\psi_{II,x}(t)\end{aligned} \tag{12.35}$$

$$\begin{aligned} \psi_{I,x}(t) &= L_1\, i_{I,x}(t) + L_{1m}\, i_{II,x}(t) \\ \psi_{I,y}(t) &= L_1\, i_{I,y}(t) + L_{1m}\, i_{II,y}(t) \end{aligned} \tag{12.36}$$

$$\begin{aligned} \psi_{II,x}(t) &= L_2\, i_{II,x}(t) + L_{1m}\, i_{I,x}(t) \\ \psi_{II,y}(t) &= L_2\, i_{II,y}(t) + L_{1m}\, i_{I,y}(t) \end{aligned} \tag{12.37}$$

The torque equation is transformed like follows:

$$\begin{aligned} T(t) &= -\frac{3}{2} p\ \mathrm{Re}\left\{ j\, \underline{\psi}_{II}(t)\, \underline{i}_{II}^{*}(t) \right\} \\ &= -\frac{3}{2} p\ \mathrm{Re}\left\{ j \left[\psi_{II,x}(t) - j\, \psi_{II,y}(t) \right] \left[i_{II,x}(t) + j\, i_{II,y}(t) \right] \right\} \\ &= -\frac{3}{2} p\ \mathrm{Re}\left\{ j \left[\psi_{II,x}(t) i_{II,x}(t) + \psi_{II,y}(t) i_{II,y}(t) \right] - \right. \\ &\qquad \left. \left[\psi_{II,x}(t) i_{II,y}(t) - \psi_{II,y}(t) i_{II,x}(t) \right] \right\} \\ &= \frac{3}{2} p \left[\psi_{II,x}(t) i_{II,y}(t) - \psi_{II,y}(t) i_{II,x}(t) \right] \end{aligned} \tag{12.38}$$

As the coordinate system rotates with the angular frequency of the rotor flux (please refer to the choice of $\alpha(t)$ in Eq. (12.1)), the above given description in components is representing the stator and rotor MMF decomposition in direct axis (d-axis, y-component) and quadrature axis (q-axis, x-component) with respect to the rotor flux. As this is a flux-oriented coordinate system, the y- and x-components are named in the following d-component (direct component) and q-component (quadrature component).[21]

This decomposition into d- and q-components leads to a clear *decoupling* and by *clever control* enables the impression of suitable phase currents that the following aims are achievable:

- the rotor flux linkage in the quadrature axis is zero ($\psi_{II,q} = 0$)
- the rotor flux linkage in the direct axis only depends on the magnetizing current ($\psi_{II,d} \sim i_{\mu,d}$)

[21] The identification with “d“ and “q“ is just a different naming of the components, that usually is introduced for flux-oriented coordinate systems.

- the torque is then only generated by the perpendicular components of rotor flux and stator current: $T \sim \psi_{II,d} i_{I,q}$.

The described operation is called "field-oriented". An observer, stationary to the system rotating with $\alpha(t)$, sees the same field distribution and torque generation like for the DC-machine. Simple relations for the control variables rotor flux and active stator current are obtained, that can be adjusted independently from each other like for the DC-machine.

In the following it will be described, how to reach the above mentioned aims (for the sake of clarity the explicit description of the time dependencies is avoided in the following equations). For the rotor flux linkages it is required in direct axis and in quadrature axis, respectively:

$$\begin{aligned} \psi_{II,d} &= L_2 i_{II,d} + L_{1m} i_{I,d} \\ &\overset{!}{=} L_{1m} i_{\mu,d} \end{aligned} \tag{12.39}$$

$$\begin{aligned} \psi_{II,q} &= L_2 i_{II,q} + L_{1m} i_{I,q} \\ &\overset{!}{=} 0 \end{aligned} \tag{12.40}$$

Here $i_{\mu,d}$ is a magnetizing current defined being proportional to the rotor flux linkage. From the requirement to the rotor flux linkages it follows for the rotor currents:

$$\begin{aligned} i_{II,d} &= \frac{L_{1m}}{L_2}\left(i_{\mu,d} - i_{I,d}\right) \\ i_{II,q} &= \frac{L_{1m}}{L_2}\left(-i_{I,q}\right) \end{aligned} \tag{12.41}$$

and for the angular frequencies:

$$\begin{aligned} &\frac{d\gamma}{dt} = \omega_{mech} = p\Omega, \qquad \frac{d\alpha}{dt} = \omega_{CS} = \omega_{\mu} = \omega_{mech} + \omega_R \\ &\Rightarrow \quad \frac{d(\alpha - \gamma)}{dt} = \omega_R \end{aligned} \tag{12.42}$$

As the rotor windings are short-circuited (e.g. for the squirrel-cage rotor) the respective voltage equations are:

$$
\begin{aligned}
0 &= R_2 i_{II,q} + \frac{d\psi_{II,q}}{dt} + \frac{d(\alpha-\gamma)}{dt}\psi_{II,d} \\
0 &= R_2 i_{II,d} + \frac{d\psi_{II,d}}{dt} - \frac{d(\alpha-\gamma)}{dt}\psi_{II,q}
\end{aligned}
\tag{12.43}
$$

Introducing the above equations for the currents, flux linkages, and angular frequencies this leads to:

$$
\begin{aligned}
0 &= R_2 \frac{L_{1m}}{L_2}\left(-i_{I,q}\right) + 0 + \omega_R L_{1m} i_{\mu,d} \\
0 &= R_2 \frac{L_{1m}}{L_2}\left(i_{\mu,d} - i_{I,d}\right) + L_{1m}\frac{di_{\mu,d}}{dt} - 0
\end{aligned}
\tag{12.44}
$$

With the rotor time constant $\tau_2 = \dfrac{L_2}{R_2}$ the rotor voltage equations in field-oriented coordinates are obtained:

$$
\begin{aligned}
\omega_R &= \frac{i_{I,q}}{\tau_2\, i_{\mu,d}} = \omega_\mu - \omega_{mech} \\
\tau_2 \frac{di_{\mu,d}}{dt} &+ i_{\mu,d} = i_{I,d}
\end{aligned}
\tag{12.45}
$$

The torque equation in field-oriented coordinates is:

$$
\begin{aligned}
T &= \frac{3}{2}p\left(\psi_{II,q} i_{II,d} - \psi_{II,d} i_{II,q}\right) = -\frac{3}{2}p\psi_{II,d}\frac{L_{1m}}{L_2}\left(-i_{I,q}\right) \\
&= \frac{3}{2}pL_{1h} i_{\mu,d}\frac{L_{1m}}{L_2} i_{I,q} = \frac{3}{2}p\frac{L_{1m}^2}{L_2} i_{\mu,d}\, i_{I,q} \\
&= \frac{3}{2}p\frac{L_{1m}}{1+\sigma_2} i_{\mu,d}\, i_{I,q}
\end{aligned}
\tag{12.46}
$$

For the stator flux linkages, after introducing the rotor currents, the following is true:

$$\begin{aligned}\psi_{I,d} &= L_1 i_{I,d} + L_{1m} i_{II,d} \\ &= L_1 i_{I,d} + L_{1m}\frac{L_{1m}}{L_2}\left(i_{\mu,d} - i_{I,d}\right) \\ &= \frac{L_{1m}^2}{L_2} i_{\mu,d} + \left(L_1 - \frac{L_{1m}^2}{L_2}\right) i_{I,d}\end{aligned} \tag{12.47}$$

$$\begin{aligned}\psi_{I,q} &= L_1 i_{I,q} + L_{1m} i_{II,q} \\ &= L_1 i_{I,q} + L_{1m}\frac{L_{1m}}{L_2}\left(-i_{I,q}\right) \\ &= \left(L_1 - \frac{L_{1m}^2}{L_2}\right) i_{I,q}\end{aligned} \tag{12.48}$$

Now the stator voltage equations can be transformed:

$$\begin{aligned}u_{I,q} &= R_1\, i_{I,q} + \frac{d\psi_{I,q}}{dt} + \omega_\mu \psi_{I,d} \\ &= R_1\, i_{I,q} + \frac{d}{dt}\left(\left(L_1 - \frac{L_{1m}^2}{L_2}\right) i_{I,q}\right) + \omega_\mu\left(\frac{L_{1m}^2}{L_2} i_{\mu,d} + \left(L_1 - \frac{L_{1m}^2}{L_2}\right) i_{I,d}\right) \\ &= R_1\, i_{I,q} + \left(L_1 - \frac{L_{1m}^2}{L_2}\right)\frac{di_{I,q}}{dt} + \omega_\mu \frac{L_{1m}^2}{L_2} i_{\mu,d} + \omega_\mu\left(L_1 - \frac{L_{1m}^2}{L_2}\right) i_{I,d}\end{aligned} \tag{12.49}$$

$$\begin{aligned}u_{I,d} &= R_1\, i_{I,d} + \frac{d\psi_{I,d}}{dt} - \omega_\mu \psi_{I,q} \\ &= R_1\, i_{I,d} + \frac{d}{dt}\left(\frac{L_{1m}^2}{L_2} i_{\mu,d} + \left(L_1 - \frac{L_{1m}^2}{L_2}\right) i_{I,d}\right) - \omega_\mu\left(L_1 - \frac{L_{1m}^2}{L_2}\right) i_{I,q} \\ &= R_1\, i_{I,d} + \frac{L_{1m}^2}{L_2}\frac{di_{\mu,d}}{dt} + \left(L_1 - \frac{L_{1m}^2}{L_2}\right)\frac{di_{I,d}}{dt} - \omega_\mu\left(L_1 - \frac{L_{1m}^2}{L_2}\right) i_{I,q}\end{aligned} \tag{12.50}$$

With the stator time constant $\tau_1 = \frac{L_1}{R_1}$ and the total leakage coefficient (please refer to Eq. (4.75))

$$\begin{aligned}\sigma &= 1 - \frac{1}{(1+\sigma_1)(1+\sigma_2)} = 1 - \frac{L_{1m}^2}{(1+\sigma_1)L_{1m}(1+\sigma_2)L_{1m}} \\ &= 1 - \frac{L_{1m}^2}{L_1\, L_2}\end{aligned} \tag{12.51}$$

the stator voltage equations in field-oriented coordinates are obtained:

$$\sigma\tau_1 \frac{di_{I,q}}{dt} + i_{I,q} = \frac{u_{I,q}}{R_1} - \omega_\mu \sigma\tau_1 i_{I,d} - (1-\sigma)\,\tau_1 \omega_\mu i_{\mu,d} \tag{12.52}$$

$$\sigma\tau_1 \frac{di_{I,d}}{dt} + i_{I,d} = \frac{u_{I,d}}{R_1} + \omega_\mu \sigma\tau_1 i_{I,q} - (1-\sigma)\,\tau_1 \frac{di_{\mu,d}}{dt} \tag{12.53}$$

By means of these equations the block diagram of the induction machine in field-oriented coordinates can be deduced (Fig. 12.8). For the sake of clarity the five equations (two stator voltage equations, two rotor voltage equations, and the torque equation) are highlighted in grey. The coordinate transformation and the torque balance can be described by the block diagrams shown in Fig. 12.9.

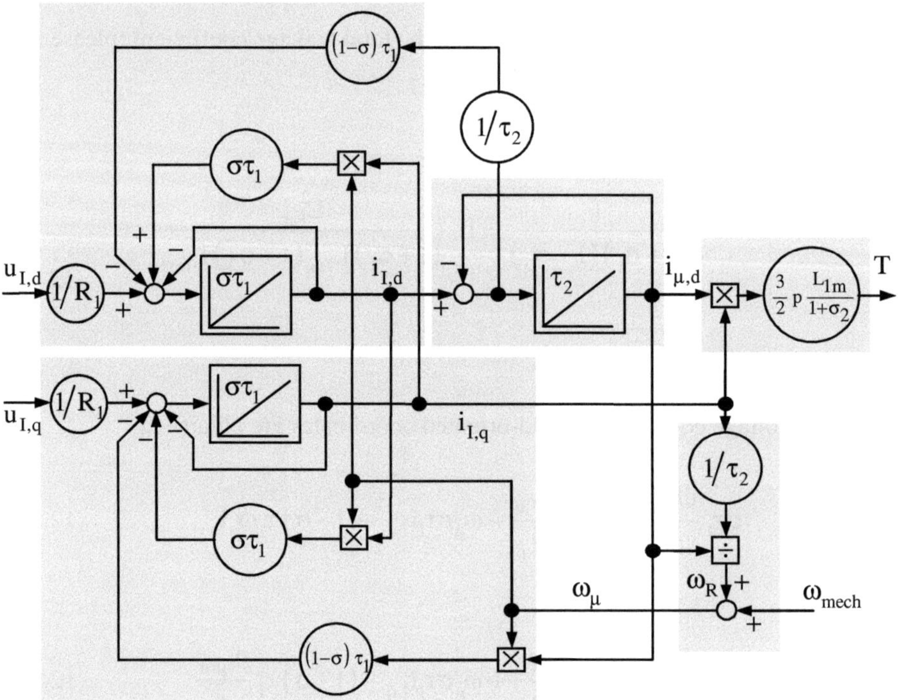

Fig. 12.8. Block diagram of the induction machine in field-oriented coordinates.

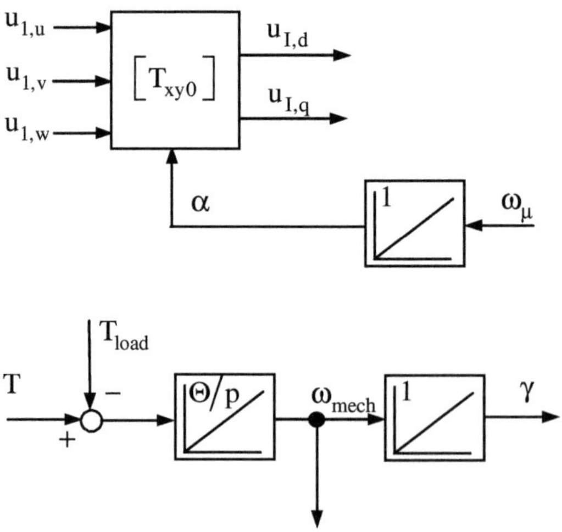

Fig. 12.9. Block diagrams of coordinate transformation and torque balance.

The equation

$$\tau_2 \frac{di_{\mu,d}}{dt} + i_{\mu,d} = i_{I,d} \tag{12.54}$$

shows that the direct component of the stator current ($i_{I,d}$) determines the magnitude of the rotor flux (that is proportional to $i_{\mu,d}$, see above). Like for the field winding of the DC-machine a large time constant (the rotor time constant τ_2) is relevant. Therefore, the magnitude of the rotor flux is not suitable for realizing fast control actions.

The equation

$$\omega_R = \frac{i_{I,q}}{\tau_2 \, i_{\mu,d}} = \omega_\mu - \omega_{mech} \tag{12.55}$$

shows that the angular frequency of the slip (angular frequency of the rotor currents ω_R) is determined by the quadrature component of the stator current ($i_{I,q}$) and the magnitude of the rotor flux ($\sim i_{\mu,d}$). The angular frequency of the rotor flux is calculated from the angular frequency of the slip and the mechanical angular frequency of the rotor.

The equation

$$T = \frac{3}{2} p \frac{L_{1m}^2}{L_2} i_{\mu,d} \, i_{I,q} \tag{12.56}$$

describes the torque generation. Analogously to the DC-machine the torque is produced by the direct axis flux ($\sim i_{\mu,d}$) and the quadrature component of the stator current ($i_{I,q}$). If (like required) $\psi_{II,d} = L_{1m} i_{\mu,d} = \text{const.}$ is true, then the torque T and the angular frequency of the slip ω_R are directly proportional to the quadrature component of the stator current $i_{I,q}$.

The equations

$$\sigma\tau_1 \frac{di_{I,q}}{dt} + i_{I,q} = \frac{u_{I,q}}{R_1} - \omega_\mu \sigma \tau_1 i_{I,d} - (1-\sigma)\tau_1 \omega_\mu i_{\mu,d} \tag{12.57}$$

$$\sigma\tau_1 \frac{di_{I,d}}{dt} + i_{I,d} = \frac{u_{I,d}}{R_1} + \omega_\mu \sigma\tau_1 i_{I,q} - (1-\sigma)\tau_1 \frac{di_{\mu,d}}{dt} \tag{12.58}$$

are completing the machine model concerning the interaction of stator voltages and stator currents. Regarding the stator current components the induction machine acts like a first-order delay element (PT1 element) with the time constant $\sigma\tau_1$ and the gain $1/R_1$. The stator current components are coupled by the right-hand terms in the above equations. The expressions $\omega_\mu\sigma\tau_1 i_{I,q}$ and $\omega_\mu\sigma\tau_1 i_{I,d}$ are the rotatory induced voltages, that are caused by the currents of the respective different axis. $(1-\sigma)\tau_1 \frac{di_{\mu,d}}{dt}$ is the transformatory induced voltage that occurs by changing the magnetizing current. $(1-\sigma)\tau_1\omega_\mu i_{\mu,d}$ is the rotatory induced voltage of the main field.

These equations and the deduced block diagram (see Fig. 12.8) describe the induction machine equivalently to the equations in Sects. 12.1 and 12.2. Switching the mains voltage suddenly to the machine at zero speed the same acceleration characteristics (and the same transient characteristics when suddenly loading the machine) like in Sect. 12.2 are obtained!

The advantage of the field-oriented description for the dynamic operation of the induction machine will become clear in the following Sect. 12.4.

12.4 Field-Oriented Control of Induction Machines with Impressed Stator Currents

The advantage of the description shown in the last section is that now the same control strategy like for the separately excited DC-machine can be applied to the induction machine, by which the induction machine is qualified to be applied as a highly dynamic drive: The magnetizing current and consequently the rotor flux shall be hold constantly at their nominal values, the torque shall be adjusted only by means of the quadrature component of the stator current.[22] To reach this the transformed stator currents must be *independently controllable* in direct and quadrature axis. This kind of control (field-oriented control, FOC) has been developed end of the 1960s, begin of the 1970s by Karl Hasse and Felix Blaschke independently from each other.

The above described controllability is enabled by

---

[22] Here it is to be regarded that the stator flux increases with the load, therefore saturation may occur in the stator.

- power electronic devices with high switching frequency (for small power about 20kHz) and
- short sampling intervals for the control (for small power some 100μs for the current control, for the speed control some ms).

If these conditions are valid, it can be assumed that the stator currents are impressed. Then the equations of the relations between stator voltages and stator currents can be omitted because these are handled intrinsically in the power electronic converter. The respective block diagram is shown in Fig. 12.10.

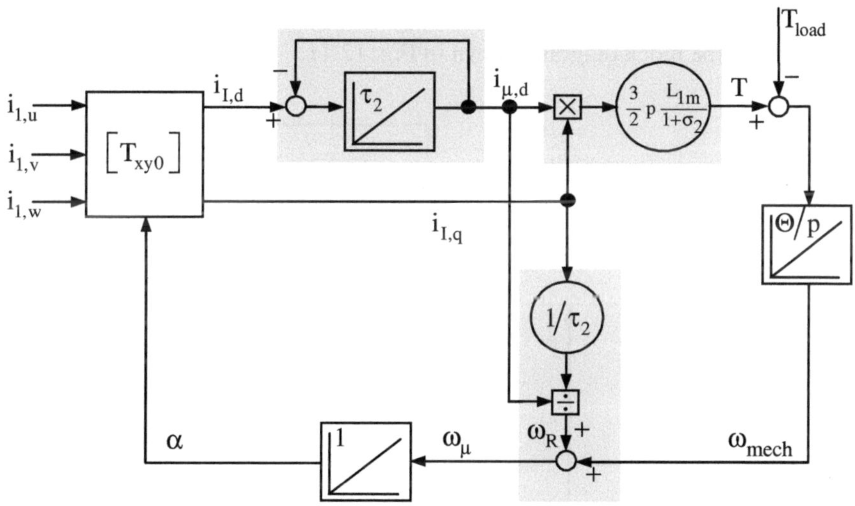

Fig. 12.10. Block diagram of the induction machine in field-oriented coordinates with impressed stator currents.

The block diagram of the induction machine in field-oriented coordinates with impressed stator currents corresponds to the block diagram of the separately excited DC-machine with neglected armature time constant (please refer to Sect. 10.1). This means that the torque production at constant rotor flux follows the quadrature component of the stator current $i_{I,q}$ without any delay and the rotor flux is controllable only by the direct component of the stator current.

Consequently, the aim of a highly dynamic drive system is reached in principle. But if the induction machine shall be controlled in field-oriented coordinates, it is necessary to know the instantaneous amplitude and phase of the rotor flux. Having a squirrel-cage rotor the rotor currents and voltages cannot be measured and the measurement of the air-gap flux, which is just an approximation, is very costly and susceptible to faults.

However, amplitude and phase of the rotor flux can be calculated from measured values of the stator currents and the speed, by evaluating the rotor voltage equations of the induction machine. This is called “flux model“:

$$
\begin{aligned}
&\tau_2 \frac{di_{\mu,d}}{dt} + i_{\mu,d} = i_{1,d} \\
&\frac{i_{1,q}}{\tau_2\, i_{\mu,d}} + \omega_{mech} = \omega_{\mu} = \frac{d\alpha}{dt}
\end{aligned}
\tag{12.59}
$$

Then there is the block diagram shown in Fig. 12.11.

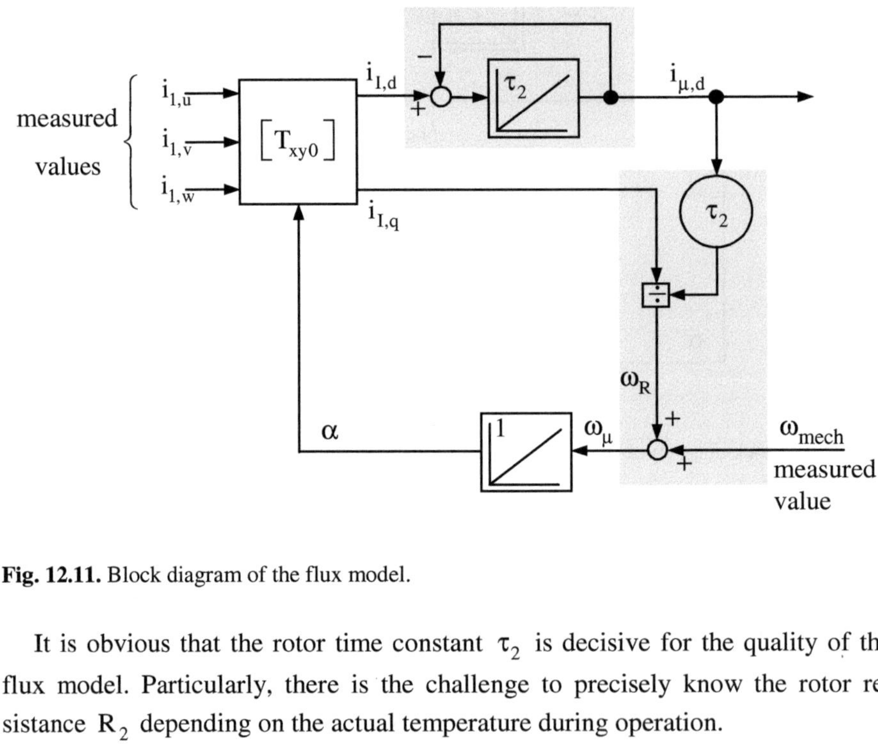

Fig. 12.11. Block diagram of the flux model.

It is obvious that the rotor time constant τ_2 is decisive for the quality of the flux model. Particularly, there is the challenge to precisely know the rotor resistance R_2 depending on the actual temperature during operation.

Once the rotor time constant τ_2 is known, the induction machine can be controlled highly dynamic. For a drive system with speed controller, torque controller, and flux controller the block diagram in Fig. 12.12 is obtained.

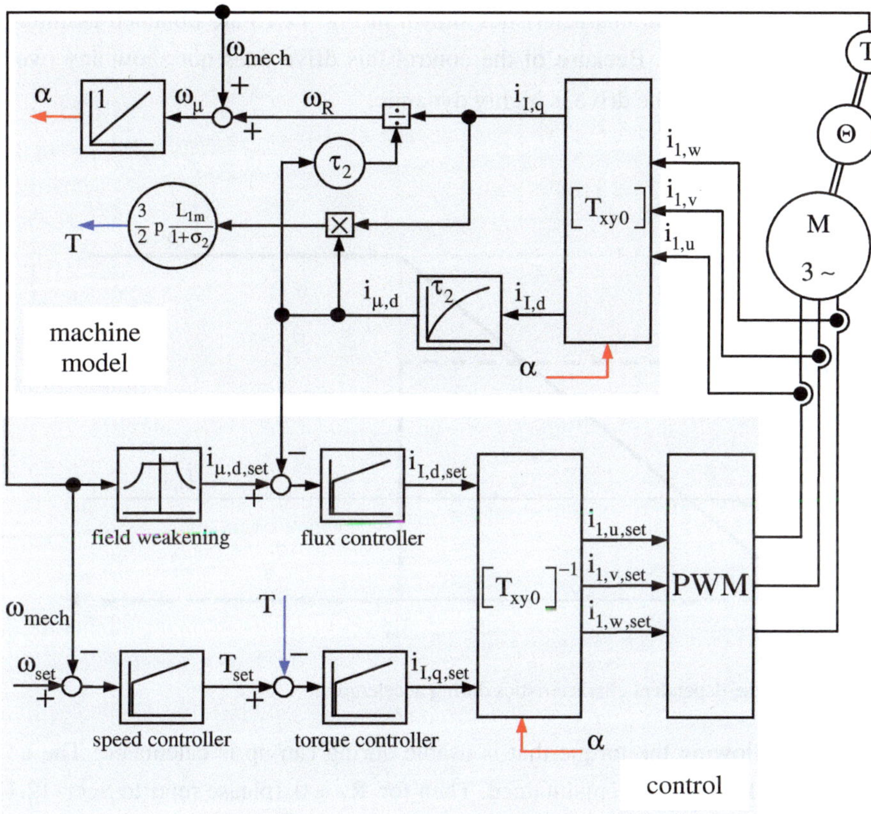

Fig. 12.12. Block diagram of the field-oriented controlled induction machine with impressed currents.

If at the time $t = 0$ a step function $\omega = \omega_1 = \omega_{set}$ is applied to the machine at zero speed, the phase currents $i_{1,u}$, $i_{1,v}$ and $i_{1,w}$ are impressed *by the control* via the power electronic converter (named "PWM" in the block diagram) in such a way, that

- the magnetizing current $i_{\mu,d}$ increases with the rotor time constant $\tau_2 = \dfrac{(1+\sigma_2)L_{1m}}{R_2'} = \dfrac{L_2'}{R_2'} = \dfrac{L_2}{R_2}$ to its nominal value and
- the acceleration happens nearly linearly according to the impressed quadrature current $i_{I,q} = i_{I,q,max}$ during the run-up time $\tau_\Theta = \dfrac{\Theta}{p}\dfrac{\omega_1}{T_{max}}$.

The time-dependent characteristics shown in Fig. 12.13 are obtained assuming that $\tau_2 \ll \tau_\Theta$ is true. Because of the control this drive does not show any overshoot or oscillation; the drive is highly dynamic.

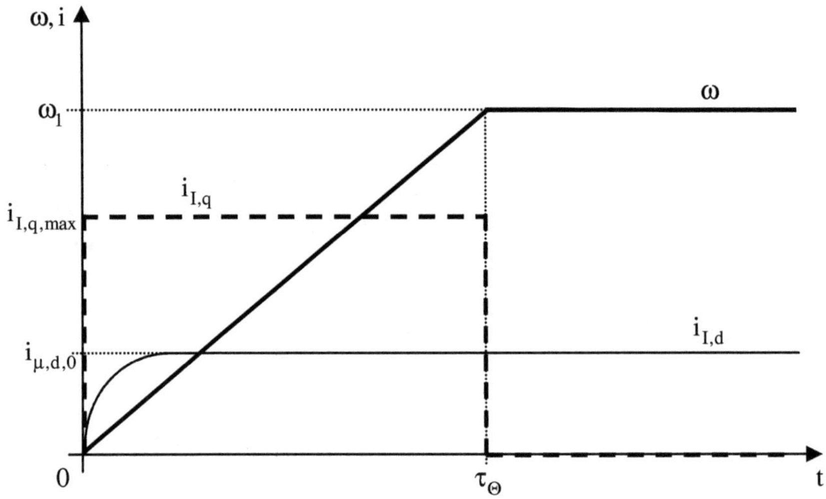

Fig. 12.13. Time-dependent characteristics during acceleration.

In the following the torque that is usable during run-up is calculated: The no-load flux linkage shall be maintained. Then for $R_1 \approx 0$ (please refer to Sect. 12.1) it follows:

$$\underline{i}_I(t) = -j\frac{\sqrt{2}\,U_1}{\omega_{CS}\,L_1} = i_{I,q} - j\,i_{I,d}$$
$$\Rightarrow \quad i_{I,d} = i_{\mu,d,0} = \frac{\sqrt{2}\,U_1}{\omega_{CS}\,L_1} = \sqrt{2}\,I_0 \tag{12.60}$$

This magnetizing current increases with the time constant τ_2:

$$i_{\mu,d} = i_{I,d}\left(1 - e^{-\frac{t}{\tau_2}}\right) \tag{12.61}$$

In steady-state operation the maximum torque is obtained at the pull-out operation point:

$$\omega_R = \omega_{\text{pull-out}} = \frac{R_2'\left(1+\sigma_1\right)^2}{L_1 \dfrac{\sigma}{1-\sigma}} = \frac{R_2'}{\sigma L_{1m}\left(1+\sigma_2\right)}\left(1+\sigma_2\right)\left(1+\sigma_1\right)\left(1-\sigma\right)$$

$$= \frac{1}{\sigma\tau_2}\left(1+\sigma_2\right)\left(1+\sigma_1\right)\left(1-\left(1-\frac{1}{\left(1+\sigma_1\right)\left(1+\sigma_2\right)}\right)\right) = \frac{1}{\sigma\tau_2} \tag{12.62}$$

For the quadrature current (i.e. the torque producing current) it follows (at field-oriented control and $\omega_R = 1/\sigma\tau_2$):

$$i_{I,q} = \omega_R \tau_2 i_{\mu,d} = \frac{1}{\sigma}\sqrt{2}\, I_0 \tag{12.63}$$

The usable torque is then:

$$T = \frac{3}{2} p \frac{L_{1m}}{1+\sigma_2} i_{\mu,d} i_{I,q} = \frac{3}{2} p \frac{L_{1m}}{1+\sigma_2} \sqrt{2}\, I_0 \frac{1}{\sigma} \sqrt{2}\, I_0$$

$$= \frac{3p}{\omega_1} X_1 I_0^2 \frac{1-\sigma}{\sigma} = \frac{3p}{\omega_1} \frac{U_1^2}{X_1 \dfrac{\sigma}{1-\sigma}} = 2\, T_{\text{pull-out}} \tag{12.64}$$

During the acceleration with field-oriented control at $\omega_R = 1/\sigma\tau_2$ the double torque (against the steady-state operation at symmetric, fixed mains) is obtained.

Having the same machine data as for the example in Sect. 12.2 (fast acceleration and sudden load change at fixed mains), the time-dependent characteristics shown in Fig. 12.14 are obtained (just the acceleration is shown; therefore the time is limited to an interval from 0s to 0.12s).

The flux generating current $i_{\mu,d}$ (shown in red in Fig. 12.14) is increased quite slowly in this example because of the relatively large time constant τ_2. However, the torque generating current $i_{I,q}$ (shown in blue in Fig. 12.14) is switched from its nominal value to zero after about 0.097s, because the run-up period is already finished.

Figures 12.15 and 12.16 show the torque and the speed during this acceleration period (the scale of the vertical axes are the same like in Sect. 12.2). The torque is increased analogously to the magnetizing current $i_{\mu,d}$ as long as $i_{I,q} > 0$ is true. The acceleration time is considerably shorter (compared to the operation at fixed mains) and there are no oscillations.

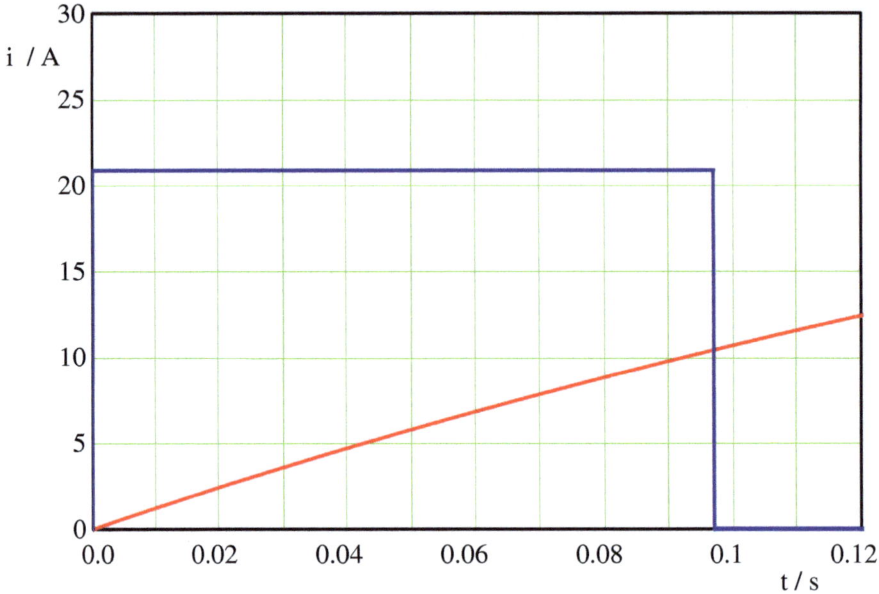

Fig. 12.14. Flux generating current (*red*) and torque generating current (*blue*) during acceleration of the field-oriented controlled induction machine.

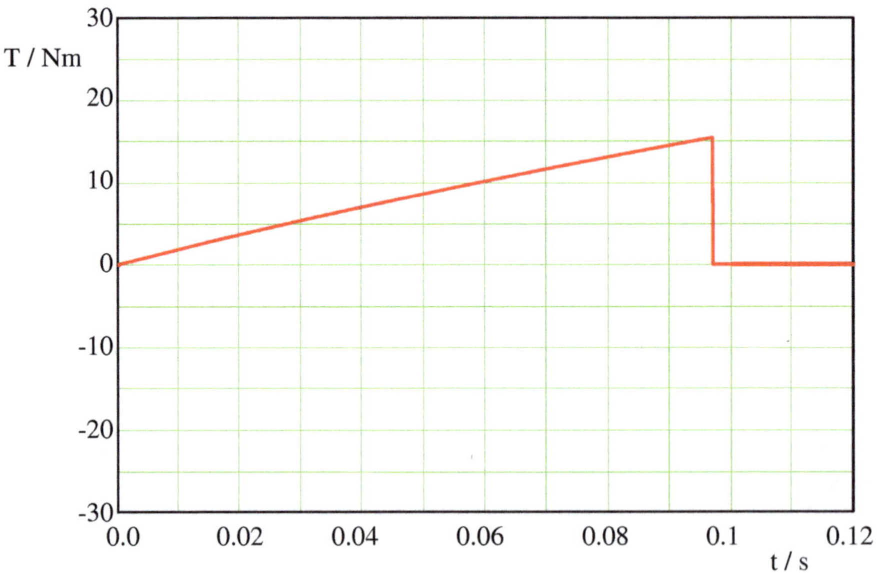

Fig. 12.15. Time-dependent torque during acceleration of the field-oriented controlled induction machine.

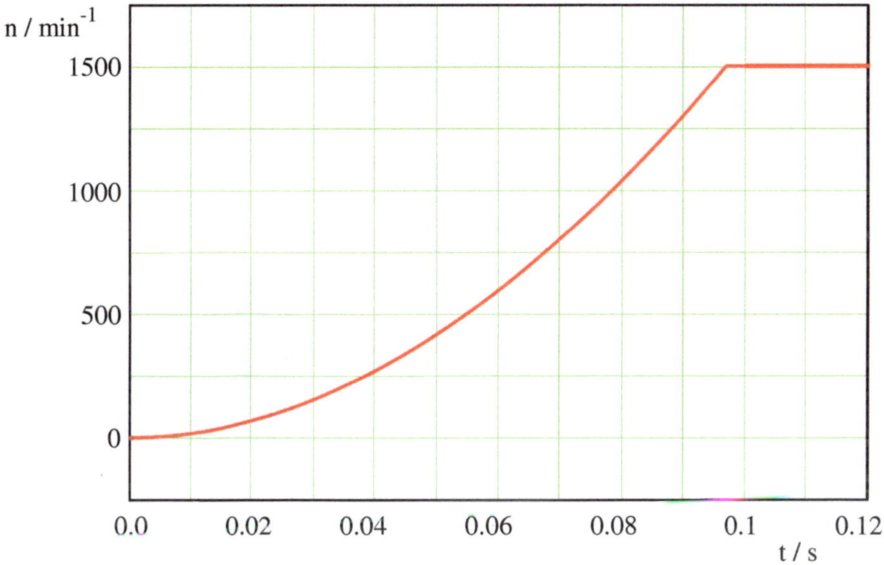

Fig. 12.16. Time-dependent speed during acceleration of the field-oriented controlled induction machine.

Figure 12.17 shows the time-dependent characteristics of speed, torque producing quadrature current $i_{I,q}$ and field producing magnetizing current $i_{\mu,d}$.

- The acceleration period, finished after 0.097s, is characterized by $i_{I,q} = i_{I,q,max}$ and a very steep speed increase (the magnetizing current is increased quite slowly because of the large time constant τ_2; this is true even for the further simulation time).
- After finishing the acceleration and until switching on the load at the time 0.6s the quadrature current is set to $i_{I,q} = 0$, then the speed is constant.
- After switching on the load the influence of the control can be recognized clearly: small speed changes provoked by the control of the torque producing quadrature current component $i_{I,q}$ are noticeable.

For a better comparison the speed-time-characteristics for acceleration at fixed mains (red line) and for acceleration with field-oriented control (blue line) are presented once again in Fig. 12.18. The entire acceleration period is shown from 0s to 0.6s. The improvement of the dynamic behavior against the operation at fixed mains (please refer to Sect. 12.2) is impressive. The additional effort to realize this operational behavior mainly is a powerful controller and the power electronic converter.

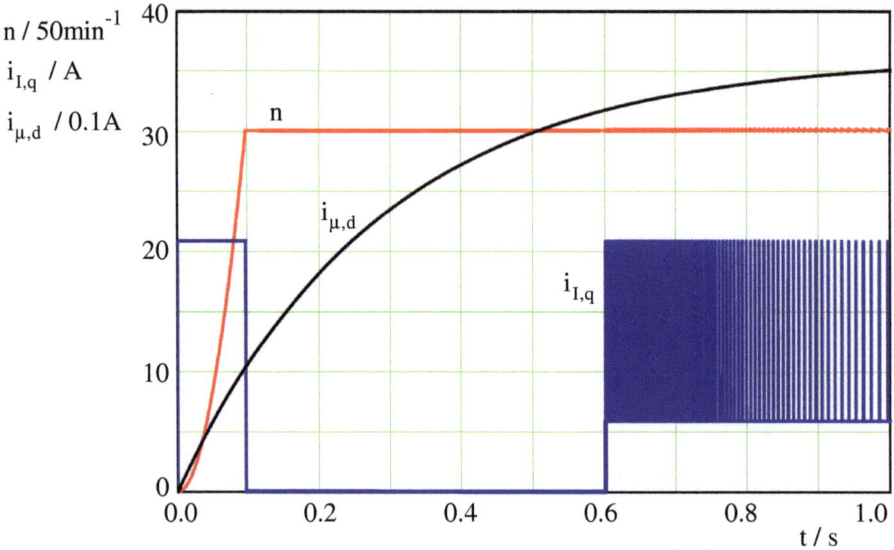

Fig. 12.17. Time-dependent characteristics during acceleration of the field-oriented controlled induction machine: speed (*red*), torque generating current (*blue*), flux generating current (*black*).

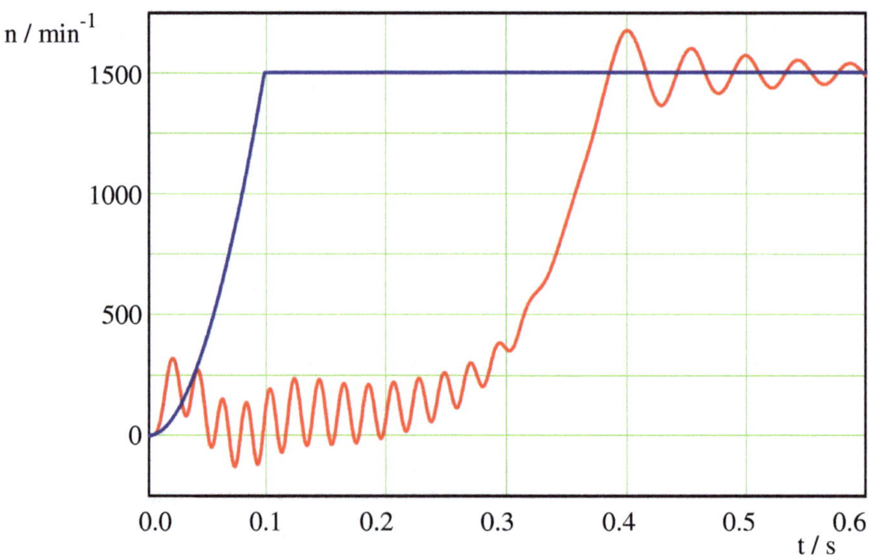

Fig. 12.18. Speed versus time characteristics of the induction machine: operation at fixed mains supply (*red*) and operation with field-oriented control (*blue*).

The time-dependent characteristics of the three phase currents are obtained from the currents $i_{I,d}$ and $i_{I,q}$ by means of reverse transformation. With the deduction presented above it follows:

$$i_{I,d} = \sqrt{2}\, I_0 = \text{const.}$$

$$i_{I,q}(t) = \begin{cases} \frac{1}{\sigma}\sqrt{2}\, I_0, & \text{for } 0\text{s} \le t \le 0.097\text{s} \\ 0, & \text{for } 0.097\text{s} \le t \le 0.6\text{s} \end{cases} \tag{12.65}$$

$$I_0 = \frac{U_1}{X_1}$$

Further there is (please refer to Sect. 11.3):

$$\begin{aligned} i_{1,u} &= i_{I,q}\cos(\alpha) + i_{I,d}\sin(\alpha) \\ i_{1,v} &= i_{I,q}\cos\left(\alpha - \frac{2\pi}{3}\right) + i_{I,d}\sin\left(\alpha - \frac{2\pi}{3}\right) \\ i_{1,w} &= -i_{1,u} - i_{1,v} \end{aligned} \tag{12.66}$$

with the time-dependent angle

$$\begin{aligned} \alpha(t) &= \int_0^t \omega_\mu(\tilde{t})\,d\tilde{t} \quad \text{with} \quad \omega_\mu(\tilde{t}) = \omega_{mech}(\tilde{t}) + \omega_R \\ \omega_R &= 1/\sigma\tau_2 \end{aligned} \tag{12.67}$$

Inserting these equations, the solution for the first 0.097s of the acceleration period is:

$$\begin{aligned} i_{1,u} &= \sqrt{2}\, I_0\left[\frac{1}{\sigma}\cos(\alpha) + \sin(\alpha)\right] \\ i_{1,v} &= \sqrt{2}\, I_0\left[\frac{1}{\sigma}\cos\left(\alpha - \frac{2\pi}{3}\right) + \sin\left(\alpha - \frac{2\pi}{3}\right)\right] \\ i_{1,w} &= -i_{1,u} - i_{1,v} \end{aligned} \tag{12.68}$$

These equations are valid as long as $i_{I,d}$ and $i_{I,q}$ are maintained constantly onto their respective maximum values ($\sqrt{2}\,I_0$ and $\sqrt{2}\,I_0/\sigma$). As soon as the desired speed is reached the current component $i_{I,q}$ is switched to zero (see Fig. 12.17), and therefore even the acceleration torque gets zero (the magnetizing condition of the machine remains unchanged, i.e. $i_{I,d} = \sqrt{2}\,I_0 = \text{const.}$, the magnetizing current $i_{\mu,d}$ is increased with the time constant τ_2). To get the phase currents by means of reverse transformation for the entire simulation time, the time-dependent current component $i_{I,q} = i_{I,q}(t)$ has to be considered. It follows:

$$
\begin{aligned}
i_{1,u} &= i_{I,q}(t)\cos(\alpha) + i_{I,d}\sin(\alpha) \\
i_{1,v} &= i_{I,q}(t)\cos\left(\alpha - \frac{2\pi}{3}\right) + i_{I,d}\sin\left(\alpha - \frac{2\pi}{3}\right) \\
i_{1,w} &= -i_{1,u} - i_{1,v}
\end{aligned}
\tag{12.69}
$$

The time-dependent characteristics of the three phase currents, calculated by these equations, are illustrated in Fig. 12.19 for the starting period (the scale of the vertical axis is the same like in Sect. 12.2).

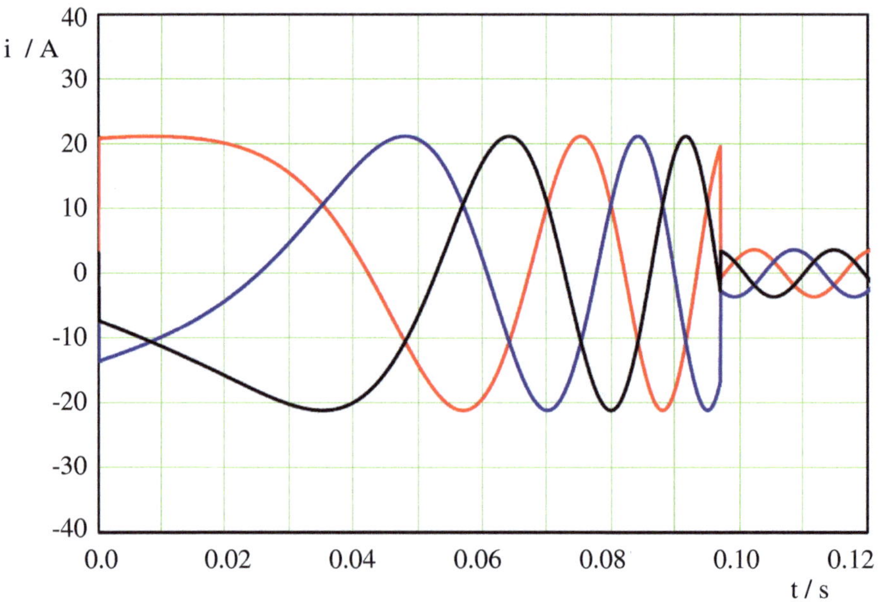

Fig. 12.19. Time-dependent currents in phase u (*red*), phase v (*blue*), and phase w (*black*) during acceleration of the field-oriented controlled induction machine.

In contrast to the acceleration at fixed mains (see Fig. 12.2) the amplitudes of the three phase currents are identical and (during acceleration and during steady-state operation) constant in time. In addition, the maximum phase current is lower than for the uncontrolled acceleration.

The change of frequency of the phase currents is characteristic for the influence of the power electronic converter, this would not be possible at fixed mains (constant voltage concerning amplitude and frequency).

When finishing the acceleration period (at about 0.097s) the current component $i_{I,q}$ is set to zero. This is noticeable in the characteristics of the phase currents by the simultaneous change of amplitude and phase; then the frequency is not changed any longer. During the entire operation the slip is maintained at the pull-out slip of the steady-state operation ($\omega_R = 1/\sigma\tau_2$); the torque is adjusted by the current (more precisely: the current component $i_{I,q}$).

Figure 12.20 shows the stator current space vector at field-oriented control (in red) and the circle diagram of the current amplitude in steady-state operation (blue curve). Even with this graph the differences to the dynamic operation at fixed mains supply (please refer to Fig. 12.6) become obvious.

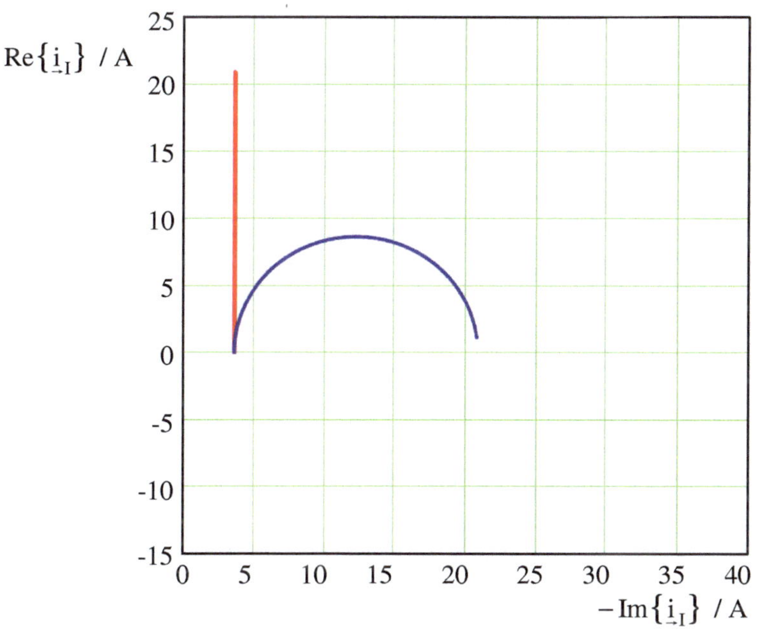

Fig. 12.20. Circle diagram of the induction machine: steady-state operation (*blue, current amplitude*) and during dynamic acceleration with field-oriented control (*red, current space vector*).

12.5 Field-Oriented Control of Induction Machines with Impressed Stator Voltages

Up to now the field-oriented control (FOC) of induction machines has been regarded assuming that current injecting electronic power converters with high switching frequency and sufficient voltage reserve are available as well as fast microcontrollers for calculating the control algorithms. Regarding electrical drives with a power of up to some kW this is given (servo drives with transistorized inverters and switching frequencies up to about 20 kHz). For larger drives often pulse-width modulated inverters with intermediate DC-voltage and switching frequencies of a few kHz are used. Then the above mentioned conditions are no longer fulfilled. This means that the stator voltage equations have to be regarded in addition ($\tau_1 = L_1/R_1$):

$$\sigma\tau_1 \frac{di_{I,d}}{dt} + i_{I,d} = \frac{1}{R_1}\left(u_{I,d} + \omega_\mu \sigma L_1 i_{I,q} - (1-\sigma) L_1 \frac{di_{\mu,d}}{dt}\right) \tag{12.70}$$

$$\sigma\tau_1 \frac{di_{I,q}}{dt} + i_{I,q} = \frac{1}{R_1}\left(u_{I,q} - \omega_\mu \sigma L_1 i_{I,d} - (1-\sigma) L_1 \omega_\mu i_{\mu,d}\right) \tag{12.71}$$

Both control paths are coupled via the stator currents, therefore they are not independent from each other. However, a decoupling is required so that the current controllers can be adjusted independently. This can be achieved if negative compensation voltages are added to the output voltages of the controllers ($u_{C,d}$ and $u_{C,q}$) in such a way that the coupling voltages are zero. Now the controllers see decoupled paths. For the compensation usually it is assumed that the rotor flux linkage is constant, i.e. $di_{\mu,d}/dt = 0$. Then it follows:

$$\begin{aligned} u_{C,d} - \omega_\mu \sigma L_1 i_{I,q} &= u_{I,d} \\ u_{C,q} + \omega_\mu \sigma L_1 i_{I,d} + (1-\sigma) L_1 \omega_\mu i_{\mu,d} &= u_{I,q} \end{aligned} \tag{12.72}$$

Further:

$$\begin{aligned} \sigma\tau_1 \frac{di_{I,d}}{dt} + i_{I,d} &= \frac{1}{R_1} u_{C,d} \\ \sigma\tau_1 \frac{di_{I,q}}{dt} + i_{I,q} &= \frac{1}{R_1} u_{C,q} \end{aligned} \tag{12.73}$$

The block diagram of this decoupling network is shown in Fig. 12.21.

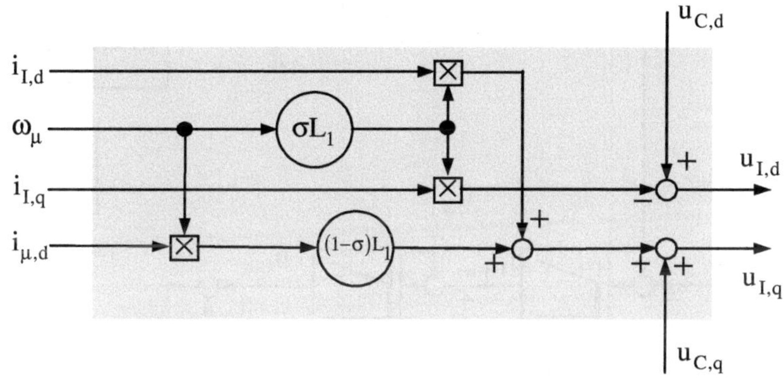

Fig. 12.21. Block diagram of the decoupling network.

Figure 12.22 shows the entire diagram of a field-oriented controlled induction machine with impressed voltages realized by a power electronic converter ("PWM"). For the currents in direct axis and quadrature axis a cascaded control is applied respectively; the control parameters can be adjusted independently by means of the decoupling network. The instantaneous values of rotor flux amplitude and phase, which are required for the control, are calculated by means of the flux model.

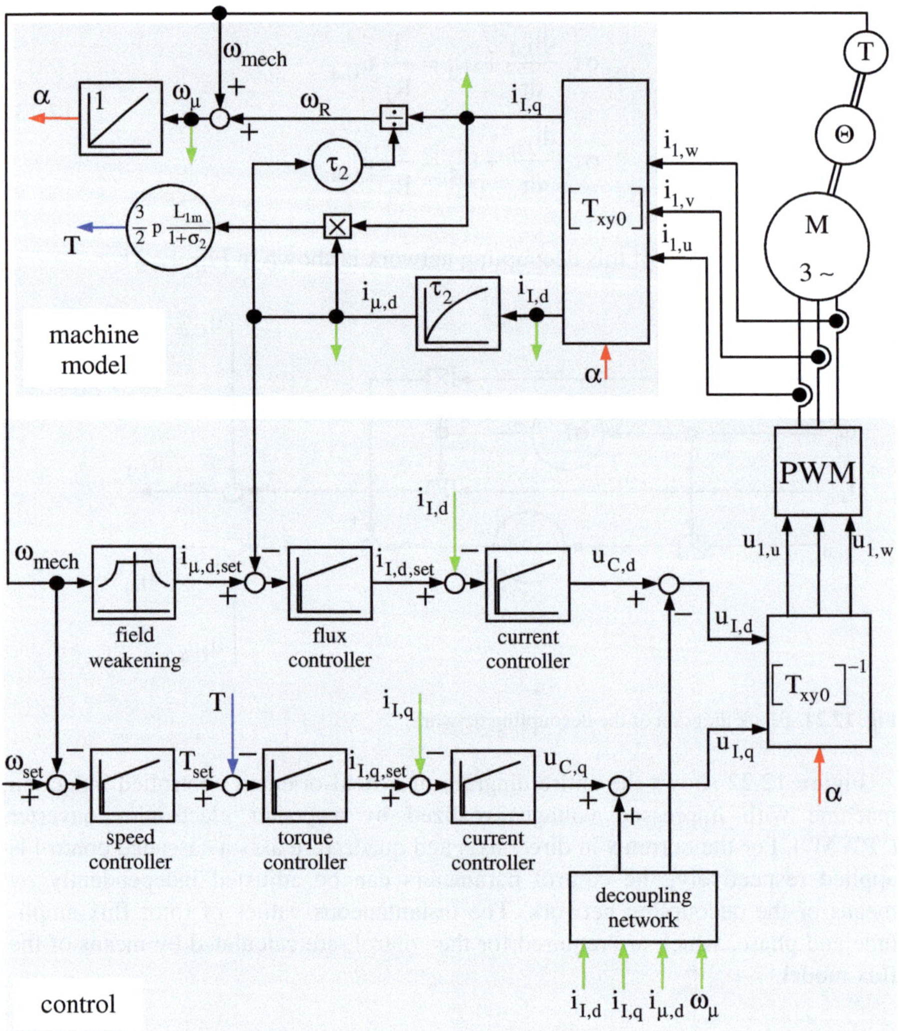

Fig. 12.22. Block diagram of the field-oriented controlled induction machine with impressed voltages.

12.6 Field-Oriented Control of Induction Machines without Mechanical Sensor (Speed or Position Sensor)

The mechanical speed has to be known for the field-oriented control (FOC) of induction machines: This value is necessary for the speed control as well as for the coordinate transformation (the angle α is calculated by means of the mechanical speed).

However, mechanical speed sensors have some disadvantages that preferably should be avoided:

- vulnerability against outside impacts (forces, torques, temperatures, dirt)
- costs
- space consumption
- necessity of a free shaft extension

Therefore it is desirable to compute the speed from the measured terminal values of the machine (often this method is denominated as "sensorless" speed control). The method can be explained by means of the block diagram of the induction machine shown in Fig. 12.23.

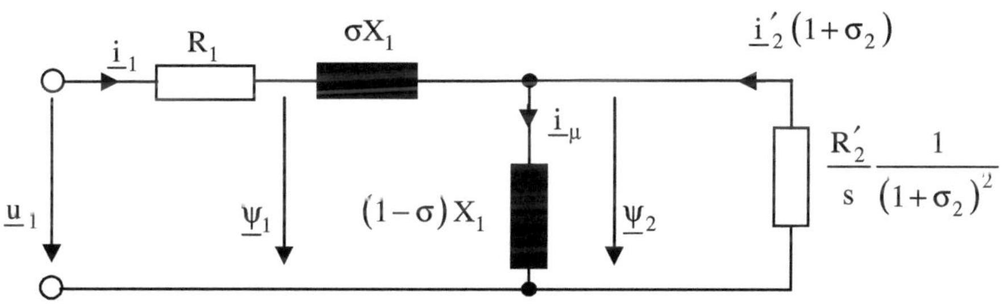

Fig. 12.23. Block diagram of the induction machine.

The stator flux linkage is:

$$\underline{\psi}_1 = \int(\underline{u}_1 - R_1\underline{i}_1)\,dt + \underline{\psi}_{1,0} \tag{12.74}$$

and the rotor flux linkage is:

$$\underline{\psi}_2 = \underline{\psi}_1 - \sigma L_1 \underline{i}_1 \tag{12.75}$$

Consequently the amplitude and phase of the rotor flux linkage are known:

- The phase is the angle α, which is necessary for the coordinate transformation. By differentiating the angular frequency ω_μ is obtained; together with the value for ω_R this is used for the speed control.
- The amplitude of the rotor flux linkage is already known from Sect. 12.4 (flux model) and is therefore not necessary at this moment.

- The above equations for calculating the phase of the rotor flux linkage are evaluated for a two-phase system, after measuring the three-phase values (phase currents and voltages) and subsequent coordinate transformation.
- The evaluation of the integral is difficult for small voltages or small frequencies (preciseness and size of the time interval); i.e. for very small speed this method is quite incorrect. The method is reliable from about 5% of the nominal speed onwards.

The flux model from Sect. 12.4 (that solely is based on current measurement) showed the disadvantage of temperature dependency (dependency of the rotor time constant τ_2); the model presented in this section possesses the disadvantage of inaccuracy at small speed. By clever combination the area of reliable operation without mechanical speed sensor can be considerably increased.

12.7 Direct Torque Control

The direct torque control (DTC) has been developed independently, nearly simultaneously and in similar form at the beginning of the 1980s in Germany (Manfred Depenbrock) and in Japan (Isao Takahashi and Toshihiko Noguchi) for induction machines. In the meantime this method also is applied to different rotating field machines. The principle of DTC is that by choosing the phase voltages the flux and the torque are directly influenced.

To explain this, a simple switch-model for the power electronic converter and a machine with Y-connected phases is regarded. The diagram is shown in Fig. 12.24.

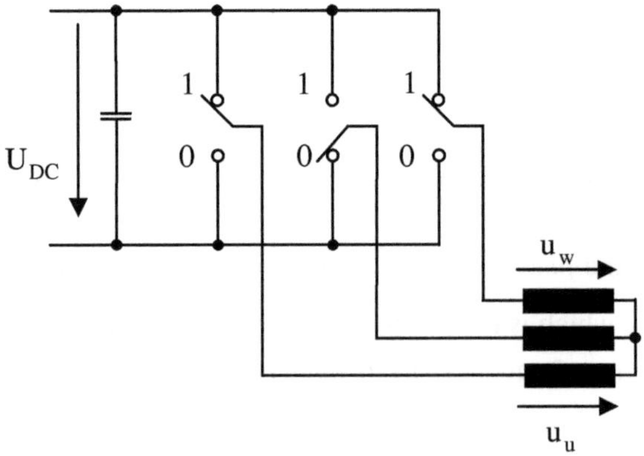

Fig. 12.24. Diagram of the inverter-fed induction machine.

The three switches with two different switching positions each ("0" and "1") define eight voltages that are illustrated in Fig. 12.25; the voltages $\underline{u}_0(000)$ and $\underline{u}_7(111)$ are called zero-vectors and they are drawn in the origin of the coordinate system:

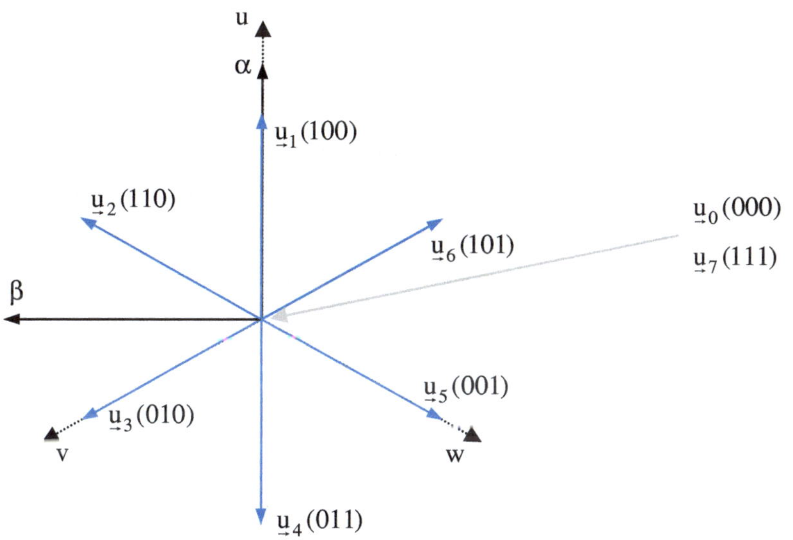

Fig. 12.25. Possible voltage vectors of the inverter-fed induction machine.

If only these positions of the power electronic switches are allowed, there is – for any point in time (except for the positions "000" und "111") – a series connection of one machine phase with the parallel connection of the other two phases (like it is shown in Fig. 12.26).

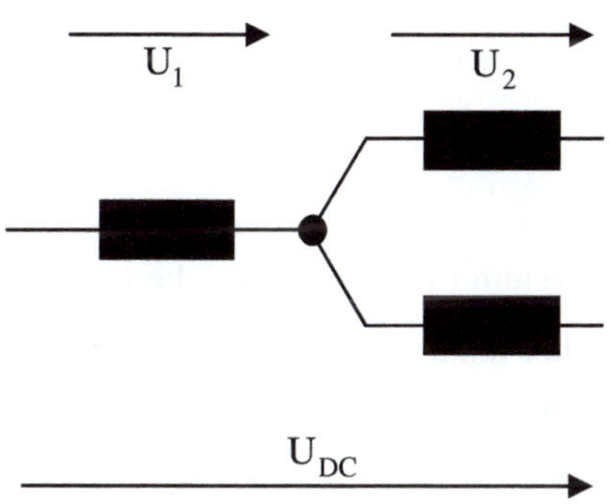

Fig. 12.26. Connection of phases of the inverter-fed induction machine at any point in time.

With the phase impedance Z and the total current I it follows:

$$\begin{aligned} U_{DC} &= U_1 + U_2 \\ &= Z\,I + \frac{Z\,Z}{Z+Z} I = \left(1+\frac{1}{2}\right) Z\,I \\ &= \frac{3}{2} U_1 \end{aligned} \tag{12.76}$$

Consequently there is always one phase with a voltage drop of 2/3 of the intermediate voltage U_{DC}, and the voltage drop across the parallel connected phases is 1/3 of the intermediate voltage. Then the voltage space vector $\vec{\underline{u}}_1$ becomes (see the definition of the complex voltage space vector in Sect. 11.3):

$$\begin{aligned} \vec{\underline{u}}_1 &= \frac{2}{3}\left(u_u(t) + \underline{a}\, u_v(t) + \underline{a}^2\, u_w(t)\right) e^{-j\alpha(t)} \\ &= \frac{2}{3}\left(\frac{2}{3} U_{DC} - \frac{1}{3} U_{DC}\left(\underline{a} + \underline{a}^2\right)\right) e^{j0} \\ &= \frac{2}{3}\left(\frac{2}{3} U_{DC} + \frac{1}{3} U_{DC}\right) \\ &= \frac{2}{3} U_{DC} \end{aligned} \tag{12.77}$$

The other voltage space vectors can be calculated analogously. Summarizing, the phase voltages of the machine can be described by space vectors like follows:

$$\vec{\underline{u}}_\nu = \begin{cases} \frac{2}{3} U_{DC}\, e^{j(\nu-1)\frac{\pi}{3}} & \text{if } \nu = 1, \cdots, 6 \\ 0 & \text{if } \nu = 0, 7 \end{cases} \tag{12.78}$$

By choosing a stationary coordinate system ($\alpha(t) = \text{const.}$, here as a special case $\alpha(t) = 0$) it follows for the stator flux linkage from the stator voltage equation in space vector notation (see Sect. 11.6):

$$\vec{\underline{u}}_I = R_I\, \vec{\underline{i}}_I + \frac{d}{dt} \vec{\underline{\psi}}_I \quad \Rightarrow \quad \vec{\underline{\psi}}_I = \int \left(\vec{\underline{u}}_I - R_I\, \vec{\underline{i}}_I\right) dt \tag{12.79}$$

If the stator resistance R_I can be neglected, the stator voltage space vectors $\underline{\vec{u}}_1$ to $\underline{\vec{u}}_6$ are causing a continuous motion of the stator flux space vector, whereas the stator voltage space vectors $\underline{\vec{u}}_0$ and $\underline{\vec{u}}_7$ are stopping the stator flux space vector. If the stator voltage space vectors $\underline{\vec{u}}_1$ to $\underline{\vec{u}}_6$ are switched just once per period (which is called "block-mode operation"), the stator flux space vector moves on a hexagon. Consequently the first task (adjusting the flux of the machine) is fulfilled (see Fig. 12.27).

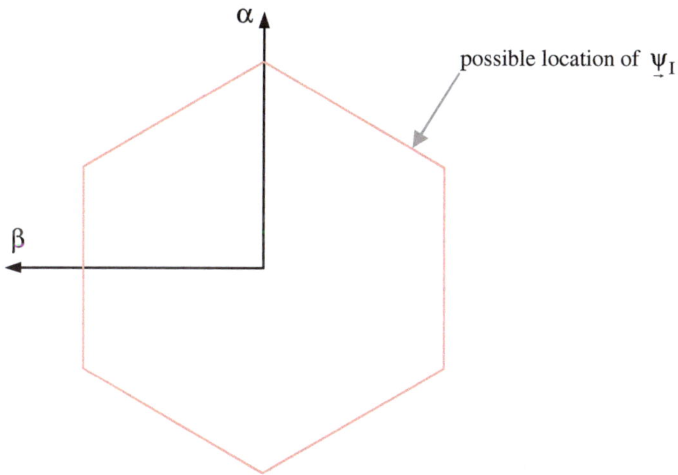

Fig. 12.27. Possible locations of the stator flux space vector.

The torque can be calculated like it is shown in Sect. 11.9:

$$T = \frac{3}{2} p \operatorname{Im}\left\{ \underline{\vec{i}}_I \, \underline{\vec{\psi}}_I^* \right\} \tag{12.80}$$

With

$$\begin{aligned} \underline{\vec{\psi}}_I &= L_1 \, \underline{\vec{i}}_I + L_{1m} \, \underline{\vec{i}}_{II} = L_1 \, \underline{\vec{i}}_I + L_{1m} \frac{L_2}{L_2} \underline{\vec{i}}_{II} \\ &= \frac{L_{1m}}{L_2} \left(L_2 \, \underline{\vec{i}}_{II} + L_{1m} \, \underline{\vec{i}}_I \right) + L_1 \, \underline{\vec{i}}_I - \frac{L_{1m}}{L_2} L_{1m} \, \underline{\vec{i}}_I \\ &= \frac{L_{1m}}{L_2} \underline{\vec{\psi}}_{II} + \left((1+\sigma_1) - \frac{1}{1+\sigma_2} \right) L_{1m} \, \underline{\vec{i}}_I \end{aligned} \tag{12.81}$$

and further

$$\begin{aligned}\underline{\psi}_{I} &= \frac{L_{1m}}{L_2}\underline{\psi}_{II} + \left(1 - \frac{1}{(1+\sigma_2)(1+\sigma_1)}\right) L_1\, \underline{i}_I \\ &= \frac{L_{1m}}{L_2}\underline{\psi}_{II} + \sigma\, L_1\, \underline{i}_I\end{aligned} \tag{12.82}$$

it follows

$$T = \frac{3}{2}\frac{p}{\sigma L_1}\,\mathrm{Im}\left\{\left(\underline{\psi}_I - \frac{L_{1m}}{L_2}\underline{\psi}_{II}\right)\underline{\psi}_I^*\right\} \tag{12.83}$$

With

$$\mathrm{Im}\left\{\underline{\psi}_I\ \underline{\psi}_I^*\right\} = 0 \tag{12.84}$$

it can be deduced finally

$$\begin{aligned}T &= -\frac{3}{2}\frac{p}{\sigma L_1}\frac{L_{1m}}{L_2}\,\mathrm{Im}\left\{\underline{\psi}_{II}\ \underline{\psi}_I^*\right\} \\ &= -\frac{3}{2}p\frac{1-\sigma}{\sigma L_{1m}}\,\mathrm{Im}\left\{\underline{\psi}_{II}\ \underline{\psi}_I^*\right\}\end{aligned} \tag{12.85}$$

Consequently, the torque generation is determined by the amplitudes of the stator flux linkage, the rotor flux linkage, and the relative phase shift of both.

In the following it is assumed that the speed and the amplitude of the rotor flux linkage is constant during one switching condition of the stator voltage space vector. Then the stator flux linkage and the torque are adjusted directly by the choice of the stator voltage space vector. For this choice of the stator voltage space vector – required in the actual operating condition of the machine – the following steps have to be performed:

- division of the α - β -plane into sectors;
- calculation, in which sector the actual stator flux linkage is located;
- evaluation, if the stator flux linkage and the torque have to be increased or decreased;
- adjusting the resulting stator voltage space vector.

The sectors in the α - β -plane can be chosen e.g. like it is shown in Fig. 12.28.

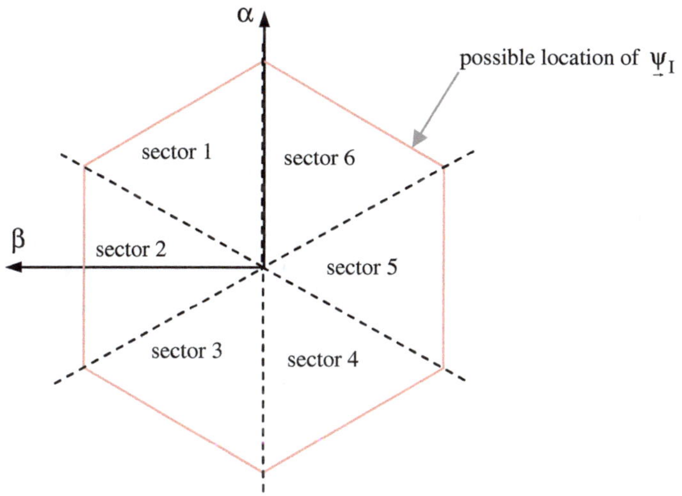

Fig. 12.28. Definition of the sectors in the α-β-plane.

By means of a machine model the instantaneous location and amplitude of the stator flux linkage and the value of the torque can be calculated at any time. By comparison with the respective set values differences are obtained that are fed to a hysteresis controller each. From the instantaneous location of the stator flux linkage and the necessity to increase or decrease the stator flux linkage and the torque (outputs of the hysteresis controllers "1" or "-1"), the next stator voltage space vector is obtained by means of a table. Respective switching signals s_u, s_v and s_w activate the power electronic switches of the converter. The corresponding block diagram is given in Fig. 12.29.

The according table for choosing the stator voltage space vector is shown in Table 12.1.

Table 12.1. Switching table for the direct torque control.

ϕ	τ	sector 1	sector 2	sector 3	sector 4	sector 5	sector 6
1	1	$\underline{u}_2$	$\underline{u}_3$	$\underline{u}_4$	$\underline{u}_5$	$\underline{u}_6$	$\underline{u}_1$
1	-1	$\underline{u}_6$	$\underline{u}_1$	$\underline{u}_2$	$\underline{u}_3$	$\underline{u}_4$	$\underline{u}_5$
-1	1	$\underline{u}_3$	$\underline{u}_4$	$\underline{u}_5$	$\underline{u}_6$	$\underline{u}_1$	$\underline{u}_2$
-1	-1	$\underline{u}_5$	$\underline{u}_6$	$\underline{u}_1$	$\underline{u}_2$	$\underline{u}_3$	$\underline{u}_4$

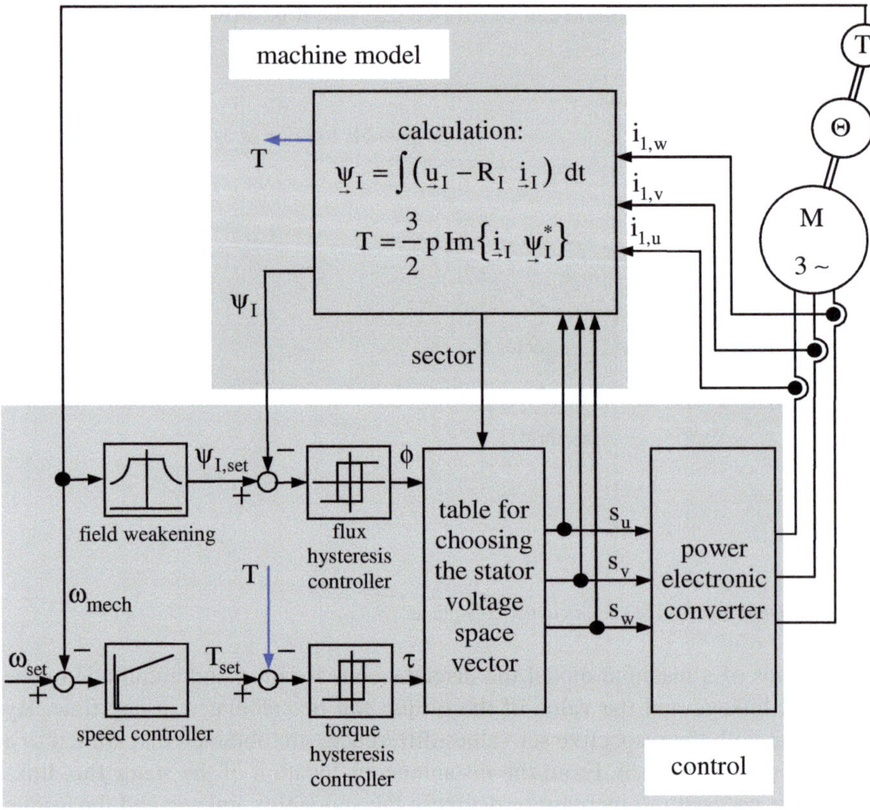

Fig. 12.29. Block diagram of the inverter-fed induction machine with direct torque control.

The main advantages of the direct torque control against the field-oriented control are:

- The calculation load in the microcontroller is much lower, because no coordinate transformation is required.
- Flux and torque are adjusted by means of simple hysteresis controllers; there is no need for current controllers or for pulse width modulation. Consequently the switching frequency of the power electronic switches is quite low.
- There is only low sensitivity against varying rotor parameters, because only flux and torque calculation is needed.
- For torque and flux control no knowledge of the speed is required; having the field-oriented control this was necessary to calculate the angle α.
- Depending on the preciseness of the machine model even for the speed control the speed sensor can be omitted.
- Generally, the torque control using DTC is faster than using the field-oriented control.

However, disadvantages of the DTC are:

- Because of the missing current controllers there is no possibility of active forming of the current waveform. This results in considerable deviations from the ideal sinusoidal function and therefore leads to increased losses of the induction machine.
- Having non-sinusoidal currents the preciseness of the flux and torque calculation strongly depends on the sampling interval and preciseness of the current measurement as well as on the cycle period of the controller.
- The torque ripple depends on the current waveform and the chosen widths of the hysteresis controllers; usually this torque ripple is larger than for the field-oriented control. This results in mechanical load and acoustic noise.
- The switching frequency of the power electronic devices is not fixed and it changes with the speed of the machine. Consequently, even the switching losses of the power electronic devices are speed-dependent.

Generally the DTC is characterized by simplicity, robustness, low switching losses, and fast torque control. These features are especially interesting for variable-speed drives with high power.

The DTC described until now can be extended, e.g. by

- consideration of the measured intermediate voltage when calculating the stator voltage space vector;
- high-frequent switching (e.g. PWM) and utilization of all voltage vectors $\underset{\rightarrow}{u}_0$ to $\underset{\rightarrow}{u}_7$ (by this and with increasing frequency the motion of the flux space vector can be approximated more and more to a circle);
- increasing the number of sectors;
- hysteresis controller with three steps (additional step "0", i.e. no change of flux or torque).

By these means the technical features (e.g. current waveform and torque ripple) are improved, but the effort is increased.

12.8 References for Chapter 12

Blaschke F (1973) Das Verfahren der Feldorientierung zur Regelung der Drehfeldmaschine. Dissertation Technische Universitaet Braunschweig

Boldea I, Tutelea L (2010) Electric machines. CRC Press, Boca Raton

Buja GS, Kazmierkowski P (2004) Direct torque control of PWM inverter-fed AC motors – a survey. IEEE Transactions on Industrial Electronics, 51:744-757

Casadei D, Profumo F, Serra G, Tani A (2002) FOC and DTC: two viable schemes for induction motors torque control. IEEE Transactions on Power Electronics, 17:779-787

DeDoncker RW, Pulle DWJ, Veltman A (2011) Advanced eldectrical Drives. Springer-Verlag, Berlin

Depenbrock M (1988) Direct self-control (DSC) of inverter-fed induction machine. IEEE Transactions on Power Electronics, 3:420-429

Hasse K (1969) Zur Dynamik drehzahlgeregelter Antriebe mit stromrichtergespeisten Asynchron-Kurzschlußläufermotoren. Dissertation Technische Universitaet Darmstadt

Krishnan R (2001) Electric motor drives. Prentice Hall, London

Li Y (2010) Direct torque control of permanent magnet synchronous machine. Shaker-Verlag, Aachen

Nasar SA (1970) Electromagnetic energy conversion devices and systems. Prentice Hall, London

Schröder D (1995) Elektrische Antriebe 2. Springer-Verlag, Berlin

Takahashi I, Noguchi T (1986) A new quick-response and high-efficiency control strategy of an induction motor. IEEE Transactions on Industry Application, 22:820-827

White DC, Woodson HH (1958) Electromechanical energy conversion. John Wiley & Sons, New York

13 Dynamic Operation of Synchronous Machines

13.1 Oscillations of Synchronous Machines, Damper Winding

In this section the behavior of the synchronous machine will be regarded, if the rotor angle ϑ is changed by small values $\Delta\vartheta$ from the operation point (index "0"). As only small changes are considered, the description of the synchronous machine in steady-state operation (Chap. 5) will be used. There is:

$$\vartheta = \vartheta_0 + \Delta\vartheta \tag{13.1}$$

In steady-state operation the external driving torque of the turbine is equal to the torque of the synchronous machine in every operating point:

$$T_{ext} = T_{pull-out} \sin(\vartheta_0), \qquad T_{pull-out} = \frac{3p}{\omega_1} \frac{U_{N,phase} U_P}{X} \tag{13.2}$$

The torque of the synchronous generator and the acceleration torque are:

$$T_{gen} = T_{pull-out} \sin(\vartheta) \tag{13.3}$$

$$T_a = \Theta \frac{d\Omega}{dt} \tag{13.4}$$

with Θ being the inertia of all rotating masses and $\Omega \neq 2\pi n_0$ being the speed of the synchronous machine: [23]

$$\Omega = 2\pi n_0 + \frac{d\vartheta/p}{dt} \tag{13.5}$$

---

[23] In the following it will be assumed that the stator current angular frequency ω_1 always is adapted to the speed of the machine. Otherwise the frequency condition for generating a constant torque (see Chap. 3 "Rotating Field Theory") would not be fulfilled and a pure oscillating torque would occur. The influence of this frequency change on other data (e.g. the pull-out torque) is neglected because just small changes are regarded; in addition the friction is neglected.

The torque balance

$$T_{ext} - T_{gen} = T_a \tag{13.6}$$

leads to the following differential equation:

$$T_{pull-out} \sin(\vartheta_0) - T_{pull-out} \sin(\vartheta) = \Theta \frac{d\Omega}{dt} = \frac{\Theta}{p} \frac{d^2\vartheta}{dt^2} \tag{13.7}$$

This differential equation will be linearized by a Taylor expansion and truncation after the first term:

$$\begin{aligned} & f(x + \Delta x) = f(x) + \frac{f'(x)}{1!} \Delta x + \cdots \\ \Rightarrow \quad & \sin(\vartheta) = \sin(\vartheta_0 + \Delta\vartheta) \approx \sin(\vartheta_0) + \Delta\vartheta \cos(\vartheta_0) \end{aligned} \tag{13.8}$$

Moreover there is:

$$\frac{d^2\vartheta}{dt^2} = \frac{d^2(\vartheta_0 + \Delta\vartheta)}{dt^2} = \frac{d^2\Delta\vartheta}{dt^2} \tag{13.9}$$

Then the differential equation becomes:

$$\begin{aligned} & T_{pull-out} \sin(\vartheta_0) - T_{pull-out} \left(\sin(\vartheta_0) + \Delta\vartheta \cos(\vartheta_0)\right) = \frac{\Theta}{p} \frac{d^2\Delta\vartheta}{dt^2} \\ \Rightarrow \quad & \frac{\Theta}{p} \frac{d^2\Delta\vartheta}{dt^2} + T_{pull-out} \Delta\vartheta \cos(\vartheta_0) = 0 \end{aligned} \tag{13.10}$$

With the synchronizing torque in the operating point $T_{syn,0} = T_{pull-out} \cos(\vartheta_0)$ it can be deduced:

$$\frac{d^2\Delta\vartheta}{dt^2} + \frac{T_{syn,0}}{\Theta/p} \Delta\vartheta = 0 \tag{13.11}$$

The solution of this differential equation is an undamped harmonic oscillation:

$$\Delta\vartheta = \sin\left(\Omega_{e,0} t\right) \tag{13.12}$$

with the mechanical resonance frequency (eigenfrequency)[24]

$$\Omega_{e,0} = 2\pi f_{e,0} = \sqrt{\frac{T_{syn,0}}{\Theta/p}} \tag{13.13}$$

Most often the frequency of this mechanical oscillation of the synchronous machine is in the range of $f_{e,0} = 1 \ldots 2\,\mathrm{Hz}$.

During operation of the synchronous machine oscillations can be evoked by electrical or mechanical load changes, which are accompanied by current oscillations. Especially for drives with a non-constant torque (e.g. diesel engine or piston compressor) these oscillations may reach critically high values, if the excitation is near to the eigenfrequency. It is also possible that different generators may excite each other until they fall out of synchronism.

For damping of these oscillations all synchronous machines are equipped with a damper winding. The effect of such a damper winding is comparable with the squirrel-cage of an induction machine.

For high-speed generators with cylindrical rotor damper bars are inserted into the slots of the rotor in addition to the excitation winding; these damper bars are short-circuited at their axial ends (even electrically conductive slot wedges may be used as damper bars). Solid rotors have a damping effect as well, because eddy currents may develop. For salient-pole synchronous machines there are additional slots with bars in the poles; then again the bars are short-circuited at their axial ends.

The calculation of the damper winding can be started from the equations of the induction machine (the torque is negative as it decelerates the machine):

$$\frac{T_D}{T_{pull-out,IM}} = \frac{-2}{\dfrac{s}{s_{pull-out}} + \dfrac{s_{pull-out}}{s}}, \quad T_{pull-out,IM} = \frac{3p}{\omega_1} \frac{U_1^2}{2X_1 \dfrac{\sigma}{1-\sigma}}$$

$$s_{pull-out} = \frac{R_2'\left(1+\sigma_1\right)^2}{X_1 \dfrac{\sigma}{1-\sigma}} \tag{13.14}$$

Near to the synchronous speed (just small changes $\Delta\vartheta$ are regarded) there is:

---

[24] In the mechanical analogon the synchronizing torque corresponds to the spring stiffness, the inertia divided by the number of pole pairs corresponds to the mass.

$$\frac{s}{s_{pull-out}} \ll \frac{s_{pull-out}}{s} \tag{13.15}$$

Consequently it follows:

$$T_D \approx -T_{pull-out,IM} \frac{2s}{s_{pull-out}} \tag{13.16}$$

The slip can be described as:

$$s = \frac{\Omega_0 - \Omega}{\Omega_0} = \frac{\Omega_0 - \left(\Omega_0 + \frac{d\vartheta/p}{dt}\right)}{\Omega_0} = -\frac{1}{p\Omega_0}\frac{d\vartheta}{dt} = -\frac{1}{p\Omega_0}\frac{d\Delta\vartheta}{dt} \tag{13.17}$$

Therefore the damping torque becomes:

$$T_D = \frac{2T_{pull-out,IM}}{s_{pull-out}p\Omega_0}\frac{d\Delta\vartheta}{dt} = D\frac{d\Delta\vartheta}{dt} \tag{13.18}$$

Introducing this damping torque into the differential equation it follows:

$$\frac{d^2\Delta\vartheta}{dt^2} + \frac{D}{\Theta/p}\frac{d\Delta\vartheta}{dt} + \frac{T_{syn,0}}{\Theta/p}\Delta\vartheta = 0 \tag{13.19}$$

The solution of this differential equation is a damped oscillation of the following kind:

$$\Delta\vartheta = e^{-\frac{t}{\tau_D}} \sin\left(\Omega_e t\right) \tag{13.20}$$

with the mechanical resonance frequency

$$\Omega_e = \sqrt{\Omega_{e,0}^2 - \frac{1}{\tau_D^2}} \tag{13.21}$$

the damping

$$D = \frac{2T_{pull-out,IM}}{s_{pull-out} p\Omega_0} = \frac{2\frac{3p}{\omega_1}\frac{U_1^2}{2X_1\frac{\sigma}{1-\sigma}}}{\frac{R_2'(1+\sigma_1)^2}{X_1\frac{\sigma}{1-\sigma}}p\Omega_0} = \frac{\frac{3p}{\omega_1}U_1^2}{R_2'(1+\sigma_1)^2 p\Omega_0} \tag{13.22}$$

$$= \frac{3p}{\omega_1^2}\frac{U_1^2}{R_2'(1+\sigma_1)^2}$$

and the time constant

$$\tau_D = \frac{2\Theta}{pD} \tag{13.23}$$

For maximizing the effect of the damper winding (and therefore damping the oscillations caused by load changes most quickly) the time constant τ_D must be small or the damping D must be large. This means that the resistance R_2' must be small. Consequently there is high copper mass and costs needed for the damper cage.

Besides the damping of oscillations the damper winding has two additional tasks:

- From asymmetric loading an opposite rotating field with harmonic oscillations in stator voltage and stator current is generated, which causes additional iron losses and copper losses. In the damper winding currents are evoked that (according to Lenz' Law) act against their cause. Therefore, these harmonic oscillations and the additional losses are strongly reduced.
- With sufficient heat capacity of the damper winding the synchronous machine may be started with the damper cage like an induction machine. Because of the large slip during run-up the rotating field of the stator would induce high voltages in the not-connected excitation winding; therefore the excitation winding is short-circuited at first. Reaching the no-load speed the excitation voltage is switched on and the machine synchronizes suddenly. This is accompanied by current pulses and oscillating torque components, so that this kind of run-up can be applied only for small power machines.

13.2 Steady-State Operation of Non Salient-Pole Synchronous Machines in Space Vector Notation

Starting with the equations in space vector notation (which are also valid for the dynamic operation of electrical machines) the steady-state operation of the symmetric synchronous machine with non salient-pole (cylindrical) rotor shall be calculated in this section.

The air-gap is assumed being constant; the rotor shall be symmetric, i.e. there are two identical windings shifted electrically by 90°. The excitation winding is supplied with DC current via slip rings, the damper winding is short-circuited.

According to Sects. 11.6 and 11.9 it follows in space vector notation:

$$\underrightarrow{u}_I(t) = R_I\ \underrightarrow{i}_I(t) + \frac{d\underrightarrow{\psi}_I(t)}{dt} + j\frac{d\alpha}{dt}\underrightarrow{\psi}_I(t) \tag{13.24}$$

$$\underrightarrow{u}_{II}(t) = R_{II}\ \underrightarrow{i}_{II}(t) + \frac{d\underrightarrow{\psi}_{II}(t)}{dt} + j\frac{d(\alpha-\gamma)}{dt}\underrightarrow{\psi}_{II}(t) \tag{13.25}$$

$$T(t) = \frac{3}{2}p\,\mathrm{Im}\left\{\underrightarrow{i}_I(t)\,\underrightarrow{\psi}^*_I(t)\right\} \tag{13.26}$$

In the following a coordinate system is chosen that rotates in synchronism with the rotor: $\omega_{CS} = d\alpha/dt = d\gamma/dt = \omega_{mech}$. This coordinate system is rotor flux oriented, consequently the axes are nominated with “d“ and “q“ (instead of “y“ and “x“). The stator systems are then called I,d and I,q, the rotor systems II,d and II,q. The splitting up of the above shown complex equations into their components gives (analogously to Sect. 12.3):

$$\begin{aligned} u_{I,q}(t) &= R_I\ i_{I,q}(t) + \frac{d\psi_{I,q}(t)}{dt} + \omega_{CS}\ \psi_{I,d}(t) \\ u_{I,d}(t) &= R_I\ i_{I,d}(t) + \frac{d\psi_{I,d}(t)}{dt} - \omega_{CS}\ \psi_{I,q}(t) \end{aligned} \tag{13.27}$$

$$\begin{aligned} u_{II,q}(t) &= R_2' \, i_{II,q}(t) + \frac{d\psi_{II,q}(t)}{dt} + (\omega_{CS} - \omega_{mech})\psi_{II,d}(t) \\ u_{II,d}(t) &= R_2' \, i_{II,d}(t) + \frac{d\psi_{II,d}(t)}{dt} - (\omega_{CS} - \omega_{mech})\psi_{II,q}(t) \end{aligned} \tag{13.28}$$

$$\begin{aligned} T(t) &= \frac{3}{2} p \, \mathrm{Im}\left\{ \underline{i}_I(t) \underline{\psi}_I^*(t) \right\} \\ &= \frac{3}{2} p \, \mathrm{Im}\left\{ \left[i_{I,q}(t) - j \, i_{I,d}(t) \right] \left[\psi_{I,q}(t) + j \, \psi_{I,d}(t) \right] \right\} \\ &= \frac{3}{2} p \left[\psi_{I,d}(t) i_{I,q}(t) - \psi_{I,q}(t) i_{I,d}(t) \right] \end{aligned} \tag{13.29}$$

During steady-state operation there are no changes of the flux linkages ($d\psi/dt = 0$) and the speed is constant ($\omega_{mech} = d\gamma/dt = \text{const.}$). Now the excitation winding is laid into the II, d -axis and the (short-circuited) damper winding into the II, q -axis. For most synchronous machines (in particular for large generators) the Ohmic resistance of the stator winding can be neglected. This is done in the following.

Then the above set of equations becomes ($\omega_1 = \omega_{CS} = d\alpha/dt = \text{const.}$):

$$\begin{aligned} u_{I,q}(t) &= \omega_1 \, \psi_{I,d}(t) \\ u_{I,d}(t) &= -\omega_1 \, \psi_{I,q}(t) \end{aligned} \tag{13.30}$$

$$\begin{aligned} 0 &= R_2' \, i_{II,q}(t) \\ u_{II,d}(t) &= R_2' \, i_{II,d}(t) \end{aligned} \tag{13.31}$$

$$T(t) = \frac{3}{2} p \left[\psi_{I,d}(t) i_{I,q}(t) - \psi_{I,q}(t) i_{I,d}(t) \right] \tag{13.32}$$

As the space vector theory has been developed in the energy consumption system, this is the set of equations for the synchronous machine in the energy consumption system. Usually the synchronous machine is described in the energy generation system (because generally this machine type is used as a generator, see

Chap. 5), therefore in the following the synchronous machine shall be described in terms of the energy generation system. To reach this the voltages $u_{I,d}$ and $u_{I,q}$ are changed in sign. Even the torque equation has to be attached with a negative sign, see below. The two stator voltage equations become:

$$\begin{aligned} u_{I,q}(t) &= -\omega_1\ \psi_{I,d}(t) \\ &= -\omega_1\ L_1\ i_{I,d}(t) - \omega_1\ L_{1m}\ i_{II,d}(t) \\ u_{I,d}(t) &= \omega_1\ \psi_{I,q}(t) \\ &= \omega_1\ L_1\ i_{I,q}(t) + \omega_1\ L_{1m}\ i_{II,q}(t) \end{aligned} \tag{13.33}$$

The initial value of the rotating coordinate system shall be:

$$\alpha(t) = \omega_1 t + \alpha_0 \tag{13.34}$$

The reverse transformation of the stator voltage equations

$$\begin{gathered} \begin{bmatrix} u_{1,u}(t) \\ u_{1,v}(t) \end{bmatrix} = \left[T_{xy}\right]^{-1} \begin{bmatrix} u_{I,q}(t) \\ u_{I,d}(t) \end{bmatrix}, \\ \left[T_{xy}\right]^{-1} = \begin{bmatrix} \cos(\alpha) & \sin(\alpha) \\ \cos\left(\alpha - \frac{2\pi}{3}\right) & \sin\left(\alpha - \frac{2\pi}{3}\right) \end{bmatrix} \end{gathered} \tag{13.35}$$

then gives:

$$\begin{aligned} u_{1,u}(t) &= u_{I,q}(t)\cos(\alpha) + u_{I,d}(t)\sin(\alpha) \\ &= u_{I,q}(t)\cos(\omega_1 t + \vartheta) + u_{I,d}(t)\sin(\omega_1 t + \vartheta) \end{aligned} \tag{13.36}$$

Considering

$$\begin{aligned} u_{1,u}(t) &= \mathrm{Re}\left\{\sqrt{2}\underline{U}_1 e^{j\omega_1 t}\right\} \\ \cos(\omega_1 t + \vartheta) &= \mathrm{Re}\left\{e^{j\omega_1 t} e^{j\vartheta}\right\} \\ \sin(\omega_1 t + \vartheta) &= \mathrm{Im}\left\{e^{j\omega_1 t} e^{j\vartheta}\right\} = \mathrm{Re}\left\{-j e^{j\omega_1 t} e^{j\vartheta}\right\} \end{aligned} \tag{13.37}$$

it follows:

$$\mathrm{Re}\left\{\sqrt{2}\underline{U}_1 e^{j\omega_1 t}\right\} = u_{I,q}\,\mathrm{Re}\left\{e^{j\omega_1 t}e^{j\vartheta}\right\} + u_{1,d}\,\mathrm{Re}\left\{-je^{j\omega_1 t}e^{j\vartheta}\right\} \tag{13.38}$$

As there are no currents induced into the damper winding in steady-state operation, there is $i_{II,q} = 0$. Consequently:

$$\begin{aligned}\mathrm{Re}\left\{\sqrt{2}\underline{U}_1 e^{j\omega_1 t}\right\} &= -\omega_1 \psi_{I,d}\,\mathrm{Re}\left\{e^{j\omega_1 t}e^{j\vartheta}\right\} + \omega_1 \psi_{1,q}\,\mathrm{Re}\left\{-je^{j\omega_1 t}e^{j\vartheta}\right\} \\ &= -\omega_1\left(L_1 i_{I,d} + L_{1m} i_{II,d}\right)\mathrm{Re}\left\{e^{j\omega_1 t}e^{j\vartheta}\right\} \\ &\quad + \omega_1 L_1 i_{I,q}\,\mathrm{Re}\left\{-je^{j\omega_1 t}e^{j\vartheta}\right\}\end{aligned} \tag{13.39}$$

The left side of this equation describes a harmonic oscillation of a single frequency. This has to be true even for the right side of the equation. Then the equality of both sides is unchanged if on both sides the respective imaginary components are added. It follows:

$$\begin{aligned}&\sqrt{2}\underline{U}_1 e^{j\omega_1 t} = -\omega_1\left(L_1 i_{I,d} + L_{1m} i_{II,d}\right)\left\{e^{j\omega_1 t}e^{j\vartheta}\right\} + \omega_1 L_1 i_{I,q}\left\{-je^{j\omega_1 t}e^{j\vartheta}\right\} \\ &\Rightarrow\quad \underline{U}_1 = \frac{1}{\sqrt{2}}\left[j\,\omega_1 L_1\left(j\,i_{I,d} - i_{I,q}\right)e^{j\vartheta} + j\,\omega_1 L_{1m}\left(j\,i_{II,d}\right)e^{j\vartheta}\right]\end{aligned} \tag{13.40}$$

Now the following currents are defined:

$$\underline{I}_{1,q} = \frac{i_{I,q}}{\sqrt{2}} e^{j\vartheta} \tag{13.41}$$

$$\underline{I}_{1,d} = -j\frac{i_{I,d}}{\sqrt{2}} e^{j\vartheta} \tag{13.42}$$

$$\underline{I}_2' = -j\frac{i_{II,d}}{\sqrt{2}} e^{j\vartheta} \tag{13.43}$$

Then the above voltage equation becomes:

$$\underline{U}_1 = j\omega_1 L_1\left(-\underline{I}_{1,d} - \underline{I}_{1,q}\right) - j\omega_1 L_{1m}\underline{I}_2' \tag{13.44}$$

With

$$\underline{I}_1 = \underline{I}_{1,q} + \underline{I}_{1,d} \tag{13.45}$$

and

$$\underline{U}_P = -j\omega_1 L_{1m} \underline{I}_2' \tag{13.46}$$

it follows further:

$$\underline{U}_P = \underline{U}_1 + j\omega_1 L_1 \underline{I}_1 \tag{13.47}$$

This is the well-known voltage equation of the synchronous machine with cylindrical rotor (please refer to Sect. 5.1). Choosing the terminal voltage being real

$$\underline{U}_1 = U_1 \tag{13.48}$$

the phasor diagram in the well-known form (see Sect. 5.1 and Fig. 13.1) can be drawn:

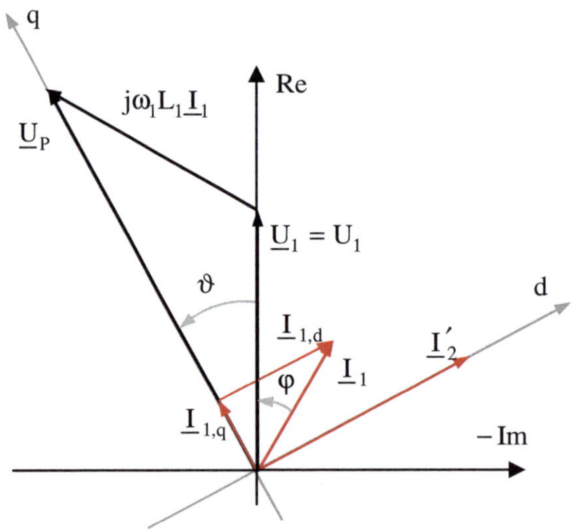

Fig. 13.1. Phasor diagram of the synchronous machine in steady-state operation.

Splitting the stator current into d-axis component and q-axis component leads to the flux generating and torque generating parts. This becomes obvious by transforming the torque equation:

$$
\begin{aligned}
T &= -\frac{3}{2}p\left[\psi_{I,d}(t)\, i_{I,q}(t) - \psi_{I,q}(t)\, i_{I,d}(t)\right] \\
&= -\frac{3}{2}p\Bigg[\left[L_1\, i_{I,d}(t) + L_{1m}\, i_{II,d}(t)\right] i_{I,q}(t) \\
&\qquad - \Bigg[L_1\, i_{I,q}(t) + L_{1m}\, \underbrace{i_{II,q}(t)}_{=0}\Bigg] i_{I,d}(t)\Bigg] \\
&= -\frac{3}{2}p\Big[\left[L_1\, i_{I,d}(t)\, i_{I,q}(t) + L_{1m}\, i_{II,d}(t)\, i_{I,q}(t)\right] \\
&\qquad - \left[L_1\, i_{I,q}(t)\, i_{I,d}(t)\right]\Big] \\
&= -\frac{3}{2}p\, L_{1m}\, i_{II,d}(t)\, i_{I,q}(t) \\
&= -\frac{3}{2}p\, L_{1m}\left[\frac{\sqrt{2}}{-j}\underline{I}_2'\, e^{-j\vartheta}\right]\left[\sqrt{2}\, \underline{I}_{1,q}\, e^{-j\vartheta}\right] \\
&= -j\, 3\frac{p}{\omega_1} X_{1m}\, \underline{I}_2'\, \underline{I}_{1,q}\, e^{-j2\vartheta} = 3\frac{p}{\omega_1}\underline{U}_P\, \underline{I}_{1,q}\, e^{-j2\vartheta} \\
&= 3\frac{p}{\omega_1} U_P\, e^{j\vartheta}\, I_{1,q}\, e^{j\vartheta}\, e^{-j2\vartheta} = 3\frac{p}{\omega_1} U_P\, I_{1,q} \\
&= 3\frac{p}{\omega_1} U_P\, I_1 \cos(\vartheta + \varphi) \\
&= 3\frac{p}{\omega_1} U_P\, I_1 \cos(\delta_G)
\end{aligned}
\tag{13.49}
$$

The angle δ_G is called load angle (angle between stator current and internal voltage of the machine, please refer to Sect. 5.1).

With the relations

$$U_P \cos(\vartheta + \varphi) = U_1 \cos(\varphi) \tag{13.50}$$

and

$$U_P \sin(\vartheta) = X_1 I_1 \cos(\varphi) \tag{13.51}$$

it follows further:

$$\begin{aligned} T &= 3\frac{p}{\omega_1} U_1\, I_1 \cos(\varphi) \\ &= 3\frac{p}{\omega_1}\frac{U_P\, U_1}{X_1}\sin(\vartheta) \end{aligned} \tag{13.52}$$

The equations obtained for calculating the torque of the synchronous machine with cylindrical rotor are already known from Chap. 5:

- Torque from internal voltage, phase voltage and rotor angle

$$T = 3\frac{p}{\omega_1}\frac{U_P\, U_1}{X_1}\sin(\vartheta) \tag{13.53}$$

- Torque from phase voltage, phase current and phase angle

$$T = 3\frac{p}{\omega_1} U_1\, I_1 \cos(\varphi) \tag{13.54}$$

- Torque from internal voltage, phase current and load angle

$$T = 3\frac{p}{\omega_1} U_P\, I_1 \cos(\delta_G) \tag{13.55}$$

13.3 Sudden Short-Circuit of Non Salient-Pole Synchronous Machines

13.3.1 Fundamentals

As an example for the dynamic operation of the synchronous machine the sudden three-phase short-circuit of a non salient-pole synchronous generator at no-load and nominal excitation will be calculated (no-load means: open terminals and rotor mechanically driven at synchronous speed). Sudden short-circuit is the transient process that is occurring directly after short-circuiting of the stator terminals (in contrast to the permanent short-circuit that is present when all transient processes are subsided); i.e. it is the transient phase between steady-state no-load operation to steady-state short-circuit operation.

For the sake of simplification the following approximations are introduced:

- The speed of the rotor shall remain constant at synchronous speed during the transient operation (the generator further shall be driven mechanically with synchronous speed).
- The non salient-pole machine shall be symmetric with two identical rotor windings shifted electrically by 90°. The excitation winding is supplied with DC current via two slip rings; the damper winding, lying in the quadrature axis, is short-circuited.
- A rotating coordinate system with $d\alpha/dt = \omega_{CS} = \omega_1 = \omega_{mech} = d\gamma/dt$ (like in Sect. 13.2) is chosen. The stator is composed of systems I,d and I,q, the rotor shows the systems II,d (excitation winding) and II,q (damper winding).
- The initial condition is defined by the switching moment: $\alpha(t) = \omega_1 t + \varepsilon$. Here ε is an (at this time) arbitrary phase angle.

13.3.2 Initial Conditions for t = 0

The original state before the sudden short-circuit is the no-load operation at nominal excitation. In this steady-state operation the Ohmic resistances of the stator winding generally can be neglected: $R_1 = 0$ (this assumption will be abolished later).

In the following the values just before the moment of switching get the additional index "0". For the initial conditions it follows:

- The stator currents are zero, because the terminals are not connected; the damper current is zero, because there is a steady-state operation at synchronous speed.

$$
\begin{aligned}
i_{I,d,0} &= i_{I,q,0} = i_{II,q,0} = 0 \\
i_{II,d,0} &= \frac{u_{II,d,0}}{R_2'}
\end{aligned}
\tag{13.56}
$$

- For the flux linkages the following is true:

$$
\begin{aligned}
\psi_{I,d,0} &= L_1\ i_{I,d,0} + L_{1m}\ i_{II,d,0} = L_{1m}\ i_{II,d,0} \\
\psi_{I,q,0} &= L_1\ i_{I,q,0} + L_{1m}\ i_{II,q,0} = 0
\end{aligned}
\tag{13.57}
$$

$$
\begin{aligned}
\psi_{II,d,0} &= L_2'\ i_{II,d,0} + L_{1m}\ i_{I,d,0} = L_2'\ i_{II,d,0} \\
\psi_{II,q,0} &= L_2'\ i_{II,q,0} + L_{1m}\ i_{I,q,0} = 0
\end{aligned}
\tag{13.58}
$$

- The stator voltages are (calculation in the energy generation system):

$$
\begin{aligned}
-u_{I,q,0} &= R_1\ i_{I,q,0} + \frac{d\psi_{I,q,0}}{dt} + \omega_1\ \psi_{I,d,0} = \omega_1\ \psi_{I,d,0} \\
-u_{I,d,0} &= R_1\ i_{I,d,0} + \frac{d\psi_{I,d,0}}{dt} - \omega_1\ \psi_{I,q,0} = -\omega_1\ \psi_{I,q,0}
\end{aligned}
\tag{13.59}
$$

Introducing the flux linkages into the stator voltage equations gives:

$$
\begin{aligned}
u_{I,q,0} &= -\omega_1\ L_{1m}\ i_{II,d,0} \\
u_{I,d,0} &= 0
\end{aligned}
\tag{13.60}
$$

and further:

$$\begin{aligned} u_{I,q,0} &= -\omega_1\ L_{1m} \frac{i_{II,d,0}}{\sqrt{2}} \sqrt{2} \\ &= \sqrt{2}\ U_{P,0} = \sqrt{2}\ U_{1,N} = \sqrt{2}\ \omega_1\ L_{1m}\ I_{F,0} \end{aligned} \tag{13.61}$$

From this last equation it follows:

$$i_{II,d,0} = \frac{-U_{1,N}\sqrt{2}}{\omega_1 L_{1m}} \tag{13.62}$$

Now the voltage of phase u will be calculated by reverse transformation. As there is a symmetric operation before the moment of switching, the other two phase voltages are symmetrical to the first one; therefore it is not necessary to calculate them separately.

$$\begin{aligned} u_{1,u,0} &= u_{I,q,0} \cos(\alpha) + u_{I,d,0} \sin(\alpha) \\ &= \sqrt{2}\ U_{1,N} \cos(\alpha) + 0 \\ &= \sqrt{2}\ U_{1,N} \cos(\omega_1 t + \varepsilon) \end{aligned} \tag{13.63}$$

From this equation it can be deduced that the phase angle ε introduced before characterizes the moment of switching:

- For $\varepsilon = 0$ and time $t = 0$ the flux linkage in phase u is zero, i.e. the peak value of the voltage is induced.
- For $\varepsilon = \pi/2$ and time $t = 0$ the flux linkage in phase u is maximum, i.e. the induced voltage is zero.

13.3.3 Set of Equations for t > 0

Now the above introduced approximation of neglecting the Ohmic resistances of the stator winding is abolished ($R_1 \neq 0$), because these resistances are responsible for the subsiding of the stator currents (damping characteristic).

The stator voltage equations are now in matrix notation (because of the short-circuit of the terminals this is identical for the energy consumption system and the energy generation system):

$$\begin{bmatrix} u_{I,d} \\ u_{I,q} \end{bmatrix} = \begin{bmatrix} 0 \\ 0 \end{bmatrix} = R_1 \begin{bmatrix} i_{I,d} \\ i_{I,q} \end{bmatrix} + \frac{d}{dt} \begin{bmatrix} \psi_{I,d} \\ \psi_{I,q} \end{bmatrix} + \omega_1 \begin{bmatrix} -\psi_{I,q} \\ \psi_{I,d} \end{bmatrix} \tag{13.64}$$

The rotor voltage equations become (as the excitation does not change with time, and it is further true $d\alpha/dt = \omega_{CS} = \omega_1 = \omega_{mech} = d\gamma/dt$):

$$\begin{bmatrix} u_{II,d} \\ u_{II,q} \end{bmatrix} = \begin{bmatrix} R_2' i_{II,d,0} \\ 0 \end{bmatrix} = \begin{bmatrix} -R_2' \dfrac{U_{1,N}\sqrt{2}}{\omega_1 L_{1m}} \\ 0 \end{bmatrix} \tag{13.65}$$
$$= R_2' \begin{bmatrix} i_{II,d} \\ i_{II,q} \end{bmatrix} + \frac{d}{dt} \begin{bmatrix} \psi_{II,d} \\ \psi_{II,q} \end{bmatrix}$$

According to the space vector theory the torque is calculated like (here in the energy consumption system):

$$T = \frac{3}{2} p \left[\psi_{I,d}\, i_{I,q} - \psi_{I,q}\, i_{I,d} \right] \tag{13.66}$$

The flux linkages are:

$$\begin{bmatrix} \psi_{I,d} \\ \psi_{I,q} \end{bmatrix} = L_1 \begin{bmatrix} i_{I,d} \\ i_{I,q} \end{bmatrix} + L_{1m} \begin{bmatrix} i_{II,d} \\ i_{II,q} \end{bmatrix} \tag{13.67}$$

and

$$\begin{bmatrix} \psi_{II,d} \\ \psi_{II,q} \end{bmatrix} = L_2' \begin{bmatrix} i_{II,d} \\ i_{II,q} \end{bmatrix} + L_{1m} \begin{bmatrix} i_{I,d} \\ i_{I,q} \end{bmatrix} \tag{13.68}$$

or summarized:

$$\begin{bmatrix} \psi_{I,d} \\ \psi_{I,q} \\ \psi_{II,d} \\ \psi_{II,q} \end{bmatrix} = L_{1m} \underbrace{\begin{bmatrix} 1+\sigma_1 & 0 & 1 & 0 \\ 0 & 1+\sigma_1 & 0 & 1 \\ 1 & 0 & 1+\sigma_2 & 0 \\ 0 & 1 & 0 & 1+\sigma_2 \end{bmatrix}}_{[L]} \begin{bmatrix} i_{I,d} \\ i_{I,q} \\ i_{II,d} \\ i_{II,q} \end{bmatrix} \tag{13.69}$$

The currents are calculated by means of the inverted matrix $[L]^{-1}$:

$$\begin{bmatrix} i_{I,d} \\ i_{I,q} \\ i_{II,d} \\ i_{II,q} \end{bmatrix} = \frac{1-\sigma}{\sigma L_{1m}} \underbrace{\begin{bmatrix} 1+\sigma_2 & 0 & -1 & 0 \\ 0 & 1+\sigma_2 & 0 & -1 \\ -1 & 0 & 1+\sigma_1 & 0 \\ 0 & -1 & 0 & 1+\sigma_1 \end{bmatrix}}_{[L]^{-1}} \begin{bmatrix} \psi_{I,d} \\ \psi_{I,q} \\ \psi_{II,d} \\ \psi_{II,q} \end{bmatrix} \tag{13.70}$$

This set of equations together with the torque equation and the flux equations can be solved numerically. For easier programming the flux equations are transformed like follows:

$$\begin{aligned} \frac{d}{dt}\psi_{I,d} &= -R_1\, i_{I,d} + \omega_1\, \psi_{I,q} \\ \frac{d}{dt}\psi_{I,q} &= -R_1\, i_{I,q} - \omega_1\, \psi_{I,d} \\ \frac{d}{dt}\psi_{II,d} &= -R_2' \frac{U_{1,N}\sqrt{2}}{\omega_1 L_{1m}} - R_2'\, i_{II,d} \\ \frac{d}{dt}\psi_{II,q} &= -R_2'\, i_{II,q} \end{aligned} \tag{13.71}$$

The currents are obtained by means of reverse transformation. According to the requirements the excitation current is located in the II, d -axis and the damper current in the II, q -axis; therefore the rotor currents transformed to the stator winding are:

$$\begin{bmatrix} i_F \\ i_D \end{bmatrix} = \begin{bmatrix} i_{II,d} \\ i_{II,q} \end{bmatrix} \tag{13.72}$$

The stator phase currents are calculated like follows (please refer to Sect. 11.3):

$$\begin{gathered} \begin{bmatrix} i_{1,u} \\ i_{1,v} \end{bmatrix} = \left[T_{xy}\right]^{-1} \begin{bmatrix} i_{I,q} \\ i_{I,d} \end{bmatrix}, \\ \left[T_{xy}\right]^{-1} = \begin{bmatrix} \cos(\alpha) & \sin(\alpha) \\ \cos\left(\alpha - \frac{2\pi}{3}\right) & \sin\left(\alpha - \frac{2\pi}{3}\right) \end{bmatrix} \\ i_{1,w} = -i_{1,u} - i_{1,v} \end{gathered} \tag{13.73}$$

With $\alpha(t) = \omega_1 t + \varepsilon$ it follows further:

$$\begin{aligned} i_{1,u} &= i_{I,q} \cos(\omega_1 t + \varepsilon) + i_{I,d} \sin(\omega_1 t + \varepsilon) \\ i_{1,v} &= i_{I,q} \cos\left(\omega_1 t + \varepsilon - \frac{2\pi}{3}\right) + i_{I,d} \sin\left(\omega_1 t + \varepsilon - \frac{2\pi}{3}\right) \\ i_{1,w} &= -i_{1,u} - i_{1,v} \end{aligned} \tag{13.74}$$

Using some approximation (which generally are fulfilled) the set of equations can be solved even analytically. The main advantage is that even qualitative predictions are possible and the influence of the different parameters can be investigated on principle. In the following, this analytical solution is not shown in detail, just the results are given.

The envelopes of the different time-dependent characteristics are:

- current in phase u

$$i_{1,u,min}(t) = -\sqrt{2}\,U_1 \left[\frac{1}{X_1} + \left(\frac{1}{X_{1,stall}} - \frac{1}{X_1} \right) e^{-\frac{t}{\tau_{F,stall}}} + \frac{1}{X_{1,stall}} \sin(\varepsilon)\, e^{-\frac{t}{\tau_{1,stall}}} \right]$$

$$i_{1,u,max}(t) = -\sqrt{2}\,U_1 \left[-\frac{1}{X_1} - \left(\frac{1}{X_{1,stall}} - \frac{1}{X_1} \right) e^{-\frac{t}{\tau_{F,stall}}} + \frac{1}{X_{1,stall}} \sin(\varepsilon)\, e^{-\frac{t}{\tau_{1,stall}}} \right] \tag{13.75}$$

$$i_{1,u,min}(t) \le i_{1,u}(t) \le i_{1,u,max}(t)$$

- permanent short-circuit current (i.e. short-circuit current after subsiding of the transient processes) in phase u

$$i_{1,u,min,perm} = -\frac{\sqrt{2}\,U_1}{X_1}$$
$$i_{1,u,max,perm} = \frac{\sqrt{2}\,U_1}{X_1} \tag{13.76}$$
$$i_{1,u,min,perm} \le i_{1,u,perm} \le i_{1,u,max,perm}$$

- for $t \to \infty$ there is:

$$i_{1,u,min}(t \to \infty) \to i_{1,u,min,perm}; \qquad i_{1,u,max}(t \to \infty) \to i_{1,u,max,perm} \tag{13.77}$$

- for $t = 0$ there is:

$$
\begin{aligned}
i_{1,u,min}(t=0) &= -\frac{\sqrt{2}\,U_1}{X_{1,stall}}\left[1+\sin(\varepsilon)\right] \\
i_{1,u,max}(t=0) &= -\frac{\sqrt{2}\,U_1}{X_{1,stall}}\left[-1+\sin(\varepsilon)\right]
\end{aligned} \tag{13.78}
$$

- excitation current

$$
\begin{aligned}
i_{F,min}(t) &= i_{F,0}\left[1+\frac{1-\sigma}{\sigma}e^{-\frac{t}{\tau_{F,stall}}}-\frac{1-\sigma}{\sigma}e^{-\frac{t}{\tau_{1,stall}}}\right] \\
i_{F,max}(t) &= i_{F,0}\left[1+\frac{1-\sigma}{\sigma}e^{-\frac{t}{\tau_{F,stall}}}+\frac{1-\sigma}{\sigma}e^{-\frac{t}{\tau_{1,stall}}}\right] \\
i_{F,min}(t) &\le i_F(t) \le i_{F,max}(t)
\end{aligned} \tag{13.79}
$$

- damper current

$$
\begin{aligned}
& i_{D,min}(t) = -i_{F,0}\frac{1-\sigma}{\sigma}e^{-\frac{t}{\tau_{1,stall}}}; \qquad i_{D,max}(t) = i_{F,0}\frac{1-\sigma}{\sigma}e^{-\frac{t}{\tau_{1,stall}}} \\
& i_{D,min}(t) \le i_D(t) \le i_{D,max}(t)
\end{aligned} \tag{13.80}
$$

- torque

$$
\begin{aligned}
T_{min}(t) &= -\frac{3p}{\omega_1}\frac{U_1^2}{\sigma X_1}e^{-\frac{t}{\tau_{F,stall}}}e^{-\frac{t}{\tau_{1,stall}}} \\
T_{max}(t) &= \frac{3p}{\omega_1}\frac{U_1^2}{\sigma X_1}e^{-\frac{t}{\tau_{F,stall}}}e^{-\frac{t}{\tau_{1,stall}}} \\
T_{min}(t) &\le T(t) \le T_{max}(t)
\end{aligned} \tag{13.81}
$$

In these equations the time constants and the reactances are:

- no-load time constant of the rotor winding

$$\tau_F = \frac{(1+\sigma_2)L_{1m}}{R_2'} \tag{13.82}$$

- short-circuit (stall) time constant of the rotor winding

$$\tau_{F,stall} = \sigma\tau_F \tag{13.83}$$

- short-circuit (stall) time constant of the stator winding

$$\tau_{1,stall} = \sigma\frac{L_1}{R_1} \tag{13.84}$$

- synchronous reactance

$$X_1 = \omega_1 L_1 \tag{13.85}$$

- short-circuit (stall) reactance

$$X_{1,stall} = \sigma X_1 \tag{13.86}$$

It is obvious from the above equations that the short-circuit values $\tau_{F,stall}$, $\tau_{1,stall}$ and $X_{1,stall}$ determine the transient change between sudden short-circuit and permanent short-circuit.

For high-speed generators the following orders of magnitude are quite usual:

$$\begin{aligned} x_{1,stall} &= \sigma x_1 = \sigma\frac{X_1}{U_N/I_N} = 0.15\cdots 0.25 \\ \tau_{F,stall} &= 0.5\cdots 2\text{s} \\ \tau_{1,stall} &= 60\cdots 250\text{ms} \end{aligned} \tag{13.87}$$

Now the time-dependent characteristics of stator current (in phase U), excitation current, damper current, and torque shall be illustrated. For this a high-speed generator with the following data is calculated:

$$\begin{aligned} &U_{1,N} = 11\text{kV}, \quad && I_{1,N} = 758\text{A}, \quad && P_N = 20\text{MW} \\ &X_1 = 21.77\Omega, \quad && \tau_{F,\text{stall}} = 0.2\text{s}, \quad && \tau_{1,\text{stall}} = 104\text{ms} \\ &T_N = 127\text{kNm} && && \end{aligned} \tag{13.88}$$

13.3.4 Maximum Voltage Switching

As it can be deduced from the above equations for the enveloping characteristics, the moment of switching (for maximum voltage switching there is $\varepsilon = 0$) has no influence onto the maximum values of torque, excitation current, and damper current. However, there is an influence onto the phase current. In the following Figs. 13.2 to 13.6 all currents are given in kA, the torque is given in kNm and the time in s.

Figure 13.2 shows the characteristic of the stator current in phase u (red) together with its medium value (magenta), the envelopes of the sudden short-circuit current (black dotted) and the envelopes of the permanent short-circuit current (blue dotted). The maximum value of the permanent short-circuit current is

$$i_{1,u,\max,\text{perm}} = \frac{\sqrt{2}\, U_1}{X_1} \tag{13.89}$$

and this is for the considered machine 715A (at a nominal value of 758A). The current for sudden short-circuit is

$$i_{1,u,\max}(t=0) = -\frac{\sqrt{2}\, U_1}{X_{1,\text{stall}}}\left[-1+\sin(\varepsilon)\right] \tag{13.90}$$

and this is for the considered machine for maximum voltage switching ($\varepsilon = 0$) 4764A (a factor of $1/\sigma$ larger than the permanent short-circuit current). Consequently the sudden short-circuit current is several times larger than the nominal current. The mean value of the current for maximum voltage switching is zero.

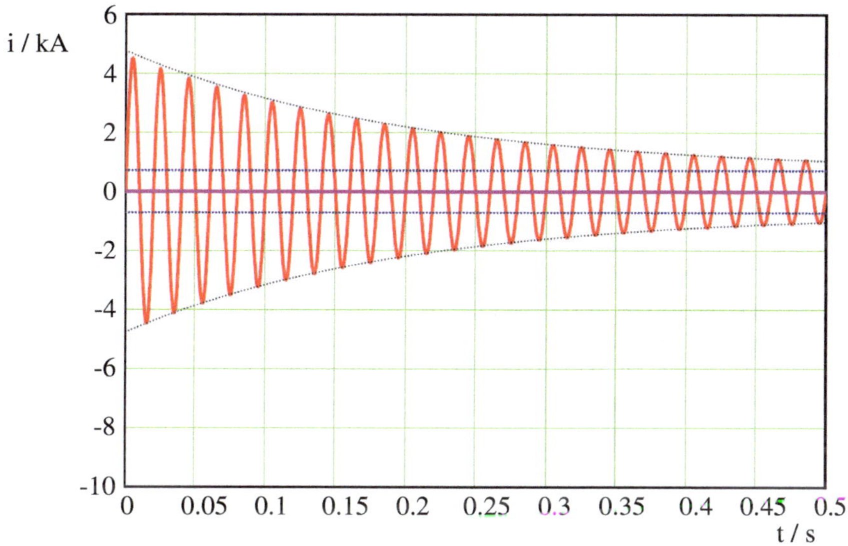

Fig. 13.2. Stator current in phase u (*red*) together with its medium value (*magenta*), the envelopes of the sudden short-circuit current (*black dotted*) and the envelopes of the permanent short-circuit current (*blue dotted*).

The time dependent characteristics of torque, excitation current, and damper current are shown in Figs. 13.3 to 13.6 (in red each); the respective enveloping functions are shown as dotted black characteristics.

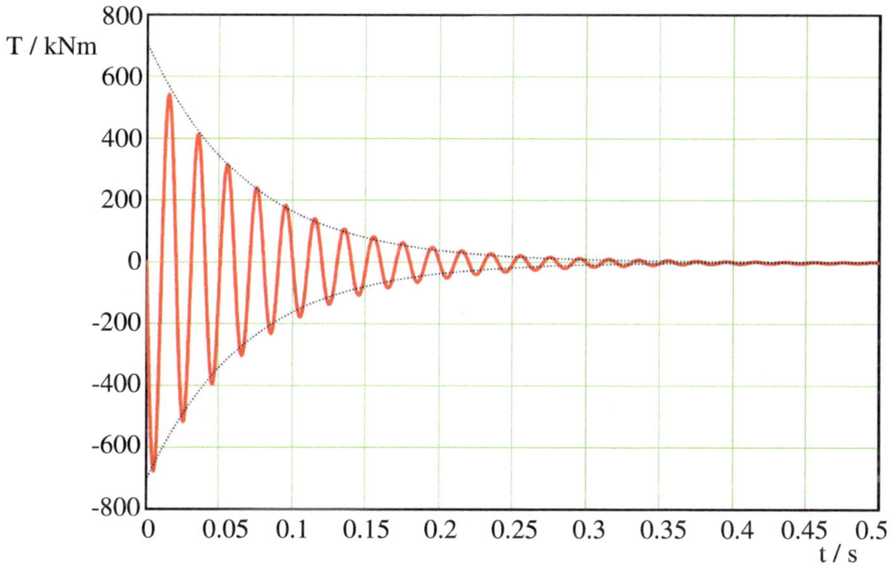

Fig. 13.3. Torque (*red*) together with the envelopes of the sudden short-circuit torque (*black dotted*).

A torque is generated that oscillates with the mains frequency and that subsides with time. The maximum value is

$$T_{max}(t=0) = \frac{3p}{\omega_1}\frac{U_1^2}{\sigma X_1} \qquad (13.91)$$

and for the regarded generator amounts to 707.7kNm (at a nominal torque of 127kNm; i.e. about a factor of 5.6 higher than the nominal torque). When constructing synchronous machines this high mechanical load has to be considered.

The permanent short-circuit stator currents generate losses in the stator resistances. These losses have to be covered by the driving torque (the power loss has to be equalized by the input power). As the speed has been assumed being constant, the acceleration torque is zero. Therefore, the torque balance gives

$$T = -T_{ext} \qquad (13.92)$$

As the external torque has to cover the losses, it follows

$$T_{ext} = \frac{P_{loss}}{2\pi n} = \frac{3R_1\left(i_{1,u,max,perm}\right)^2}{2\pi n} \qquad (13.93)$$

and consequently (inserting the numbers from above)

$$T = -\frac{3R_1\left(i_{1,u,max,perm}\right)^2}{2\pi n} \approx -488\text{Nm} \qquad (13.94)$$

This low torque cannot be recognized in the above Fig. 13.3. But evaluating the same simulation like before with a different scale of the vertical axis results in Fig. 13.4, which supports the above calculation.

The currents $i_{I,d}$, $i_{I,q}$, $i_{II,d}$, $i_{II,q}$ are independent from the moment of switching ε. Therefore, even the time-dependent characteristics of the torque (Figs. 13.3 and 13.4) and the time-dependent characteristics of the rotor currents (excitation current in Fig. 13.5 and damper current in Fig. 13.6) are independent from the moment of switching. The excitation current as well as the damper current reach multiples of their nominal values during sudden short-circuit (similar to the phase current).

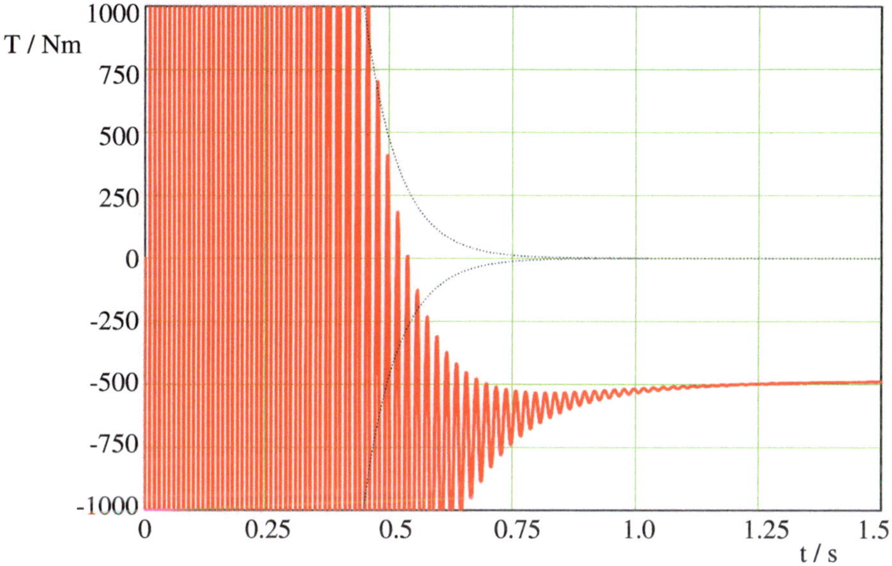

Fig. 13.4. Torque (*red*) together with the envelopes of the sudden short-circuit torque (*black dotted*); zoomed view compared to Fig. 13.3 concerning amplitude and time.

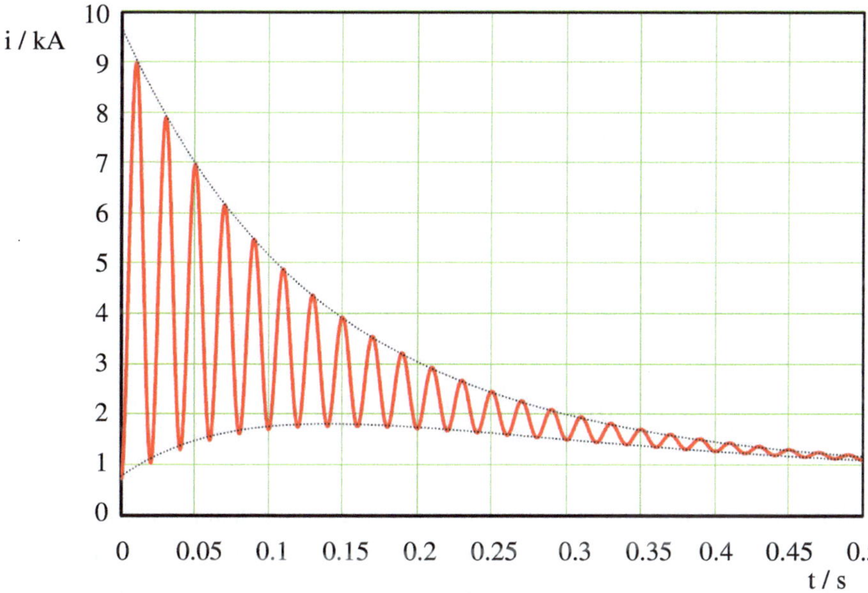

Fig. 13.5. Excitation current (*red*) together with the envelopes of the sudden short-circuit excitation current (*black dotted*).

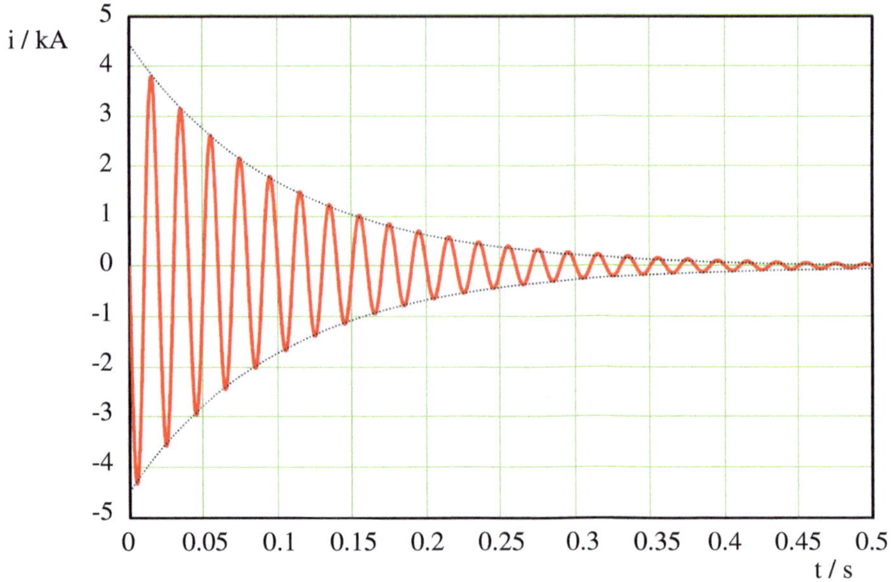

Fig. 13.6. Damper current (*red*) together with the envelopes of the sudden short-circuit damper current (*black dotted*).

13.3.5 Zero Voltage Switching

In contrast to the rotor currents and the torque the stator phase current depends on the moment of switching ε. Switching at a time instant where the voltage is zero ($\varepsilon = \pi/2$), the minimal and maximal values of the sudden short-circuit current are

$$\begin{aligned} i_{1,u,min}(t=0) &= -\frac{\sqrt{2}\,U_1}{X_{1,stall}}\left[1+\sin(\varepsilon)\right] = -2\frac{\sqrt{2}\,U_1}{X_{1,stall}} \\ i_{1,u,max}(t=0) &= -\frac{\sqrt{2}\,U_1}{X_{1,stall}}\left[-1+\sin(\varepsilon)\right] = 0 \end{aligned} \tag{13.95}$$

Comparing the amplitudes of the extreme values of the stator phase current for $t=0$, it is true that for $\varepsilon = \pi/2$ the double is reached than for the case $\varepsilon = 0$: 9528A (factor $2/\sigma$ higher than the permanent short-circuit current). In this case the sudden short-circuit current for $\varepsilon = \pi/2$ excesses the nominal value (758A) by more than a factor of 12.

The mean value of the phase current for $t = 0$ is $\dfrac{-\sqrt{2}\ U_1}{X_{1,\text{stall}}}$ and subsides to zero with the short-circuit time constant of the stator winding $\tau_{1,\text{stall}}$ (see Fig. 13.7).

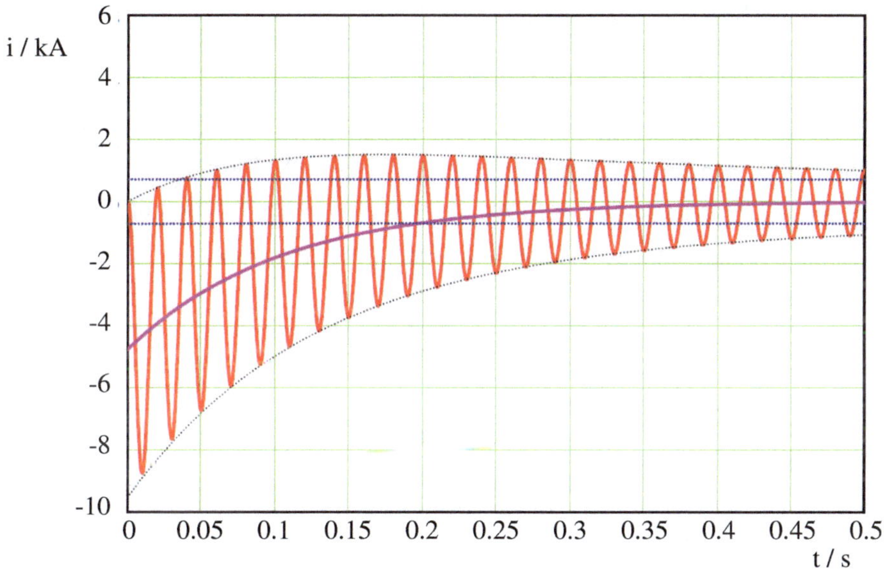

Fig. 13.7. Stator current in phase u (*red*) together with its medium value (*magenta*), the envelopes of the sudden short-circuit current (*black dotted*) and the envelopes of the permanent short-circuit current (*blue dotted*).

13.3.6 Sudden Short-Circuit with Changing Speed and Rough Synchronization

So far the sudden short-circuit of the non salient-pole machine has been calculated assuming that the speed is constant. This assumption shall be retracted now. For the following calculation it is assumed that:

- At the time $t = 0\text{s}$ the machine operates with synchronous speed.
- At the time $t = 0\text{s}$ the machine will be short-circuited, zero voltage switching for phase u ($\varepsilon = \pi/2$) is assumed (please refer to the preceding section).
- The external driving torque T_{ext} shall be zero for the entire regarded time period.

- The speed decreases after switching on the short-circuit, because the short-circuit currents generate losses in the Ohmic resistances.
- The slope of the speed decrease depends on the inertia; this is assumed being $\Theta = 2.5 \cdot 10^3 \, \mathrm{kgm}^2$.
- At the time $t = 0.2\mathrm{s}$ the machine is switched to the mains again (still with $T_{ext} = 0$). As generally not all synchronizing conditions (please refer to Sect. 5.3) are fulfilled, a rough synchronization happens.

The stator terminal voltages in d-q-representation *after* the rough synchronization are time-dependent, because there is a time-dependent instantaneous angle ς between the q-axis (rotating with $\omega_{mech} = d\gamma/dt$) and the location of the voltage $\underline{U}_1$ (which rotates with ω_1): $d\varsigma/dt = \omega_1 - \omega_{mech}$ (see Fig. 13.8).

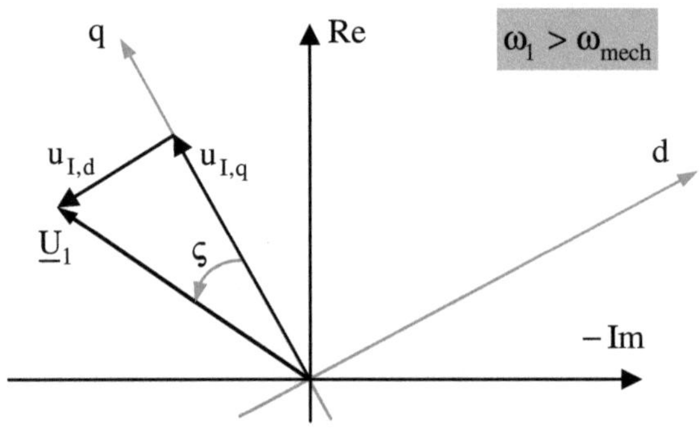

Fig. 13.8. Terminal voltages in d-q-representation after rough synchronization.

$$\begin{aligned}\begin{bmatrix} -u_{I,d} \\ -u_{I,q} \end{bmatrix} &= \begin{bmatrix} U_{1,N}\sqrt{2}\sin(\varsigma+\varepsilon) \\ -U_{1,N}\sqrt{2}\cos(\varsigma+\varepsilon) \end{bmatrix} \\ &= R_1 \begin{bmatrix} i_{I,d} \\ i_{I,q} \end{bmatrix} + \frac{d}{dt}\begin{bmatrix} \psi_{I,d} \\ \psi_{I,q} \end{bmatrix} + \frac{d\gamma}{dt}\begin{bmatrix} -\psi_{I,q} \\ \psi_{I,d} \end{bmatrix} \\ &= R_1 \begin{bmatrix} i_{I,d} \\ i_{I,q} \end{bmatrix} + \frac{d}{dt}\begin{bmatrix} \psi_{I,d} \\ \psi_{I,q} \end{bmatrix} + \omega_{mech}\begin{bmatrix} -\psi_{I,q} \\ \psi_{I,d} \end{bmatrix}\end{aligned} \tag{13.96}$$

It follows:

$$\frac{d}{dt}\begin{bmatrix} \psi_{I,d} \\ \psi_{I,q} \end{bmatrix} = -R_1 \begin{bmatrix} i_{I,d} \\ i_{I,q} \end{bmatrix} + \omega_{mech}\begin{bmatrix} \psi_{I,q} \\ -\psi_{I,d} \end{bmatrix} + U_{1,N}\sqrt{2}\begin{bmatrix} \sin(\varsigma+\varepsilon) \\ -\cos(\varsigma+\varepsilon) \end{bmatrix} \tag{13.97}$$

Figures 13.9 to 13.13 on the one hand show the influence of the now time-dependent speed, on the other hand the impact of the rough synchronization at the time $t = 0.2\text{s}$ is illustrated (like before the currents are represented in kA, the torque in kNm; the scale of the speed axis is in min^{-1}).

Figure 13.9 shows the time-dependent characteristic of the phase current. The small deviations compared to the respective figure in Sect. 13.3.5 (Fig. 13.7) in the first 0.2s, provoked by the slightly decreasing speed (time-dependent characteristic of the speed see Fig. 13.13), are only noticeable by detailed analysis. The rough synchronization again evokes a transient operation with remarkable current oscillations. As the machine is supplied by the mains voltage after this rough synchronization, the final steady-state value of the phase current is no longer the permanent short-circuit current, but it is equal to zero (because for these calculations the machine was assumed being unloaded).

Even for the torque there is a transient in the time-dependent characteristic at time $t = 0.2\text{s}$ coming from the rough synchronization, see Fig. 13.10. It is obvious that the torque is negative directly after the short circuit ($t = 0\text{s}$), directly after the rough synchronization ($t = 0.2\text{s}$) it is positive. Of course this depends on the relative position between the rotating stator field and the rotor at the time instant of rough synchronization.

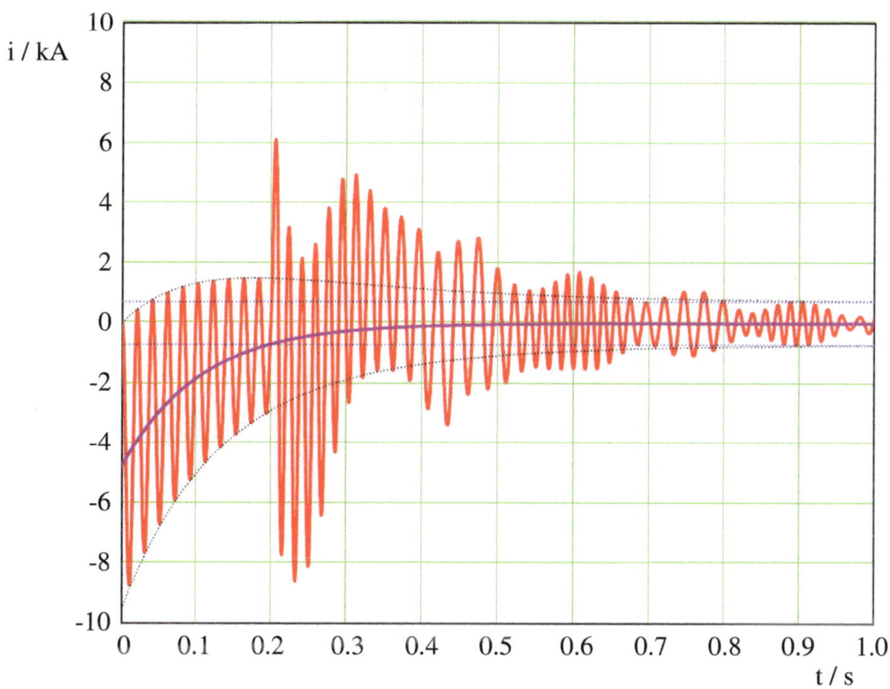

Fig. 13.9. Stator current in phase u (*red*) together with its medium value (*magenta*), the envelopes of the sudden short-circuit current (*black dotted*) and the envelopes of the permanent short-circuit current (*blue dotted*).

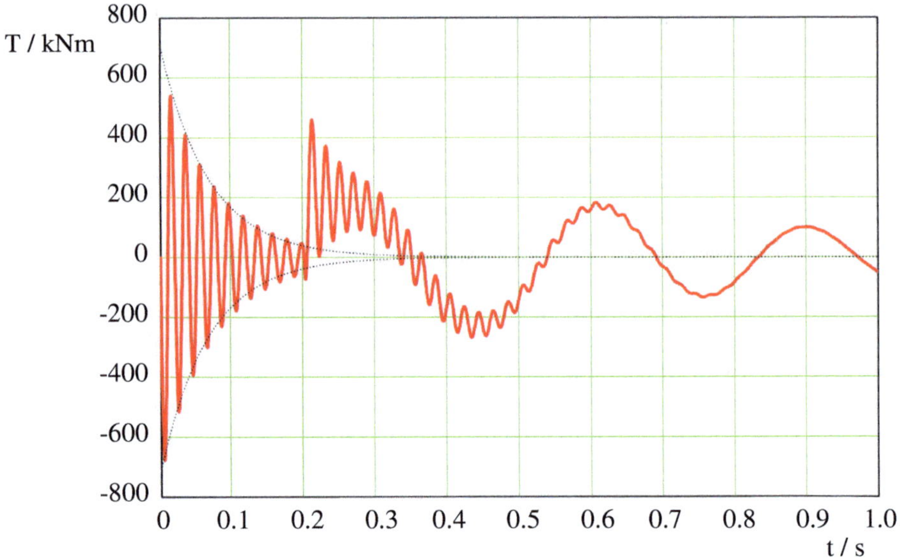

Fig. 13.10. Torque (*red*) together with the envelopes of the sudden short-circuit torque (*black dotted*).

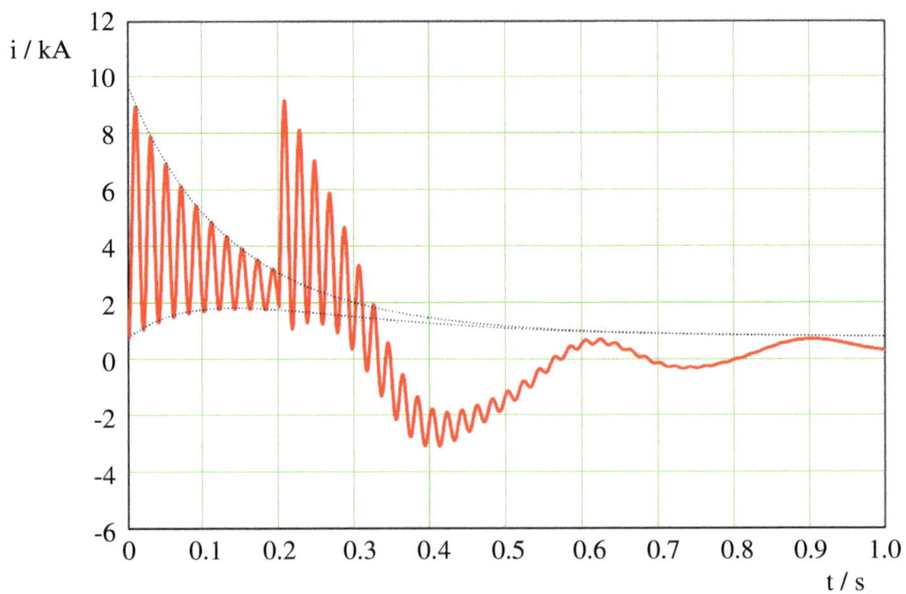

Fig. 13.11. Excitation current (*red*) together with the envelopes of the sudden short-circuit excitation current (*black dotted*).

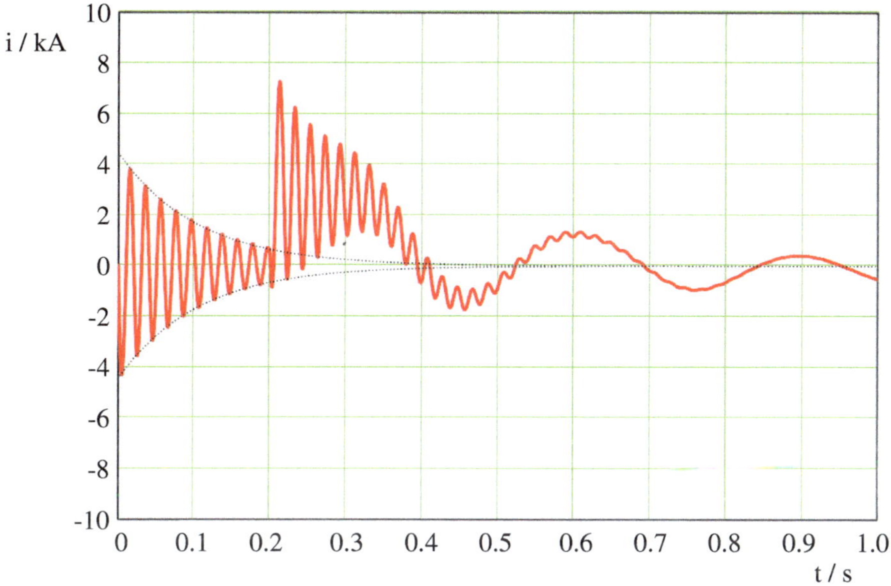

Fig. 13.12. Damper current (*red*) together with the envelopes of the sudden short-circuit damper current (*black dotted*).

Even the excitation current (Fig. 13.11) shows the transient effect after rough synchronization, the final steady-state value is the respective nominal value. The maximum value of the excitation current after the rough synchronization has about the same value like after the short-circuit.

The amplitude of the damper current after rough synchronization is even a factor of 1.5 larger than after the short-circuit; the final steady-state value is zero for each transient operation, see Fig. 13.12.

Finally the speed-time-characteristic is shown in Fig. 13.13 (the synchronous speed for this machine is $1{,}500\text{min}^{-1}$). This speed-time-characteristic correlates with the torque-time-characteristic shown in Fig. 13.10: after the short-circuit the torque acts decelerating, after the rough synchronization it accelerates the rotor.

During short-circuit (0s to 0.2s) the mean value of the speed in decreased, because the phase currents generate losses in the Ohmic resistances; this energy is taken from the reduction of rotational kinetic energy of the rotor (friction was neglected). After the rough synchronization the machine is accelerated again and the speed oscillates transiently to the synchronous speed; this is done by means of the damper cage analogously to the operation of an induction machine. As in this simulation the entire load torque was assumed being zero (and therefore friction was neglected), the machine reaches the synchronous speed.

In reality the synchronous speed has to be adjusted by increasing the external driving torque (compensation of friction and other load torques) and stabilized by control (avoiding the overshoot).

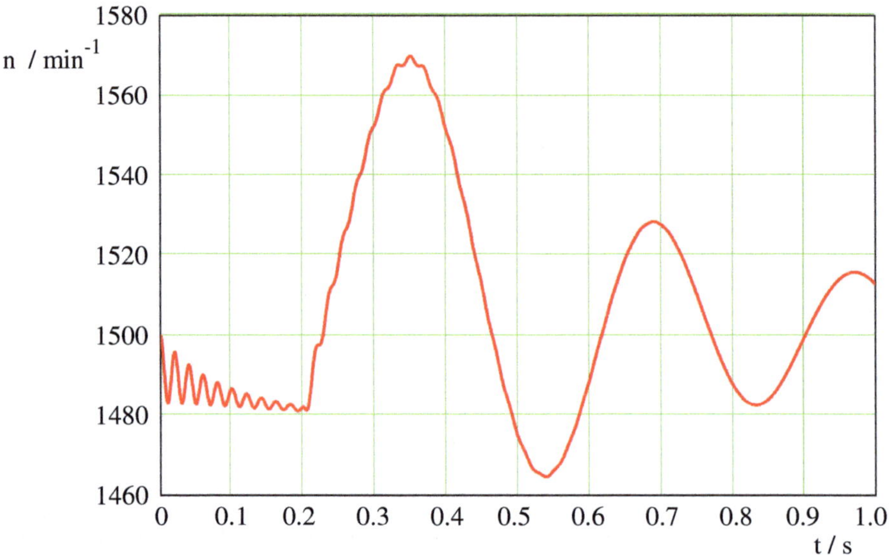

Fig. 13.13. Speed-time-characteristic.

13.3.7 Physical Explanation of the Sudden Short-Circuit

In this section the physical conditions during sudden short-circuit of a synchronous machine will be explained in principle.

- During no-load operation the excited and rotating rotor generates a rotating flux distribution and consequently sinusoidal terminal voltages ($u_{1,u} = \sqrt{2}\, U \sin(\omega t)$); this is illustrated in Fig. 13.14.

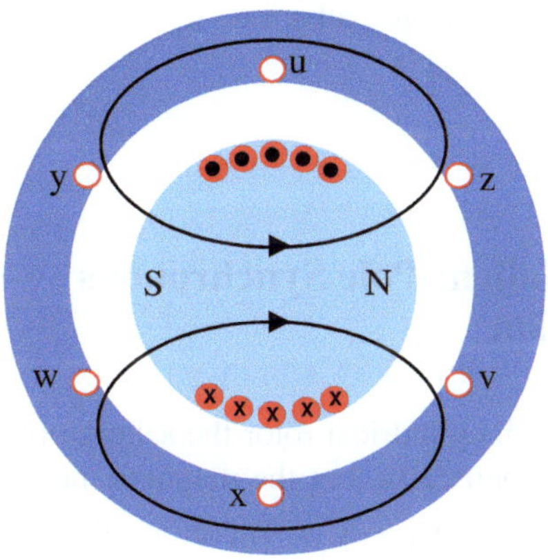

Fig. 13.14. Principle situation just after the sudden short-circuit.

- If the terminals of the machine are short-circuited, $u \equiv 0$ is true. Neglecting the Ohmic resistances ($R_1 = 0$) the stator flux linkage becomes: $d\psi/dt = 0$, therefore $\psi = \text{const.}$
- The stator flux linkage remains constant because of the short-circuited terminals, therefore the situation shown in Fig. 13.15 is valid after half a period:

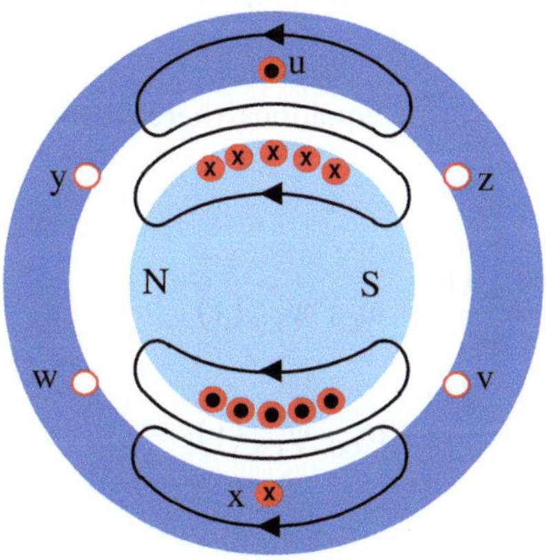

Fig. 13.15. Principle situation half a period after the sudden short-circuit.

The flux is pushed to a magnetically poor conducting leakage path, i.e. the stator winding has to deliver a MMF (i.e. a respective current must be conducted) that is able to drive the flux even along the leakage path. This demonstrates why the current has to be larger by a factor of $1/\sigma$ compared to the permanent short-circuit (during permanent short-circuit the flux is conducted across stator and rotor again and no longer along the leakage paths, i.e. L_1 is relevant and no longer σL_1).

13.4 Steady-State Operation of Salient-Pole Synchronous Machines in Space Vector Notation

In contrast to the synchronous machine with cylindrical rotor the salient-pole synchronous machine is not constructed symmetrically, but the magnetic permeance is different in direct axis and quadrature axis ($X_d > X_q$). Therefore, the transformation may not be done arbitrarily, but it has to be oriented to the asymmetric part (rotor): only then constant mutual inductivities are obtained.

Figure 13.16 is taken from Sect. 5.5. Here the replaced winding system of the salient-pole synchronous machine is designed in such a way that all windings are located in the d-axis (direct axis, axis of the excitation current) or perpendicular to this (q-axis, quadrature axis). Voltages and currents are shown here for the energy consumption system.

Additionally it shall now be considered that a damper winding is present in the rotor having a direct component with the current i_D and a quadrature component with the current i_Q.

Analogously to Sect. 13.2 the following set of equations (here in the energy generation system) is obtained with $\omega_{CS} = \omega_{mech}$:

$$\begin{aligned} -u_{I,q}(t) &= R_1\, i_{I,q}(t) + \frac{d\psi_{I,q}(t)}{dt} + \omega_{CS}\, \psi_{I,d}(t) \\ -u_{I,d}(t) &= R_1\, i_{I,d}(t) + \frac{d\psi_{I,d}(t)}{dt} - \omega_{CS}\, \psi_{I,q}(t) \end{aligned} \tag{13.98}$$

$$\begin{aligned} 0 &= R_D\, i_D(t) + \frac{d\psi_D(t)}{dt} \\ 0 &= R_Q\, i_Q(t) + \frac{d\psi_Q(t)}{dt} \\ -u_F(t) &= R_F\, i_F(t) + \frac{d\psi_F(t)}{dt} \end{aligned} \tag{13.99}$$

$$T(t) = -\frac{3}{2} p\left[\psi_{I,d}(t) i_{I,q}(t) - \psi_{I,q}(t) i_{I,d}(t)\right] \tag{13.100}$$

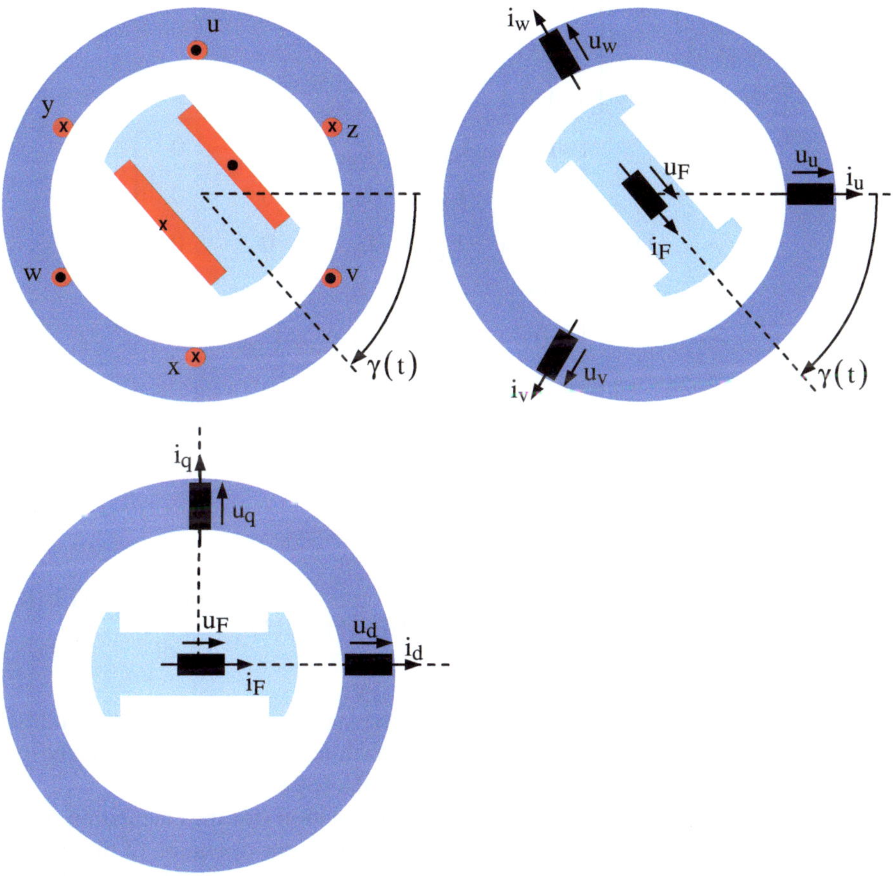

Fig. 13.16. Sketch of the salient-pole synchronous machine: original system (*above left*), replacement (*above right*) and two-phase replacement (*below*).

Setting up the equations for the flux linkages the correct consideration of the sign is important. In the direct axis the stator winding and the excitation winding are magnetizing in the same direction, if they conduct positive current. However the damper winding magnetizes in opposite direction. In the quadrature axis the stator winding and the damper winding are magnetizing in opposite direction. There is no magnetic coupling between the direct axis and the quadrature axis. The following matrix equation is obtained for the flux linkages:

$$\begin{bmatrix} \psi_{I,d}(t) \\ \psi_F(t) \\ \psi_D(t) \\ \psi_{I,q}(t) \\ \psi_Q(t) \end{bmatrix} = [L] \begin{bmatrix} i_{I,d}(t) \\ i_F(t) \\ i_D(t) \\ i_{I,q}(t) \\ i_Q(t) \end{bmatrix}$$

$$[L] = \begin{bmatrix} L_d & L_{dF} & -L_{dD} & 0 & 0 \\ L_{dF} & L_F & -L_{DF} & 0 & 0 \\ -L_{dD} & -L_{DF} & L_D & 0 & 0 \\ 0 & 0 & 0 & L_q & -L_{qQ} \\ 0 & 0 & 0 & -L_{qQ} & L_Q \end{bmatrix} \tag{13.101}$$

The external driving torque T_{ext} has to overcome the electrodynamic torque T of the machine and the acceleration torque T_{acc}:

$$\begin{aligned} T_{ext}(t) &= T(t) + T_{acc}(t) \\ &= -\frac{3}{2} p \left[\psi_{I,d}(t) i_{I,q}(t) - \psi_{I,q}(t) i_{I,d}(t) \right] + \frac{\Theta}{p} \frac{d^2\gamma}{dt^2} \end{aligned} \tag{13.102}$$

With the equations for the flux linkages, the stator voltage equations, and the torque equation the salient-pole synchronous machine is described completely in a coordinate system rotating in synchronism with the rotor.

Starting from these equations in space vector notation (for the calculation of the dynamic behavior) firstly as a special case the steady-state operation of the salient-pole synchronous machine is regarded. In *steady-state operation* there is no time-dependent change of the flux linkages ($d\psi/dt = 0$) and the speed is constant ($d\gamma/dt = \omega_{mech} = \text{const.}$). In addition, in the following the stator resistance will be neglected ($R_1 = 0$). Then the set of equations becomes:

$$\begin{aligned} u_{I,q}(t) &= -\omega_{CS}\ \psi_{I,d}(t) \\ u_{I,d}(t) &= \omega_{CS}\ \psi_{I,q}(t) \end{aligned} \tag{13.103}$$

$$\begin{aligned} 0 &= i_D(t) \\ 0 &= i_Q(t) \\ u_F(t) &= -R_F\ i_F(t) \end{aligned} \tag{13.104}$$

$$T_{ext}(t) = T(t) = -\frac{3}{2} p\left[\psi_{I,d}(t) i_{I,q}(t) - \psi_{I,q}(t) i_{I,d}(t)\right] \tag{13.105}$$

$$\begin{aligned} \psi_{I,d}(t) &= L_d\ i_{I,d}(t) + L_{dF}\ i_F(t) \\ \psi_{I,q}(t) &= L_q\ i_{I,q}(t) \end{aligned} \tag{13.106}$$

All currents and voltages are now DC values; therefore in the following an explicit time-dependency is to be omitted. The other flux linkages are not interesting any longer. Inserting the flux linkages gives (with $\omega_{CS} = \omega_{mech}$):

$$\begin{aligned} u_{I,d} &= \omega_{mech}\ L_q\ i_{I,q} \\ u_{I,q} &= -\omega_{mech}\left(L_d\ i_{I,d} + L_{dF}\ i_F\right) \\ u_F &= -i_F\ R_F \\ T &= \frac{3}{2}\frac{p}{\omega_{mech}}\left(i_{I,d}\ u_{I,d} + i_{I,q}\ u_{I,q}\right) \end{aligned} \tag{13.107}$$

Using an analogous calculation like in Sect. 13.2 the reverse transformation is performed; for the rotating coordinate system the angle $\alpha(t)$ is chosen as:

$$\alpha(t) = \omega_1 t + \alpha_0, \qquad \alpha_0 = \vartheta \tag{13.108}$$

$$\begin{aligned} u_{1,u}(t) &= u_{I,q}(t)\cos(\alpha) + u_{I,d}(t)\sin(\alpha) \\ &= u_{I,q}(t)\cos(\omega_1 t + \vartheta) + u_{I,d}(t)\sin(\omega_1 t + \vartheta) \end{aligned} \tag{13.109}$$

and further (analogous to the calculation in Sect. 13.2):

$$\underline{U}_1 = \frac{1}{\sqrt{2}}\left[j\,\omega_1 \left(j\,L_d\; i_{I,d} - L_q\; i_{I,q} \right) e^{j\vartheta} + j\,\omega_1\; L_{dF} \left(j\,i_F \right) e^{j\vartheta} \right] \quad (13.110)$$

With the definitions from Sect. 13.2

$$\underline{I}_{1,q} = \frac{i_{I,q}}{\sqrt{2}} e^{j\vartheta} \quad (13.111)$$

$$\underline{I}_{1,d} = -j \frac{i_{I,d}}{\sqrt{2}} e^{j\vartheta} \quad (13.112)$$

$$\begin{aligned} \underline{U}_P &= -j\omega_1 L_{dF} \left(-j \frac{i_F}{\sqrt{2}} e^{j\vartheta} \right) \\ &= -\omega_1 L_{dF} \frac{i_F}{\sqrt{2}} e^{j\vartheta} \end{aligned} \quad (13.113)$$

the stator voltage equation becomes

$$\underline{U}_1 = -j\,\omega_1\; L_d\; \underline{I}_{1,d} - j\,\omega_1\; L_q\; \underline{I}_{1,q} + \underline{U}_P \quad (13.114)$$

Omitting the index “1“, it follows further with the direct axis reactance $X_d = \omega L_d$ and the quadrature axis reactance $X_q = \omega L_q$:

$$\underline{U} + jX_q \underline{I}_q + jX_d \underline{I}_d = \underline{U}_P \quad (13.115)$$

With this the phasor diagram of the salient-pole synchronous machine can be drawn (please compare to Sect. 5.5), see Fig. 13.17. A transformation of the voltage equation delivers:

$$\begin{aligned} &\underline{U} + jX_q \underline{I}_q + jX_d \underline{I}_d = \underline{U}_P, \qquad \underline{I} = \underline{I}_d + \underline{I}_q \\ \Rightarrow\quad &\underline{U} + jX_q \underline{I} + j\left(X_d - X_q \right) \underline{I}_d = \underline{U}_P \end{aligned} \quad (13.116)$$

This leads to the next phasor diagram (Fig. 13.18) that often is used in practice.

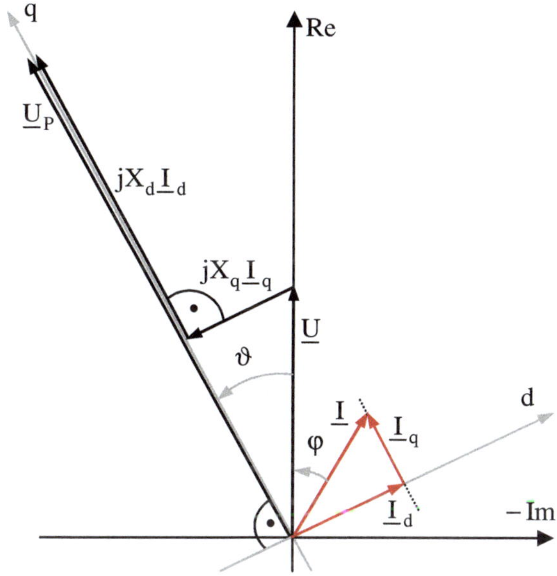

Fig. 13.17. Phasor diagram of salient-pole synchronous machines.

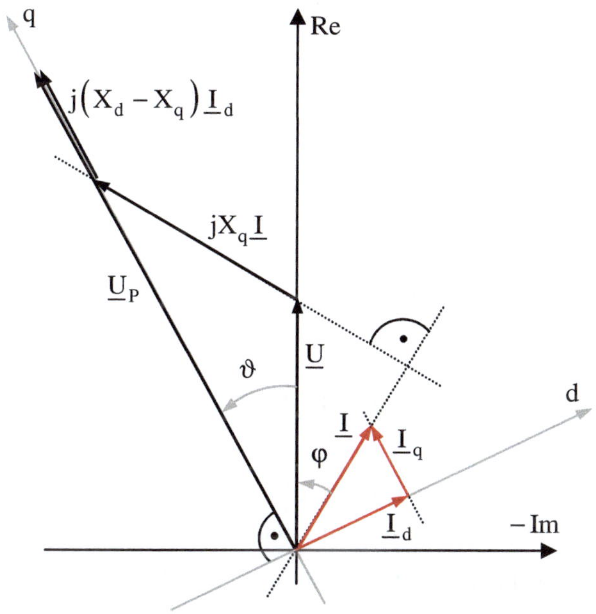

Fig. 13.18. Alternative phasor diagram of salient-pole synchronous machines.

This phasor diagram can be developed from the machine data like follows:

- Given are U, I, and φ, e.g. as measurement values.
- The voltage $\underline{U} = U$ is chosen being real.
- The current $\underline{I}$ is drawn considering the phase angle φ.
- Perpendicular to this phasor $\underline{I}$ the phasor $jX_q\underline{I}$ is drawn with its base at the top of the phasor $\underline{U} = U$.
- The direction of the internal voltage $\underline{U}_P$ is defined by the origin of the coordinate system and the tip of the phasor $jX_q\underline{I}$. With this even the angle ϑ and the direction of the q-axis are fixed.
- Perpendicular to the q-axis the d-axis is located. Now the current $\underline{I}$ can be divided into the components $\underline{I}_d$ and $\underline{I}_q$.
- At the tip of the phasor $jX_q\underline{I}$ the base of the phasor $j\left(X_d - X_q\right)\underline{I}_d$ is located; by this the value of the internal voltage $\underline{U}_P$ is known (base of $\underline{U}_P$ in the origin, tip of $\underline{U}_P$ identical with the tip of the phasor $j\left(X_d - X_q\right)\underline{I}_d$).

The salient-pole synchronous machine differs from the synchronous machine with cylindrical rotor by the different reactances in d- and q-axis. For $X_d = X_q = X$ the above given equations and phasor diagrams deliver the respective description of the synchronous machine with cylindrical rotor.

For the torque the equation

$$T = \frac{3}{2}\frac{p}{\omega}\left(i_d\ u_d + i_q\ u_q\right) \tag{13.117}$$

holds true. Inserting the rms-values I_d, U_d, I_q, U_q instead of the DC values, it follows:

$$T = \frac{3p}{\omega}\left(I_d\ U_d + I_q\ U_q\right) \tag{13.118}$$

Consequently, the active power of direct axis and quadrature axis are added.

From the above phasor diagram the following relations can be deduced (U_d and U_q come from the segmentation of U according to d- and q-axis):

$$I_d = \frac{U_P - U\cos(\vartheta)}{X_d}, \qquad U_d = U\sin(\vartheta) \tag{13.119}$$

$$I_q = \frac{U\sin(\vartheta)}{X_q}, \qquad U_q = U\cos(\vartheta) \tag{13.120}$$

Inserting this, the torque becomes:

$$\begin{aligned} T &= \frac{3p}{\omega}\left(I_d\, U_d + I_q\, U_q\right) \\ &= \frac{3p}{\omega}\left(\frac{U_P - U\cos(\vartheta)}{X_d} U\sin(\vartheta) + \frac{U\sin(\vartheta)}{X_q} U\cos(\vartheta)\right) \\ &= \frac{3p}{\omega}\left(\frac{U_P U}{X_d}\sin(\vartheta) + U^2\left(\frac{1}{X_q} - \frac{1}{X_d}\right)\cos(\vartheta)\sin(\vartheta)\right) \\ &= \frac{3p}{\omega}\left(\frac{U U_P}{X_d}\sin(\vartheta) + \frac{U^2}{2}\left(\frac{1}{X_q} - \frac{1}{X_d}\right)\sin(2\vartheta)\right) \end{aligned} \tag{13.121}$$

This result is already known from Chap. 5: the first summand corresponds to the torque of the non salient-pole synchronous machine and it depends on the excitation, the second summand is the so-called reaction torque (reluctance torque) and it is *not dependent on the excitation* (it is merely generated by the difference in the magnetic permeance in d- and q-axis). Because of this reluctance torque the pull-out torque is reached at a rotor angle less than $\pi/2$, see Fig. 13.19 (and Fig. 5.23). This figure shows the ratio of torque and pull-out torque ($T_{ratio} = T/T_{pull-out}$) versus the rotor angle ϑ for different excitations U_P/U:

- red: $U_P/U = 0$
- blue: $U_P/U = 0.5$
- green: $U_P/U = 1.0$
- magenta: $U_P/U = 2.0$
- black: torque of the non-salient-pole synchronous machine for $U_P/U = 1.0$ (for comparison reasons)

In addition to the shift of the pull-out torque to smaller rotor angles it can be deduced that the salient-pole synchronous machine delivers a higher pull-out torque for the same excitation current, because of the additional reluctance torque component (assumed is $X_{d,\text{salient-pole}} = X_{\text{non-salient-pole}}$). In addition it becomes obvious that for the assumed relation $X_d = 2X_q$ the pull-out torque without excitation (i.e. the reluctance pull-out) is just half of the pull-out torque of the non-salient-pole machine.

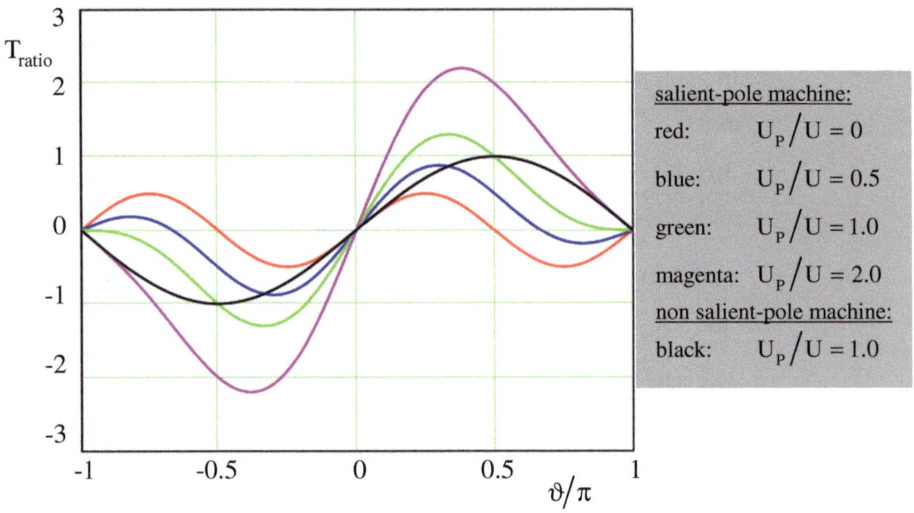

Fig. 13.19. Ratio of torque and pull-out torque versus rotor angle for different excitations.

13.5 Sudden Short-Circuit of Salient-Pole Synchronous Machines

13.5.1 Initial Conditions for t = 0

With some approximations the sudden short-circuit of the salient-pole synchronous machine can be calculated analytically. As there is a considerable calculation effort, in the following only the numerical solution is regarded.

Firstly the initial conditions have to be determined. The situation before the sudden short-circuit shall be the no-load operation with nominal excitation (this means that the excitation winding is supplied by its nominal voltage and the rotor rotates with synchronous speed). The values before the switching moment are nominated in the following with the additional index "0". The initial conditions then are:

- The stator currents are zero, because the terminals are not connected; the damper current is zero, because there is a steady-state operation at synchronous speed.

$$i_{d,0} = i_{q,0} = i_{D,0} = i_{Q,0} = 0 \tag{13.122}$$

- The stator voltages are (calculation in the energy generation system):

$$\begin{aligned} u_{d,0} &= \omega\psi_{q,0} = \omega L_q i_{q,0} = 0 \\ u_{q,0} &= -\omega\psi_{d,0} = -\omega\left(L_d i_{d,0} + L_{dF} i_{F,0}\right) = -\omega L_{dF} i_{F,0} = \sqrt{2}\, U_{1,N} \end{aligned} \tag{13.123}$$

- From the last equation it follows:

$$i_{F,0} = -\frac{\sqrt{2}\, U_{1,N}}{\omega L_{dF}} \tag{13.124}$$

- In addition it is true:

$$\begin{aligned} u_{F,0} &= R_F\, i_{F,0} \\ T_{ext,0} &= \frac{3}{2}\frac{p}{\omega_{mech}}\left(i_{d,0}\, u_{d,0} + i_{q,0}\, u_{q,0}\right) = 0 \end{aligned} \tag{13.125}$$

Similar to the calculation of the non salient-pole machine the angle $\alpha(t) = \omega_1 t + \varepsilon$ is chosen. Then it follows by means of reverse transformation:

$$\begin{aligned} u_{1,u,0} &= u_{I,q,0} \cos(\alpha) + u_{I,d,0} \sin(\alpha) \\ &= \sqrt{2}\, U_{1,N} \cos(\alpha) + 0 \\ &= \sqrt{2}\, U_{1,N} \cos(\omega_1 t + \varepsilon) \end{aligned}$$

The angle ε therefore characterizes the moment of switching (like for the calculation of the non salient-pole machine):

- For $\varepsilon = 0$ and time $t = 0$ the flux linkage in phase u is zero, i.e. the peak value of the voltage is induced.
- For $\varepsilon = \pi/2$ and time $t = 0$ the flux linkage in phase u is maximum, i.e. the induced voltage is zero.

13.5.2 Set of Equations for t > 0

Transforming the set of equations of the salient-pole synchronous machine (see Sect. 13.4) into the description in state space, it can be solved with numerical methods on a digital computer. The set of equations in the energy generation system and for $\omega_{CS} = \omega_{mech}$ is (the explicit time-dependency and the index "I" are omitted for the sake of simplicity):

$$
\begin{aligned}
\frac{d\psi_d}{dt} &= -u_d - i_d\, R_1 + \omega\, \psi_q \\
\frac{d\psi_q}{dt} &= -u_q - i_q\, R_1 - \omega\, \psi_d \\
\frac{d\psi_F}{dt} &= -u_F - i_F\, R_F \\
\frac{d\psi_D}{dt} &= -i_D\, R_D \\
\frac{d\psi_Q}{dt} &= -i_Q\, R_Q \\
\frac{d\omega}{dt} &= \frac{p}{\Theta}\left[T_{ext} - \frac{3}{2} p \left(i_d \psi_q - i_q \psi_d \right) \right]
\end{aligned}
\tag{13.126}
$$

The relation between the currents and the flux linkages is given by the inductivity matrix $[L]$.

$$\begin{bmatrix} i_d \\ i_F \\ i_D \\ i_q \\ i_Q \end{bmatrix} = [L]^{-1} \begin{bmatrix} \psi_d \\ \psi_F \\ \psi_D \\ \psi_q \\ \psi_Q \end{bmatrix} \tag{13.127}$$

with

$$[L] = \begin{bmatrix} L_d & L_{dF} & -L_{dD} & 0 & 0 \\ L_{dF} & L_F & -L_{DF} & 0 & 0 \\ -L_{dD} & -L_{DF} & L_D & 0 & 0 \\ 0 & 0 & 0 & L_q & -L_{qQ} \\ 0 & 0 & 0 & -L_{qQ} & L_Q \end{bmatrix} \tag{13.128}$$

For calculation of the sudden short-circuit the stator voltages for $t > 0$ are set to zero:

$$u_d = u_q = 0 \tag{13.129}$$

The block diagram shown in Fig. 13.20 illustrates the above set of equations.

Performing a respective simulation for calculating the sudden short-circuit of the salient-pole synchronous machine (for this the machine data have to be known), time-dependent characteristics are obtained that are similar (but not identical) to those obtained for the sudden short-circuit of the synchronous machine with cylindrical rotor:

- In the case of sudden short-circuit the stator current reaches a very high value, which subsides to the permanent short-circuit current.
- The excitation current increases to a very high value as well; subsequently it decreases analogously to the stator current.
- The transformed stator currents i_d and i_q are DC currents in steady-state operation; i_d is the reactive component (magnetizing), i_q is the active component (torque).
- The damper currents (for the regarded salient-pole synchronous machine these are i_D and i_Q) are only relevant during a small time interval directly after the switching, otherwise they are zero.

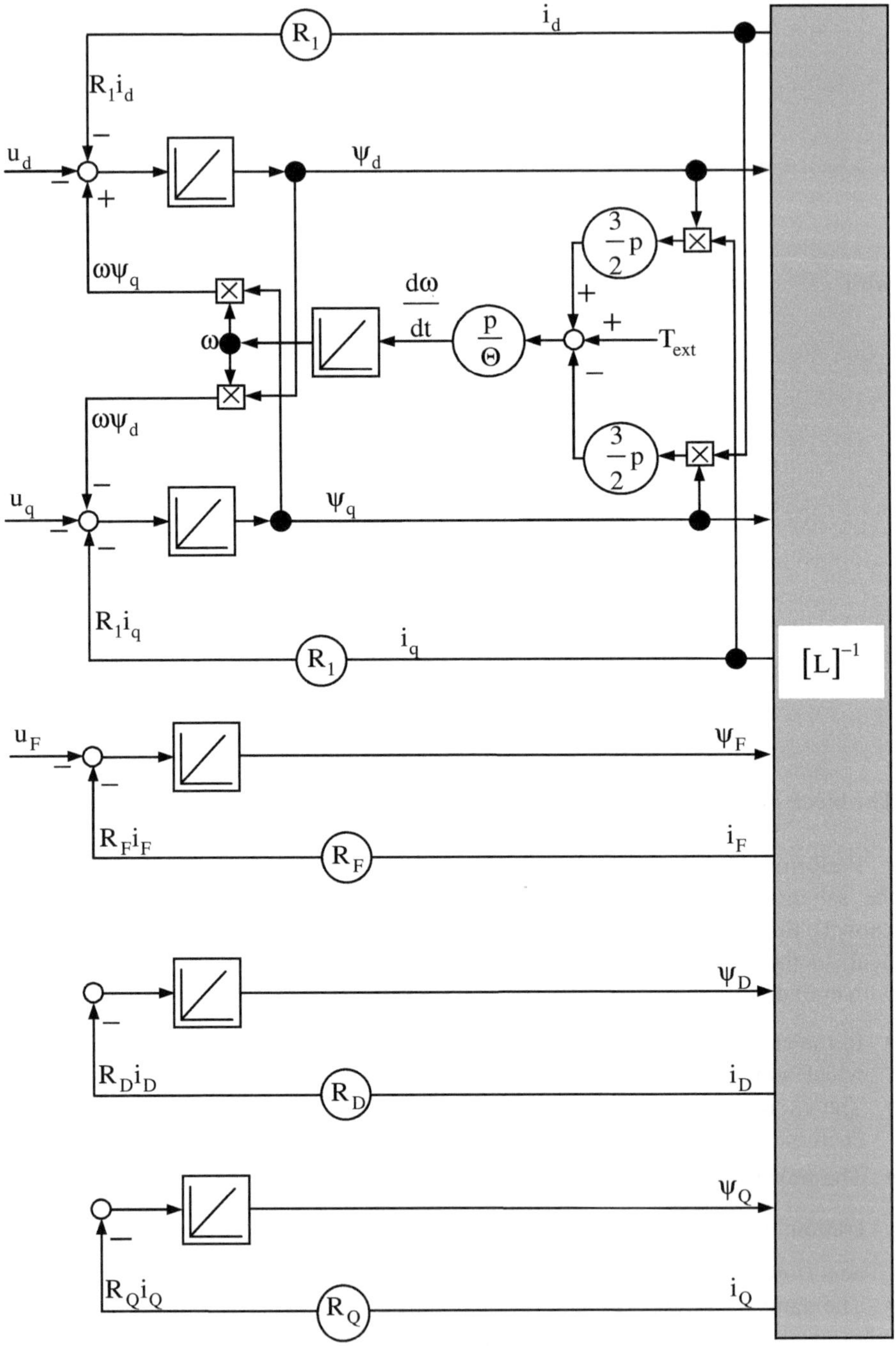

Fig. 13.20. Block diagram of the salient-pole synchronous machine.

Having a constant driving torque T_{ext} instead of a constant speed, the speed decreases after the short-circuit (analogously to the behavior of the non salient-pole machine), because the short-circuit currents generate losses in the Ohmic resistances. These losses are supplied by a reduction of the kinetic energy of the system.

The calculation of the salient-pole synchronous machine is more complex than the calculation of the synchronous machine with cylindrical rotor, because there is an additional differential equation for the damper winding.

13.6 Transient Operation of Salient-Pole Synchronous Machines

The dynamic operation can be calculated similar to the steady-state operation if some simplifying requirements are assumed. The integration of a nonlinear set of equations is avoided, but it has to be checked always, if the assumptions are valid for the regarded operation:

- The speed of the salient-pole synchronous machine is assumed being constant. The larger the inertia of the rotating masses, the better this assumption is fulfilled.
- The effect of the damper winding is not considered: Either the transient process of the damper currents subsides very quickly (considerably faster than the transient process of the stator phase currents), or there does not exist a damper winding.
- The transformatorily induced voltage components can be neglected compared to the rotatorily induced voltage components (e.g. $\frac{d\psi_d}{dt} \ll \omega\psi_q$).
- The excitation flux linkage ψ_F is constant during the entire transient period. This is valid, e.g. if the machine is equipped with a voltage controller that compensates the Ohmic voltage drop at the resistance in the excitation circuit if the current changes: $-u_F - R_F i_F = 0 = \frac{d\psi_F}{dt}$.

If these conditions are fulfilled it is called the "transient operation" of the synchronous machine (compared to the "dynamic operation" regarded in the preceding section). From the general set of equations for the salient-pole synchronous machine the following set of equations for this transient operation can be deduced:

$$
\begin{aligned}
0 &= -u_d - i_d\, R_1 + \omega\, \psi_q \\
0 &= -u_q - i_q\, R_1 - \omega\, \psi_d \\
0 &= -u_F - i_F\, R_F \\
0 &= -i_D\, R_D \\
0 &= -i_Q\, R_Q \\
0 &= \frac{p}{\Theta}\left[T_{ext} - \frac{3}{2} p\left(i_d \psi_q - i_q \psi_d\right)\right]
\end{aligned}
\tag{13.130}
$$

Neglecting the Ohmic voltage drop at the stator resistance it follows:

$$
\begin{aligned}
u_d &= +\omega\psi_q \\
u_q &= -\omega\psi_d \\
u_F &= -i_F R_F \\
i_D &= 0 \\
i_Q &= 0 \\
T_{ext} &= T = \frac{3}{2} p\left(i_d \psi_q - i_q \psi_d\right)
\end{aligned}
\tag{13.131}
$$

The flux linkages are:

$$
\begin{bmatrix} \psi_d \\ \psi_F \\ \psi_D \\ \psi_q \\ \psi_Q \end{bmatrix} = [L] \begin{bmatrix} i_d \\ i_F \\ i_D \\ i_q \\ i_Q \end{bmatrix}
\tag{13.132}
$$

with

$$[L] = \begin{bmatrix} L_d & L_{dF} & -L_{dD} & 0 & 0 \\ L_{dF} & L_F & -L_{DF} & 0 & 0 \\ -L_{dD} & -L_{DF} & L_D & 0 & 0 \\ 0 & 0 & 0 & L_q & -L_{qQ} \\ 0 & 0 & 0 & -L_{qQ} & L_Q \end{bmatrix} \tag{13.133}$$

Consequently (with $i_D = i_Q = 0$) there is:

$$\begin{aligned} \psi_F &= L_F\, i_F + L_{dF}\, i_d \\ \psi_d &= L_d\, i_d + L_{dF}\, i_F \\ \psi_q &= L_q\, i_q \end{aligned} \tag{13.134}$$

The set of equations for this transient operation has the same form like the set of equations for the steady-state operation, but there is one main difference:

- steady-state operation: excitation current $i_F = \text{const.}$
- transient operation: excitation flux $\psi_F = \text{const.}$

Because of the constant excitation flux it follows (the index "0" characterizes the steady-state situation before the switching):

$$\begin{aligned} \psi_F &= L_F\, i_F + L_{dF}\, i_d \\ &= \psi_{F,0} = L_F\, i_{F,0} + L_{dF}\, i_{d,0} \\ \Rightarrow\quad & i_F + \frac{L_{dF}}{L_F} i_d = i_{F,0} + \frac{L_{dF}}{L_F} i_{d,0} \\ \Rightarrow\quad & i_F = i_{F,0} + \frac{L_{dF}}{L_F}\left(i_{d,0} - i_d\right) \end{aligned} \tag{13.135}$$

Therefore the excitation current i_F changes with the stator current i_d. For the internal voltage there is:

$$u_P = -\omega\, L_{dF}\, i_F \tag{13.136}$$

With this an additional difference between steady-state and transient operation becomes obvious:

- steady-state operation: internal voltage $u_P = \text{const.}$
- transient operation: internal voltage $u_P \neq \text{const.}$

Further:

$$\begin{aligned} u_P &= -\omega\, L_{dF}\, i_F \\ &= -\omega\, L_{dF} \left[i_{F,0} + \frac{L_{dF}}{L_F} \left(i_{d,0} - i_d \right) \right] \\ &= -\omega\, L_{dF}\, i_{F,0} - \omega\, L_{dF} \frac{L_{dF}}{L_F} \left(i_{d,0} - i_d \right) \\ &= u_{P,0} + \omega\, L_{dF} \frac{L_{dF}}{L_F} \left(i_d - i_{d,0} \right) \end{aligned} \tag{13.137}$$

With

$$\sigma_{dF} = 1 - \frac{L_{dF}^2}{L_d\, L_F} \tag{13.138}$$

it follows:

$$\begin{aligned} u_P &= u_{P,0} + \omega\, L_{dF} \frac{L_{dF}}{L_F} \left(i_d - i_{d,0} \right) \\ &= u_{P,0} + \left(1 - \sigma_{dF} \right) \omega\, L_d \left(i_d - i_{d,0} \right) \end{aligned} \tag{13.139}$$

The flux linkage in the d-axis becomes:

$$\begin{aligned} \psi_d &= L_d\, i_d + L_{dF}\, i_F \\ &= L_d\, i_d + L_{dF} \left[i_{F,0} + \frac{L_{dF}}{L_F} \left(i_{d,0} - i_d \right) \right] \\ &= L_{dF}\, i_{F,0} + L_d\, i_d \left(1 - \frac{L_{dF}^2}{L_d\, L_F} \right) + \frac{L_{dF}^2}{L_d\, L_F} L_d\, i_{d,0} \\ &= L_{dF}\, i_{F,0} + \sigma_{dF}\, L_d\, i_d + \left(1 - \sigma_{dF} \right) L_d\, i_{d,0} \end{aligned} \tag{13.140}$$

Consequently the voltage equation of the quadrature axis becomes:

$$\begin{aligned} u_q &= -\omega\,\psi_d \\ &= -\omega\left[L_{dF}\, i_{F,0} + \sigma_{dF}\, L_d\, i_d + (1-\sigma_{dF}) L_d\, i_{d,0}\right] \\ &= -\omega\, L_{dF}\, i_{F,0} - \omega\,\sigma_{dF}\, L_d\, i_d - \omega(1-\sigma_{dF}) L_d\, i_{d,0} \end{aligned} \tag{13.141}$$

With the internal voltage before the switching moment $u_{P,0} = -\omega\, L_{dF}\, i_{F,0}$ and the transient reactance $X_d' = \sigma_{dF}\,\omega\, L_d$ the voltage in the quadrature axis is:

$$\begin{aligned} u_q &= -\omega\, L_{dF}\, i_{F,0} - \omega\,\sigma_{dF}\, L_d\, i_d - \omega(1-\sigma_{dF}) L_d\, i_{d,0} \\ &= u_{P,0} - X_d'\, i_d - (1-\sigma_{dF}) X_d\, i_{d,0} \qquad \text{with} \quad X_d = \omega L_d \end{aligned} \tag{13.142}$$

A transformation (all constant values before the switching moment are written on the left side) gives:

$$u_{P,0} - (1-\sigma_{dF}) X_d\, i_{d,0} = u_q + X_d'\, i_d = u_P' \tag{13.143}$$

The value $u_P' = u_q + X_d' i_d$ therefore is a constant during the transient operation (analogously to the value $u_P = u_q + X_d\, i_d$, which is constant for steady-state operation).

For the internal voltage it holds:

$$\begin{aligned} u_P &= u_{P,0} + (1-\sigma_{dF})\,\omega\, L_d\left(i_d - i_{d,0}\right) \\ &= u_P' + (1-\sigma_{dF}) X_d\, i_d \end{aligned} \tag{13.144}$$

The voltage in the direct axis is unchanged:

$$u_d = +\omega\,\psi_q = X_q\, i_q \tag{13.145}$$

Now the description can be changed from space vectors to rms-values by using the respective equations of the preceding sections. The voltage equations for the transient operation then are:

$$\begin{aligned} U'_P &= U_{P,0} - (1-\sigma_{dF}) X_d \, I_{d,0} \\ &= U_q + X'_d \, I_d = \text{const.} \\ U_P &= U'_P + (1-\sigma_{dF}) X_d \, I_d \\ U_d &= X_q \, I_q \end{aligned} \tag{13.146}$$

For the rms-values (and for the phasor diagram developed from the above equations, see Fig. 13.21) it holds: During transient operation no longer U_P is constant (like in steady-state operation), but U'_P is constant. The value of U'_P can be calculated from the conditions just before the switching moment.

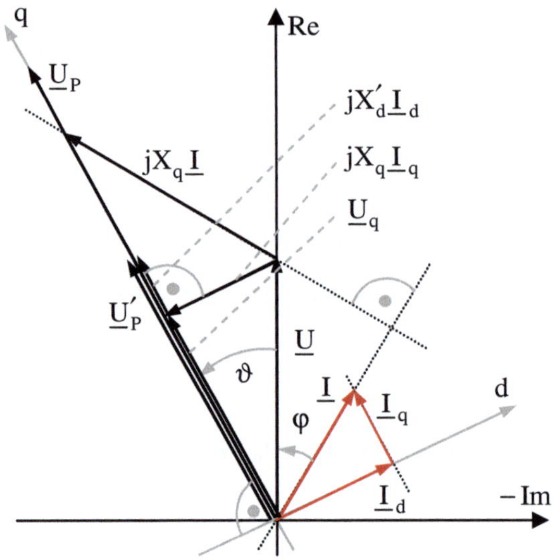

Fig. 13.21. Phasor diagram of the salient-pole synchronous machine in transient operation.

For the torque it holds during the transient operation:

$$T = \frac{3p}{\omega} \left(I_d U_d + I_q U_q \right) \tag{13.147}$$

From the phasor diagram the following relations can be deduced:

$$\begin{aligned} I_d &= \frac{U'_P - U_q}{X'_d} & \qquad I_q &= \frac{U_d}{X_q} \\ U_d &= U \sin(\vartheta) & \qquad U_q &= U \cos(\vartheta) \end{aligned} \tag{13.148}$$

By insertion it follows:

$$\begin{aligned} T &= \frac{3p}{\omega}\left[\frac{U_P' - U\cos(\vartheta)}{X_d'} U\sin(\vartheta) + \frac{U\sin(\vartheta)}{X_q} U\cos(\vartheta)\right] \\ &= \frac{3p}{\omega}\left[\frac{U_P' U}{X_d'}\sin(\vartheta) + \left(\frac{1}{X_q} - \frac{1}{X_d'}\right) U^2 \sin(\vartheta)\cos(\vartheta)\right] \\ &= \frac{3p}{\omega}\left[\frac{U_P' U}{X_d'}\sin(\vartheta) + \left(\frac{1}{X_q} - \frac{1}{X_d'}\right)\frac{U^2}{2}\sin(2\vartheta)\right] \end{aligned} \tag{13.149}$$

With the nominal torque of the synchronous machine (please refer to Chap. 5)

$$T_N = \frac{3U_N I_N \cos(\varphi_N)}{\omega/p} \tag{13.150}$$

the ratio of the torque during transient operation to the nominal torque of the synchronous machine becomes:

$$\begin{aligned} \frac{T}{T_N} &= \frac{\frac{3p}{\omega}\left[\frac{U_P' U}{X_d'}\sin(\vartheta) + \left(\frac{1}{X_q} - \frac{1}{X_d'}\right)\frac{U^2}{2}\sin(2\vartheta)\right]}{\frac{3U_N I_N \cos(\varphi_N)}{\omega/p}}, \quad U = U_N \\ &= \frac{\frac{U_P'}{X_d'}\sin(\vartheta) + \left(\frac{1}{X_q} - \frac{1}{X_d'}\right)\frac{U_N}{2}\sin(2\vartheta)}{I_N\cos(\varphi_N)} \end{aligned} \tag{13.151}$$

With

$$\begin{aligned} \frac{U_N}{I_N} &= X_N \qquad & \frac{U_P'}{U_N} &= u_P' \\ \frac{X_d'}{X_N} &= x_d' \qquad & \frac{X_q}{X_N} &= x_q \end{aligned} \tag{13.152}$$

it follows in normalized description:

$$T_{ratio} = \frac{T}{T_N} = \frac{1}{\cos(\varphi_N)}\left[\frac{u_P'}{x_d'}\sin(\vartheta) + \left(\frac{1}{x_q} - \frac{1}{x_d'}\right)\frac{1}{2}\sin(2\vartheta)\right] \quad (13.153)$$

Relating the torque to the pull-out torque of a respective non-salient-pole synchronous machine, Fig. 13.22 is obtained (this figure can be compared directly to the figures shown in Sects. 13.4 and 13.5; $U_P/U = 0.5$ has been chosen). It becomes obvious that during transient operation a considerably higher pull-out torque is achieved than during steady-state operation. Moreover the pull-out torque is reached for a rotor angle $\vartheta > \pi/2$, whereas for the steady-state operation the pull-out torque is reached for $\vartheta < \pi/2$. This effect comes from the fact that during transient operation the excitation flux is kept constant by increasing the excitation current (and therefore even the torque) to a considerably higher value. After subsiding of the time-dependent characteristics the new steady-state operation is obtained.

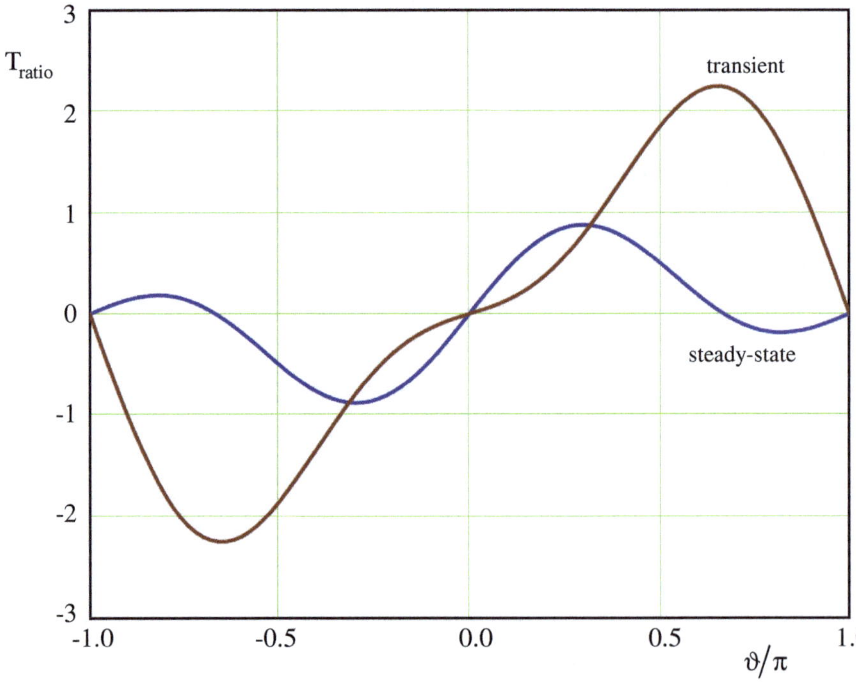

Fig. 13.22. Torque (normalized to the pull-out torque) versus rotor angle characteristic of the salient-pole synchronous machine in steady-state and transient operation.

Having the synchronous machine supplied from the mains the transient operation can be evoked by two different load changes:

- *Electrical load change:* A switching action in the supplying mains suddenly changes mains voltage and mains reactance. For calculating U'_P the values of the steady-state operation just before the switching moment have to be taken. For the time period after the switching moment the new values of mains voltage and mains reactance have to be considered. Because of U'_P being constant the currents I_d und I_q change nearly according to a step-function. Therefore the machine generates a different torque than before the switching moment. Because of the inertia the rotor moves with synchronous speed; consequently a torque difference occurs resulting in a movement $\vartheta(t)$ of the rotor angle.
- *Mechanical load change:* When the driving torque is changed suddenly, mains voltage and mains reactance remain unchanged. Directly after this disturbance the same currents like before are flowing, consequently the torque generated by the synchronous machine is unchanged. Because of the change of the driving torque there is a torque difference resulting in a movement $\vartheta(t)$ of the rotor angle. Subsequently even the currents and the torque of the synchronous machine are changed.

13.7 References for Chapter 13

Boldea I, Tutelea L (2010) Electric machines. CRC Press, Boca Raton

DeDoncker RW, Pulle DWJ, Veltman A (2011) Advanced electrical drives. Springer-Verlag, Berlin

Nasar SA (1970) Electromagnetic energy conversion devices and systems. Prentice Hall, London

Schröder D (1995) Elektrische Antriebe 2. Springer-Verlag, Berlin

White DC, Woodson HH (1958) Electromechanical energy conversion. John Wiley & Sons, New York

14 Dynamic Operation and Control of Permanent Magnet Excited Rotating Field Machines

14.1 Principle Operation

As already described in Chap. 6, the principle operation of the permanent magnet excited rotating field machine is like follows:

The synchronous machine contains permanent magnets to generate the excitation field, but there is no starting cage present. The three-phase machine is supplied by an inverter which realizes a three-phase current system. As a main difference to what is described in Chap. 6, here no limiting assumptions concerning the time-dependency of the currents is made (especially there is no need for sinusoidal currents).

The fundamental frequency of the supplying three-phase system determines the frequency of the rotating magneto-motive force and therefore even the rotor speed. The rotating magneto-motive force together with the field of the permanent magnet rotor generates the torque. Mostly, this torque shall be as smooth as possible. The rotation of the rotating stator field is realized depending on the rotor position by means of the inverter in such a way, that the electrical angle between rotating magneto-motive force of the stator and the rotor field is $\pi/2$ (i.e. $\vartheta = -\varphi$). With this the load angle in the energy consumption system already defined in Sect. 6.3 becomes $\delta_M = -\delta_G = -\vartheta - \varphi = 0$.

The rotor position can be measured by using sensors or it can be deduced from the terminal voltages and/or terminal currents.

An operation is obtained that does no longer correspond to the synchronous machine, but to the DC-machine:

- DC-machine: magneto-motive force of the armature and excitation field build an electrical angle of $\pi/2$; this adjustment is performed mechanically by means of the commutator.
- Synchronous machine: the rotor angle ϑ and the phase angle φ are adjusted depending on the operation point; there is no active influence on the phase shift between magneto-motive force of the stator and excitation field.
- Electronically commutated permanent magnet excited rotating field machine: magneto-motive force of the stator and rotor field build an electrical angle of $\pi/2$; this adjustment is performed electronically by means of the inverter.

14.2 Set of Equations for the Dynamic Operation

In Fig. 14.1 a two-poles machine is shown; the ratio of pole arc per pole pitch is $\alpha_i < 1$. Moreover a magnetic asymmetry with $L_d \neq L_q$ is present. This is caused by the fact that the geometric air-gap along the circumference is constant whereas the magnetic effective air-gap is not constant (magnet materials have a relative permeability of about $\mu_{r,PM} \approx 1$; compared to this iron shows a very high relative permeability: $\mu_{r,Fe} \gg 1$). The special case that the magnets are placed onto the surface of a cylindrical iron rotor is included in the following description, if $L_d = L_q = L$ is used.

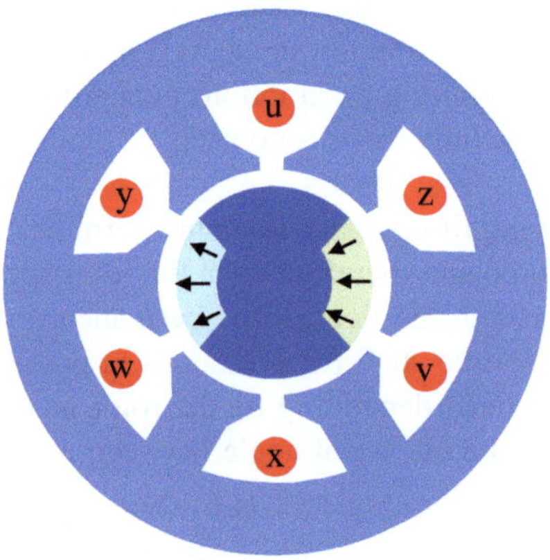

Fig. 14.1. Sketch of a two-poles permanent magnet excited rotating field machine.

Assuming that there is no damper winding also means that the rotor is manufactured from iron sheets (solid iron is electrically conductive and therefore would have a damping effect) and that rare-earth magnets (SmCo or NdFeB) have to be made of several isolated parts, because even these materials are electrically conductive (this is not the case for the much cheaper ferrite magnets).

For such a permanent magnet excited rotating field machine just a system of three windings has to be considered (as the coordinate system is oriented to the rotor flux, the axes are nominated with “d“ and “q“):

- I, d : stator direct axis (reactive current, flux generating)
- I, q : stator quadrature axis (active current, torque generating)
- II, d : rotor direct axis (permanent magnets)
- II, q : rotor quadrature axis (no winding)

As the permanent magnets are magnetizing in the direct axis, i.e. in the direct axis there is the large iron-iron-distance between stator and rotor, it follows: $L_d < L_q$.
The electrical angular frequency of the rotor is:

$$\omega_{mech} = \frac{d\gamma}{dt} = 2\pi pn \tag{14.1}$$

The following coordinate system is chosen:

$$\omega_{CS} = \frac{d\alpha}{dt} = \frac{d\gamma}{dt} = \omega_{mech} = \omega \tag{14.2}$$

Analogously to Chap. 6 the energy consumption system is used here, because this machine topology mainly is used as a motor.

By means of the space vector theory for rotating field machines the following set of equations is obtained (for the permanent magnets there does not exist a voltage equation; the constant rotor flux evoked from the permanent magnets is considered by the constant substitutive excitation current $i_{II,d,0}$ in the equation of the stator flux linkage):[25]

$$\begin{aligned}
u_{I,d} &= R_1 i_{I,d} + \frac{d\psi_{I,d}}{dt} - \omega\psi_{I,q} \\
u_{I,q} &= R_1 i_{I,q} + \frac{d\psi_{I,q}}{dt} + \omega\psi_{I,d} \\
\psi_{I,d} &= (1+\sigma_1) L_{md} i_{I,d} + L_{md} i_{II,d,0} \\
\psi_{I,q} &= (1+\sigma_1) L_{mq} i_{I,q} \\
T &= \frac{3}{2} p\left(i_{I,q}\psi_{I,d} - i_{I,d}\psi_{I,q}\right) = T_{load} + \frac{\Theta}{p}\frac{d\omega}{dt}
\end{aligned} \tag{14.3}$$

[25] This set of equations is very similar to that of the salient-pole synchronous machine (see Sect. 13.5), but with two main differences: firstly, here the energy consumption system is used, and secondly for the permanent magnet excited machine $L_d < L_q$ is true, whereas for the salient-pole synchronous machine $L_d > L_q$ is true. This comes from the fact that the magnetically effective air-gap (the iron-iron distance between stator and rotor) for the salient-pole synchronous machine is smaller in the d-axis than in the q-axis; for the permanent magnet excited machine however it is vice versa.

No-load operation with nominal voltage and nominal frequency is characterized by:

$$i_{I,d} = i_{I,q} = 0 \tag{14.4}$$

Then it follows for the flux linkages

$$\begin{aligned} \psi_{I,d} &= L_{md} i_{II,d,0} \\ \psi_{I,q} &= 0 \end{aligned} \tag{14.5}$$

and for the voltages

$$\begin{aligned} u_{I,d} &= 0 \\ u_{I,q} &= \omega \psi_{I,d} = \omega L_{md} i_{II,d,0} \end{aligned} \tag{14.6}$$

If sinusoidal time-dependencies can be assumed, it follows with

$$U_{1,N} = \frac{u_{I,q}}{\sqrt{2}} \tag{14.7}$$

for the substitutive excitation current

$$\begin{aligned} u_{I,q} &= \sqrt{2}\, U_{1,N} = \omega L_{md} i_{II,d,0} \\ \Rightarrow \quad i_{II,d,0} &= \frac{\sqrt{2}\, U_{1,N}}{\omega L_{md}} \end{aligned} \tag{14.8}$$

The substitutive internal voltage is

$$u_P = \omega L_{md} i_{II,d,0} \tag{14.9}$$

Now the flux linkages are inserted into the voltage equations and the torque equation. It follows:

$$
\begin{aligned}
u_{I,d} &= R_1 i_{I,d} + \frac{d}{dt}\left[(1+\sigma_1)L_{md} i_{I,d} + L_{md} i_{II,d,0}\right] - \omega(1+\sigma_1)L_{mq} i_{I,q} \\
u_{I,q} &= R_1 i_{I,q} + \frac{d}{dt}\left[(1+\sigma_1)L_{mq} i_{I,q}\right] + \omega\left[(1+\sigma_1)L_{md} i_{I,d} + L_{md} i_{II,d,0}\right] \\
T &= \frac{3}{2} p\left(i_{I,q}\left[(1+\sigma_1)L_{md} i_{I,d} + L_{md} i_{II,d,0}\right] - i_{I,d}(1+\sigma_1)L_{mq} i_{I,q}\right) \\
&= T_{load} + \frac{\Theta}{p}\frac{d\omega}{dt}
\end{aligned}
\tag{14.10}
$$

Further

$$
\begin{aligned}
u_{I,d} &= R_1 i_{I,d} + (1+\sigma_1)L_{md}\frac{d}{dt} i_{I,d} - \omega(1+\sigma_1)L_{mq} i_{I,q} \\
u_{I,q} &= R_1 i_{I,q} + (1+\sigma_1)L_{mq}\frac{d}{dt} i_{I,q} + \omega\left[(1+\sigma_1)L_{md} i_{I,d} + L_{md} i_{II,d,0}\right] \\
T &= \frac{3}{2} p\, i_{I,q}\left[\left((1+\sigma_1)L_{md} - (1+\sigma_1)L_{mq}\right) i_{I,d} + L_{md} i_{II,d,0}\right] \\
&= T_{load} + \frac{\Theta}{p}\frac{d\omega}{dt}
\end{aligned}
\tag{14.11}
$$

With

$$
L_d = (1+\sigma_1)L_{md}, \qquad L_q = (1+\sigma_1)L_{mq}
\tag{14.12}
$$

it follows:

$$
\begin{aligned}
u_{I,d} &= R_1 i_{I,d} + L_d\frac{d}{dt} i_{I,d} - \omega L_q i_{I,q} \\
u_{I,q} &= R_1 i_{I,q} + L_q\frac{d}{dt} i_{I,q} + \omega\left[L_d i_{I,d} + L_{md} i_{II,d,0}\right] \\
T &= \frac{3}{2} p\, i_{I,q}\left[L_{md} i_{II,d,0} - \left(L_q - L_d\right) i_{I,d}\right] = T_{load} + \frac{\Theta}{p}\frac{d\omega}{dt}
\end{aligned}
\tag{14.13}
$$

Introducing the time constants

$$\tau_d = \frac{L_d}{R_1}, \qquad \tau_q = \frac{L_q}{R_1} \tag{14.14}$$

the following differential equations are obtained:

$$\begin{aligned}
\tau_d \frac{d}{dt} i_{I,d} + i_{I,d} &= \frac{u_{I,d}}{R_1} + \omega\tau_q i_{I,q} \\
\tau_q \frac{d}{dt} i_{I,q} + i_{I,q} &= \frac{u_{I,q}}{R_1} - \omega\tau_d \left[i_{I,d} + \frac{i_{II,d,0}}{1+\sigma_1} \right] \\
\frac{\Theta}{p}\frac{d\omega}{dt} &= \frac{3}{2} p L_d \left[\frac{i_{II,d,0}}{1+\sigma_1} - \left(\frac{L_q}{L_d} - 1 \right) i_{I,d} \right] i_{I,q} - T_{load}
\end{aligned} \tag{14.15}$$

where rotor position and speed are linked by:

$$\frac{d\gamma}{dt} = \frac{d\alpha}{dt} = \omega = 2\pi pn \tag{14.16}$$

With this set of differential equations the permanent magnet excited rotating field machine is described completely.

In the base speed region the machine is operated in such a way that stator MMF and rotor field (excitation field, $L_{md} i_{II,d,0}$) are perpendicular to each other. Then the stator current component in the direct axis has to be zero ($i_{I,d} = 0$); the stator current component in the quadrature axis $i_{I,q}$ is the torque generating component. The voltages in the direct axis and in the quadrature axis can be deduced from the above equations.

However, if a control method with $i_{I,d} = i_{I,d,\infty} \neq 0$ is used (e.g. with $i_{I,d} < 0$ for field weakening), stator MMF and rotor field are no longer perpendicular to each other (this operation mode will be explained in detail in the next Sect. 14.3). For the direct axis voltage equation in steady-state operation (i.e. no more change of the currents $i_{I,d}$ and $i_{I,q}$) there is:

$$i_{I,d,\infty} = \frac{u_{I,d}}{R_1} + \omega\tau_q i_{I,q} \tag{14.17}$$

Then for the dynamic operation it follows:

$$\tau_d \frac{d}{dt} i_{I,d} + i_{I,d} = i_{I,d,\infty} \tag{14.18}$$

The solution of this differential equation is well-known, the current characteristic is:

$$i_{I,d} = i_{I,d,\infty} \left(1 - e^{-\frac{t}{\tau_d}} \right) \tag{14.19}$$

The block diagram of the controlled permanent magnet excited rotating field machine can be deduced from the above equations; it is shown in Fig. 14.2.

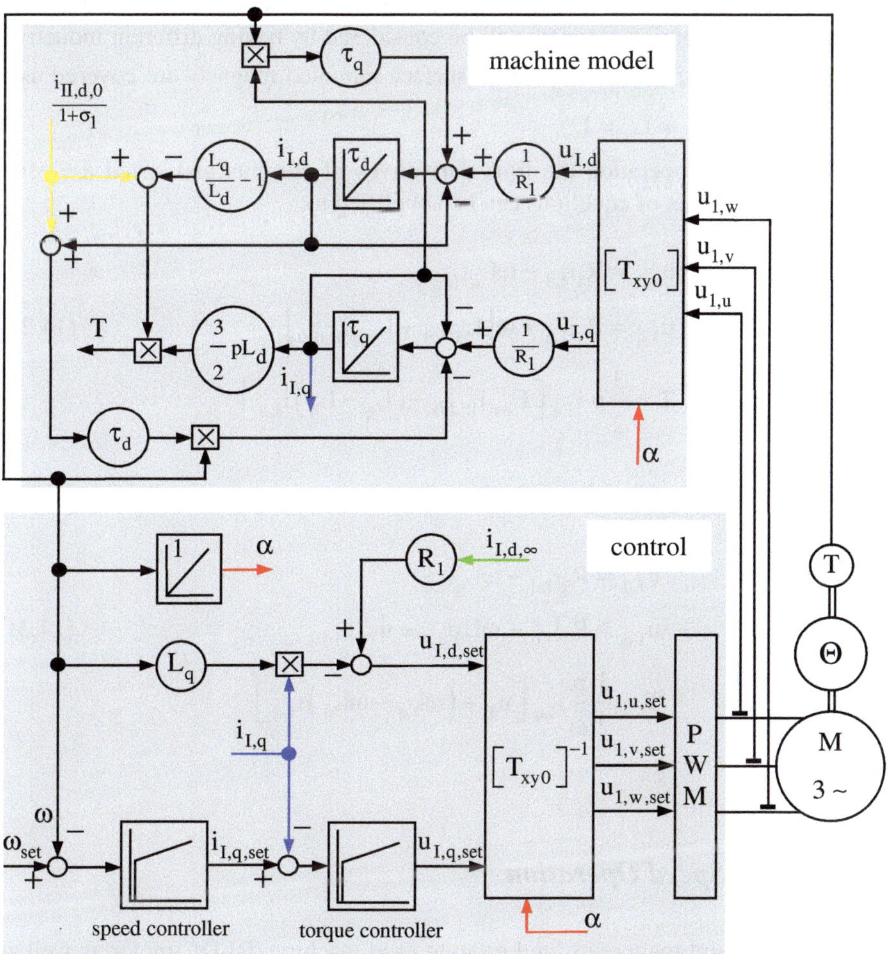

Fig. 14.2. Block diagram of the controlled permanent magnet excited rotating field machine.

14.3 Steady-State Operation

14.3.1 Fundamentals

In the following the steady-state operation of permanent magnet excited rotating field machines will be calculated. As in principle they show the same motor construction, permanent magnet synchronous machines (with sinusoidal currents) and brushless DC-motors (with step-by-step constant currents) will be regarded simultaneously. Because not always sinusoidal currents are used, the space vector theory has to be used even for the steady-state operation. In addition, the general case of machines with buried magnets will be considered by having different inductivities in d-axis and q-axis as $L_d < L_q$ (surface mounted magnets are covered as a special case for $L_d = L_q = L$).

In steady-state operation the time derivatives of currents and speed are zero. Consequently the set of equations can be simplified to:

$$\begin{aligned} u_{I,d} &= R_1 i_{I,d} - \omega L_q i_{I,q} \\ u_{I,q} &= R_1 i_{I,q} + \omega \left[L_d i_{I,d} + L_{md} i_{II,d,0} \right] \\ T &= \frac{3}{2} p\, i_{I,q} \left[L_{md} i_{II,d,0} - \left(L_q - L_d \right) i_{I,d} \right] \end{aligned} \quad (14.20)$$

With $u_P = \omega L_{md} i_{II,d,0}$ it follows further:

$$\begin{aligned} u_{I,d} &= R_1 i_{I,d} - \omega L_q i_{I,q} \\ u_{I,q} &= R_1 i_{I,q} + \omega L_d i_{I,d} + u_P \\ T &= \frac{3}{2} \frac{p}{\omega} i_{I,q} \left[u_P - \left(\omega L_q - \omega L_d \right) i_{I,d} \right] \end{aligned} \quad (14.21)$$

14.3.2 Base Speed Operation

For the permanent magnet excited rotating field machine (BLDC-motor as well as PMSM) the positive q-direction is defined in the real axis and the positive d-direction in the negative imaginary axis (please note that this d-q coordinate system has the positive horizontal axis opposite to the positive horizontal axis of the complex plane).

During base speed operation the angle between the magneto-motive force of the stator and the rotor field is 90° electrically. The rotor field is drawn in the direct axis (negative imaginary axis), see also Sect. 14.2; the magneto-motive force of the stator merely is composed of the current $i_I = i_{I,q} = i_{I,N}$ (quadrature axis, real axis), the current in the direct axis is $i_{I,d} = 0$. Then the load angle becomes: $\delta_M = -\vartheta - \varphi = 0$. The described operation condition usually is reached with a power factor of about $\cos\varphi \approx 0.8$ and without magnetic asymmetry ($L_d = L_q = L$). The set of equations then becomes

$$\begin{aligned} u_{I,d} &= -\omega L i_{I,q} \\ u_{I,q} &= R_1 i_{I,q} + u_P \\ T &= \frac{3}{2}\frac{p}{\omega} i_{I,q}\, u_P \end{aligned} \tag{14.22}$$

The equations of the machine are illustrated in the vector diagram of Fig. 14.3.

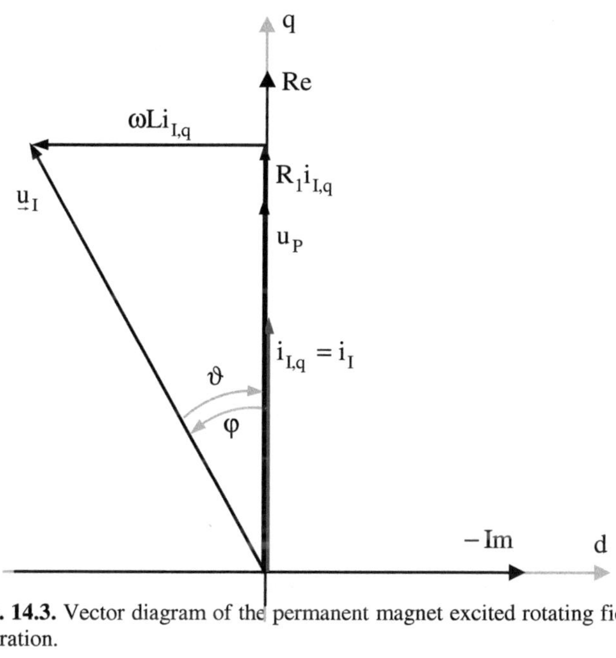

Fig. 14.3. Vector diagram of the permanent magnet excited rotating field machine in base speed operation.

In this vector diagram[26] the voltage space vector is (please refer to Sect. 11.3)

$$\begin{aligned} \underline{u}_I &= \mathrm{Re}\{\underline{u}_I\} + j\ \mathrm{Im}\{\underline{u}_I\} \\ &= u_{I,q} - j\, u_{I,d} \\ &= R_1 i_{I,q} + u_P - j\left(-\omega L i_{I,q}\right) \\ &= R_1 i_{I,q} + u_P + j\omega L i_{I,q} \end{aligned} \tag{14.23}$$

14.3.3 Operation with Leading Load Angle and without Magnetic Asymmetry

During operation with leading load angle δ_M and without magnetic asymmetry it is still true: $L_d = L_q = L$. But now, in addition to the torque producing current $i_{I,q}$ (quadrature axis), a negative current component $i_{I,d} < 0$ is supplied to the direct axis. As this negative d-axis current is opposite to the rotor field (in the negative imaginary axis), it has a demagnetizing effect (therefore, this is called "field weakening"). Consequently, $i_{I,q}$ has to be decreased, so that the total current i_I is not exceeding the nominal value (to avoid overheating). Therefore:

$$i_{I,d} < 0, \qquad i_{I,q} \le i_{I,N}, \qquad i_I = i_{I,N} \tag{14.24}$$

The set of equations becomes

$$\begin{aligned} u_{I,d} &= R_1 i_{I,d} - \omega L i_{I,q} \\ u_{I,q} &= R_1 i_{I,q} + \omega L i_{I,d} + u_P \\ T &= \frac{3}{2}\frac{p}{\omega} i_{I,q}\, u_P \end{aligned} \tag{14.25}$$

[26] Please note that complex phasors were introduced as the non time-dependent components of the complex description of sinusoidally time-dependent variables (please refer to Sect. 1.6). In this chapter, explicitly non-sinusoidal time-dependencies of the variables are permissible. Therefore, the respective illustration is called vector diagram and not phasor diagram.

Because of this additional current in the negative d-axis $i_{I,d}$ the angle between the magneto-motive force of the stator ($\underline{i}_I$) and the rotor field (in the negative imaginary axis) is enlarged to more than 90° electrically. Consequently, the angle φ between voltage and current decreases, i.e. the power factor is changed into the direction $\cos\varphi \to 1$. Moreover, for $i_{I,d} \neq 0$ even for the load angle holds true $\delta_M \neq 0$. These characteristics are illustrated in the vector diagram shown in Fig. 14.4.

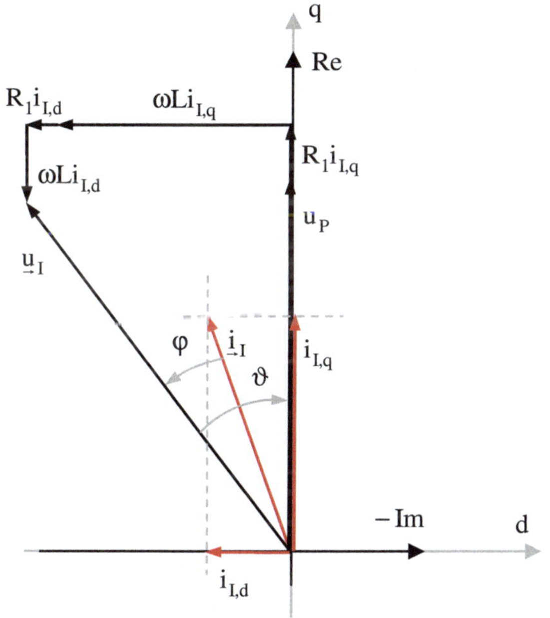

Fig. 14.4. Vector diagram of the permanent magnet excited rotating field machine with leading load angle and without magnetic asymmetry.

Neglecting the Ohmic voltage drop in the voltage equations, the current in the negative d-axis $i_{I,d}$ has the following consequences:

- The current $i_{I,q}$ decreases, because the total current is limited.
- Consequently the absolute value of the voltage $u_{I,d} = -\omega L i_{I,q}$ decreases as well.
- Then the voltage $u_{I,q} = \omega L i_{I,d} + u_P$ may increase because of $|\underline{u}_I| = \sqrt{(u_{I,q})^2 + (u_{I,d})^2} \leq |\underline{u}_{I,N}|$. As furthermore $\omega L i_{I,d}$ and u_P are oppo-

site to each other, the voltage u_P increases. Because the magnetization is constant, this can only be realized by increasing the speed.

- From the torque equation $T = \frac{3}{2}\frac{p}{\omega} i_{I,q}\ u_P$ it is obvious that the torque decreases: $\frac{u_P}{\omega}$ describes the magnetization and this is constant; the lower current $i_{I,q}$ leads to a lower torque.

Therefore, a typical field weakening operation is obtained (lower torque at higher speed; please compare e.g. with the field weakening operation of the induction machine).

If the lowering of the torque shall be omitted, an additional torque component has to be generated to compensate for this effect. This is reached by an additional reluctance torque component by introducing a magnetic asymmetry.

14.3.4 Operation with Leading Load Angle and Magnetic Asymmetry

During operation with leading load angle δ_M and with magnetic asymmetry it is now $L_d < L_q$. Like in the preceding section the following holds:

- In addition to the torque generation current $i_{I,q}$ (quadrature axis) a current in the negative d-axis $i_{I,d} < 0$ is applied; for the load angle $\delta_M \neq 0$ holds true.
- This current in the negative d-axis has a demagnetizing effect.
- The torque generation current $i_{I,q}$ has to be decreased, so that the total current i_I is not exceeding the nominal value (to avoid overheating).

The according vector diagram is shown in Fig. 14.5.

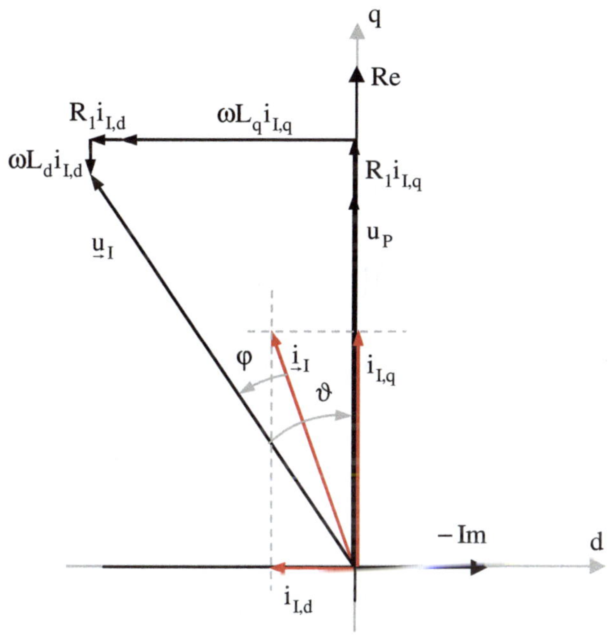

Fig. 14.5. Vector diagram of the permanent magnet excited rotating field machine with leading load angle and with magnetic asymmetry.

The torque can be calculated from:

$$T = \frac{3}{2}\frac{p}{\omega} i_{I,q}\left[u_P - \left(\omega L_q - \omega L_d\right) i_{I,d}\right] \tag{14.26}$$

At constant speed the torque can be increased to the initial value by means of the positive reluctance torque component $-\frac{3}{2}\frac{p}{\omega}\left(\omega L_q - \omega L_d\right) i_{I,d}\ i_{I,q}$. Depending on the design of the electromagnetic circuit the torque may be even higher than initially. This can be used in two different ways:

- increase of the speed;
- reduction of the current.

By suitable electromagnetic design of the machine the current can be reduced and simultaneously the power factor $\cos\varphi$ can be improved. This leads to lower losses in the machine and the supplying inverter (because of the lower current level) as well as to a lower apparent power of the inverter (because of the improved power factor).

14.3.5 Torque Calculation from Current Loading and Flux Density

With $u_P = \omega \, \Psi_P$ the torque equation becomes

$$\begin{aligned} T &= \frac{3}{2}\frac{p}{\omega} i_{I,q}\left[\omega\Psi_P - \left(\omega L_q - \omega L_d\right) i_{I,d}\right] \\ &= \frac{3}{2} p \, i_I \cos\left(\delta_M\right)\left[\Psi_P - \left(L_q - L_d\right)\left\{-i_I \sin\left(\delta_M\right)\right\}\right] \\ &= \frac{3}{2} p \, i_I \cos\left(\delta_M\right)\Psi_P + \frac{3}{2} p \, i_I^2 \cos\left(\delta_M\right)\sin\left(\delta_M\right)\left(L_q - L_d\right) \end{aligned} \tag{14.27}$$

With $\cos\left(\delta_M\right)\sin\left(\delta_M\right) = \frac{1}{2}\sin\left(2\delta_M\right)$ it follows further

$$T = \frac{3}{2} p \, i_I \cos\left(\delta_M\right)\Psi_P + \frac{3}{4} p \, i_I^2 \sin\left(2\delta_M\right)\left(L_q - L_d\right) \tag{14.28}$$

The flux linkage of the permanent magnet field can be calculated from the flux density amplitude of the working wave and the effective number of turns ($w_{eff} = w \, \xi$; the factor $2/\pi$ is the result of integrating the assumed sinusoidal flux waveform over the pole area, i.e. over a half period):

$$\begin{aligned} \Psi_P &= w_{eff} B \frac{2\pi r \ell}{2p} \frac{2}{\pi} \\ &= w_{eff} B \frac{2 r \ell}{p} \end{aligned} \tag{14.29}$$

The amplitude of the current loading of the working wave can be calculated by means of the amplitude of the current i_I (the factor 2 in the following equation comes from the fact that each turn is composed of forward and return conductor):

$$A = \frac{m \; 2 w_{eff}}{2\pi r} i_I \tag{14.30}$$

With the number of phases $m = 3$ it follows:

$$i_I = A \frac{\pi r}{3 w_{eff}} \tag{14.31}$$

In total, for the torque this results in:

$$\begin{aligned} T = & \frac{3}{2} p\, A \frac{\pi r}{3 w_{eff}} \cos(\delta_M)\, w_{eff}\, B \frac{2 r \ell}{p} + \\ & \frac{3}{4} p \left(A \frac{\pi r}{3 w_{eff}} \right)^2 \sin(2\delta_M)(L_q - L_d) \\ = & \pi r^2 \ell\, A\, B \cos(\delta_M) + \\ & \frac{3}{4} p \left(A \frac{\pi r}{3 w_{eff}} \right)^2 (L_q - L_d) \sin(2\delta_M) \end{aligned} \tag{14.32}$$

Applying the calculation of the inductivity from Sect. 4.2 to the calculation of L_d and L_q results in

$$\begin{aligned} L_d &= (1+\sigma_1) \frac{3}{2} \mu_0 \left(\frac{w_{eff}}{p} \right)^2 \frac{4}{\pi} \frac{\ell r}{\delta_d} \\ L_q &= (1+\sigma_1) \frac{3}{2} \mu_0 \left(\frac{w_{eff}}{p} \right)^2 \frac{4}{\pi} \frac{\ell r}{\delta_q} \end{aligned} \tag{14.33}$$

where δ_d and δ_q are the magnetically effective air-gaps in d-axis and q-axis, respectively. Inserting this to the above torque equation gives

$$\begin{aligned} T = & \pi r^2 \ell\, A\, B \cos(\delta_M) + \frac{3}{4} p \left(A \frac{\pi r}{3 w_{eff}} \right)^2 \cdot \\ & (1+\sigma_1) \frac{3}{2} \mu_0 \left(\frac{w_{eff}}{p} \right)^2 \frac{4}{\pi} \ell r \left(\frac{1}{\delta_q} - \frac{1}{\delta_d} \right) \sin(2\delta_M) \\ = & \pi r^2 \ell\, A\, B \cos(\delta_M) + \pi r^2 \ell\, A^2 (1+\sigma_1) \frac{\mu_0}{2p} r \left(\frac{1}{\delta_q} - \frac{1}{\delta_d} \right) \sin(2\delta_M) \end{aligned} \tag{14.34}$$

As this has been deduced on the basis of the space vector theory, this result is valid for every moment in time, even for supplying the machine with non-sinusoidal currents. The torque is made of two components, the permanent magnet excited torque and the reluctance torque. These components are:

$$T_{PM} = \pi r^2 \ell \; A \; B \cos(\delta_M) \tag{14.35}$$

$$T_{Rel} = \pi r^2 \ell \; A^2 (1+\sigma_1) \frac{\mu_0}{2p} r \left(\frac{1}{\delta_q} - \frac{1}{\delta_d} \right) \sin(2\delta_M) \tag{14.36}$$

The permanent magnet excited torque is proportional to the bore volume ($\pi r^2 \ell$), to the current loading (A), to the flux density (B), and to the cosine of the load angle δ_M. For $\delta_M = 0$ the maximum permanent magnet excited torque is reached, which is the usual operating condition in the base speed region (except for the so-called MTPA control, see Sect. 14.4.2).

The reluctance torque as well is proportional to the bore volume ($\pi r^2 \ell$). In addition, main influencing factors are the squared current loading (A^2), the difference of the inverse values of q-axis and d-axis air-gaps $\left(\frac{1}{\delta_q} - \frac{1}{\delta_d} \right)$, and the sine of the double load angle. For machines without magnetic asymmetry (i.e. machines with surface mounted magnets where $L_d = L_q$ is true) or without field weakening ($i_{I,d} = 0$, i.e. load angle $\delta_M = 0$) this torque component is zero.

14.4 Limiting Characteristics and Torque Control

14.4.1 Limiting Characteristics

From the two stator voltage equations

$$\begin{aligned} u_{I,d} &= R_1 i_{I,d} - \omega L_q i_{I,q} \\ u_{I,q} &= R_1 i_{I,q} + \omega L_d i_{I,d} + u_P \end{aligned} \tag{14.37}$$

it follows by neglecting the Ohmic resistance ($R_1 = 0$) and with $u_P = \omega\Psi_P$

$$
\begin{aligned}
u_{I,d} &= -\omega L_q i_{I,q} \\
u_{I,q} &= \omega L_d i_{I,d} + \omega\Psi_P
\end{aligned}
\tag{14.38}
$$

Now, the current and voltage limits, depending on the perpendicular current components $i_{I,d}$ and $i_{I,q}$, will be computed. The current limit is a circular function

$$
i_{max}^2 \geq i^2 = i_{I,d}^2 + i_{I,q}^2 \tag{14.39}
$$

and for the voltage limit there is an elliptic function

$$
\begin{aligned}
u_{max}^2 \geq u^2 &= u_{I,d}^2 + u_{I,q}^2 \\
&= \left(\omega L_q i_{I,q}\right)^2 + \left(\omega L_d i_{I,d} + \omega\Psi_P\right)^2 \\
&= \omega^2\left[\left(L_q i_{I,q}\right)^2 + \left(L_d i_{I,d} + \Psi_P\right)^2\right] \\
\Rightarrow \quad \frac{u_{max}^2}{\omega^2} &\geq \left(L_q i_{I,q}\right)^2 + \left(L_d i_{I,d} + \Psi_P\right)^2
\end{aligned}
\tag{14.40}
$$

Figure 14.6 illustrates these limits in the $i_{I,d}$ - $i_{I,q}$ -plane.

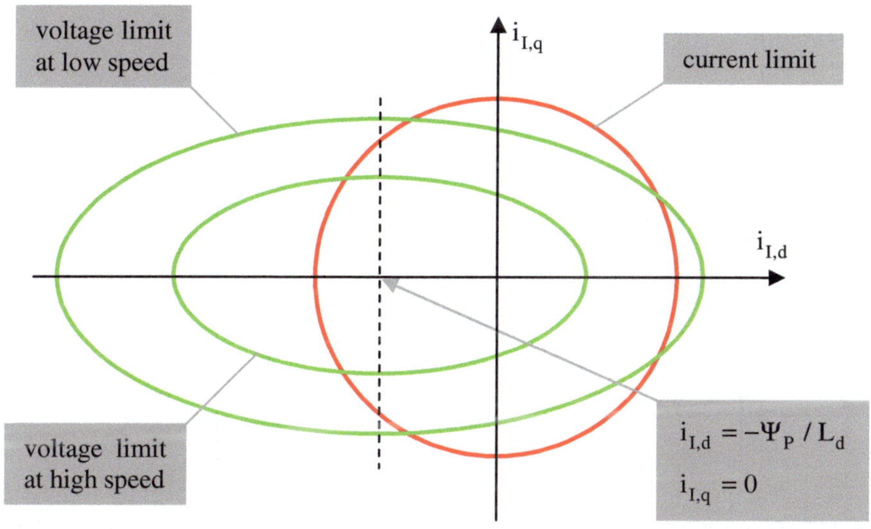

Fig. 14.6. Voltage and current limits.

The characteristics of these operation limits are:

- The voltage limit changes to circles (instead of ellipses) shifted to the left from the origin, if surface mounted magnets ($L_d = L_q$) are used.
- The voltage limit is speed dependent: with increasing speed the possible area of operation is narrowing increasingly.
- The current limit is independent from machine topology and speed.

14.4.2 Torque Control

The torque equation can be transformed like follows:

$$
\begin{aligned}
T &= \frac{3}{2}\frac{p}{\omega} i_{I,q}\left[u_P - \left(\omega L_q - \omega L_d\right) i_{I,d}\right] \\
&= \frac{3}{2}\frac{p}{\omega} i_{I,q}\left[\omega \Psi_P - \left(\omega L_q - \omega L_d\right) i_{I,d}\right] \\
&= \frac{3}{2} p\, i_{I,q}\left[\Psi_P - \left(L_q - L_d\right) i_{I,d}\right]
\end{aligned}
\tag{14.41}
$$

From this equation Fig. 14.7 with characteristics of equal torque value (iso-torque characteristics) in the $i_{I,d}$ - $i_{I,q}$ -plane is deduced for machines with surface mounted magnets and machines with buried magnets.

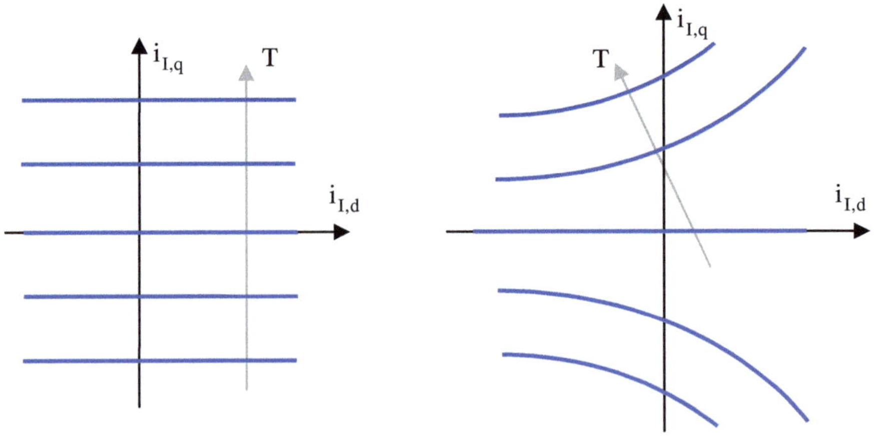

Fig. 14.7. Torque characteristics for machines with surface mounted magnets (*left*) and machines with buried magnets (*right*).

From Fig. 14.7 it is obvious that torque control is much easier for machines with surface mounted magnets ($L_d = L_q$, here a linear relation exists between the current $i_{I,q}$ and the torque) than for machines with buried magnets ($L_d < L_q$, for these machines $i_{I,d}$ and $i_{I,q}$ have to be controlled simultaneously, in addition the relation is non-linear).

Now, the question shall be answered how to choose the current components $i_{I,d}$ and $i_{I,q}$ to reach the required torque with minimum total current (minimum losses, maximum efficiency).[27] In literature, this is often referred to as "MTPA - maximum torque per ampere" control. However, more precisely it should be called "MTPC - maximum torque per current" control. For this, three cases are distinguished:

1. *Low speed, i.e. voltage limit is not relevant*
   For a certain torque the total current is then minimal if the current vector and the vector $\mathrm{grad}(T)$ have the same direction. In Fig. 14.8 this is given on the dark blue curve. Therefore, this curve gives the optimum operating points of the machine.

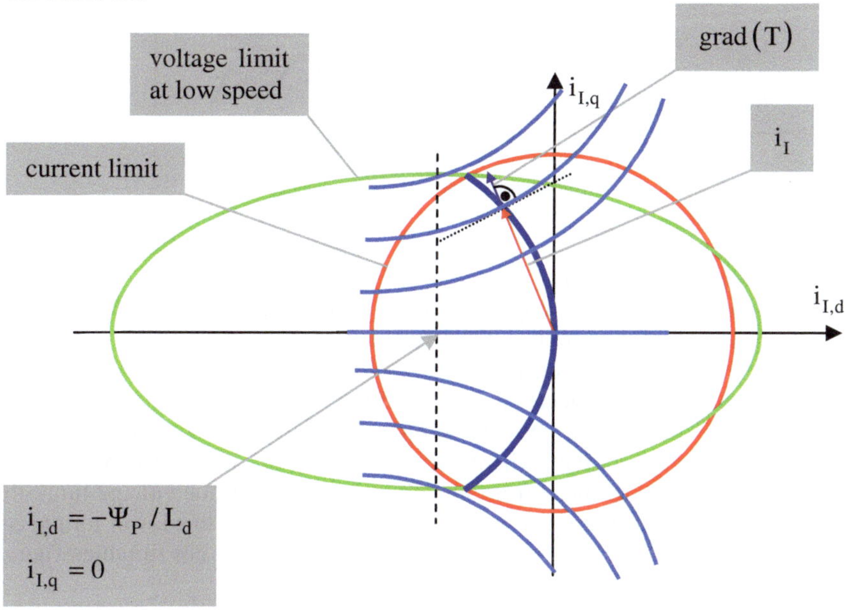

Fig. 14.8. Optimum torque characteristic at low speed.

[27] Minimum total current leads to minimum losses and maximum efficiency, if the iron losses and permanent magnet losses are neglected. This approximation is valid for low speed (i.e. low frequency), at high speed iron and permanent magnet losses may even become dominant.

2. *Medium speed (low field weakening); i.e. voltage limit is relevant*
Because of the voltage limitation the maximum torque is lower than in the case above; moreover, the optimum curve for reaching the required torque (dark blue curve) partly proceeds along the voltage limit curve, see Fig. 14.9.

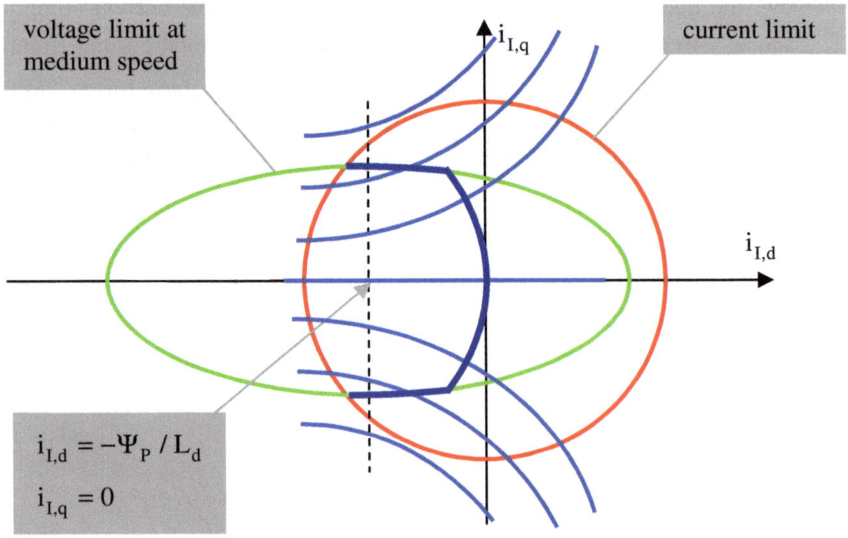

Fig. 14.9. Optimum torque characteristic at medium speed.

3. *High speed (strong field weakening); i.e. current limit is not relevant*
At strong field weakening only the voltage limit is relevant; the limit of the maximum total current is not longer reachable.
The additional black characteristic in Fig. 14.10 features that limitation, for which the torque is decreasing again if the current is further increased and the voltage is hold on its limit. With other words: the black characteristic shows those operating points, where the required torque is reached with minimum flux (as can be deduced from the ellipse equation for the voltage limit, the ellipses are the operating points with constant flux). As this lower torque can be reached even with lower current, such an operation is not suitable. Therefore, this operating area is excluded.

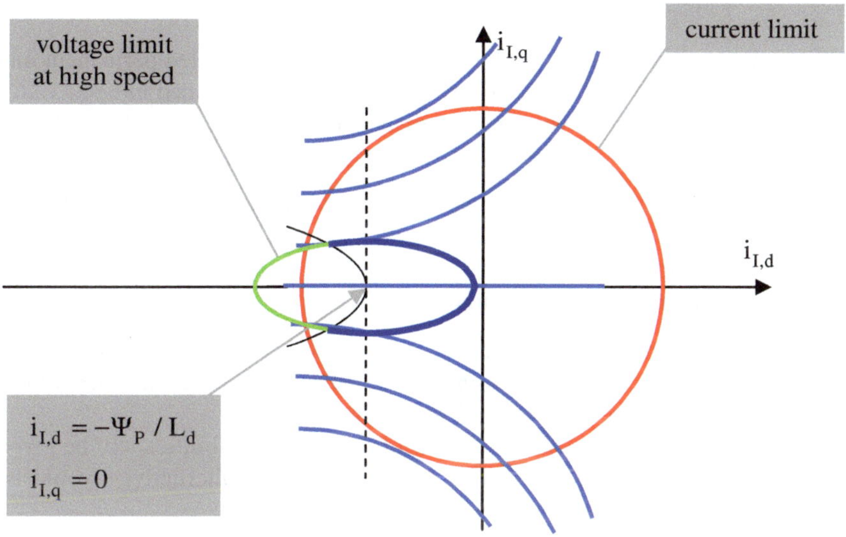

Fig. 14.10. Optimum torque characteristic at high speed.

4. *Summary*
The suitable operation area of the machine according to the above discussion is marked in grey in Fig. 14.11.

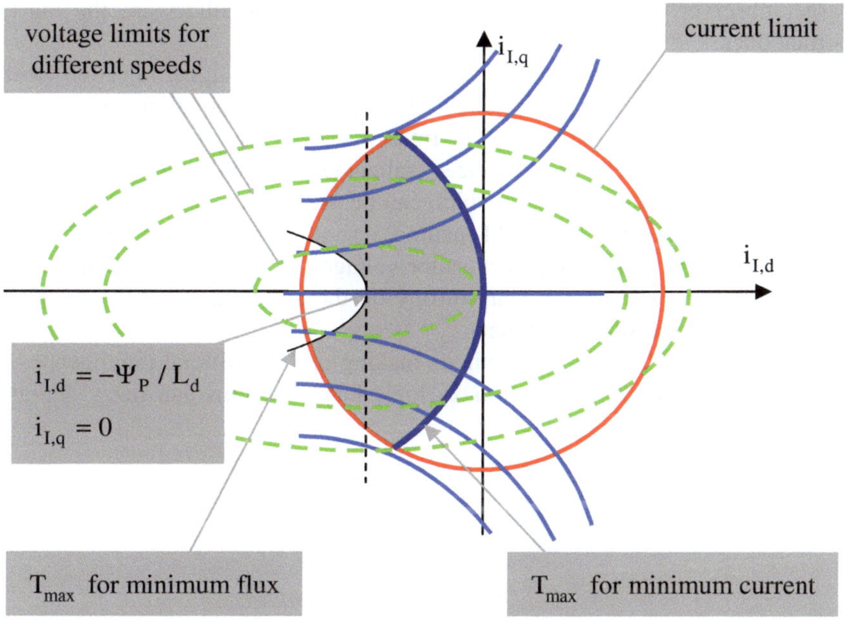

Fig. 14.11. Optimum torque characteristic for the entire speed range.

14.5 Control without Mechanical Sensor

The mechanical speed has to be known for the speed controlled permanent magnet excited machine, see Sect. 14.2. However, as already mentioned in Sect. 12.6, mechanical speed sensors have some disadvantages that preferably should be avoided:

- vulnerability against outside impacts (forces, torques, temperatures, dirt)
- costs
- space consumption
- necessity of a free shaft extension

Therefore, even for the permanent magnet excited machine it is desirable to calculate the speed from the measured terminal values of the machine (often this method is referred to as "sensorless" speed control).

In the following just a short overview of different possibilities will be given for the sake of completeness, a detailed description of the alternatives would be far beyond the scope of this book.

A first group of methods deals with direct electrical measurement (DC-voltage, DC-current, phase voltages, and / or phase currents) and a calculation of the required values by means of a nonadaptive machine model (similar to what is described in Sect. 12.6 for the induction machine).

These calculations can be improved by adaptive methods. Among these "Model Reference Adaptive Systems (MRAS)", observer based estimators (e.g. Luenberger observer, sliding mode observer), and Kalman filters are the most important ones.

A third group makes use of machine saliency and / or signal injection (with rotating or alternating carrier). In many AC machines, the position dependence is a feature of the rotor. In the case of the interior PMSM there is a measurable spatial variation of inductances or resistances (saliencies) in the d- and q-axis due to geometrical and saturation effects, which can be used for the estimation of rotor position (e.g. used for the INFORM-method). Another method to estimate the rotor position is to add a high frequency stator voltage or current component and evaluate the effects of the machine anisotropy on the amplitude of the corresponding stator voltage or current component.

Finally there is a group of methods making use of artificial intelligence. Most prominent are neural networks, fuzzy logic based systems and fuzzy neural networks.

14.6 References for Chapter 14

Benjak O, Gerling D (2010) Review of position estimation methods for PMSM drives without a position sensor, part I: Nonadaptive Methods. In: International Conference on Electrical Machines (ICEM), Rome, Italy

Benjak O, Gerling D (2010) Review of position estimation methods for PMSM drives without a position sensor, part II: Adaptive Methods. In: International Conference on Electrical Machines (ICEM), Rome, Italy

Benjak O, Gerling D (2010) Review of position estimation methods for PMSM drives without a position sensor, part III: Methods based on Saliency and Signal Injection. In: International Conference on Electrical Machines and Systems (ICEMS), Incheon, South Korea

Dajaku G (2006) Electromagnetic and thermal modeling of highly utilized PM machines. Shaker-Verlag, Aachen

Fitzgerald AE, Kingsley C, Umans SD (1983) Electric machinery. Mc-Graw Hill Book Company, New York

Krause PC (1986) Analysis of electric machinery. Mc-Graw Hill Book Company, New York

Krishnan R (2001) Electric motor drives. Prentice Hall, London

Krishnan R (2010) Permanent magnet synchronous and brushless DC motor drives. CRC Press, Boca Raton

Kwak SJ, Kim KJ, Jung HK (2004) The characteristics of the magnetic saturation in the interior permanent magnet synchronous motor. In: International Conference on Electrical Machines (ICEM) Cracow, Poland

Li Y (2010) Direct torque control of permanent magnet synchronous machine. Shaker-Verlag, Aachen

Meyer M (2010) Wirkungsgradoptimierte Regelung hoch ausgenutzter Permanentmagnet-Synchronmaschinen im Antriebsstrang von Automobilen. Dissertation Universitaet Paderborn

Schroedl M, Lambeck M (2003) Statistic properties of the INFORM method for highly dynamic sensorless control of PM motors down to standstill. In: Annual Conference of the IEEE Industrial Electronics Society (IECON) Roanoke, USA

15 Concentrated Windings

15.1 Conventional Concentrated Windings

Electrical machines with non-overlapping concentrated windings have become an increasingly popular alternative to machines with distributed windings for certain applications. Concentrated winding machines (characterized by the fact that each coil is wound around a single tooth) have potentially more compact designs compared to the conventional machine designs with distributed windings, due to shorter and less complex end-windings. With such windings, the volume of copper used in the end-windings can be reduced in significant proportions, in particular if the axial length of the machine is small. Consequently, lower costs and lower losses can be expected. Even the process of manufacturing the coils is simplified, resulting in a very cost-effective solution. In addition such a winding design is better qualified for safety critical applications, because phase-to-phase short-circuits become very unlikely. The photographs (Fig. 15.1) illustrate these differences exemplarily.

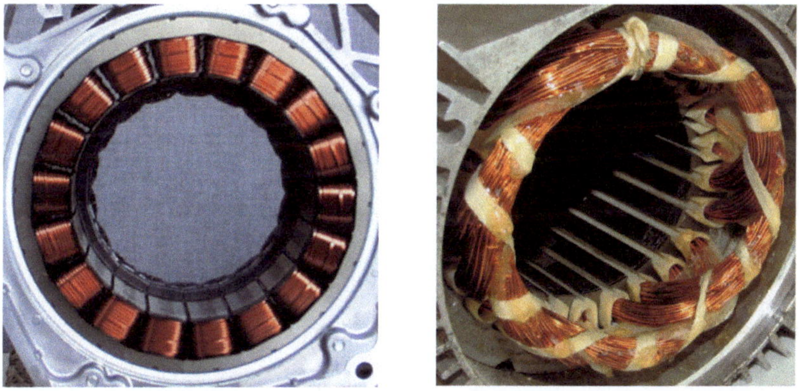

Fig. 15.1. Photographs of different winding topologies (*left*: concentrated winding, *right*: distributed winding).

There is a large variety of possibilities to realize an electrical machine with concentrated windings, e.g.

- coils wound around every tooth (often referred to as "two-layer winding") or coils wound around every other tooth ("single-layer winding");
- different number of teeth side by side with coils of the same phase;
- torque producing rotating field wave (in the following called "working wave") being the fundamental or a higher harmonic.

Some of these windings can be calculated according to Sect. 3.4, considering the concentrated winding as a special case of a distributed winding with an extreme short-pitch factor.

In the following, the concentrated winding will be analyzed analogously to Sect. 3.3 (supplying the machine with a symmetric current system and calculating the time-dependent MMF distribution). Afterwards, for this MMF distribution a Fourier analysis will be performed, which gives the harmonics of the MMF distribution.

Because of the large variety of possibilities, this will be done in the following by looking at the example of a double-layer three-phase machine with twelve stator slots, having two teeth side by side with coils of the same phase. This winding layout is shown in Fig. 15.2.

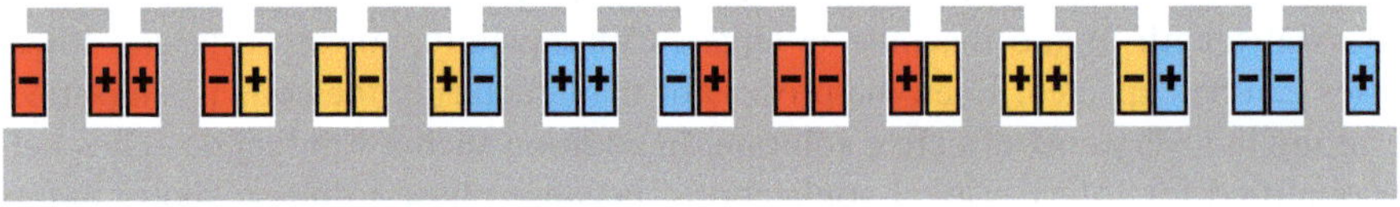

Fig. 15.2. Winding layout of a machine with concentrated coils ("wound-off" representation of the stator lamination); *red*: phase u, *yellow*: phase v, *blue*: phase w.

The MMF distribution for $\omega t = 0$ and $\omega t = \pi/2$ is illustrated in Fig. 15.3.

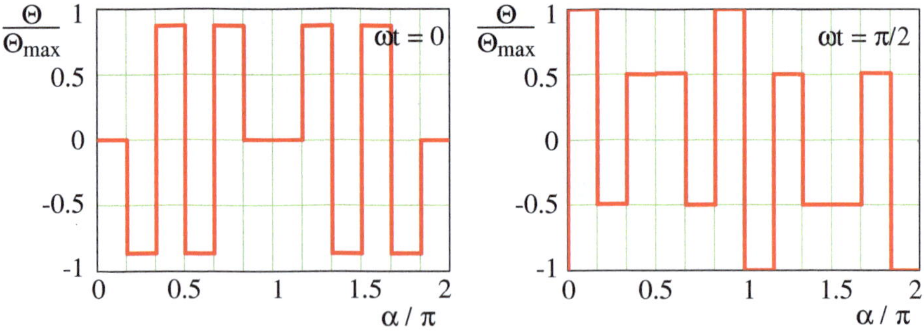

Fig. 15.3. MMF distribution of the winding shown in Fig. 15.2 versus circumference coordinate, functions normalized to a maximum value of 1; $\omega t = 0$ (*left*), $\omega t = \pi/2$ (*right*).

The harmonic analysis, which results in a time-independent characteristic for the given example, is presented in Fig. 15.4. A large number of MMF harmonics with high amplitude becomes obvious. The presence of these harmonics is the main disadvantage of concentrated windings, as they cause a variety of problems (mainly additional losses and acoustic noise).

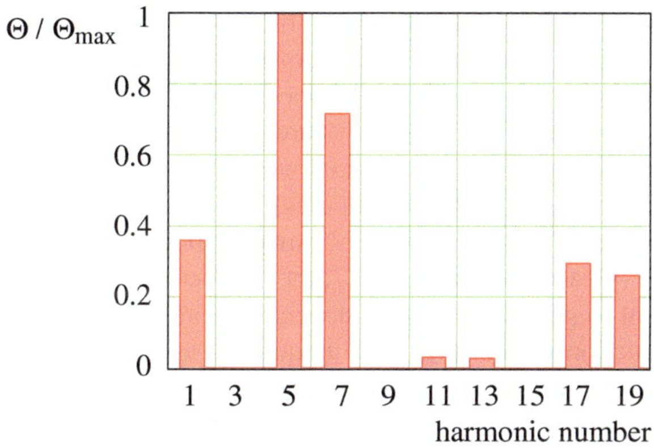

Fig. 15.4. Harmonic analysis of the MMF distribution (normalized amplitude versus harmonic number) of the winding shown in Fig. 15.2.

For calculating the torque of an electrical machine, not the amplitude of the MMF working wave is essential, but the amplitude of the current loading working wave (please refer to Sect. 14.3.5). From Sect. 3.5 it can be deduced that the current loading waves can be calculated from the spatial derivative of the MMF waves. Having a Fourier analysis of the MMF distribution means that all MMF waves are harmonic ones. Therefore, the respective spatial derivative means that the current loading waves are harmonic as well (with a phase shift of 90° electrically) and the amplitudes are proportional to the amplitudes of the MMF waves multiplied by the respective harmonic number. The result of the harmonic analysis of the current loading distribution is shown in Fig. 15.5.

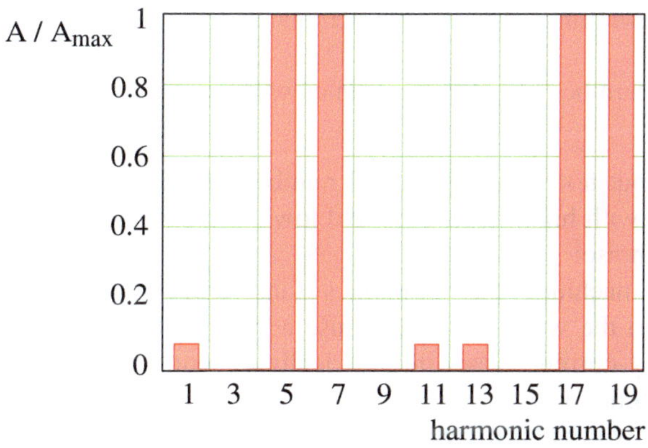

Fig. 15.5. Harmonic analysis of the current loading distribution (normalized amplitude versus harmonic number) of the winding shown in Fig. 15.2.

It is obvious that the waves with the harmonic numbers 5 and 7 may be used for torque production. Usually, higher harmonics are not used because the slot opening effect (that was not regarded in these calculations) increasingly reduces the amplitude with rising harmonic number. Using the waves with harmonic number 5 or 7 as working wave means that 10 or 14 poles are generated (e.g. a wave with harmonic number 5 contains 5 complete sinusoidal waves per circumference, which means that there are 5 north poles and 5 south poles).

Using a permanent magnet machine, the number of magnet poles in the rotor determines the working wave (please refer to Sect. 3.7: the number of poles of stator and rotor must be identical to generate a time-independent torque). Using this kind of winding with an induction motor (e.g. with a squirrel-cage rotor) torque generation will be very problematic, because the rotor adapts itself to any pole number of the stator. As the stator generates several pole numbers this will result in a very poor torque output.

If the 5th harmonic is used as working wave, the number of slots per phase per pole is according to Eq. (3.3)

$$q = \frac{N_1}{2pm} = \frac{12}{10 \cdot 3} = 0.4 \tag{15.1}$$

If the 7th harmonic is used as working wave, the number of slots per phase per pole is

$$q = \frac{N_1}{2pm} = \frac{12}{14 \cdot 3} \approx 0.29 \tag{15.2}$$

Therefore, these kinds of windings are called "fractional slot winding".

The existence of a large number of MMF harmonics with high amplitudes has some severe disadvantages:

- harmonics with similar, but not identical ordinal number and high amplitudes cause radial force waves with high amplitudes and low ordinal number, resulting in annoying acoustic noise;
- for permanent magnet machines the harmonics (and especially the sub-harmonics with respect to the working wave) cause additional losses, namely iron core losses and eddy current losses in electrically conductive permanent magnets (like NdFeB or SmCo);
- for induction machines it will be hardly possible to generate a useful torque, as the squirrel-cage rotor adapts to any pole number of the stator.

15.2 Improved Concentrated Windings

15.2.1 Increased Number of Stator Slots from 12 to 24

Because of the severe disadvantages it is necessary to reduce the unwanted harmonics by far (at least to an amount where their disturbing effect is nearly negligible). This will be explained exemplarily using the winding described in Sect. 15.1 and a rotor with 10 poles. Therefore, all low-order harmonics except for the 5th one have to be reduced to achieve an acceptable machine behavior.

The stator MMF distribution can be calculated by the following equation:

$$\Theta(x,t) = \sum_{\nu} \frac{3}{2}\, {}^{\nu}\Theta \cos\left(\omega t - \nu \frac{\pi}{\tau_p} x + \delta_M\right)$$
$$ {}^{\nu}\Theta = \frac{8\,\hat{i}\,N}{\pi\nu}\, {}^{\nu}\xi_w \qquad (15.3)$$
$$ {}^{\nu}\xi_w = \cos\left(\nu \frac{5}{6}\frac{\pi}{2}\right) \sin\left(\nu \frac{1}{6}\frac{\pi}{2}\right)$$

where ${}^{\nu}\Theta$ is the amplitude of the νth MMF space harmonic, ${}^{\nu}\xi_w$ is the winding factor, $\hat{i}$ is the phase current amplitude, δ_M is the load angle, ω is the angular frequency, and N is the number of turns per coil. Splitting this winding system into two identical winding systems, shifted against each other by an angle α_w, the resulting MMF distribution for the combined winding system becomes:

$$\Theta(x,t) = \sum_{\nu} \frac{3}{2}\, {}^{\nu}\xi_Z\, {}^{\nu}\Theta \cos\left(\omega t - \nu\left(\frac{\pi}{\tau_p} x - \frac{\alpha_w}{2}\right) + \delta_M\right) \qquad (15.4)$$
$$ {}^{\nu}\xi_Z = \cos\left(\nu \frac{\alpha_w}{2}\right)$$

where ${}^{\nu}\xi_Z$ is called distribution factor.

Figure 15.6 shows this distribution factor ${}^{\nu}\xi_Z$ for the first three relevant MMF harmonics ($\nu = 1, 5, 7$) as a function of the shifting angle α_w. This shifting angle

α_w is given in number of stator slots, because only these discrete values are possible for shifting the two winding systems against each other.

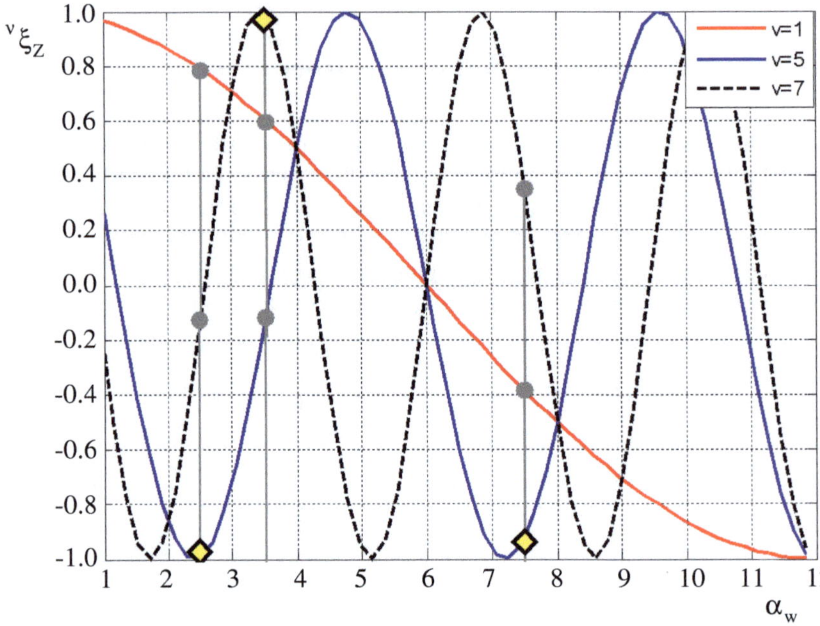

Fig. 15.6. Winding factors of the first three relevant MMF harmonics as a function of the shifting angle (measured in number of stator slots).

It can be deduced from Fig. 15.6 that reducing the 7th harmonic and maintaining the 5th harmonic results in a shift of the two winding systems of about 2.5 stator slots.[28] As mentioned before, it is only possible to shift the two winding system by an integer number of stator slots. To realize the desired shift, the number of stator slots will be doubled, resulting in a shift of five stator slots. This is illustrated in Fig. 15.7.

The combination of both winding systems is shown in Fig. 15.8. Of course it becomes obvious that the resulting winding is no longer purely concentrated, but partly overlapping. Nevertheless, this winding has the great advantages that the unwanted 7th harmonic is reduced by far and the end winding in circumferential direction is maintained as short as for the purely concentrated winding. Even if the end winding in total gets a little bit larger compared to the purely concentrated winding (because of the partly overlapping design), the end winding Ohmic losses are reduced by far compared to a conventional overlapping winding.

---

[28] For a machine with 14 rotor poles the 7th harmonic has to be maintained and the 5th harmonic has to be reduced. This results in a shift of the two winding systems of about 3.5 stator slots.

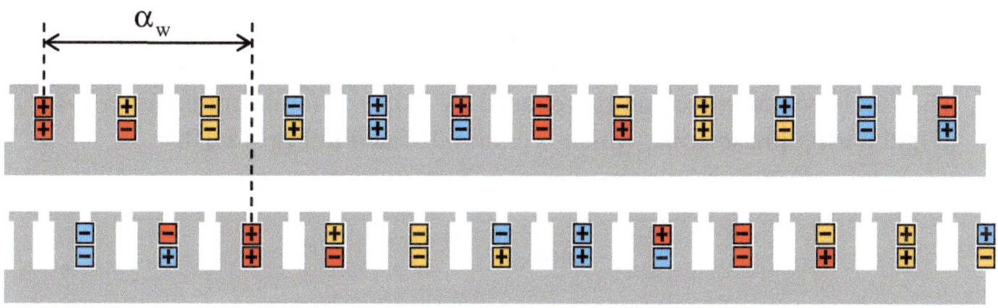

Fig. 15.7. Winding layout of a machine with two identical winding systems shifted against each other ("wound-off" representation of the stator lamination, please refer to Fig. 15.2); *red*: phase u, *yellow*: phase v, *blue*: phase w.

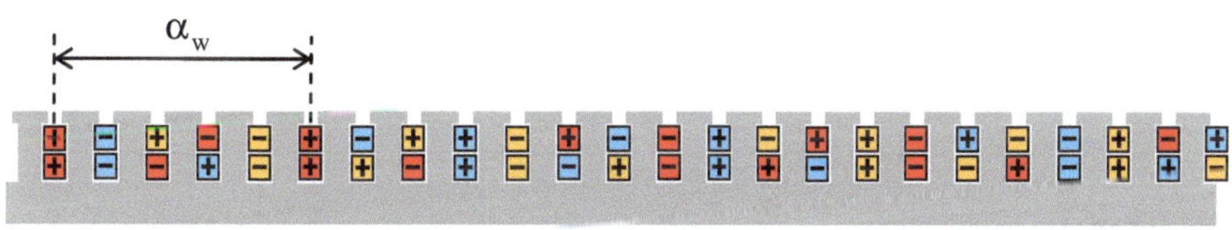

Fig. 15.8. Combination of both winding systems shown in Fig. 15.7 ("wound-off" representation of the stator lamination); *red*: phase u, *yellow*: phase v, *blue*: phase w.

The MMF harmonics of the winding layout of Fig. 15.8 is presented in Fig. 15.9. It becomes obvious that the 7th harmonic is reduced by far.

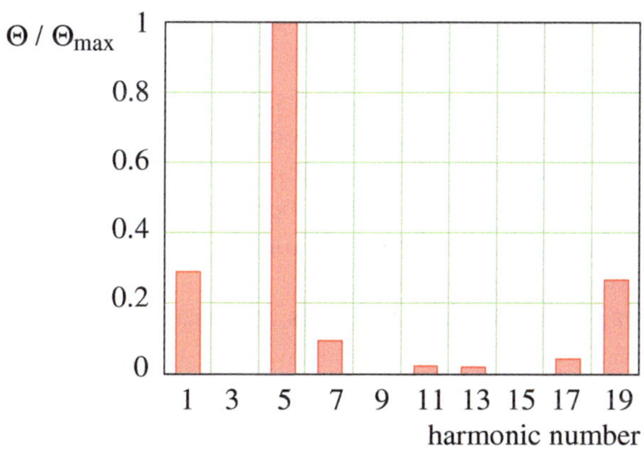

Fig. 15.9. Harmonic analysis of the MMF distribution (normalized amplitude versus harmonic number) of the winding shown in Fig. 15.8.

Now, two additional measures are introduced to further reduce the 7th harmonic and to reduce the fundamental wave:

1. Slightly different tooth widths are used to shift both winding systems a little bit more than the 2.5 slots shown in Fig. 15.6, please refer to Fig. 15.10 for a principle sketch. This results in completely reducing the 7th harmonic.[29]
2. The fundamental wave can be reduced by using different turns per coil for the neighbouring phase coils, which is also illustrated in Fig. 15.10.[30]

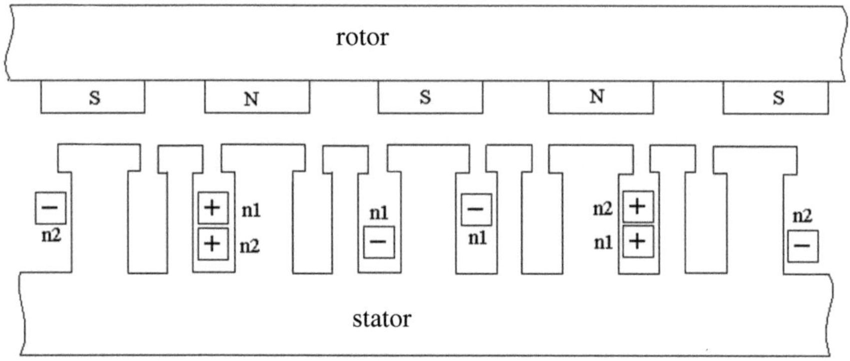

Fig. 15.10. Principle winding and lamination layout for reduction of the fundamental and 7th harmonic ("wound-off" representation of stator and rotor; only one phase shown for the sake of clarity; n1 and n2 denote different number of turns per coil).

With suitable design and optimization concerning the reduction of the fundamental harmonic and the 7th harmonic, an MMF distribution with a very low number of harmonics can be achieved. Such a design is illustrated in Fig. 15.11, showing the spectrum of the stator MMF distribution. The main advantages of such a winding design are:

- Low Ohmic stator losses because of short end windings.
- Low eddy current rotor losses because of low harmonic content in the MMF spectrum (especially low fundamental MMF harmonic).
- Low radial forces (as a reason for acoustic noises) because of low harmonic MMF waves near to the harmonic number of the working wave.
- Low torque ripple (please refer to Fig. 15.12).
- Low production costs because of short end windings and nearly concentrated coils.

[29] Of course, this additional feature can even be used differently. Possible alternatives are the maximizing of the 5th harmonic (which is the working wave) or the additional reduction of further harmonics that are not shown in Fig. 15.6.

[30] An alternative to reduce the fundamental wave is using coil windings with different turns per coil side; this implies that a single coil has to be connected from both axial ends of a radial flux machine (however, the phase winding may be still connected from one side).

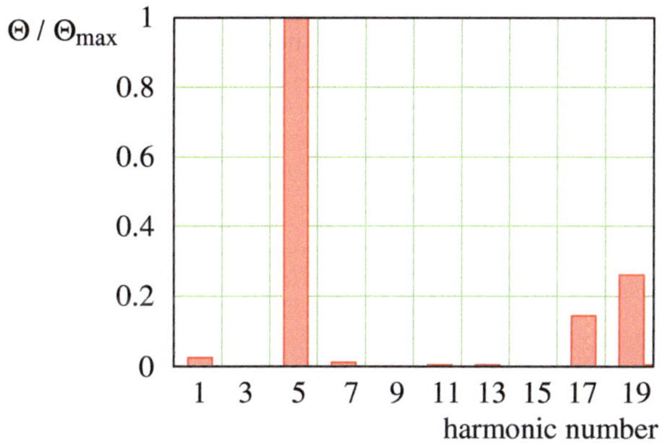

Fig. 15.11. Harmonic analysis of the MMF distribution (normalized amplitude versus harmonic number) of the optimized winding shown principally in Fig. 15.10 for one phase.

This kind of winding can be characterized by:

$$q = \frac{N_1}{2pm} = \frac{24}{10 \cdot 3} = 0.8 \tag{15.5}$$

If the 7th harmonic is used as the working wave, the following is obtained:

$$q = \frac{N_1}{2pm} = \frac{24}{14 \cdot 3} \approx 0.57 \tag{15.6}$$

The main advantages of this kind of winding are illustrated in Figs. 15.12 to 15.15, presenting a comparison with a conventional distributed wound permanent magnet machine (8 poles, 48 stator slots, $q = 2$, short-pitch of one stator slot) for the application of an automotive traction drive.

In Fig. 15.12 the comparison of the torque ripple at low speed is shown, if no skewing is used. It is obvious that for many applications the large torque ripple of distributed wound machines is not acceptable, therefore e.g. rotor magnet skewing has to be introduced (accompanied by the severe disadvantages of lower mean torque and higher costs).

The reduced Ohmic losses in the stator and the reduced eddy current losses in the rotor lead to different efficiencies in the torque-speed-plane, see Figs. 15.13 and 15.14. These results are compared by means of calculating the efficiency difference in all operating points, shown in Fig. 15.15. It becomes obvious that for nearly the entire operating area the new 24 slots / 10 poles winding is advantageous compared to a conventional distributed wound machine. Applications that are mostly used in low-load operating points (like e.g. automotive traction drives, industrial pumps and fans) benefit best from this development.

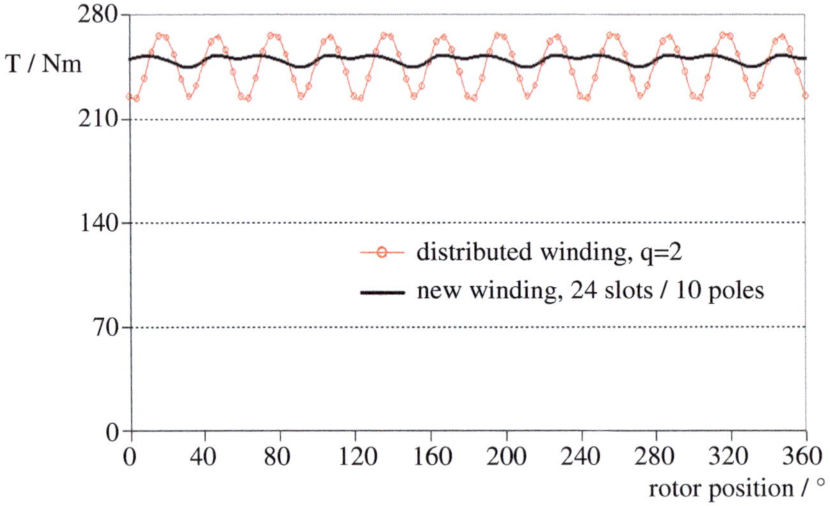

Fig. 15.12. Torque versus rotor position: Comparison of torque characteristics of a 24 slots / 10 poles permanent magnet machine and a conventional distributed wound permanent magnet machine.

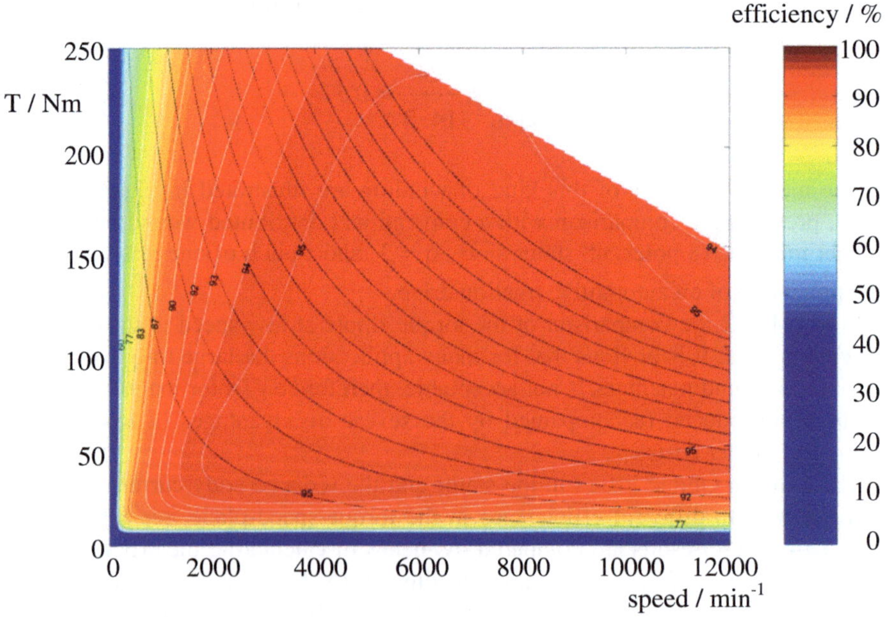

Fig. 15.13. Efficiency of the conventional distributed wound permanent magnet machine in the torque-speed-plane.

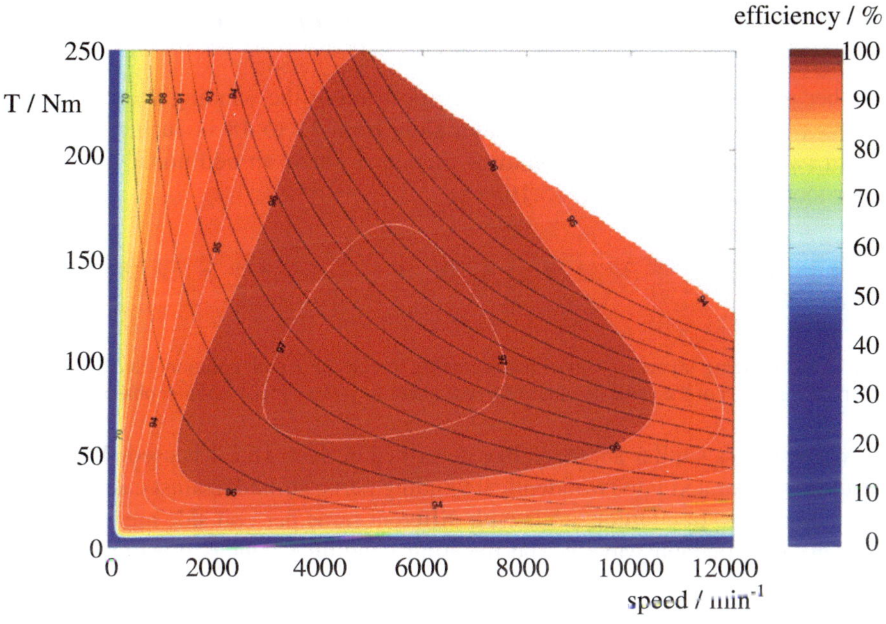

Fig. 15.14. Efficiency of the new 24 slots / 10 poles permanent magnet machine in the torque-speed-plane.

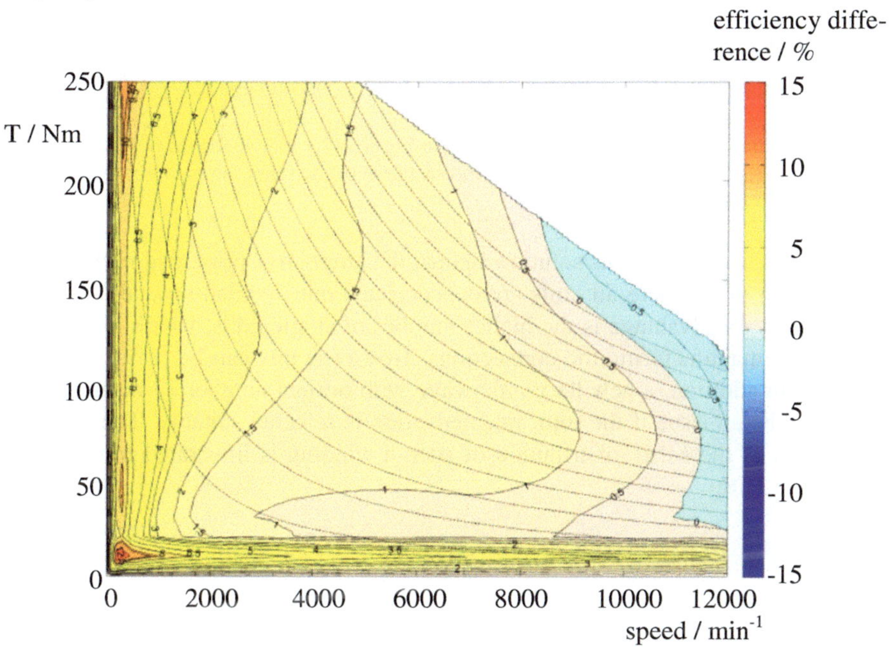

Fig. 15.15. Efficiency difference of the machines in Figs. 15.13 and 15.14 in the torque-speed-plane (positive values mean an advantage for the new 24 slots / 10 poles machine).

15.2.2 Increased Number of Stator Slots from 12 to 18

Another possibility to improve the MMF spectrum of the winding shown in Sect. 15.1 is to increase the number of stator slots from 12 to 18 in such a way, that within two neighboring stator teeth of the same phase an additional (unwound) stator tooth is placed, see Fig. 15.16.

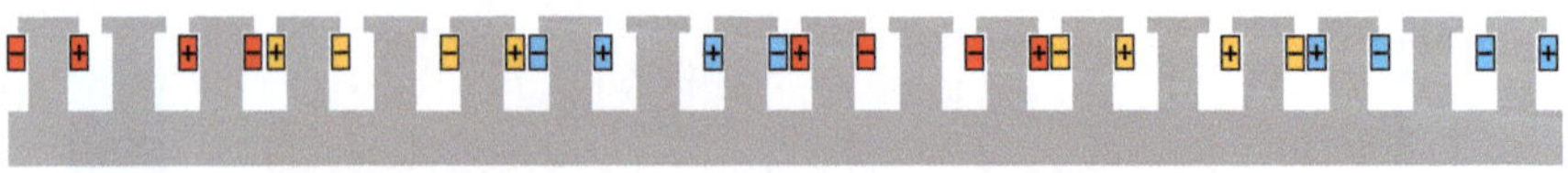

Fig. 15.16. Winding layout of a machine according to Fig. 15.2 with additional teeth ("wound-off" representation of the stator lamination); *red*: phase u, *yellow*: phase v, *blue*: phase w.

Afterwards, a similar approach like described in Sect. 15.2.1 is used: The initial winding is separated into two identical winding parts, shifted against each other by four stator slots. In addition, coils with different turns per coil side are used to reduce the fundamental wave.[31] The resulting winding layout is illustrated in Fig. 15.17.

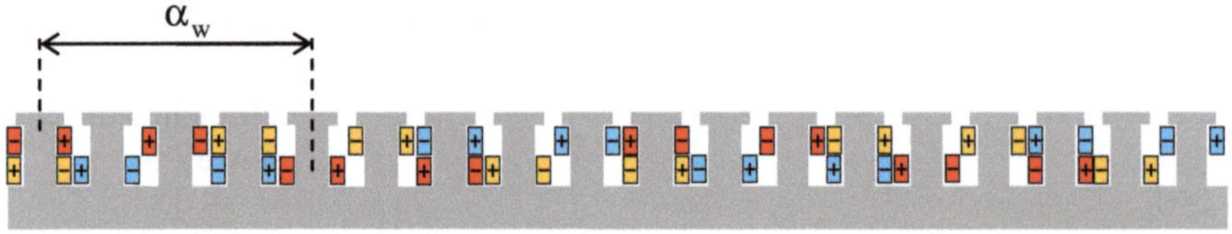

Fig. 15.17. Winding layout of a machine according to Fig. 15.12 with two identical winding systems shifted against each other ("wound-off" representation of the stator lamination); *red*: phase u, *yellow*: phase v, *blue*: phase w.

The main advantage of this solution against the alternative presented in the preceding section is that here all coils are concentrated ones wound around a single stator tooth. A disadvantage is the higher number of single coils, which increases the effort for manufacturing the connections. However, this can be realized fully automated. In addition, a new harmonic wave with ordinal number 13 occurs, but this will not be an issue if the rotor magnetization is selected suitably. The resulting spectrum of the MMF distribution of such a winding is illustrated in the following Fig. 15.18:

---

[31] Even coils with different number of turns may be used like described in the preceding section.

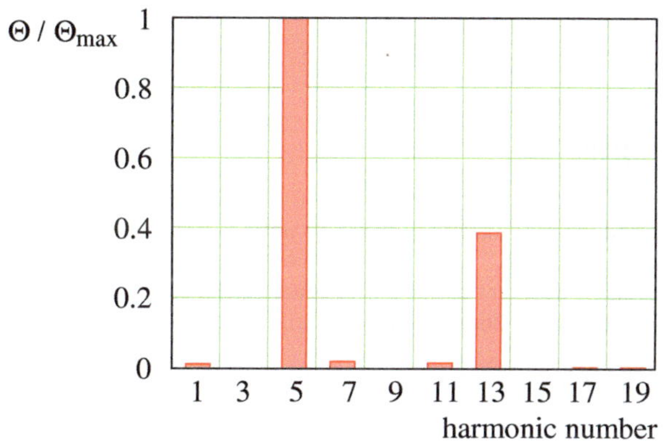

Fig. 15.18. Harmonic analysis of the MMF distribution (normalized amplitude versus harmonic number) of the optimized winding with concentrated coils (18 stator slots with purely concentrated coils).

15.2.3 Main Characteristics of the Improved Concentrated Windings

Purely concentrated windings generally are characterized by a large number of MMF harmonics with high amplitude. As considerable rotor losses and radial forces (resulting in acoustic noise) are generated by these harmonics, this kind of winding topology is unsuitable for many applications like e.g. in the automotive industry (traction, steering, and others), even if the production process of purely concentrated windings is beneficial.

For such applications alternatives are required that are advantageous compared to the well-known distributed windings (please refer to Chap. 3) which can be regarded as benchmark. The improved concentrated windings described in this chapter substantially maintain the advantages of concentrated windings like low production effort and short end windings (resulting in compact design and low Ohmic stator losses), whereas the disadvantages (high number of harmonics with large amplitude) to a large extend are eliminated. Especially the efficiency in part-load operation conditions can be improved considerably by these winding topologies, which is important for many industrial applications like pumps, blowers, traction drives, etc. Summarizing, these winding topologies are advantageous in production and operation of such electrical machines.

The positive effects of the improved concentrated winding designs were shown in this chapter using the example of permanent magnet rotors (synchronous machines as well as brushless DC machines). However, this kind of winding design is of course even applicable to electrically excited synchronous machines and induction machines.

15.3 References for Chapter 15

Dajaku G (2006) Electromagnetic and thermal modeling of highly utilized PM machines. Shaker-Verlag, Aachen

Dajaku G, Gerling D (2008) Analysis of different permanent magnet machines for hybrid vehicles application. In: ANSYS Conference & 26th CADFEM Users' Meeting, Darmstadt, Germany

Dajaku G, Gerling D (2009) Magnetic radial force density of the PM machine with 12-teeth/10-poles winding topology. In: IEEE International Electric Machines and Drives Conference (IEMDC), Miami, Florida, USA

Dajaku G, Gerling D (2010) Stator slotting effect on the magnetic field distribution of salient pole synchronous permanent-magnet machines. IEEE Transactions on Magnetics, 46:3676-3683

Dajaku G, Gerling D (2011) A novel 24-slots/10-poles winding topology for electric machines. In: IEEE International Electric Machines and Drives Conference (IEMDC), Niagara Falls, Ontario, Canada

Dajaku G, Gerling D (2011) Eddy Current Loss Minimization in Rotor Magnets of PM Machines using High-Efficiency 12-teeth/10-poles Winding Topology. In: International Conference on Electrical Machines and Systems (ICEMS), Beijing, China

Dajaku G, Gerling D (2011) Cost-Effective and High-Performance Motor Designs. In: International Electric Drives Production Conference (EDPC), Erlangen, Germany

Gerling D (2008) Analysis of the Magnetomotive Force of a Three-Phase Winding with Concentrated Coils and Different Symmetry Features. In: International Conference on Electrical Machines and Systems (ICEMS), Wuhan, China

Gerling D (2009) Influence of the stator slot opening on the characteristics of windings with concentrated coils. In: IEEE International Electric Machines and Drives Conference (IEMDC), Miami, Florida, USA

Gerling D, Dajaku G, Muehlbauer K (2010) Electric machine design tailored for powertrain optimization. In: 25th World Battery, Hybrid and Fuel Cell Electric Vehicle Symposium & Exhibition (EVS), Shenzhen, China

Ishak D, Zhu ZQ, Howe D (2006) Comparison of PM brushless motors, having either all teeth or alternate teeth wound. IEEE Transactions on Energy Conversion, 21: 95-103

Libert F, Soulard J (2004) Investigation on pole-slot combinations for permanent-magnet machines with concentrated windings. In: International Conference on Electrical Machines (ICEM), Cracow, Poland

Magnussen F, Sadarangani C (2 003)Winding factors and Joule losses of permanent magnet machines with concentrated windings. In: IEEE International Electric Machines and Drives Conference (IEMDC), Madison, Wisconsin, USA

Polinder H, Hoeijmakers MJ, Scuotto M (2007) Eddy-current losses in the solid back-iron of PM machines for different concentrated fractional pitch windings. In: IEEE International Electric Machines and Drives Conference (IEMDC), Antalya, Turkey

16 Lists of Symbols, Indices and Acronyms

16.1 List of Symbols

Table 16.1. List of Symbols.

symbol	meaning
latin letters	
A	area
A	current loading
a	complex operator
a	number of parallel paths
a	loss factor
a	factor
a, b, c	labeling of mains (line) phases
B	magnetic flux density
b	width
C	capacity
C	Esson's number
c	constant
D	displacement current
D	diameter
D	damping constant
d	differential operator
d	d-axis
div	divergence operator
E	electric field strength
E	energy
e	back electromotive force
e	Euler's number
F	force
f	force per surface area
f	frequency
f	field weakening factor
f	function
G	transfer function
g	numbering
H	magnetic field strength
h	height
I, i	current
Im	imaginary operator

J	electrical current density
j	imaginary unit
K	number of commutator sections
K	factor
K	gain of a PI-controller
K_C	Carter's factor
k	motor constant
k	numbering of slots
L	inductivity
L	Laplace operator
ℓ	length
ℓ	numbering of layers
ℓ	complex operator
m	number of phases
m	mass
N	number of slots
N	speed in the Laplace domain
n	speed
n	numbering
n1, n2	number of turns per coil side
P, p	active (real) power
p, n, 0	positive, negative, zero component
p	number of pole pairs
Q	reactive (wattless) power
q	number of slots per pole per phase
q	q-axis
R	resistance
R, r	radius
r	ratio
Re	real operator
rot	rotation operator
S	apparent power
s	distance
s	slip
s	Laplace variable
s	switching signal for power electronic device
T	torque
T, t	time
U, u	voltage
U_P	internal machine voltage (open circuit voltage)
u, v, w	labeling of machine phases
u	number of coils side-by-side in a single slot
u	transmission ratio

v	velocity
V	volume
W, W′	energy, co-energy
W	input quantity
w	energy density
w	number of turns
X	reactance
x	circumference direction (x-direction)
x, y, z	return wires of phases u, v, w
Y	conductance
Y	output quantity
y	radial direction (y-direction)
y_B	distance between brushes
Z	impedance
Z	disturbance quantity
z	axial direction (z-direction)
z	total number of conductors in all slots
greek letters	
α	mechanical angle
α_i	pole arc as a fraction of pole pitch
β	electrical angle
γ	electric conductivity
γ	angle
Δ	difference
∂	partial differential operator
δ	air-gap width
δ	load angle
$\varepsilon, \varepsilon_0$	dielectric constant, dielectric constant of vacuum
ε	angle
Φ, ϕ	magnetic flux
ϕ	output of flux hysteresis controller
φ	phase angle
Ψ, ψ	flux linkage
ξ	winding factor
ζ	parameter
λ	wave length
μ, μ_0	permeability, permeability of vacuum
μ	numbering
ν	harmonic number (order)

η	efficiency
ρ	electric charge
ρ	specific resistance
ρ	specific weight
ρ	angle
σ	leakage coefficient
τ	dimension in circumference direction (x- or α -direction)
τ	time constant
τ	output of torque hysteresis controller
τ_p	pole pitch
ϑ	rotor angle
Θ	magneto-motive force
Θ	inertia
Ω	mechanical angular frequency
ω	angular frequency

16.2 List of Indices

Table 16.2. List of Indices.

index	meaning
0	zero component
0	steady-state
0, 1, 2…	numbering
1, 2	stator, rotor
I, II	stator, rotor
A	armature
a	auxiliary
a	acceleration
a, b, c…	numbering
air	air
alt	alternating
B	brush
bar	rotor bar
C	coercive force
C	commutation, commutator
C	compensation winding
C	constant factor
C	coupling
C	characteristic in the Heyland-diagram
C	controller
CP	commutation pole
CS	coordinate system
Cu	copper
coil	coil
D	disturbance changes
D	damping
d	d-axis
DC	direct current (intermediate circuit)
e	eigen
edd	eddy current
eff	effective
el	electric
end	end
endw	end winding
F	field (exciting) winding
Fe	iron
fric	friction
G	generator

gen	generator
harm	harmonic
Hi	high
hys	hysteresis
I	inductivity
I	current
i	induced
i	current
i	internal
IM	induction motor
in	inside
k	numbering
kin	kinetic
limit	limit
lin	linear (straight)
line	line, mains
load	load
loss	losses
m	main
m	mean
M	magnet
M	motor
mag	magnetic
max	maximum
mech	mechanical
min	minimum
N	nominal
n	speed
non salient-pole	non salient-pole
off	off
on	on
op	operation
opt	optimum
out	outside
p	parallel
perm	permanent
phase	phase
PM	permanent magnet
pull-out	pull-out
q	q-axis
R	remanence
R	rotor
R	resistance
r	relative

ratio	ratio
real	real part of a complex number
Rel	reluctance
res	resultant
ring	rotor ring
rot	rotational
rotor	rotor
S	short-pitch
S	stator
S	setpoint changes
S	small
s	series
salient-pole	salient-pole
set	set value
skew	skewing
slot	slot
SO	slot opening
st	starting
stall	stand-still, short-circuit
stator	stator
syn	synchronizing
tot	total
u	voltage
u, v, w	labeling of phases
w	winding
wire	wire
x	circumference direction (x-direction)
x	x-direction of a two-phase-system
Y	star connection
y	radial direction (y-direction)
y	y-direction of a two-phase-system
Z	distribution (zoning)
z	axial direction (z-direction)
α	α -direction of a two-phase-system
β	β -direction of a two-phase-system
Δ	delta connection
δ	air-gap
μ	magnetic
ν	numbering index
Θ	nominal starting
Σ	sum
σ	leakage
$\emptyset$	diameter

∞	operating point with (ideally) infinite slip
∞	infinite time

16.3 List of Acronyms

Table 16.3. List of Acronyms.

acronym	meaning
AC	alternating current
BLDC	brushless DC
DC	direct current
DTC	direct torque control
EC	electronically commutated
FEM	finite element method
FOC	field-oriented control
INFORM	indirect flux estimation by online reactance measurement
IPM	interior permanent magnet machine
MMF	magneto-motive force
MRAS	model reference adaptive system
MTPA	maximum torque per ampere
MTPC	maximum torque per current
PI	proportional-integral
PMSM	permanent magnet synchronous machine
PWM	pulse width modulation
SMPM	surface mounted permanent magnet
SPM	surface permanent magnet
SR	switched reluctance
SRM	switched reluctance motor

Index

A

air-gap 37, 90, 102, 116, 127, 135, 159, 191, 196, 220, 244, 298, 426
air-gap field 125, 151, 171, 193, 228, 250
Ampere's Law 2, 5
armature reaction 80, 193
armature winding 45
asynchronous 135
auxiliary winding 252

B

block-mode operation 239, 363
brushes 37, 46
brushless DC-motor 223, 432

C

Carter's factor 116
cascaded control 288
coarse synchronization 396
co-energy 19, 234, 236, 238
commutation 40, 84
commutation poles 85
commutator 37
commutator segments 37, 82
compensation winding 82
complex plane 28, 147
concentrated winding 226, 449
coupling factor 174
critical damping 286
current control 288
current loading 49, 91, 114, 298, 438, 451

D

damper winding 371
d-axis 209, 322, 337, 378, 432
DC-machine 37, 273
demagnetization 68
direct torque control 360
distributed winding 104
distribution factor 110
disturbance changes 283
dynamo-electrical principle 71

E

efficiency 62, 162, 443
electrical angle 42, 90
electrical braking 269
energy 16, 48, 128, 144, 234, 263, 267, 273, 312, 399, 415
Esson's number 50

F

Faraday's Law 3, 7
field calculation 297
field weakening 181, 434
field-oriented control 344
finite element method 297
flux density 7, 14, 39, 49, 65, 93, 114, 175, 298, 438
flux linkage 14, 120, 140, 223, 237, 303, 318, 438
flux model 346
fractional slot winding 452

G

generator 38, 41, 58, 71, 150, 183, 189, 369

H

harmonic 96, 102, 142, 170, 298, 449
Heyland-diagram 148
hysteresis controller 365

I

induced voltage 11, 45, 51, 120
induction machine 135, 325
inductivity 142
inertia 258
interior permanent magnet machines 219
internal machine voltage 192
inverter 79, 179, 183, 220, 223, 239, 356, 425
iron losses 34
isolated operation 205

K

Kloss's Law 155

L

Laplace transformation 277
leakage 142
Lenz's Law 10
Leonard-converter 79
load angle 194, 433, 434, 436

losses 16, 42, 48, 132, 153, 234, 267, 312, 392, 399, 437, 443, 449

M

magnetic circuit 6, 17
magnetic field 5, 17, 32, 37, 80, 90, 117
magnetizing current 164, 337, 344
magneto-motive force 5, 93, 114
maximum torque per ampere 443
maximum voltage switching 390
Maxwell's equations 1
mutual inductivity 32

N

non salient-pole 209, 214, 374
non-linearity 239

O

operation limits 204
optimum of magnitude 292

P

permanent magnet 63
permanent magnet excited rotating field machine 425
permanent magnet synchronous machine 228, 432
permeability 2, 52, 63, 220, 426
phase angle 28, 194
phasor 28, 136, 192, 225, 323, 408
PI-controller 288
pole pairs 42, 89
pole pitch 42
power 29
  active power 29
  apparent power 29
  reactive power 29
power factor 160
Poynting's vector 15
pull-out torque 150, 199
pulse width modulation 366
pulsed operation 239

Q

q-axis 209, 322, 337, 378, 432
quasi steady-state 264

R

reluctance 92, 116, 196, 214, 231, 409, 436, 440
resistance 141
rotating wave 102
rotor angle 193

S

salient-pole 196, 209, 402
Sankey-diagram 132
saturation 17, 101, 238, 298
sector 364
sensorless speed control 359
setpoint changes 278
shaded-pole motor 253
short-circuit current 57
short-pitch factor 14, 110
short-pitch winding 104
single-layer winding 449
Single-Phase Induction Machine 250
single-phase machines 247
skewing 170
skewing factor 174
skin effect 175
slip 122
slot opening factor 112
space vector 299
speed control 59, 74, 179, 187, 288
squirrel cage rotor 166
Stability 257
stall current 57
star-delta-switching 182
Steinmetz equation 35
surface mounted permanent magnet machines 219
switched reluctance machine 232
symmetrical optimum 293
synchronization 197
synchronizing torque 199
synchronous machine 189, 369
synchronous reluctance machine 231
synchronous speed 133, 135, 189, 196

T

three-phase 30, 31, 99, 135, 189, 220, 247
torque 47, 55, 128, 152, 198, 222, 234, 248, 317, 438
torque ripple 457
torque-speed-characteristic 158, 333
transformer voltage 8
two-layer winding 42, 449

U

unipolar 238
universal motor 247
utilization factor 50

V

voltage of movement 9

W

wind power plant 183
working wave 449

Z

Zero voltage switching 394

MIX
Papier aus verantwortungsvollen Quellen
Paper from responsible sources
FSC® C105338

If you have any concerns about our products,
you can contact us on
ProductSafety@springernature.com

In case Publisher is established outside the EU,
the EU authorized representative is:
Springer Nature Customer Service Center GmbH
Europaplatz 3, 69115 Heidelberg, Germany

Printed by Libri Plureos GmbH
in Hamburg, Germany